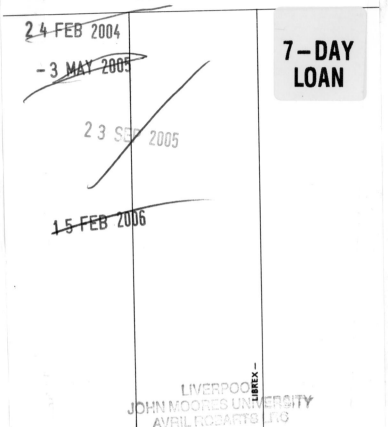

Pharmaceutical Gene Delivery Systems

DRUGS AND THE PHARMACEUTICAL SCIENCES

Executive Editor

James Swarbrick

PharmaceuTech, Inc.
Pinehurst, North Carolina

Advisory Board

DRUGS AND THE PHARMACEUTICAL SCIENCES

A Series of Textbooks and Monographs

ADDITIONAL VOLUMES IN PREPARATION

Pharmaceutical Gene Delivery Systems

edited by
Alain Rolland
Vical, Inc.
San Diego, California, U.S.A.

Sean M. Sullivan
University of Florida
Gainesville, Florida, U.S.A.

MARCEL DEKKER, INC.   NEW YORK · BASEL

Library of Congress Cataloging-in-Publication Data
A catalog record for this book is available from the Library of Congress.

ISBN: 0-8247-4235-4
This book is printed on acid-free paper.

Headquarters
Marcel Dekker, Inc.
270 Madison Avenue, New York, NY 10016
tel: 212-696-9000; fax: 212-685-4540

Eastern Hemisphere Distribution
Marcel Dekker AG
Hutgasse 4, Postfach 812, CH-4001 Basel, Switzerland
tel: 41-61-260-6300; fax: 41-61-260-6333

World Wide Web
http://www.dekker.com

The publisher offers discounts on this book when ordered in bulk quantities. For more information, write to Special Sales/Professional Marketing at the headquarters address above.

Current printing (last digit):
10 9 8 7 6 5 4 3 2 1

Preface

Gene therapy is a rapidly advancing field with great potential for the treatment of genetic and acquired systemic diseases. In developing pharmaceutical gene therapies, the disease, the therapeutic genes, and the gene delivery system need to be taken into consideration. The disease areas that have been most investigated in gene therapy to date are cancer and cardiovascular, pulmonary, and infectious diseases. Therapeutic genes for each of these diseases have been identified and are either in the clinic or due to enter the clinic. As future generation products are developed, the search for better therapeutic genes for the diseases currently being treated and genes to treat other diseases will continue. What remains to be identified is the gene delivery system needed to control the spatial and temporal modulation of the gene function. Furthermore, the duration, fidelity, regulation, and level of gene expression will be essential features to control by the design of specific gene expression systems.

Present delivery systems can be divided into virus-based, plasmid-based, and composites of both virus-based and plasmid-based systems.

Virus-based gene delivery systems comprise viruses that have been engineered not to replicate but to deliver genes to cells for expression. The viruses offer several advantages with regard to long-term expression, gene transfer efficiency, and expression of therapeutic genes. The limitations are (1) amenability of the viruses to pharmaceutical scale manufacture, (2) the restriction of viral infection to those cells expressing a receptor, and (3) the immunogenicity of the viral proteins. Viral systems do have applications for life-threatening diseases that do not require large quantities of material to be administered. Preclinical research is focused on alleviating some of these limitations with adenovirus and adeno-associated virus being the lead candidates.

Plasmid-based gene delivery systems for so-called nonviral gene therapy are being developed as an alternative to the virus-based systems. Plasmid is propagated in and isolated from bacteria. The isolated plasmid by itself can be administered locally, yielding expression in the injected tissue. Formulating the plasmid with synthetic gene delivery systems can increase the transfection efficiency. These gene delivery systems comprise mainly lipids, peptides, or polymers. The delivery systems protect the plasmid from degrading enzymes prior to internalization, facilitate cellular uptake by endocytosis, and can promote endocytic vacuole release of the plasmid into the cytoplasm. Some of the technologies developed in the early stages of the field are presently used in the clinic. The technology is limited to local administration, yields transfection of a small portion of cells, and duration of expression is at best on the order of days to weeks. The disease indications (primarily cancer and cardiovascular and infectious diseases) to be treated by these technologies conform to these specifications.

This book is divided into the following sections: an introduction providing an overview of gene therapy, including a brief history summarizing the field of plasmid and virus-based gene delivery; a description of the present studies in the clinics covering the highlighted technology for each clinical application; a review of the plasmid-based expression systems; a summary of the present plasmid-based gene delivery technologies; the gene therapy applications, both preclinical and clinical; and new technologies for expansion of the applications to new diseases.

It is critical for the field of gene therapy to transition from a proof of concept to a pharmaceutical product at the beginning of the new millenium. A successful outcome will result in a new clinical modality that represents a revolutionary approach to medicine. One immediate benefit will be to produce a continuous level of therapeutic protein, avoiding the characteristic peak and trough behavior of intermittent administrations. Physicians will have the capability to turn genes on or off on demand, producing a therapy that can treat the disease rather than the symptoms and with minimal side effects.

Alain Rolland
Sean M. Sullivan

Contents

Contributors

Eric W. F. W. Alton, M.D. Professor, Department of Gene Therapy, Imperial College at the National Heart and Lung Institute, and Royal Brompton Hospital, London, England

Jonathan Black, Ph.D. Research Microbiologist, Lineberger Comprehensive Cancer Center, University of North Carolina at Chapel Hill, Chapel Hill, North Carolina, U.S.A.

Jonathan L. Bramson, Ph.D. Assistant Professor, Department of Pathology and Molecular Medicine, McMaster University, Hamilton, Ontario, Canada

Barrie J. Carter, Ph.D. Chief Scientific Officer, Targeted Genetics Corporation, Seattle, Washington, U.S.A.

Seng H. Cheng, Ph.D. Vice President, Genetic Diseases Science, Genzyme Corporation, Framingham, Massachusetts, U.S.A.

Olivier Danos, Ph.D. Director, Scientific Department, Genethon, Evry, France

Jane C. Davies, M.B.Ch.B., M.D. Senior Lecturer and Honorary Consultant, Department of Gene Therapy, Imperial College at the National Heart and Lung Institute, and Royal Brompton Hospital, London, England

G. William Demers, Ph.D. Department of Pharmacology, Canji Inc., San Diego, California, U.S.A.

Duncan M. Geddes, M.D., F.R.C.P.   Professor, Department of Respiratory Medicine, Imperial College at the National Heart and Lung Institute, and Royal Brompton Hospital, London, England

Mikko O. Hiltunen, M.D.   A. I. Virtanen Institute, University of Kuopio, Kuopio, Finland

John A. Howe, Ph.D.   Principal Scientist, Department of Molecular Biology and Biochemistry, Canji Inc., San Diego, California, U.S.A.

Leaf Huang, Ph.D.   Joseph Koslow Professor, Department of Pharmaceutical Sciences, and Director, Center for Pharmacogenetics, University of Pittsburgh School of Pharmacy, Pittsburgh, Pennsylvania, U.S.A.

Yasufumi Kaneda, M.D., Ph.D.   Professor, Division of Gene Therapy Science, Department of Molecular Therapeutics, Graduate School of Medicine, Osaka University, Osaka, Japan

Ralf Kircheis, M.D., Ph.D.   Project Leader, Ingeneon AG, Vienna, Austria

Vivian Wai-Yan Lui, Ph.D.   Research Associate, Department of Cell Biology, Duke University Medical Center, Durham, North Carolina, U.S.A.

Fiona MacLaughlin, Ph.D.   Valentis, Inc., The Woodlands, Texas, U.S.A.

Jeffrey L. Nordstrom, Ph.D.   Senior Director, Valentis, Inc., Burlingame, California, U.S.A.

Jean-Christophe Pagès, M.D., Ph.D.   Department of Biochemistry, Centre Hospitalier Régional Universitaire de Tours, Tours, France

Robin J. Parks, Ph.D.   Center for Molecular Medicine, Ottawa Hospital Research Institute, Ottawa, Ontario, Canada

Murali Ramachandra, Ph.D.   Senior Principal Scientist, Molecular Biology and Biochemistry, Canji Inc., San Diego, California, U.S.A.

Alain Rolland, Pharm.D., Ph.D.   Vice President, Product Development, Vical, Inc., San Diego, California, U.S.A.

Sean M. Sullivan, Ph.D.   Associate Professor, Department of Pharmaceutics, University of Florida, Gainesville, Florida, U.S.A.

Francis C. Szoka, Jr., Ph.D.   Professor, Department of Biopharmaceutical Sciences and Pharmaceutical Chemistry, University of California at San Francisco, San Francisco, California, U.S.A.

Mikko P. Turunen, Ph.D.   A. I. Virtanen Institute, University of Kuopio, Kuopio, Finland

Lisa S. Uyechi-O'Brien, Ph.D.   Associate Scientist, Analytical and Formulation Development, Onyx Pharmaceuticals, Richmond, California, U.S.A.

Jean-Michel Vos, Ph.D.*   Lineberger Comprehensive Cancer Center and Department of Biochemistry and Biophysics, University of North Carolina at Chapel Hill, Chapel Hill, North Carolina, U.S.A.

Ernst Wagner, Ph.D.   Professor, Department of Pharmaceutical Biology-Biotechnology, Centre for Drug Research, Ludwig-Maximilians-University, Munich, Germany

Lionel Wightman, Ph.D.   Business Support Biotech, Boehringer Ingelheim Austria, Vienna, Austria

Nelson S. Yew, Ph.D.   Senior Scientist, Gene Transfer Research, Genzyme Corporation, Framingham, Massachusetts, U.S.A.

Seppo Ylä-Herttuala, M.D., Ph.D.   Professor, Department of Molecular Medicine, A. I. Virtanen Institute, University of Kuopio, Kuopio, Finland

---

* Deceased.

Pharmaceutical Gene Delivery Systems

1

Introduction to Gene Therapy and Guidelines to Pharmaceutical Development

Sean M. Sullivan
University of Florida, Gainesville, Florida, U.S.A.

I. WHAT IS GENE THERAPY?

Medicine is entering a new era for treating diseases that will enable physicians to treat the cause of a disease rather than the symptoms. Human genetic disease is the result of mutation or deletion of genes that impair normal metabolic pathways, ligand/receptor function, regulation of cell cycle, or structure and function of cytoskeletal or extracellular proteins. Diseases that are suitable for treatment by gene therapy can be divided into genetic and acquired diseases. Genetic diseases are typically caused by a single gene mutation or deletion. The acquired diseases are those for which no single gene has been identified as the only cause of the disease state. However, expression of a single gene delivered by a gene delivery system to the correct cell type(s) can potentially lead to the elimination of the disease state. Some examples for each type of disease are listed in Tables 1 and 2.

The genetic diseases are what sparked the initiative to create such gene-based therapeutics. Prior to this endeavor, there were no alternatives to correcting a genetic disorder. Therapies were developed that transiently alleviated the symptoms of the disease. Through the efforts of gene therapy, the correction of the mutation and the subsequent return to normalcy is now feasible. Fischer's Laboratory in France [2] recently achieved the demonstration of this principle.

Patients with severe combined immunodeficiency lack functional T cells and natural killer cells (NKs) due to mutations in the γc chain cytokine receptor subunit for interleukin (e.g., IL-2, -4, -7, -9 and -15) receptors. Ex vivo gene

Table 1   Candidate Genetic Diseases for Gene Therapy

Disease	Defect	Incidence	Target
Severe combined immunodeficiency	Adenosine Deaminase	Rare	Bone marrow or T cells
Hemophilia	Clotting Factors VIII and IX	1/10,000	Liver and muscle
Familial hypercholesterolemia	Low-density lipoprotein receptor	1/1,000,000	Liver
Cystic fibrosis	Loss of cystic fibrosis transmembrane conductance regulator (CFTR)	1/3000	Lung
Hemoglobinopathies	α and β Globin	1/600	Red blood cell precursors
Gaucher's disease	Glucocerebrosidase	1/450 Ashkenazi Jews	Liver
Inherited emphysema	α_1 Antitrypsin	1/3500	Lung, liver
Muscular dystrophy	Dystrophin	1/3500 males	Muscle

Table 2   Candidate-Acquired Disease for Gene Therapy

Disease	Defect	Incidence	Target
Cancer	Defects in tumor suppressors and/or presence of oncogenic factors	1 million/yr	Liver, lung, brain, pancreas, breast, prostate, kidney
Neurological diseases (Parkinson's and Alzheimer's)	Neurotransmitter release; structural defect of β-amyloid protein	1 million Parkinson's 4 million Alzheimer's in United States	Brain, neurons, glial cells, Schwann cells
Cardiovascular diseases	Defect in blood vessels	13 million in United States	Arteries, vascular endothelial cells
Infectious diseases	Suppressed immune system (HIV), liver destruction (HBV)	Increasing numbers	T cells, liver, macrophages, antigen-presenting cells

Source: Ref. 1.

transfer was employed to correct this genetic disease by delivering the wild-type sequence for the γc chain cytokine receptor subunit to hematopoietic stem cells using a nonreplicating murine retrovirus that has the therapeutic gene encoded within the viral genome [3,4]. Pagès and Danos describe the properties and utilities of this virus in much greater detail in Chapter 9. Preclinical results in γc chain cytokine receptor subunit–deficient mice showed that gene expression could be restored resulting in normal T-cell and NK cell development. Also, gene transfer to canine bone marrow resulted in long-term expression. Since this is a chronic disease, long-term gene expression is required.

A clinical trial was initiated in which hematopoietic stem cells from two infants were infected with a retrovirus carrying the γc chain cytokine receptor subunit. A 30% infection efficiency was obtained, and the patient's blood was analyzed on a regular basis for receptor expression in appropriate cell types and immunological function. After 10 months, the results showed that T-cell and NK cell counts were comparable to matched normal patients and the T and NK cells were found to be completely functional. This represents the first successful human gene therapy trial and validates the field. Eleven infants have been treated to date, all yielding seminormal immune systems. However, 2 infants have developed rare forms of leukemia. One infant is in remission and the other is being treated.

An extension of this success is being observed in in vivo human gene therapy trials for treatment of hemophilia. Adeno-associated virus has been delivered intramuscularly to patients lacking functional clotting Factor IX [5]. Carter describes the features and utility of this virus in Chapter 8. The Phase I clinical trial showed that the therapy was well tolerated. In addition, a reduction in the number of bleeding episodes for the treated patients was also observed.

In vivo gene therapy is a more acceptable clinical and pharmaceutical approach, because the therapeutic (a gene and a delivery system) is directly administered to the patient via conventional routes and methods, rather than a service that requires isolating the patient's cells, introducing the gene into those cells, and then reintroducing the modified cells back into the patient. Ex vivo gene therapy (as used with success in France for X-SCID patients) is not a desirable product for pharmaceutical companies but may be a service that can be provided by private or medical institutions. The listed acquired diseases should all be amenable to in vivo gene therapy.

The types of gene delivery systems are not restricted to viral vectors but can also be the simpler administration of the gene itself. The gene can be administered as a circular double-stranded DNA that is propagated in bacteria. This is termed *naked DNA*, and it was first applied in 1990 by Drs. Wolff and Felgner [6]. Intramuscular administration of such plasmid DNA yielded gene expression in skeletal muscle fibers. The transfection efficiency is relatively low, but current therapeutic applications include the development of genetic vaccines [7] and expression of angiogenic factors, such as vascular endothelium–derived growth factor

(VEGF) and fibroblast-derived growth factor (FGF), for the stimulation of new blood vessel growth in peripheral vascular disease and ischemic heart disease [8].

Gene transfer efficiency with a plasmid can be increased by formulating the plasmid in a number of synthetic gene delivery systems or applying external stimuli, such as electrical pulses (i.e., electroporation) [9–13] or ultrasound (i.e., sonoporation) [14]. MacLaughlin describes electroporation of plasmid in detail in Chapter 10. These methods are effective for local administration (e.g., intramuscular or intratumoral). Systemic administration requires the formulation of plasmid with cationic lipids, cationic polymers, or a combination of positively charged condensing agents with negatively charged liposomes [15]. Huang describes this latter system in more detail in Chapter 11. It is the surface charge interaction of the transfection complexes with the cells that results in the binding, internalization, and ultimately the transfection of endothelial cells after systemic administration. The transfection complexes can be further modified by covalent attachment of molecules such as polyethylene glycol (PEG) to the surface of the complexes to increase their circulation half-life [16,17]. In addition, targeting ligands, in the form of peptides, carbohydrates, or antibodies, can also be attached to the complex surface to bind cell surface antigens or receptors expressed on specific cell types to achieve site-directed gene transfer. This feature of the gene delivery systems is not restricted to nonviral systems but can also be applied to viral systems, such as retrovirus, adenovirus, and adeno-associated virus [18].

Successful demonstration of nonviral, cationic lipid-based gene transfer has been achieved clinically for the treatment of restenosis following balloon angioplasty [19]. The strategy was to administer a plasmid formulated with cationic liposomes composed of DOTMA/DOPE using a dispatch catheter following plaque removal by balloon angioplasty. The plasmid encoded the vascular endothelial growth factor ($VEGF_{165}$). Expression of VEGF-induced endothelial cell nitric oxide synthetase resulted in the inhibition of neointimal invasion of the blood vessel by smooth muscle due to nitric oxide production. Patients were monitored for status of the ballooned vessel and also production of new blood vessels as a result of VEGF expression. A comparison of adenovirus to the nonviral DNA approach showed that both gene delivery vehicles yielded similar results. A more detailed description of the technology and the design of the clinical trial can be found in Chapter 13 by Ylä-Herttuala.

Subsequent chapters describe various gene delivery systems and their therapeutic applications. This brief summary of the recent success in gene therapy was selected to introduce the exciting promise of gene therapy as a viable therapeutic for the future. What is not readily available in the literature is information regarding the processes by which a gene-based therapeutic is turned into a pharmaceutical product. The last part of this chapter will be devoted to product development of a gene-based therapeutic up to the filing of an Investigational New Drug application (IND). It is not meant to be a complete map to an all encompassing depiction of the transition from the preclinical proof of concept to clinical trials, but it will offer insights from personal experience, discussions with colleagues directly involved in the drug development process, and is supplemented with published information from the Food and Drug Administration (FDA).

II. DEVELOPMENT OF A GENE THERAPY PHARMACEUTICAL

Marketing of a gene-based therapeutic requires approval by the FDA. The drug sponsor must show, through supporting scientific evidence, that the therapy is safe and effective for its intended use. The sponsor must also demonstrate that the processes and controls used in manufacturing are properly controlled and validated to ensure the quality of the product.

The Center for Drug Evaluation and Research (CDER) is involved in New Drug Development and Review, Generic Drug Review, Over-the-Counter Drug Review, and Post Drug Approval Activities. The Center does not review gene therapy submissions. However, the approval process for gene-based therapeutics is similar and information obtained from the Center can be applied. For example, Figure 1 outlines the New Drug Development time course. The responsibilities

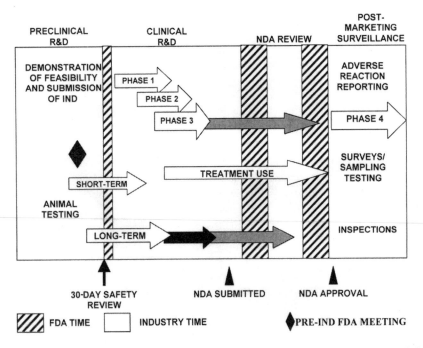

Figure 1 Gene therapy-based drug development time course. (Adapted from Ref. 20.)

of the sponsor are depicted in arrows, the white background is industrial time, and FDA review times are depicted in diagonal line with marked bars. There are several important features in the depicted timeline, the first one being the meetings between the sponsor and the FDA. The first of these meetings takes place upon demonstration of the proof of concept and the sponsor's decision to develop the product. There should be a preclinical development plan in place to review with the FDA. Timing for this meeting is critical, because it should not be conducted prematurely, nor should it be conducted too far into the development stage to avoid performance of unnecessary studies, and also to identify deficiencies in the design of the plan.

Items not displayed in the outline shown in Figure 1 are the continued processes after the drug candidate has entered Phase I clinical trials. This continuum includes, modifications to the manufacturing process, formulation optimization, and development and validation of quality control assays.

A misconception of an IND application is that it corresponds to an approved document that has to be signed by the FDA. In fact, the only approved, signed document by the FDA is a New Drug Application (NDA) that is submitted upon demonstration of the safety and effectiveness of the drug from the results obtained in the clinical trials.

The sponsor submits the IND application. The FDA notifies the sponsor that it has received the submission. Lack of a response within a 30-day period from the time of the notification allows the sponsor to proceed with the clinical trials. Technically, submission of an IND application is a request by the sponsor to be exempt from the federal statute that prohibits an unapproved drug from being shipped in interstate commerce. Only federally approved drugs are allowed to be shipped or distributed across state lines. The sponsor will usually conduct the clinical trials in several clinical centers in different states, thus requiring the authorization to ship the investigational drug to these centers. An exemption from this legal requirement is achieved by submission and approval of the IND application. However, in reviewing the IND document, the FDA ensures that the investigational drug is safe and the quality is suitable for testing in human clinical trials.

The FDA has two centers for the review of investigational drugs: Center for Drug Evaluation and Research (CDER) and Center for Biologics Evaluation and Research (CBER). CDER regulates conventional nonbiologically derived investigational drugs and CBER regulates biopharmaceuticals (e.g., recombinant proteins, monoclonal antibodies) and human gene therapies, which fall under the legal definition of a "biologic." Figure 1 serves as a general outline for the preclinical and clinical development of gene therapy products. The pharmacology and safety concerns of these products are significantly different from conventional drugs and also are different from biological drugs, such as recombinant proteins, blood products, and vaccines. One of the most important features is the pre-IND meeting with the FDA. The purpose is to review preclinical results and outlines for experiments to complete the IND submission, including the safety/toxicology and clinical plans. This step is important to ensure that the preclinical information can meet the requirements for IND filing.

Up until January 2000, no product-related serious adverse events had been reported in approximately 350 human gene therapy clinical trials. However, investigation of the death of a patient being treated for a rare liver disorder resulted in the FDA placing a clinical hold on all gene therapy clinical trials. The results from the investigation identified several deficiencies in conducting the clinical trials, in the recruitment and information of patients, and in the reporting of adverse events. As of April 2002, the FDA requires that the sponsor submit the IND documentation to the recombinant DNA Advisory committee [RAC; currently the Office of Biotechnology (OBA)] before submitting it to the FDA. The OBA's major role is to evaluate all protocols for clinical trials involving the transfer of genetically modified tissue into humans. The committee also considers safety standards, potential hazards, and methods for monitoring and minimizing risks associated with recombinant DNA research and advises the National Institutes of Health (NIH) director and staff on related issues and activities.

The OBA requires that investigational review boards (IRBs) from each institution conducting the proposed gene therapy clinical trials approve the protocol prior to submission of the IND application. Hence, in establishing timelines for product development, these considerations need to be taken into account. Presumably as more safe and efficacious gene therapies emerge from the current clinical trials, the time constraints placed on the review process may be reduced.

A. Product Discovery

The product discovery stage or "feasibility study" requires demonstration of pharmacological activity, or efficacy, in an appropriate animal model with expression of a gene that could correct the disease state and assessment of the safety profile of the drug product. Before initiating the animal studies, the therapeutic gene is cloned into a vector that can be delivered by the gene delivery system. The sequence is checked to ensure that it is correct. It is recommended that the entire expression cassette be checked to ensure proper expression of the gene. This includes not only the open reading frame of the gene but the promoter (including the enhancer), the intron/exon, the g-methyl cap sequence, the 5″ untranslated region, the Kozak sequence for identifying the initiation codon for translation, the 3′ untranslated region, and the poly adenylation sequence. Any ATG initiation sequences in the 5′ untranslated region upstream of the Kozak sequence can lead to false initiation of the transcript. The gene delivery system is then tested for gene transfer in tissue culture. Expression of the gene product is verified usually by Western blot. If it is a secreted protein, both the cell pellet and the extracellular media are assayed to ensure that the gene product is secreted. Finally, the biological activity of the expressed gene product is characterized. In planning ahead, the design of these identity tests should be amenable for development into quality control tests for release of manufactured product upon demonstration of feasibility. Upon validation of the gene sequence, expression of the correct gene product, and demonstration of biological activity, in vivo studies are initiated in an appropriate animal model.

The animal model should be a relative facsimile of the disease state in humans such that demonstration of efficacy in the animal model could parallel that observed in the clinic. The relevance also pertains to maximal tolerated dose and minimal effective dose yielding the therapeutic index for the therapy. These results can serve to define the initial dosing range for a Phase I clinical trial.

There are accepted models, for instance, for different types of cancer, cardiovascular disorders, or infectious diseases. A gene delivery system needs to be decided upon based on gene transfer efficiency, desired transfected cell type, duration of expression, administration route, and single-dose or multiple-dose regimen, as well as balance of risk versus therapeutic activity.

Once these factors have been decided upon, in vivo expression studies are initiated to show that the desired cell type in the diseased state is either transfected or impacted by gene expression from another cell type. An example would be the local expression of a cytokine gene or tumor suppressor gene in a tumor following intratumoral administration, or the intramuscular administration of an antiangiogenic agent resulting in subsequent secretion into the blood stream, thus acting distally on the tumor relative to the gene product production site. Once biological activity has been established, an efficacy study is then conducted showing that the diseased state can be cured.

B. Preclinical Development

1. Dose Optimization

Upon completion of feasibility studies, detailed pharmacokinetic and pharmacodynamic studies are required to determine the dosing schedule and to determine the minimal effect dose and maximal tolerated dose. The pharmacokinetic profile can be better defined as biodistribution with an acute time point for rapid clearance and a later time point for residual clearance; for example, 2 days versus 30 days. This would be applicable for a locally administered gene delivery system. For a systemically administered gene-based medicine, the time points would have to be more extensive and may require multiple dosing followed by administration depending on the type of gene delivery system, the administered dose, and the number of injections. Duration of expression is also part of the therapeutic profile. This also provides an estimate for the frequency of administration needed to yield an efficacious end result. The pharmacokinetic profile can be obtained using blood and tissue samples and analyzing by quantitative polymerase chain reaction (qPCR). The same samples used for the DNA analysis can also be used for determination of expression by assaying for RNA by reverse transcriptase–polymerase chain reaction (RT-PCR). The primary organs of interest are the injection site and typically the lung, liver, heart, spleen, muscle, kidney, bone marrow, and gonads. Gonadal tissue is of particular importance to the FDA. The requirement is that no transgene be vectorially transferred to offspring.

2. Stability

The manufactured gene-based product should ideally have a shelf life sufficient to ensure adequate manufacturing and use in toxicology and/or clinical trials (e.g., 1–2 years). However, for some gene-based medicines, this is not possible or conditions to achieve this are still under development while completing the preclinical development data package. Hence, the minimal shelf life stability should be from time of manufacture to completion of the first half of the Phase I clinical trials. It is not feasible to carry out real-time shelf life stability prior to submission of the IND. Accelerated stability studies utilize incubation of the product at several temperatures, and product stability is monitored as a function of time yielding an Arrhenius plot from which the slope is the reciprocal of the activation energy. The question then becomes which parameters to monitor as stability indicators. Ideally, the most informative assay would be an in vivo gene transfer assay, applicable for all gene-based products. However, use of this assay for stability studies is impractical because of the time requirement, degree of variability, and the potential for loss of replicates due to uncontrolled variables, such as infection.

The next preferred assay would be a cell-based bioactivity (potency) or gene expression assay. This is also a tedious and variable assay requiring the characterization and banking of cells, also displaying a potentially high degree of variability compared to other analytical methods. A cell-based assay should not be used alone but in conjunction with analytical techniques such as refractive index, appearance, particle size, pH, enzyme-linked immunosorbent assay (ELISA) for retention of receptor-binding domain for the case of virus-based gene delivery, high-performance liquid chomatography (HPLC) analysis of DNA for ratio of supercoiled to open circle for nonviral gene delivery, electrophoretic or mass spectroscopy analysis of protein degradation, HPLC analysis of cationic lipids, polymers, and polypeptides. Results from the biological assay can then be used to validate the use of the physical and chemical assays for stability testing, thus eliminating the need for the cell-based assay for future stability studies and release assay for a manufactured lot.

3. Safety/Toxicity

The design of the safety/toxicity studies depends on several parameters: (1) the disease to be treated, (2) the potential side effects of the gene product either due to expression in non-diseased tissue or to overexpression, and (3) the gene delivery system. Before initiating the design of the safety/toxicity studies, the following questions should be answered:

- Is a single administration sufficient for treatment or will multiple administrations be required?
- What is the duration of expression and what is the level of gene expression?
- What are the target organs for gene expression?
- If side effects are observed, what is their severity and are they reversible?

These questions are not meant to be inclusive. However, they should address the desired performance of the gene therapy product and potential side effects derived from the gene delivery system, expression of the therapeutic gene, and the administration route. Ideally, these studies can be conducted in conjunction with efficacy studies. However, owing to sampling times and dosage, it is often not possible to achieve this goal.

Specific points to consider are that the duration of gene expression determines the dosing schedule for the product, and the safety/toxicity dosing schedule should parallel the pharmacological dosing schedule. The assays should follow Good Laboratory Practice (GLP) procedures to assess safety and toxicity, including, for instance, histopathology of target organs along with other assays, such as blood chemistry to monitor liver damage. Board certified pathologists, either from a Contract Research Organization (CRO) or in-house personnel can conduct the organ pathology examination. The duration of expression will determine when to assay the animals. The time points can be divided into an acute phase in which tissues are harvested shortly after the last administration of the gene delivery system, such as 2–7 days, and a recovery phase where the tissue is harvested 1 month after administration. The safety/toxicity design will be dependent upon the disease to be treated.

Another point to consider is that the gene product should be active in the safety/toxicity animal model. For example, human IL-1 is active in mice and consequently the human sequence can be used for mouse safety/toxicity studies. There are genes in which the activity of the gene product is species specific (e.g., IL-12, interferon-α). Hence the safety/toxicity studies require expression of the species-specific sequence.

Under Good Manufacturing Practice (GMP) compliance, this will require the manufacture of a separate batch of the gene therapy product for each species. For example, in a nonviral setting, a bacterial master cell bank will be required for each new sequence. The DNA sequence of the manufactured plasmid will have to be verified. If an ELISA is used to assay for the transgene product, the assay will have to be developed and validated. If a bioassay is required for release, then a species-specific cell type may be required for release of the product.

Proper selection of suitable animals for the safety/toxicity studies is essential. The FDA requires that safety/toxicity studies be conducted in two animal species for an IND submission. One assumption is that primates should be one of these species. However, this may not always be the most suitable species. Primates are expensive both to purchase and house. They are high maintenance and require anesthesia for most procedures. Their immune system is not as well characterized as some rodent models, and the primate viruses are poorly characterized and may interfere with the study. Mice, dogs, rabbits, and pigs are more commonly used, and once again species selection will ultimately depend on several parameters, including the disease, expressed transgene, and route of administration. For example, the vascular system of pigs resembles the human vascular system compared to other animals. Hence, pigs would be better suited for demonstration of side effects for excessive expression of the gene product. Before initiating the safety/toxicity studies, it is advisable to have a preliminary meeting with the FDA and request a preliminary review of the study protocol to avoid later surprises upon the IND submission.

4. Scale Up and Validation of Manufacturing

A decision needs to be made early in the development of the gene therapy product, often as early as the feasibility period, as to whether manufacturing will be conducted internally or licensed out to a contract facility. There are generally some internal capabilities to supply the research teams with material for future development. At that stage, the manufacturing facilities and processes can be "GMP-like," without the necessary validation. There are advantages and disadvantages to both strategies. Cost is typically the first consideration. Establishing a certified GMP manufacturing process requires a large up front expense starting with something as simple as the supply of water for injection. In addition to building a manufacturing facility and certifying the manufacturing process, quality assurance needs to be established with auditors and creation of standard operating procedures (SOPs) for all aspects of the manufacture, quality control, and stability analysis.

The manufacturing process will most likely be unique to a specific product with regard to formulation development at the very least. Different administration routes require the addition of additional excipients. Development of a lyophilized product may also require specific cryoprotectants and other additives. Hence, the manufacturing process will often be codeveloped with a CRO. The advantage is that CROs already possess the instrumentation and possibly the quality assurance personnel, although in-house quality assurance personnel should coordinate activities with the CRO. A disadvantage of the CRO is loss of control over scheduling and cost. This requires a very high degree of planning and scheduling to ensure that lack of time slots does not slow the development of the manufacturing process. It is advisable to identify a CRO with which a good working relationship can be established. One additional concern is the stability of the CRO, especially if the sponsor is solely reliant on the expertise of the CRO for manufacturing.

5. Methods Development and Quality Control

The purpose of quality control is to ensure that the product can be manufactured reproducibly within a set of specifications that can be used to release the product for human administration. The assays will measure gene sequence and DNA content. Regardless of the manufacturing location, establishment of quality control will be required to release the finished product for clinical trials. These assays and release specifications can begin to be set with the safety/toxicity manufactured material. Assays for release can follow standard physical inspection such as appearance (clear or cloudy, color), pH, particle size (indicative of aggregation: this may not be applicable for low-titer viruses, such as retrovirus or lentivirus, or noncondensed, nonviral products), or refractive index. Early in the development process, bioassays involving in vitro cell culture or even in vivo expression study can be performed in parallel with physical measurements. Positive or negative correlations may be observed between physical/chemical analysis of the gene delivery system and its biological activity. Observed correlations from several GMP manufactured lots of the gene delivery system may yield substitution of the physical–chemical analytical methods for the in vitro or in vivo biological activity assays for quality control release assays. Cell-based quality control assays have far greater requirements and are more prone to variation compared to chemical and physical analytical methods. Also, the time required to perform a chemical or physical assay is generally far less than that required for a biological assay.

C. IND Submission

The flowchart in Figure 2 depicts the IND review process upon submission to the FDA. This diagram is from the Center for Drug Evaluation and Research Division (CDER) of the FDA and the review categories will somewhat differ for Biologicals. Specifically, the Chemistry Review section reviews all chemical aspects of the new drug entity such as manufacturing, stability, quality control, and quality assurance. These aspects of the gene-based pharmaceuticals will be reviewed by a similar committee who has expertise in biochemistry, virology, and microbiology for evaluation of the manufacture of the gene-based medicine and, for example, establishing validated quality control assays and determining the storage shelf half-life.

The medical/clinical reviewers evaluate the clinical section of the submission, such as the safety of the clinical protocols. The protocols are evaluated to

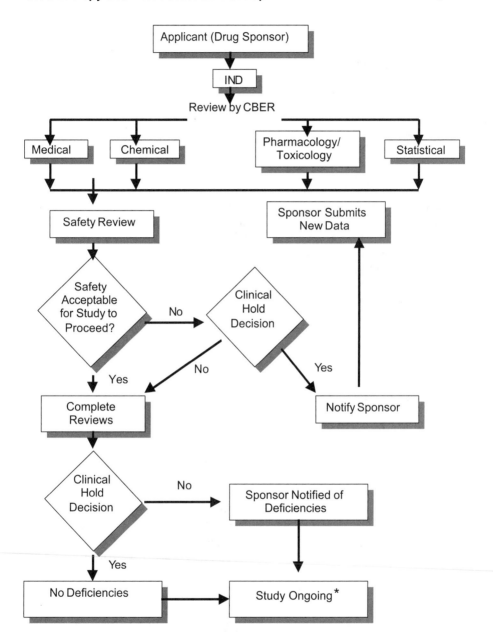

Figure 2 FDA IND review process. (http://www.fda.gov/cder/handbook/ind.htm).
*While sponsor answers any deficiencies.

determine if the participants are at unnecessary risk and will the design of the study yield results that can evaluate the effectiveness of the gene therapy.

The pharmacology/toxicology review section will evaluate the tolerability studies conducted in animal models and determine whether the results reflect the potential impact on humans. There will be additional safety concerns that are common to gene therapy–based investigational therapies. For example, one concern of the FDA is gene transfer to germ cells that could result in vertical gene transfer to offspring. The purpose is to contain the observation of any adverse effects to the patient and not have to be concerned with long-term development effects on the offspring.

Completion of the review by the first set of committees sends the application to be reviewed for safety. The submission progresses to completion if there are no concerns regarding the safety and study design to show drug product efficacy. However, if safety issues are raised, the FDA may place a clinical hold on the submission and the sponsor has a set time frame to respond to the concerns. Once the sponsor has submitted the response to the cited concerns by the FDA and they are favorably reviewed, the process can proceed. As stated previously, the review process takes 30 days from the notification date to the sponsor that the FDA has received the submission. If the sponsor receives no requests for information within this 30-day process, the sponsor can proceed with the clinical trials.

D. Clinical Trials

The protocols for all three phases of the clinical trials must be submitted in the IND document. The description of the clinical trials is also included in the Investigator's Brochure of the IND submission (Section 5). Also included is the information regarding the clinical facilities and the individual Institutional Review Boards (IRBs). Phase I clinical trials are designed to show that the new drug product or therapy is safe and well tolerated. Second, it will identify whether the patients display any adverse effects and at what dose. Since this first study involves the introduction of a new drug candidate into humans, the patients are closely monitored for any effects exerted on them as the result of the administration of the drug product. Dosing is initiated at a low dose deemed safe based on the preclinical results and often at a nonobservable effect level. The dose is escalated until a maximal tolerated dose (MTD) is determined. Phase I study design also includes the characterization of transgene expression, such as secreted transgene products assayed from the blood or locally expressed in biopsied tissues following local administration. This will serve to show that the gene transfer was successful, determine the variation in therapeutic gene expression levels from patient to patient, and ultimately determine the duration of expression. This latter point will facilitate the determination of dosing frequency for Phases II and III. Phase I studies can be conducted in diseased patients or in healthy volunteers depending on the disease. The total numbers of patients can range from 15 to 80 depending upon the disease and clinical endpoints, and trials are generally conducted at one or more clinical centers depending upon availability of patients.

Once the new therapy has been shown to be safe and well tolerated with an assessment of MTD, Phase II clinical trials are conducted to determine the effectiveness of the therapy in patients with the disease or condition and to determine the common short-term side effects associated with the therapy. These studies are not only designed to evaluate the effectiveness of the therapy but also to determine the minimal effective dose. The number of patients is still relatively small, not exceeding a few hundred patients and trials are conducted at a few centers.

Pivotal Phase III clinical trials are expanded controlled and uncontrolled studies performed at multiple centers with the total study treating several hundred to several thousand subjects. The increased enrollment will yield necessary information to allow the overall effectiveness and safety of the therapy to be statistically determined.

The FDA's primary objectives in reviewing the IND documentation are to assure the safety and rights of the subjects. The central theme of the initial IND submission should be on the general investigational plan and the protocols for specific human studies. Subsequent amendments to the IND document, either in response to inquiries by the FDA or modifications to the clinical trial protocols as the result of additional information, should build logically on previous submissions and should be supported by additional information, including the results of animal toxicology studies or other human studies as appropriate.

The field of gene therapy has great potential for patients, resulting in either cure or prevention of disease. In the early years of gene therapy, there may have been a tendency to progress too rapidly experimental gene therapies from the bench to the clinic. What is currently being observed is the transition from getting the therapy into the clinic to getting the therapy successfully out of the clinic and into the population. There will be false starts in this endeavor, but the question will no longer be "Will a gene therapy product ever be approved for treatment of a disease?" but rather "When will it be an approved therapy commercially available for every patient?"

REFERENCES

1. Verma IM, Somia N. Gene therapy—promises, problems and prospects. Nature 1997; 389(6648):239–242.
2. Cavazzana-Calvo M, Hacein-Bey S, de-Saint-Basile G, Gross F, Yvon E, Nusbaum P, Selz F, Hue C, Certain S, Casanova J-L, Bousso P, Deist F-L, Fischer A. Gene therapy of human severe combined immunodeficiency (SCID)–X1 disease. Science, 2000; 288(5466):669–672.

3. Hacein-Bey H, Cavazzana-Calvo M, Le Deist F, Dautry-Varsat A, Hivroz C, Riviere I, Danos O, Heard JM, Sugamura K, Fischer A, De Saint Basile G. Gamma-c gene transfer into SCID X1 patients' B-cell lines restores normal high-affinity interleukin-2 receptor expression and function. Blood 1996; 87(8):3108–3116.

4. Hacein-Bey S, Gross F, Nusbaum P, Hue C, Hamel Y, Fischer A, Cavazzana-Calvo M. Optimization of retroviral gene transfer protocol to maintain the lymphoid potential of progenitor cells. Hum Gene Ther 2001; 12(3):291–301.

5. High KA, AAV-mediated gene transfer for hemophilia. Ann NY Acad Sci 2001; 953:64–74.

6. Wolff JA, Malone RW, Williams P, Chong W, Acsadi G, Jani A, Felgner PL. Direct gene transfer into mouse muscle in vivo. Science 1990; 247(4949 Pt 1):1465–1468.

7. Felgner PL. DNA vaccines. Curr Biol 1998; 8(16):R551–553.

8. Isner JM, Ropper A, Hirst K. VEGF gene transfer for diabetic neuropathy. Hum Gene Ther 2001; 12(12):1593–1594.

9. Fewell JG, MacLaughlin F, Mehta V, Gondo M, Nicol F, Wilson E, Smith LC. Gene therapy for the treatment of hemophilia B using PINC-formulated plasmid delivered to muscle with electroporation. Mol Ther 2001; 3(4):574–583.

10. Shibata MA, Morimoto J, Otsuki Y. Suppression of murine mammary carcinoma growth and metastasis by HSVtk/GCV gene therapy using in vivo electroporation. Cancer Gene Ther 2002; 9(1):16–27.

11. Cupp CL, Bloom DC. Gene therapy, electroporation, and the future of wound-healing therapies. Facial Plast Surg 2002; 18(1):53–58.

12. Yamashita Y, Shimada M, Tanaka S, Okamoto M, Miyazaki J, Sugimachi K. Electroporation-mediated tumor necrosis factor-related apoptosis-inducing ligand (TRAIL)/Apo2L gene therapy for hepatocellular carcinoma. Hum Gene Ther 2002; 13(2):275–286.

13. Nakano A, Matsumori A, Kawamoto S, Tahara H, Yamato E, Sasayama S, Miyazaki JI. Cytokine gene therapy for myocarditis by in vivo electroporation. Hum Gene Ther 2001; 12(10):1289–1297.

14. Anwer K, Kao G, Proctor B, Anscombe I, Florack V, Earls R, Wilson E, McCreery T, Unger E, Rolland A, Sullivan SM. Ultrasound enhancement of cationic lipid-mediated gene transfer to primary tumors following systemic administration. Gene Ther 2000; 7(21):1833–1839.

15. Tan Y, Whitmore M, Li S, Frederik P, Huang L. LPD nanoparticles—novel nonviral vector for efficient gene delivery. Methods Mol Med 2002; 69:73–81.

16. Shi N, Zhang Y, Zhu C, Boado RJ, Pardridge WM. Brain-specific expression of an exogenous gene after i.v. administration. Proc Natl Acad Sci USA 2001; 98(22):12754–12759.

17. Fenske DB, MacLachlan I, Cullis PR. Stabilized plasmid-lipid particles: a systemic gene therapy vector. Methods Enzymol 2002; 346:36–71.

18. Anwer K, Bailey A, Sullivan SM. Targeted gene delivery: a two-pronged approach. Crit Rev Ther Drug Carrier Syst 2000; 17(4):377–424.

19. Laitinen M, Hartikainen J, Hiltunen MO, Eranen J, Kiviniemi M, Narvanen O, Makinen K, Manninen H, Syvanne M, Martin JF, Laakso M, Ylä-Herttuala S. Catheter-mediated vascular endothelial growth factor gene transfer to human coronary arteries after angioplasty. Hum Gene Ther 2000; 11(2):263–270.

20. Lesko LJ, Rowland M, Peck CC, Blaschke TF. Optimizing the science of drug development: opportunities for better candidate selection and accelerated evaluation in humans. J Clin Pharm 2000; 40:803–814.

2
Sustaining Transgene Expression In Vivo

Nelson S. Yew and Seng H. Cheng
Genzyme Corporation, Framingham, Massachusetts, U.S.A.

I. INTRODUCTION

Preclinical studies have demonstrated the potential of gene therapy to reverse, albeit transiently, several disease phenotypes. For some indications such as cancer, these short-term effects may be sufficient. However, for a large proportion of diseases being considered for gene therapy, sustained expression of the therapeutic gene will be required. For genetic disorders such as cystic fibrosis, hemophilia, muscular dystrophy, or lysosomal storage disorders, the therapeutic protein will need to be supplied to the appropriate tissues essentially for the lifetime of the patient. Although one can repetitively administer therapy, for many current gene delivery vectors this remains problematic, and for an organ such as the brain, frequent redosing may not be feasible. If gene therapy is to become a truly practical mode of treatment, the therapeutic gene will need to be expressed for a sustained length of time. How long will depend on the disease, route of delivery, and ability to redose, but a period of several months at therapeutic levels is considered a desirable goal. One can consider three basic requirements to achieve this goal: (1) stable transduction of the target tissue, (2) maintenance of gene expression, and (3) minimizing the immune response to the vector and the transgene product. This chapter will review the ability of current viral and nonviral vectors to fulfill each of these three requirements.

II. STABILITY OF TRANSDUCED VECTOR GENOMES

The first basic requirement for sustained expression is to maintain the transduced vector within the target cells. Vector DNA can exist either in an extrachromo-somal episomal form, as occurs with adenoviral (Ad) or plasmid DNA (pDNA) vectors, or as one or more integrated copies in the genome, as occurs with adeno-associated viral (AAV) and lentiviral vectors. Is integration necessary for stable gene expression? Although there are examples of stable nonintegrating vectors in tissues, in other instances elements that promote integration, episomal replica-tion, or nuclear retention appear to be necessary to increase vector DNA stability.

A. Stability of Episomal Vector DNA

1. Adenoviral Vectors

Adenovirus is a linear double-stranded DNA virus whose genome persists as an episome in the infected cell, but the stability of the transduced DNA is highly dependent on surviving the strong host immune response to the virus. Adenoviral vectors primarily transduce the liver when delivered intravenously, where 90% of the vector DNA is lost within 24 hrs postinjection as a result of an innate immune response to the virus [1]. This is followed by a cytotoxic T-lymphocyte (CTL) response and destruction of the infected cells with additional loss of vector DNA [2,3]. However, there is a surviving population of infected cells within which adenoviral DNA can be stably maintained for at least several weeks. This stability may be aided by the presence of terminal protein that is covalently attached to the end of the adenoviral genome, which may help anchor the adenovi-ral DNA through its interaction with the nuclear scaffold [4–7]. Very long-term stability of viral DNA has been observed with adenoviral vectors deleted of all viral genes ("gutless" or helper-dependent vectors), with expression being sus-tained from several months to a year [8]. Compared to first-generation E1-deleted vectors, significantly higher levels of vector DNA were detected by Southern analysis at 8 weeks postinfection with fully deleted vectors [9]. This increased survival of vector DNA is due to the greatly decreased CTL response. Thus, it would appear that adenoviral DNA can exist stably as an episome so long as it can avoid the significant cellular immune response of the host.

2. Plasmid DNA Vectors

Transduced plasmid vectors also exist as extrachromosomal entities. An ex-tremely small proportion can be maintained stably in particular tissues, but even this amount is eventually degraded over time. pDNA is usually delivered com-plexed to cationic lipids, polycationic polymers, or other molecular conjugates or as naked DNA when injected into muscle. All of these delivery vehicles are quite inefficient, and the bulk of the injected pDNA is eliminated within the first 24 hr after administration by either the intravenous (IV), intranasal, or intramuscular (IM) route [10–12]. For example, the half-life of pDNA was observed to be less than 5 mins after IV injection of pDNA complexed with DMRIE-DOPE [13] due to the presence of serum nucleases. The pDNA that is delivered into the cell must then survive the nucleases present in the cytosol, which has been shown to degrade naked pDNA within 50–90 mins after microinjection into the cytoplasm of HeLa and COS cells [14]. Whether this time is extended when the DNA is present in a complex has not been discerned. pDNA that reaches the nucleus should conceivably be more stable, although the kinetics of DNA loss from the nucleus are actually unknown, as in vivo studies have not been able to distinguish nuclear from cytoplasmic pDNA. Total pDNA levels, though, decline much more slowly after the first few days. At day 7 after IV injection of cationic lipid–pDNA complex, pDNA could be detected by polymerase chain reaction (PCR) in all tissues, with the greatest amounts seen in muscle, spleen, liver, heart, and bone marrow [13]. At day 28 postinjection, these levels decreased further, with the largest decrease being in the liver and the smallest in the lung. At 6 months postinjection, pDNA could be detected only in the muscle [13].

There are very low levels of persistent expression from these residual amounts of pDNA at the later time points. In the lung and liver, in the absence of promoter inactivation, expression can be detected out to at least 10 weeks, and in the muscle, luciferase expression could be detected for up to 1 year postinjection [15]. With a few exceptions these levels are generally far below what is required for a therapeutic effect. So a primary issue with pDNA stability is first obtaining sufficient transduction of the target tissue and having a greater proportion of pDNA reach the nucleus. This would permit a longer period of expression at therapeutic levels as pDNA levels decline slowly over time.

3. Plasmid-Based Transposons

One approach to increase the stability of pDNA is to promote its integration into the genome. This can be achieved by adding sequence elements from transposons, which are mobile genetic elements capable of moving from one region of a chromosome to another [16]. The Tcl/mariner superfamily of transposons consists of a transposase flanked by inverted repeats (IRs) [17]. The transposase enzyme binds to the IR sequences, catalyzing excision and integration by a cut and paste mechanism. The target sequence for the transposase is a TA dinucleotide, so integration is essentially random [17]. Although vertebrate transposons have through evolution been inactivated by mutation, Ivics et al. [18] eliminated the inactivating mutations in a transposon from fish and restored its enzymatic activity. This reconstructed transposon was named Sleeping Beauty and was shown to mediate transposition in human cells. Using this system, Yant et al. [19] con-

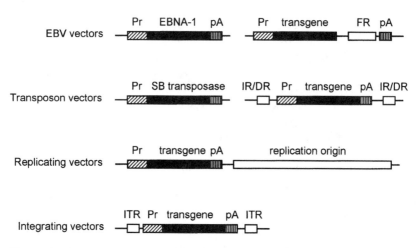

Figure 1   Strategies to increase the stability of transduced vector genomes. Boxes indicate different sequence elements, lines indicate vector backbone sequence, which could be either plasmid or virus. EBV, Epstein–Barr virus; Pr, promoter; pA, polyadenylation signal; FR, EBV family of repeats; SB, Sleeping Beauty; IR/DR, inverted repeats/direct repeats; ITR, inverted terminal repeats.

structed a plasmid containing an α-$_1$-antitrypsin (AAT) expression cassette flanked by two ITRs and a second plasmid that expressed the Sleeping Beauty transposase from the cytomegalovirus (CMV) promoter (Fig. 1). The plasmids were codelivered into the livers of mice and AAT expression in the serum was measured. The ITR–plasmid was found to be inserted into approximately 5–6% of the hepatocytes. Compared to control mice, mice that received both the ITR and transposase plasmids expressed approximately 40-fold more AAT for more than 6 months. This system also expressed therapeutic levels of Factor IX for greater than 5 months (Fig. 2) [19]. An obvious concern of such a system is safety, as there is always a risk from random integration events into the genome and an additional unknown risk from injecting millions of copies of an active transposase into the body. The stability of the transpositions also needs to be determined. Nevertheless, this system offers an effective approach to increase both pDNA transduction efficiency and stability.

4.   Replicating Vectors and Nuclear Retention

A vector that can be maintained stably as an episome would theoretically be safer than a vector that integrates randomly into the genome. In dividing cells, such as hematopoietic stem cells, regenerating vascular endothelium, and tumor cells, the vector would also need to replicate to be maintained without dilution [20].

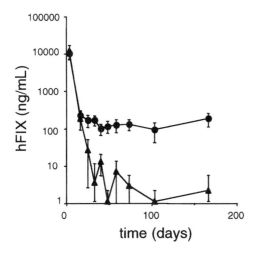

Figure 2 Long-term transgene expression mediated by Sleeping Beauty transposon vectors in mice. 25 μg of pTEF1α-FIX (contains cassette-expressing human Factor IX from the EF1α promoter flanked by inverted repeats) was injected via the tail vein with 1 μg of pCMV-SB (expressing the Sleeping Beauty transposon; filled circles) or 1 μg of pCMV–BS (control; filled triangles). Serum Factor IX concentrations were measured at different days post-injection. (From Ref. 19.)

In slowly dividing or nondividing cells, such as hepatocytes and myoblasts, the vector does not necessarily need to replicate but still must be retained efficiently in the nucleus. Vectors based on the Epstein–Barr virus (EBV) appear to provide both replicative capability and nuclear retention signals [21]. EBV plasmids contain the latent viral DNA replication origin (oriP), which consists of two parts: the family of repeats (FR) and the dyad symmetry sequence [22]. The plasmids also express the EBNA-1 nuclear antigen, which binds to the FR [23]. Expression of EBNA-1 is sufficient for replication of the plasmid, which occurs once per cell cycle. EBNA-1 also binds to the nuclear scaffold, prolonging retention of the plasmid in the nucleus [24]. When an EBV plasmid was combined with hemagglutinating virus of Japan (HVJ) liposomes and injected into the liver of mice, expression persisted out to 35 days (approximately 25% of peak expression) compared to less than 14 days for the non-EBV vector [25]. A similar system using a separate plasmid to express EBNA-1 (see Fig. 1) also showed sustained expression in the lung after IV injection of cationic lipid–plasmid complexes (approximately 10% of peak expression at 11 weeks) [26]. Although EBV plasmids have been shown to replicate in human cells, they do not appear to replicate in rodent cells, so the increased persistence of expression in the mouse liver is presumably due, at least in part, to the increased nuclear retention of the pDNA.

It may be possible to substitute the replicative and retention functions provided by EBNA-1 with mammalian replication origins and retention signals (see Fig. 1). However, unlike the relatively small and discrete origins found in viruses or yeast, the origins of higher eukaryotes are much larger and less well defined [27,28]. Nevertheless, genomic sequences containing putative replication origins have been cloned into plasmids and shown to confer the ability to replicate in tissue culture cells [29]. Nuclear retention could conceivably be achieved using matrix attachment regions (MARs), which are sequence elements that associate with the nuclear scaffold [30,31]. Further studies are needed to determine the benefit from using either mammalian origins or MARs to increase the persistence of expression in vivo.

B. Integrating Viral Vectors: AAV and Lentivirus

AAV viruses can stably integrate their genomes into the target cell, although recombinant AAV vectors may be mainly episomal. Regardless, AAV vectors can achieve very stable transgene expression. For example, intraportal delivery of an AAV–Factor IX vector into hemophilic mice or canines resulted in sustained expression for greater than 17 months and partial phenotypic correction [32–34]. IM injection of an AAV– erythropoietin vector into monkeys resulted in undiminished expression for nearly 20 weeks [35]. Injection of an AAV–tyrosine hydroxylase vector into the brain of a rat model for Parkinson's disease resulted in expression for up to 4 months and behavioral correction of the lesioned rats [36]. These and other examples of sustained expression from AAV vectors are one reason for their increasing popularity.

A second integrating vector is based on lentiviruses, most commonly human immunodeficiency virus (HIV) type 1 [37,38]. Unlike retroviruses such as those derived from murine leukemia virus, lentiviruses can transduce nondividing cells efficiently, and they can transduce a broad range of cell types when pseudotyped with the vesicular stomatitis G protein. Expression was sustained for 6 months after injection of a lentiviral vector into the striatum and hippocampus of adult rats [39], and green fluorescent protein (GFP) expression persisted for 22 weeks after transduction of CD34$^+$ hematopoietic stem cells [40,41]. Long-term expression was also seen after injection into the rat liver parenchyma and the muscle [42].

C. Simian Virus Vectors

A viral vector that may persist either episomally or possibly integrate into the genome is based on simian virus 40 (SV40), a papovavirus that contains a 5.2-kb circular double-stranded genome. Strayer et al. [43,44] inserted various transgenes in place of the large T-antigen gene to generate a replication-deficient virus. These vectors transduce CD34$^+$ hematopoietic progenitor cells in vitro very efficiently, and they were stably maintained in these cells for at least 3 months when reimplanted into mice [45–47]. IV injection of an SV40–luciferase vector resulted in sustained transgene expression in several tissues for greater than 3 months [44]. There is some evidence of random integration of these vectors, although this has not been extensively characterized.

III. MAINTENANCE OF TRANSCRIPTIONAL ACTIVITY

The vector DNA that is successfully transduced into the target tissue must also be actively transcribed and translated to achieve sustained gene expression. Downregulation of gene expression over time has been observed, and is thought to be mainly due to promoter inactivation. This phenomenon has most often been seen using the strong promoter from human CMV. Several studies using viral and nonviral vectors have shown that transgene expression from this promoter peaks at 2–3 days postadministration, and then declines to near background levels within 4 weeks [10,48,49]. What is the cause of this inactivation and why is promoter activity more persistent in certain tissues than others? There is currently little mechanistic data available to answer these questions, but investigators have explored the use of cellular promoters, synthetic promoters, and transactivating systems to increase the persistence of transcriptional activity.

A. Viral Promoters: CMV

One of the most commonly used promoters is from the immediate-early gene of human CMV (CMV IE), a member of the herpesvirus family. The popularity of the CMV promoter stems from its very strong transcriptional activity in a variety of cells [50,51]. The CMV IE promoter region extends to approximately -550 relative to the transcriptional start site and contains a high density of transcription factor–binding sites [50]. These sites include multiple 18-, 19-, and 21-bp repeated motifs that bind the transcription factors NFκB/rel, CREB/ATF, and Sp1, respectively, as well as binding sites for AP1, retinoic acid, ETS, and SRE [52]. The promoter can be activated via the protein kinase A, protein kinase C, and Ca^{2+}/calmodulin C pathways, and is also activated by heat shock, stress, phorbol esters, mitogens, tumor necrosis factor-α (TNF-α) and other viral transcription factors [52]. These multiple activation pathways account for at least part of its strong short-term activity, but may also factor in the downregulation of the promoter that is often observed in vivo. The reason for this inactivation is unclear, but perhaps could be viewed as either (1) an active repression mechanism that "turns off" transcriptional activity over time or (2) one or more factors "turns on" the promoter, and without these factors promoter activity is quite low.

Studies in transgenic mice that express a reporter gene using the CMV promoter provide some indirect evidence as to why the CMV promoter may not be appropriate for expression in certain tissues. In vitro, the promoter is transcriptionally active in virtually all cell types, and in vivo the promoter is also active in a broad range of tissues. However, in mice that express the Escherichia coli chloramphenicol acetyltransferase (CAT) or β-galactosidase (lacZ) gene from the CMV promoter, the level of transgene expression varied widely from tissue to tissue. Reports using different transgenic mice are not completely congruent, but the highest levels of expression generally were observed in organs such as the stomach, skeletal muscle, and spleen, whereas little or no expression was seen in the liver and lung, two of the main targets for gene therapy vectors [53–55]. These results suggest that either the activity from the integrated CMV promoter is being inhibited in the lung and liver, or some factor required for activity is missing from these tissues.

Cytokines can negatively affect CMV promoter activity transiently, although it is unclear if cytokines are principally involved in the continued inactivity of the promoter at later time points. Interferon-α (IFN-α), IFN-γ, and TNF-α have been shown to inhibit expression from the mouse and human CMV promoters in tissue culture cells [56–58]. β-Galactosidase (β-gal) expression from C2C12 myoblasts infected with CMV–β-gal or Rous sarcoma virus (RSV)–β-gal adenoviral vectors decreased 24–72% after addition of either IFN-γ or TNF-α, whereas addition of other proinflammatory cytokines, such as IL-1, IL-6, or transforming growth factor-β (TGF-β) had no effect on promoter activity [59]. IFN-γ and TNF-α were also shown to act synergistically, further depressing activity [59]. Administration of either adenoviral vectors or cationic lipid–pDNA complexes induces a transient inflammatory response that includes the induction of IFN-γ and TNF-α [60]. These elevated cytokine levels precede the loss of transgene expression. However, IFN-γ and TNF-α levels fall to basal levels within a few days, whereas expression from the CMV promoter remains depressed. Thus, if cytokines were principally responsible for the inactivation of the CMV promoter, their effects would have to be irreversible. Evidence for this in vivo remains to be determined.

A converse theory is that CMV promoter activity is essentially dependent on one or more transcriptional activators that are transiently induced early after vector delivery. Promoter activity then declines in concert with the decline in the levels of these activators. Loser et al. [48] have provided evidence that the activity of the CMV promoter in mouse liver is dependent on the transcription factor NFκB. Mice were given a partial hepatectomy or lipopolysaccharide (LPS) 28 days postinfection with an adenoviral vector. These treatments, which are known to activate NFκB activity in hepatocytes, reactivated the CMV promoter as evidenced by an increase in transgene RNA levels [48]. Elevated levels of nuclear-associated NFκB were detected after IV delivery of adenovirus, which

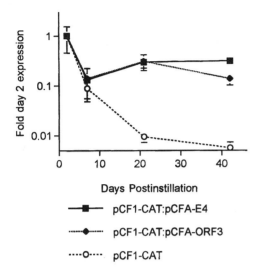

Figure 3 Effect of polypeptides from the adenoviral E4 region on the persistence of CAT expression from a CMV promoter plasmid (pCF1-CAT). Cationic lipid GL-67 was complexed with either pCF1–CAT plus pCFA–E4 (expressing the entire Ad E4 region from the CMV promoter), pCF1–CAT plus pCFA–ORF3 (expressing only ORF3), or pCF1–CAT alone. Complexes were instilled into the lungs of nude BALB/c mice. Lungs were harvested at 2, 7, 21, and 42 days postinstillation and CAT assays were performed. (From Ref. 65.)

peaked at 4 days postinfection and declined to low levels by day 14. This profile paralleled the loss of transgene expression.

Besides NFκB, an early viral protein from adenovirus can also positively regulate the CMV promoter. This discovery was made while attempting to reduce viral gene expression by deleting portions of the adenoviral genome. In immune-deficient mice, CMV-driven transgene expression can persist in both the lung and liver from adenoviral vectors containing an intact E4 region. However, expression did not persist from vectors deleted completely of E4 or vectors expressing only open reading frame 6 (ORF6) [61,62]. Sustained expression could also be achieved by coinfection of the E4-deleted virus with an E4-containing virus, showing that E4 function could be supplied in trans [61]. Thus, one or more proteins expressed from the E4 region were responsible for sustaining transgene expression. Mutation of each of the individual ORFs indicated that expression of ORF3 was required [63]. Adenoviral vectors were also constructed which contained different subsets of the ORFs. The expression of ORF3 was sufficient for sustained transgene expression from the CMV promoter in the liver, but expres-

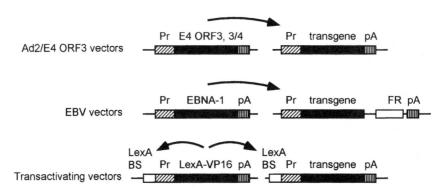

Figure 4 Strategies for transactivation of the promoter. Boxes indicate different sequence elements, lines indicate vector backbone sequence, which could be either plasmid or virus. E4 ORF3/4, adenovirus E4 open reading frames 3 or 3 and 4; Pr, promoter; pA, polyadenylation signal; FR, EBV family of repeats; BS, LexA-binding site; LexA–VP16, LexA DNA-binding domain herpes virus VP16 activation domain fusion protein.

sion of both ORF3 and ORF4 was required for sustained expression in the lung [64]. This effect could also be duplicated in the context of plasmid vectors, although in this case, expression of ORF3 alone was sufficient to increase the persistence of transgene expression in the lung (Fig. 3) [65].

There are a few examples of sustained expression from the CMV promoter without the need for any activating factor, most notably in muscle, where expression actually increases over the first 7 days and can persist for several weeks (Table 1). Factor IX expression from the CMV promoter in an AAV vector or β-gal expression in a helper-dependent Ad vector injected into the muscle also persisted for several weeks, and transgene expression from the CMV promoter in plasmid vectors injected IM was sustained. What is unique about muscle that permits long-term expression from the CMV promoter? It remains to be determined if there are particular activated factors in muscle that are absent in lung or liver.

B. Cellular Promoters

An alternative to viral promoters are cellular promoters that express endogenous genes. There is a myriad of possible promoters from which to choose, but the number that have been evaluated so far for persistent expression in vivo is relatively few (see Table 1). The advantages of using a cellular promoter are that some may be less prone to inactivation over time, particularly constitutively active, "housekeeping"-type promoters, and they may confer tissue-restricted ex-

Table 1 Representative In Vivo Studies of Various Promoters Evaluating the Persistence of Transgene Expression

Promoter[a]	Type[b]	Vector[c]	Animal	Tissue	Transgene[d]	Last timepoint	Approximate % of peak expression	Reference
AAT	C	HD Ad	mice	liver	AAT	40 weeks	100	138
Albumin	C	AAV	mice	liver	G-CSF	5 months	100	76
Albumin	C	AAV	mice (RAG-1)	liver	AAT	18 weeks	50	156
Albumin	C	Ad	mice	liver	AAT	40 days	10	124
Albumin	C	Ad	mice	liver	FVIII	12 months	2	77
EF1α	C	AAV	mice	liver	FIX	6 months	100	71
EF1α	C	Ad	mice (SCID)	liver	AAT	13 weeks	100	70
EF1α	C	HD Ad	mice	liver	epo	6 months+	85–100	72
LSP	C	AAV	mice	liver	FIX	20 weeks	100	134
CMV-actin	H	AAV	mice (RAG-1)	muscle	FIX	14 weeks	90	157
CMV-actin	H	AAV	mice (RAG-1)	liver	AAT	18 weeks	80–100	156
HBV-CMV	H	Ad	mice	liver	LDL receptor	77 days	75–100	48
CMV	V	AAV	rats	brain	β-gal	3 months		36
CMV	V	AAV	dogs	muscle	FIX	32 weeks	100	33
CMV	V	AAV	mice	muscle	β-gluc	11 weeks		158
CMV	V	plasmid	mice	muscle	luciferase	6 months	100	159
RSV	V	AAV	mice	liver	FIX	125 days	100	160
RSV	V	plasmid	mice	muscle	luciferase	19 months	20	15

[a] AAT, α₁-antitrypsin; EF1α, elongation factor 1α; LSP, liver-specific promoter containing thyroid hormone–binding globulin promoter and two copies of the α₁-microglobulin/bikunin enhancer; CMV-actin, CMV enhancer and β-actin promoter; CMV, cytomegalovirus; HBV-CMV, enhancer II from hepatitis B virus and human CMV promoter; RSV, Rous sarcoma virus LTR.

[b] C, cellular; H, hybrid; V, viral.

[c] AAV, adenoassociated virus; Ad, adenoviral; HD Ad, helper-dependent adenovirus.

[d] AAT, α₁-antitrypsin; G-CSF, granulocyte-colony stimulating factor; FVIII, Factor VIII; FIX, Factor IX; epo, erythropoietin; β-gal, β-galactosidase; β-gluc, β-glucoronidase.

pression. A disadvantage of using cellular promoters is that they are often less transcriptionally robust compared to CMV.

The promoter from the gene encoding elongation factor-1α (EF1α) is one example of a widely expressed, constitutive cellular promoter. EF1α is an abundant protein that has an essential role in protein synthesis, catalyzing the guanosine triphospate (GTP)–dependent binding of aminoacyl–tRNA to ribosomes [66]. The EF1α promoter was initially utilized for high-level expression in tissue culture cells [67–69]. This promoter has also been evaluated in the context of an Ad–AAT vector in immunodeficient SCID mice [70]. Compared to the CMV and RSV promoters, the EF1α promoter produced the highest sustained levels of AAT at 12 weeks [70]. In an AAV–Factor IX vector, the EF1α promoter expressed Factor IX for 6 months, and in a helper-dependent Ad–erythropoietin vector, the promoter expressed erythropoietin for greater than 6 months (85–100% peak expression at 6 months) [71,72].

Other groups have employed tissue-specific promoters, such as the liver-specific albumin promoter. Albumin is the most abundant protein in plasma, and is transcribed primarily in the liver. Several liver-specific elements within the promoter account for its restricted expression [73–75]. The albumin promoter was compared to other promoters such as CMV, phosphoglycerokinase (PGK), chicken β-actin, and the Moloney murine leukemia virus long terminal repeat (MMV-LTR) using AAV–AAT vectors in immunodeficient RAG-1 mice [70]. The albumin and MMV-LTR promoters gave the highest levels of expression and the levels persisted out to 18 weeks. In an AAV vector expressing granulocyte colony-stimulating factor (G-CSF), the albumin promoter expressed G-CSF for 5 months, and in an Ad–Factor VIII vector, the albumin promoter expressed Factor VIII in the liver for 1 year [76,77]. The levels of Factor VIII declined to approximately 2% of peak levels at 1 year, but were still above therapeutic levels.

C. Synthetic Promoters

The optimal promoter may be derived from neither viral nor cellular sequences, but rather constructed from a novel synthetic sequence selected for high-level, sustained expression. Such a promoter could combine the features of a high density of transcription factor–binding sites for strong activity while avoiding sequence elements responsible for promoter inactivation. Li et al. [78] randomly assembled several transcription factor–binding sites from various promoters displaying muscle-restricted expression, and then inserted these sequences 5′ of a minimal α-actin promoter. The library of random promoters was first screened in vitro, and then promising candidates were tested in mouse muscle. The best promoter was found to express luciferase at levels six to eightfold greater than that of CMV, and also exhibited muscle-specific expression. Promoters from a completely random sequence were created by Edelman et al. [79], who inserted

18-mer oligonucleotides upstream of a minimal promoter. They then constructed retroviral vectors expressing GFP from these promoters selected for expression by fluorescence activated cell sorting. The selected vectors contained promoters composed of novel composites of transcription factor–binding sites. Conceivably, one could develop an analogous screening assay to select for sustained gene expression in vivo from a library of random promoter sequences.

D. Transactivating Systems

To increase the relatively weak activity of cellular promoters, some investigators have expressed either in cis or in trans transcriptional activators that bind to a specific site just upstream of the promoter. The EBV-based vectors described earlier are an example of such a system, as EBNA-1 binding to the origin is known to activate transcription (Fig. 4) [80]. Nettlebeck et al. [81] used a chimeric transcriptional activator to boost the expression of the endothelium-specific but poorly expressing von Willebrand factor (vWF) promoter. One plasmid contained seven tandem lexA-binding sites upstream of the vWF promoter, whereas another plasmid expressed a chimeric transcription factor composed of the lexA DNA-binding domain fused to the strong viral VP16 activation domain (Fig. 4). Cotransfection of both plasmids into endothelial cells in vitro increased expression from the vWF promoter 14- to greater than 100-fold [81]. A related strategy used serum response factor (SRF), a ubiquitous transcription factor, to transactivate a muscle-specific skeletal actin promoter (skA) [82]. Cotransfection of a SRF-expressing plasmid with a skA promoter–SEAP reporter plasmid increased the level and duration of SEAP expression in the muscle. In theory, a self-perpetuating positive feedback loop could be established by having the transactivator activate its own expression in addition to the transgene by placing binding sites into the promoter expressing the transactivator (Fig. 4) [83]. Alternatively, both the transgene and transactivator could be expressed from one transcript using an internal ribosomal entry site [83].

IV. REDUCING THE IMMUNE RESPONSE TO INCREASE DURATION OF EXPRESSION

Of the three requirements for sustained gene expression, reducing the immune response to the vector and the transgene is perhaps the most consequential. This is especially true for viral vectors, as inhibiting the immune response can lead to dramatic increases in persistence of expression. This section will review strategies to suppress transiently the immune response and to reduce the immunogenicity of both viral and nonviral vectors.

A. Nature of the Immune Response to Adenoviral Vectors

The immune response to adenoviral vectors has been described to consist of three phases: (1) an early, innate response, (2) a CTL/antibody response, and (3) a gradual hypertrophy and death of transduced cells due to leaky, low-level expression of viral proteins [9]. The innate response to adenoviral vectors is one typical to any viral pathogen, characterized by a rapid induction of cytokines, such as interferons and chemokines, activation of complement, and recruitment of neutrophils and macrophages [84–89]. As a result, 90% of the adenoviral DNA is degraded within the first 24 hrs after vector administration (1). Following this early response, $CD4^+$ and $CD8^+$ T cells are activated leading to the destruction of virus-infected cells within 2–3 weeks [2,3,90,91]. Neutralizing antibodies to the capsid proteins are also generated preventing the ability to effectively readminister the vector [2,92–94]

B. Immunesuppression Strategies

1.   General Immunesuppressants

Drugs used to immunosuppress organ transplant patients are being employed to reduce the immune response to adenoviral vectors [95–100]. Examples include cyclosporin A, a compound isolated from a Tolypoiciodium fungus, and FK506 (now known as tacrolimus), a compound isolated from Streptomyces bacteria. These drugs act by disrupting intracellular signaling pathways important for T-cell proliferation. Treatment of hemophilic dogs with cyclosporin A was shown to prolong Factor IX expression from an Ad–Factor IX vector, although there was no decrease in anti-Ad antibodies and the vector could not be readministered [101]. Mice treated with FK506 resulted in prolonged dystrophin expression after IM injection of an Ad–dystrophin vector [99]. Other immune suppressants, such as deoxyspergualin, also reduced the immune response to adenovirus and prolonged expression [96,102,103]. However, these compounds are somewhat toxic to the kidney and other organs, and their use in a gene therapy regimen must be carefully evaluated. Nevertheless, their effects highlight the paramount involvement of the immune response to first-generation Ad vectors in limiting long-term expression.

2.   Inhibiting the Innate Immune Response

A second approach is to inhibit the innate immune response to the vector, and a key component of this response is the ingestion of foreign particles by macrophages. Macrophages can be depleted transiently using liposomes encapsulating dichloromethylene bisphosphonate (clodronate) [104,105]. Upon IV administration, the liposomes are phagocytosed by hepatic Kupffer cells and splenic macrophages, inducing the cells to undergo apoptosis [105]. When clodronate liposomes were injected IV 2 days prior to injecting an Ad–chloramphenicol acetyltransferase vector, the levels of CAT expressed were both increased and more persistent [106]. Without clodronate pretreatment, AAT levels from an RSV promoter–AAT Ad vector dropped a 1000-fold over 30 days, but in mice pretreated with clodronate, AAT levels dropped less than 10-fold over the same period and remained detectable out to 180 days postinjection [107]. Clodronate liposomes also prolonged Factor IX expression from an Ad–Factor IX vector, inhibited the generation of an antibody response to Factor IX, and partially inhibited the anti-Ad antibody response [108].

3. Disrupting T-cell–Antigen-Presenting Cell Interactions

Another immune target that is central to both CTL and antibody responses is T-cell activation, which requires T-cell interaction with B cells and antigen-presenting cells (APCs). This interaction can be inhibited by disrupting binding of APCs to the CD4 receptor or disrupting binding of costimulatory molecules; for example, CD40 to the CD40 ligand or B7 to CD28 (Fig. 5). In the absence of signaling through CD4 or the costimulatory molecules, T cells will become inactivated and induce a state of tolerance or anergy to the antigen. Treatment with depleting or

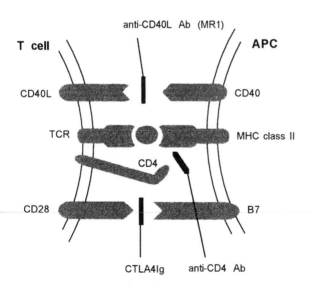

Figure 5 Approaches to inhibit the immune response to adenoviral vectors by disrupting the interaction between T cells and antigen-presenting cells (APC). TCR, T-cell receptor.

nondepleting antibodies to CD4 prior to administration of Ad vectors has been shown to prolong transgene expression [109–111]. Rats treated with a nondepleting anti-CD4 antibody prior to instillation of an Ad vector into the lung exhibited reduced lung inflammation, blocked development of anti-Ad neutralizing antibodies, and showed increased persistence of expression [111].

Other groups have used CTLA4Ig, a hybrid molecule composed of the extracellular domain of CTLA4 fused to an immunoglobulin IgGFc domain, which binds to B7 and blocks its interaction with CD28. Treatment with CTLA4Ig reduced $CD4^+$ infiltration and prolonged AAT expression from an Ad–AAT vector for greater than 5 months [112]. Nakagawa et al. [113] expressed CTLA4Ig from an adenoviral vector, and when coadministered with an Ad–β-gal vector, prolonged β-gal expression, inhibited production of anti-Ad antibodies, and permitted readministration of the vector.

Equally effective is to target the interaction between CD40 (on the APC) and CD40 ligand (on activated $CD4^+$ T cells). Mice were given intraperitoneal injections of MR1 an anti-mouse CD40L (gp39) hybridoma, approximately every other day for 6–9 days, just prior to and immediately after injection of an Ad vector. Transgene expression was more persistent only in those mice transiently immunosuppressed with MR1 (114–116). MR1 inhibited both the CTL response to Ad and the development of anti-Ad neutralizing antibodies [114,115]. MR1 also permitted a second and third administration of Ad vector, as long as the MR1 dosing regimen was rigorous enough to inhibit completely development of anti-Ad antibodies [108]. Investigators have also combined MR1 with CTLA4Ig or MR1 with clodronate liposomes [117,118].

4. Tolerization Strategies

A more difficult but potentially highly effective approach is to induce tolerance to the vector and transgene. Orally administered antigens have been shown to render patients immunologically unresponsive to the foreign protein, likely by active suppression mechanisms and clonal anergy or deletion, although the mechanism is not well understood [119,120]. Mice or rats that were fed multiple doses of ultraviolet-inactivated adenoviral particles became tolerized to the virus. Subsequent administration of Ad vectors resulted in increased persistence of expression compared to the profile seen in nontolerized animals [121–123].

C. Decreasing the Immunogenicity of the Vector

Having to immunesuppress patients transiently could be avoided if the vector could be rendered nonreactive to the host immune system. For adenoviral vectors, the following examples demonstrate that reductions in the immune response to the vector and the transgene can significantly increase sustained gene expression.

1. Use of Tissue-Specific Promoters

A simple, effective alteration is to use a tissue-specific promoter to express an antigenic transgene. Pastore et al. [124] injected C3/HeJ mice with first-generation E1-deleted Ad vectors expressing human α_1-antitrypsin from either the non-specific PGK promoter or the liver-specific albumin promoter. In mice that received the phosphoglycerate kinase (PGK) promoter vector, AAT expression declined precipitously within the first 3 weeks, and the mice developed antibodies to AAT. However, in mice that received the albumin promoter vector, AAT expression persisted for greater than 40 weeks, and the mice did not develop anti-AAT antibodies [124]. A proposed explanation for these observations is that with a liver-specific promoter, AAT was not expressed in APCs, making it less likely that an antibody response to AAT would develop. The effect may be limited to this particular transgene in C3/HeJ mice, in which AAT appears not to be highly immunogenic [124]. Nevertheless, the use of a tissue-specific promoter will likely be beneficial in most vector systems.

2. Partially and Fully Deleted Ad Vectors

Even with a tissue-specific promoter, leaky expression of viral genes from Ad vectors limits persistence of expression. E1-deleted vectors still express early and late viral genes, especially at higher multiplicities of infection [3,90]. One strategy to reduce this problem has been to delete some or all of the viral genes within the vector. Several investigators have made additional deletions in the E2a and E4 regions with varying effects on toxicity and persistence [125–136]. One of the more effective modifications deletes the polymerase gene. An E1- and polymerase-deleted Ad vector exhibited increased duration of β-gal expression in immunocompetent mice [137].

The next logical progression has been the development of "gutless" or "helper-dependent" vectors in which all the viral coding sequences are removed, retaining only the inverted terminal repeats and packaging signal. A helper virus supplies in trans the missing proteins required for replication and packaging. Several studies have shown that these vectors induce significantly less inflammation and liver toxicity compared to first-generation adenoviral vectors. Schiedner et al. [138] constructed a helper-dependent virus that contained the complete 19-kb human α_1-antitrypsin gene under the control of its own promoter. When this virus was injected into C57B1/6 mice, AAT levels plateaued at 3 weeks and remained essentially undiminished for 10 months. When this same vector was injected into baboons, expression persisted for more than 1 year in two of the three animals, with a slow decline to less than 10% of peak expression at 24 months (Fig. 6) [8]. Therapeutic levels of Factor VIII, leptin, dystrophin, and erythropoietin were also achieved for greater than 6 months in mice and canines using fully deleted vectors [72,139,140]. Maione et al. [72] showed that compared

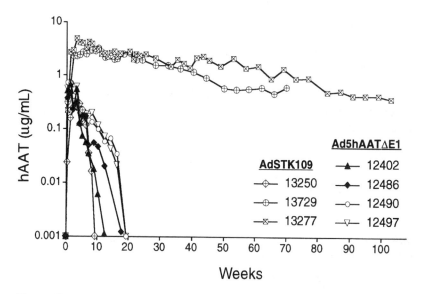

Figure 6 Expression of human α_1-antitrypsin (hAAT) in baboons after administration of Ad5hAATΔE1, a first-generation adenoviral vector, or AdSTK109, a gutless adenoviral vector. Baboons 12402 and 12486 received 6.2×10^{11} particles per kilogram (particles/kg) of Ad5hAATΔE1. Baboons 12490 and 12497 received 1.4×10^{12} particles/kg of Ad5hAATΔE1. Baboons 13250, 13277, and 13729 received 3.3×10^{11}, 3.9×10^{11}, and 3.6×10^{11} particles/kg, respectively, of AdSTK109. (From Ref. 8.)

to using a first-generation vector, a 100-fold lower dose was sufficient to achieve therapeutic levels of erythropoietin. Even though fully deleted Ad vectors are capable of inducing the formation of neutralizing antibodies to the same degree as first-generation vectors, at this lower dose neutralizing antibodies to the vector were not produced. Consequently, the deleted vector could also be readministered effectively. Thus, fully deleted vectors are a substantial improvement over first- or even second-generation Ad vectors. The major current limitation is the ability to produce sufficient amounts of these viruses, although improvements are being made.

3. Decreased Immune Response to AAV Vectors and Lentiviral Vectors

Vectors derived from AAV and lentivirus are essentially fully deleted vectors, expressing no viral proteins. This feature, more so than their ability to integrate, may explain the sustained expression from these vectors. IM injection of an

AAV–β-gal vector induced a CTL and antibody response to the viral capsid proteins. However, no CTLs or humoral response were developed to the neoantigenic β-gal protein [141]. Joos et al. [142] showed that transduction of dendritic cells is required to generate a CTL response to a foreign transgene in the muscle, and that AAV avoids this by its poor transduction of APCs. Although AAV enters the cytoplasmic compartment of APCs, they do not appear to enter the nucleus [142]. When AAV is delivered by other routes, such as intravenously, intraperitoneally, or subcutaneously, a CTL response to the transgene is generated [143].

Lentiviral vectors also show a decreased immune response compared to that seen with Ad vectors. For example, when an HIV vector was injected into the striatum and hippocampus of adult rats, there was no significant infiltration of lymphocytes or macrophages over that seen with the saline control [39]. Inflammation was also not observed after injecting an HIV vector into the mouse liver parenchyma or into the muscle [42]. However, the dose used in these studies (3×10^7 IU) was very low, which may account for the lack of a response. Nevertheless, like AAV, the decreased immune response is likely due to the absence of viral protein expression from these vectors.

4. Plasmid DNA Vectors

The immune response to nonviral vectors was thought initially not to be a significant problem, as such systems are devoid of any viral components. However, pDNA, either alone or complexed with cationic lipid, is capable of inducing a significant inflammatory response [10,60,144,145]. This is because pDNA, like bacterial DNA, contains unmethylated CpG dinucleotides in particular sequence contexts that are immunostimulatory [146,147]. This response includes the activation of B cells, natural killer cells, dendritic cells, monocytes, and macrophages, with the subsequent release of several proinflammatory cytokines (148–153).

Does this inflammatory response affect the persistence of transgene expression from plasmid vectors? There are several examples of sustained expression from naked DNA in muscle, but the inflammatory response to the pDNA backbone has been shown to decrease the longevity of expression [145]. Methylation of the plasmid was found to decrease this inflammatory response and increase persistence. The immunostimulatory CpG motifs present in the plasmid also function as effective adjuvants, stimulating an antibody response to the transgene [154]. This feature is being exploited for the development of DNA vaccines, but the production of potentially neutralizing antibodies would likely reduce persistence. Elimination of CpG motifs from pDNA has been shown to reduce its inflammatory properties [155], but the effect of these CpG reduced vectors on the duration of gene expression remains to be determined.

V. SUMMARY

An arbitrary, yet reasonable, goal applicable to many diseases targeted for gene therapy is expression at therapeutic levels over several months. In this chapter, several examples have been cited of sustained transgene expression in vivo and for a limited number of indications, the goal has been met or exceeded. AAV, lentiviral, and helper-dependent Ad vectors currently are superior to plasmid vectors or first-generation Ad vectors in achieving sustained expression. Yet there are still outstanding issues that need to be addressed. With AAV and lentiviral vectors, the levels of expression may be too low to treat some diseases effectively. With all viral vectors, redosing remains a significant problem, for even with the most persistent vectors, expression does not last forever, and effective repeated administration will be necessary. Nonetheless, these issues are no doubt solvable, and one can envision truly practical gene therapeutics for many genetic disorders in the not too distant future.

REFERENCES

1. Worgall S, Leopold PL, Wolff G, Ferris B, Van Roijen N, Crystal RG. Role of alveolar macrophages in rapid elimination of adenovirus vectors administered to the epithelial surface of the respiratory tract. Hum Gene Ther 1997; 8:1675–1684.
2. Yang Y, Jooss KU, Su Q, Ertl HC, Wilson JM. Immune responses to viral antigens versus transgene product in the elimination of recombinant adenovirus-infected hepatocytes in vivo. Gene Ther 1996; 3:137–144.
3. Yang Y, Ertl HC, Wilson JM. MHC class I-restricted cytotoxic T lymphocytes to viral antigens destroy hepatocytes in mice infected with E1-deleted recombinant adenoviruses. Immunity 1994; 1:433–442.
4. Bodnar JW, Hanson PI, Polvino-Bodnar M, Zempsky W, Ward DC. The terminal regions of adenovirus and minute virus of mice DNAs are preferentially associated with the nuclear matrix in infected cells. J Virol 1989; 63:4344–4353.
5. Fredman JN, Engler JA. Adenovirus precursor to terminal protein interacts with the nuclear matrix in vivo and in vitro. J Virol 1993; 67:3384–3395.
6. Schaack J, Ho WY, Freimuth P, Shenk T. Adenovirus terminal protein mediates both nuclear matrix association and efficient transcription of adenovirus DNA. Genes Dev 1990; 4:1197–1208.
7. Lieber A, He CY, Kay MA. Adenoviral preterminal protein stabilizes mini-adenoviral genomes in vitro and in vivo. Nat Biotechnol 1997; 15:1383–1387.
8. Morral N, O'Neal W, Rice K, et al. Administration of helper-dependent adenoviral vectors and sequential delivery of different vector serotype for long-term liver-directed gene transfer in baboons. Proc Natl Acad Sci USA 1999; 96:12816–12821.
9. Morral N, Parks RJ, Zhou H, et al. High doses of a helper-dependent adenoviral vector yield supraphysiological levels of alpha1-antitrypsin with negligible toxicity. Hum Gene Ther 1998; 9:2709–2716.

10. Li S, Huang L. In vivo gene transfer via intravenous administration of cationic lipid- protamine-DNA (LPD) complexes. Gene Ther 1997; 4:891–900.
11. Meyer KB, Thompson MM, Levy MY, Barron LG, Szoka FC, Jr. Intratracheal gene delivery to the mouse airway: characterization of plasmid DNA expression and pharmacokinetics. Gene Ther 1995; 2:450–460.
12. Levy MY, Barron LG, Meyer KB, Szoka FC, Jr. Characterization of plasmid DNA transfer into mouse skeletal muscle: evaluation of uptake mechanism, expression and secretion of gene products into blood. Gene Ther 1996; 3:201–211.
13. Lew D, Parker SE, Latimer T, et al. Cancer gene therapy using plasmid DNA: pharmacokinetic study of DNA following injection in mice. Hum Gene Ther 1995; 6:553–564.
14. Lechardeur D, Sohn KJ, Haardt M, et al. Metabolic instability of plasmid DNA in the cytosol: a potential barrier to gene transfer. Gene Ther 1999; 6:482–497.
15. Wolff JA, Ludtke JJ, Acsadi G, Williams P, Jani A. Long-term persistence of plasmid DNA and foreign gene expression in mouse muscle. Hum Mol Genet 1992; 1:363–369.
16. Haren L, Ton-Hoang B, Chandler M. Integrating DNA: transposases and retroviral integrases. Annu Rev Microbiol 1999; 53:245–281.
17. Plasterk RH, Izsvak Z, Ivics Z. Resident aliens: the Tcl/mariner superfamily of transposable elements. Trends Genet 1999; 15:326–332.
18. Ivics Z, Hackett PB, Plasterk RH, Izsvak Z. Molecular reconstruction of Sleeping Beauty, a Tcl-like transposon from fish, and its transposition in human cells. Cell 1997; 91:501–510.
19. Yant SR, Meuse L, Chiu W, Ivics Z, Izsvak Z, Kay MA. Somatic integration and long-term transgene expression in normal and haemophilic mice using a DNA transposon system. Nat Genet 2000; 25:35–41.
20. Calos MP. The potential of extrachromosomal replicating vectors for gene therapy. Trends Genet 1996; 12:463–466.
21. Sclimenti CR, Calos MP. Epstein-Barr virus vectors for gene expression and transfer. Curr Opin Biotechnol 1998; 9:476–479.
22. Reisman D, Yates J, Sugden B. A putative origin of replication of plasmids derived from Epstein-Barr virus is composed of two cis-acting components. Mol Cell Biol 1985; 5:1822–1832.
23. Ambinder RF, Shah WA, Rawlins DR, Hayward GS, Hayward SD. Definition of the sequence requirements for binding of the EBNA-1 protein to its palindromic target sites in Epstein-Barr virus DNA. J Virol 1990; 64:2369–2379.
24. Mattia E, Ceridono M, Chichiarelli S, D'Erme M. Interactions of Epstein-Barr virus origins of replication with nuclear matrix in the latent and in the lytic phases of viral infection. Virology 1999; 262:9–17.
25. Saeki Y, Matsumoto N, Nakano Y, Mori M, Awai K, Kaneda Y. Development and characterization of cationic liposomes conjugated with HVJ (Sendai virus): reciprocal effect of cationic lipid for in vitro and in vivo gene transfer. Hum Gene Ther 1997; 8:2133–2141.
26. Tu G, Kirchmaier AL, Liggitt D, et al. Non-replicating EBV-based plasmids extend gene expression and can improve gene therapy in vivo. J Biol Chem 2000; 275: 30408–30416.

27. Todorovic V, Falaschi A, Giacca M. Replication origins of mammalian chromosomes: the happy few. Front Biosci 1999; 4:D859-D868.

28. Zannis-Hadjopoulos M, Price GB. Eukaryotic DNA replication. J Cell Biochem 1999; Suppl 32–33:1–14.

29. Wohlgemuth JG, Kang SH, Bulboaca GH, Nawotka KA, Calos MP. Long-term gene expression from autonomously replicating vectors in mammalian cells. Gene Ther 1996; 3:503–512.

30. Bode J, Benham C, Knopp A, Mielke C. Transcriptional augmentation: modulation of gene expression by scaffold/matrix-attached regions (S/MAR elements). Crit Rev Eukaryot Gene Exp 2000; 10:73–90.

31. Brylawski BP, Cohen SM, Cordeiro-Stone M, Schell MJ, Kaufman DG. On the relationship of matrix association and DNA replication. Crit Rev Eukaryot Gene Exp 2000; 10:91–99.

32. Herzog RW, Yang EY, Couto LB, et al. Long-term correction of canine hemophilia B by gene transfer of blood coagulation factor IX mediated by adeno-associated viral vector [see comments]. Nat Med 1999; 5:56–63.

33. Chao H, Samulski R, Bellinger D, Monahan P, Nichols T, Walsh C. Persistent expression of canine factor IX in hemophilia B canines. Gene Ther 1999; 6:1695–1704.

34. Snyder RO, Miao C, Meuse L, et al. Correction of hemophilia B in canine and murine models using recombinant adeno-associated viral vectors. Nat Med 1999; 5:64–70.

35. Rudich SM, Zhou S, Srivastava R, Escobedo JA, Perez RV, Manning WC. Dose response to a single intramuscular injection of recombinant adeno-associated virus–erythropoietin in monkeys. J Surg Res 2000; 90:102–108.

36. Kaplitt MG, Leone P, Samulski RJ, et al. Long-term gene expression and phenotypic correction using adeno-associated virus vectors in the mammalian brain. Nat Genet 1994; 8:148–154.

37. Buchschacher GL, Jr., Wong-Staal F. Development of lentiviral vectors for gene therapy for human diseases. Blood 2000; 95:2499–2504.

38. Trono D. Lentiviral vectors: turning a deadly foe into a therapeutic agent. Gene Ther 2000; 7:20–23.

39. Blomer U, Naldini L, Kafri T, Trono D, Verma IM, Gage FH. Highly efficient and sustained gene transfer in adult neurons with a lentivirus vector. J Virol 1997; 71:6641–6649.

40. Miyoshi H, Smith KA, Mosier DE, Verma IM, Torbett BE. Transduction of human CD34+ cells that mediate long-term engraftment of NOD/SCID mice by HIV vectors. Science 1999; 283:682–686.

41. Case SS, Price MA, Jordan CT, et al. Stable transduction of quiescent CD34(+)CD38(−) human hematopoietic cells by HIV-1-based lentiviral vectors. Proc Natl Acad Sci USA 1999; 96:2988–2993.

42. Kafri T, Blomer U, Peterson DA, Gage FH, Verma IM. Sustained expression of genes delivered directly into liver and muscle by lentiviral vectors. Nat Genet 1997; 17:314–317.

43. Strayer DS. SV40 as an effective gene transfer vector in vivo. J Biol Chem 1996; 271:24741–24746.

44. Strayer DS, Milano J. SV40 mediates stable gene transfer in vivo. Gene Ther 1996; 3:581–587.

45. Strayer DS, Pomerantz RJ, Yu M, et al. Efficient gene transfer to hematopoietic progenitor cells using SV40-derived vectors. Gene Ther 2000; 7:886–895.

46. Strayer DS, Kondo R, Milano J, Duan LX. Use of SV40-based vectors to transduce foreign genes to normal human peripheral blood mononuclear cells. Gene Ther 1997; 4:219–225.

47. Strayer DS. Gene delivery to human hematopoietic progenitor cells to address inherited defects in the erythroid cellular lineage. J Hematother Stem Cell Res 1999; 8:573–574.

48. Loser P, Jennings GS, Strauss M, Sandig V. Reactivation of the previously silenced cytomegalovirus major immediate-early promoter in the mouse liver: involvement of NFkappaB. J Virol 1998; 72:180–190.

49. Yew NS, Wysokenski DM, Wang KX, et al. Optimization of plasmid vectors for high-level expression in lung epithelial cells. Hum Gene Ther 1997; 8:575–584.

50. Boshart M, Weber F, Jahn G, Dorsch-Hasler K, Fleckenstein B, Schaffner W. A very strong enhancer is located upstream of an immediate early gene of human cytomegalovirus. Cell 1985; 41:521–530.

51. Foecking MK, Hofstetter H. Powerful and versatile enhancer-promoter unit for mammalian expression vectors. Gene 1986; 45:101–105.

52. Stinski MF. Cytomegalovirus promoter for expression in mammalian cells. In: Fernandez JM, Fernandez J, Hoeffler JP, eds. Gene Expression Systems: Using Nature for the Art of Expression. San Diego: Academic Press, 1999:211–233.

53. Baskar JF, Smith PP, Nilaver G, et al. The enhancer domain of the human cytomegalovirus major immediate-early promoter determines cell type-specific expression in transgenic mice. J Virol 1996; 70:3207–3214.

54. Furth PA, Hennighausen L, Baker C, Beatty B, Woychick R. The variability in activity of the universally expressed human cytomegalovirus immediate early gene 1 enhancer/promoter in transgenic mice. Nucleic Acids Res 1991; 19:6205–6208.

55. Schmidt EV, Christoph G, Zeller R, Leder P. The cytomegalovirus enhancer: a pan-active control element in transgenic mice. Mol Cell Biol 1990; 10:4406–4411.

56. Gribaudo G, Ravaglia S, Gaboli M, Gariglio M, Cavallo R, Landolfo S. Interferon-alpha inhibits the murine cytomegalovirus immediate-early gene expression by down-regulating NF-kappa B activity. Virology 1995; 211:251–260.

57. Gribaudo G, Ravaglia S, Caliendo A, et al. Interferons inhibit onset of murine cytomegalovirus immediate-early gene transcription. Virology 1993; 197:303–311.

58. Harms JS, Splitter GA. Interferon-gamma inhibits transgene expression driven by SV40 or CMV promoters but augments expression driven by the mammalian MHC I promoter. Hum Gene Ther 1995; 6:1291–1297.

59. Qin L, Ding Y, Pahud DR, Chang E, Imperiale MJ, Bromberg JS. Promoter attenuation in gene therapy: interferon-gamma and tumor necrosis factor-alpha inhibit transgene expression. Hum Gene Ther 1997; 8:2019–2029.

60. Scheule RK, St George JA, Bagley RG, et al. Basis of pulmonary toxicity associated with cationic lipid-mediated gene transfer to the mammalian lung. Hum Gene Ther 1997; 8:689–707.

61. Armentano D, Zabner J, Sacks C, et al. Effect of the E4 region on the persistence of transgene expression from adenovirus vectors. J Virol 1997; 71:2408–2416.

62. Brough DE, Hsu C, Kulesa VA, et al. Activation of transgene expression by early region 4 is responsible for a high level of persistent transgene expression from adenovirus vectors in vivo. J Virol 1997; 71:9206–9213.

63. Armentano D, Smith MP, Sookdeo CC, et al. E4ORF3 requirement for achieving long-term transgene expression from the cytomegalovirus promoter in adenovirus vectors. J Virol 1999; 73:7031–7034.

64. Lusky M, Grave L, Dieterle A, et al. Regulation of adenovirus-mediated transgene expression by the viral E4 gene products: requirement for E4 ORF3. J Virol 1999; 73:8308–8319.

65. Yew NS, Marshall J, Przybylska M, et al. Increased duration of transgene expression in the lung with plasmid DNA vectors harboring adenovirus E4 open reading frame 3. Hum Gene Ther 1999; 10:1833–1843.

66. Riis B, Rattan SI, Clark BF, Merrick WC. Eukaryotic protein elongation factors. Trends Biochem Sci 1990; 15:420–424.

67. Kim DW, Harada T, Saito I, Miyamura T. An efficient expression vector for stable expression in human liver cells. Gene 1993; 134:307–308.

68. Kim DW, Uetsuki T, Kaziro Y, Yamaguchi N, Sugano S. Use of the human elongation factor 1 alpha promoter as a versatile and efficient expression system. Gene 1990; 91:217–223.

69. Mizushima S, Nagata S. pEF-BOS, a powerful mammalian expression vector. Nucleic Acids Res 1990; 18:5322.

70. Guo ZS, Wang LH, Eisensmith RC, Woo SL. Evaluation of promoter strength for hepatic gene expression in vivo following adenovirus-mediated gene transfer. Gene Ther 1996; 3:802–810.

71. Nakai H, Herzog RW, Hagstrom JN, et al. Adeno-associated viral vector-mediated gene transfer of human blood coagulation factor IX into mouse liver. Blood 1998; 91:4600–4607.

72. Maione D, Wiznerowicz M, Delmastro P, et al. Prolonged expression and effective readministration of erythropoietin delivered with a fully deleted adenoviral vector. Hum Gene Ther 2000; 11:859–868.

73. Zaret KS, Liu JK, DiPersio CM. Site-directed mutagenesis reveals a liver transcription factor essential for the albumin transcriptional enhancer. Proc Natl Acad Sci USA 1990; 87:5469–5473.

74. Frain M, Hardon E, Ciliberto G, Sala-Trepat JM. Binding of a liver-specific factor to the human albumin gene promoter and enhancer. Mol Cell Biol 1990; 10:991–999.

75. Maire P, Wuarin J, Schibler U. The role of cis-acting promoter elements in tissue-specific albumin gene expression. Science 1989; 244:343–346.

76. Koeberl DD, Bonham L, Halbert CL, Allen JM, Birkebak T, Miller AD. Persistent, therapeutically relevant levels of human granulocyte colony-stimulating factor in mice after systemic delivery of adeno-associated virus vectors. Hum Gene Ther 1999; 10:2133–2140.

77. Gallo-Penn AM, Shirley PS, Andrews JL, et al. In vivo evaluation of an adenoviral vector encoding canine factor VIII: high-level, sustained expression in hemophiliac mice. Hum Gene Ther 1999; 10:1791–1802.

78. Li X, Eastman EM, Schwartz RJ, Draghia-Akli R. Synthetic muscle promoters: activities exceeding naturally occurring regulatory sequences. Nat Biotechnol 1999; 17:241–245.
79. Edelman GM, Meech R, Owens GC, Jones FS. Synthetic promoter elements obtained by nucleotide sequence variation and selection for activity. Proc Natl Acad Sci USA 2000; 97:3038–3043.
80. Reisman D, Sugden B. trans Activation of an Epstein-Barr viral transcriptional enhancer by the Epstein-Barr viral nuclear antigen 1. Mol Cell Biol 1986; 6:3838–3846.
81. Nettelbeck DM, Jerome V, Muller R. A strategy for enhancing the transcriptional activity of weak cell type-specific promoters. Gene Ther 1998; 5:1656–1664.
82. Li S, MacLaughlin FC, Fewell JG, et al. Increased level and duration of expression in muscle by co-expression of a transactivator using plasmid systems. Gene Ther 1999; 6:2005–2011.
83. Veelken H, Leutgeb B, Kulmburg P, Fiebig HH, Mackensen A, Lindemann A. Enhancement of a constitutively active promoter for gene therapy by a positive feed-back transcriptional activator mechanism. Int J Mol Med 1998; 2:423–428.
84. Benihoud K, Salone B, Esselin S, et al. The role of IL-6 in the inflammatory and humoral response to adenoviral vectors. J Gene Med 2000; 2:194–203.
85. Minter RM, Rectenwald JE, Fukuzuka K, et al. TNF-alpha receptor signaling and IL-10 gene therapy regulate the innate and humoral immune responses to recombinant adenovirus in the lung. J Immunol 2000; 164:443–451.
86. Lieber A, He CY, Meuse L, Himeda C, Wilson C, Kay MA. Inhibition of NF-kappaB activation in combination with bcl-2 expression allows for persistence of first-generation adenovirus vectors in the mouse liver. J Virol 1998; 72:9267–9277.
87. Muruve DA, Barnes MJ, Stillman IE, Libermann TA. Adenoviral gene therapy leads to rapid induction of multiple chemokines and acute neutrophil-dependent hepatic injury in vivo. Hum Gene Ther 1999; 10:965–976.
88. Otake K, Ennist DL, Harrod K, Trapnell BC. Nonspecific inflammation inhibits adenovirus-mediated pulmonary gene transfer and expression independent of specific acquired immune responses. Hum Gene Ther 1998; 9:2207–2222.
89. Borgland SL, Bowen GP, Wong NC, Libermann TA, Muruve DA. Adenovirus vector–induced expression of the C-X-C chemokine IP-l0 is mediated through capsid-dependent activation of NF-kappaB. J Virol 2000; 74:3941–3947.
90. Yang Y, Wilson JM. Clearance of adenovirus-infected hepatocytes by MHC class I-restricted CD4+ CTLs in vivo. J Immunol 1995; 155:2564–2570.
91. Yang Y, Su Q, Wilson JM. Role of viral antigens in destructive cellular immune responses to adenovirus vector-transduced cells in mouse lungs. J Virol 1996; 70:7209–7212.
92. Dai Y, Schwarz EM, Gu D, Zhang WW, Sarvetnick N, Verma IM. Cellular and humoral immune responses to adenoviral vectors containing factor IX gene: tolerization of factor IX and vector antigens allows for long-term expression. Proc Natl Acad Sci USA 1995; 92:1401–1405.
93. Tripathy SK, Black HB, Goldwasser E, Leiden JM. Immune responses to transgene-encoded proteins limit the stability of gene expression after injection of replication-defective adenovirus vectors. Nat Med 1996; 2:545–550.

94. Gahery-Segard H, Farace F, Godfrin D, et al. Immune response to recombinant capsid proteins of adenovirus in humans: antifiber and anti-penton base antibodies have synergistic effect on neutralizing activity. J Virol 1998; 72:2388–2397.
95. Kuriyama S, Tominaga K, Mitoro A, et al. Immunomodulation with FK506 around the time of intravenous re-administration of an adenoviral vector facilitates gene transfer into primed rat liver. Int J Cancer 2000; 85:839–844.
96. Cichon G, Strauss M. Transient immunosuppression with 15-deoxyspergualin prolongs reporter gene expression and reduces humoral immune response after adenoviral gene transfer. Gene Ther 1998; 5:85–90.
97. Ilan Y, Jona VK, Sengupta K, et al. Transient immunosuppression with FK506 permits long-term expression of therapeutic genes introduced into the liver using recombinant adenoviruses in the rat. Hepatology 1997; 26:949–956.
98. Durham HD, Alonso-Vanegas MA, Sadikot AF, et al. The immunosuppressant FK506 prolongs transgene expression in brain following adenovirus-mediated gene transfer. Neuroreport 1997; 8:2111–2115.
99. Lochmuller H, Petrof BJ, Pari G, et al. Transient immunosuppression by FK506 permits a sustained high-level dystrophin expression after adenovirus-mediated dystrophin minigene transfer to skeletal muscles of adult dystrophic (mdx) mice. Gene Ther 1996; 3:706–716.
100. Vilquin JT, Guerette B, Kinoshita I, et al. FK506 immunosuppression to control the immune reactions triggered by first-generation adenovirus-mediated gene transfer. Hum Gene Ther 1995; 6:1391–1401.
101. Fang B, Eisensmith RC, Wang H, et al. Gene therapy for hemophilia B: host immunosuppression prolongs the therapeutic effect of adenovirus-mediated factor IX expression. Hum Gene Ther 1995; 6:1039–1044.
102. Smith TA, White BD, Gardner JM, Kaleko M, McClelland A. Transient immunosuppression permits successful repetitive intravenous administration of an adenovirus vector. Gene Ther 1996; 3:496–502.
103. Kaplan JM, Smith AE. Transient immunosuppression with deoxyspergualin improves longevity of transgene expression and ability to readminister adenoviral vector to the mouse lung. Hum Gene Ther 1997; 8:1095–1104.
104. van Rooijen N, Bakker J, Sanders A. Transient suppression of macrophage functions by liposome-encapsulated drugs. Trends Biotechnol 1997; 15:178–185.
105. Van Rooijen N, Sanders A. Liposome mediated depletion of macrophages: mechanism of action, preparation of liposomes and applications. J Immunol Methods 1994; 174:83–93.
106. Wolff G, Worgall S, van Rooijen N, Song WR, Harvey BG, Crystal RG. Enhancement of in vivo adenovirus-mediated gene transfer and expression by prior depletion of tissue macrophages in the target organ. J Virol 1997; 71:624–629.
107. Kuzmin AI, Finegold MJ, Eisensmith RC. Macrophage depletion increases the safety, efficacy and persistence of adenovirus-mediated gene transfer in vivo. Gene Ther 1997; 4:309–316.
108. Stein CS, Pemberton JL, van Rooijen N, Davidson BL. Effects of macrophage depletion and anti-CD40 ligand on transgene expression and redosing with recombinant adenovirus. Gene Ther 1998; 5:431–439.
109. Kolls JK, Lei D, Odom G, et al. Use of transient CD4 lymphocyte depletion to

prolong transgene expression of E1-deleted adenoviral vectors. Hum Gene Ther 1996; 7:489–497.

110. Poller W, Schneider-Rasp S, Liebert U, et al. Stabilization of transgene expression by incorporation of E3 region genes into an adenoviral factor IX vector and by transient anti-CD4 treatment of the host. Gene Ther 1996; 3:521–530.

111. Lei D, Lehmann M, Shellito JE, et al. Nondepleting anti-CD4 antibody treatment prolongs lung-directed E1-deleted adenovirus-mediated gene expression in rats. Hum Gene Ther 1996; 7:2273–2279.

112. Kay MA, Holterman AX, Meuse L, et al. Long-term hepatic adenovirus-mediated gene expression in mice following CTLA4Ig administration. Nat Genet 1995; 11: 191–197.

113. Nakagawa I, Murakami M, Ijima K, et al. Persistent and secondary adenovirus-mediated hepatic gene expression using adenovirus vector containing CTLA4IgG. Hum Gene Ther 1998; 9:1739–1745.

114. Yang Y, Su Q, Grewal IS, Schilz R, Flavell RA, Wilson JM. Transient subversion of CD40 ligand function diminishes immune responses to adenovirus vectors in mouse liver and lung tissues. J Virol 1996; 70:6370–6377.

115. Scaria A, St George JA, Gregory RJ, et al. Antibody to CD40 ligand inhibits both humoral and cellular immune responses to adenoviral vectors and facilitates repeated administration to mouse airway. Gene Ther 1997; 4:611–617.

116. Ziegler RJ, Yew NS, Li C, et al. Correction of enzymatic and lysosomal storage defects in Fabry mice by adenovirus-mediated gene transfer. Hum Gene Ther 1999; 10:1667–1682.

117. Kay MA, Meuse L, Gown AM, et al. Transient immunomodulation with anti-CD40 ligand antibody and CTLA4Ig enhances persistence and secondary adenovirus-mediated gene transfer into mouse liver. Proc Natl Acad Sci USA 1997; 94:4686–4691.

118. Wilson CB, Embree LJ, Schowalter D, et al. Transient inhibition of CD28 and CD40 ligand interactions prolongs adenovirus-mediated transgene expression in the lung and facilitates expression after secondary vector administration. J Virol 1998; 72:7542–7550.

119. Strober W, Kelsall B, Marth T. Oral tolerance. J Clin Immunol 1998; 18:1–30.

120. Nagler-Anderson C. Tolerance and immunity in the intestinal immune system. Crit Rev Immunol 2000; 20:103–120.

121. Ilan Y, Sauter B, Chowdhury NR, et al. Oral tolerization to adenoviral proteins permits repeated adenovirus-mediated gene therapy in rats with pre-existing immunity to adenoviruses. Hepatology 1998; 27:1368–1376.

122. Kagami H, Atkinson JC, Michalek SM, et al. Repetitive adenovirus administration to the parotid gland: role of immunological barriers and induction of oral tolerance. Hum Gene Ther 1998; 9:305–313.

123. Ilan Y, Prakash R, Davidson A, et al. Oral tolerization to adenoviral antigens permits long-term gene expression using recombinant adenoviral vectors. J Clin Invest 1997; 99:1098–1106.

124. Pastore L, Morral N, Zhou H, et al. Use of a liver-specific promoter reduces immune response to the transgene in adenoviral vectors. Hum Gene Ther 1999; 10:1773–1781.

125. Engelhardt JF, Litzky L, Wilson JM. Prolonged transgene expression in cotton rat lung with recombinant adenoviruses defective in E2a. Hum Gene Ther 1994; 5: 1217–1229.
126. Engelhardt JF, Ye X, Doranz B, Wilson JM. Ablation of E2A in recombinant adenoviruses improves transgene persistence and decreases inflammatory response in mouse liver. Proc Natl Acad Sci USA 1994; 91:6196–6200.
127. Dedieu JF, Vigne E, Torrent C, et al. Long-term gene delivery into the livers of immunocompetent mice with E1/E4-defective adenoviruses. J Virol 1997; 71: 4626–4637.
128. Gorziglia MI, Kadan MJ, Yei S, et al. Elimination of both E1 and E2 from adenovirus vectors further improves prospects for in vivo human gene therapy. J Virol 1996; 70:4173–4178.
129. Gao GP, Yang Y, Wilson JM. Biology of adenovirus vectors with E1 and E4 deletions for liver-directed gene therapy. J Virol 1996; 70:8934–8943.
130. Fang B, Wang H, Gordon G, et al. Lack of persistence of E1-recombinant adenoviral vectors containing a temperature-sensitive E2A mutation in immunocompetent mice and hemophilia B dogs. Gene Ther 1996; 3:217–222.
131. Armentano D, Sookdeo CC, Hehir KM, et al. Characterization of an adenovirus gene transfer vector containing an E4 deletion. Hum Gene Ther 1995; 6:1343–1353.
132. Yang Y, Nunes FA, Berencsi K, Gonczol E, Engelhardt JF, Wilson JM. Inactivation of E2a in recombinant adenoviruses improves the prospect for gene therapy in cystic fibrosis. Nat Genet 1994; 7:362–369.
133. Amalfitano A. Next-generation adenoviral vectors: new and improved. Gene Ther 1999; 6:1643–1645.
134. Wang Q, Greenburg G, Bunch D, Farson D, Finer MH. Persistent transgene expression in mouse liver following in vivo gene transfer with a delta E1/delta E4 adenovirus vector. Gene Ther 1997; 4:393–400.
135. Gorziglia MI, Lapcevich C, Roy S, et al. Generation of an adenovirus vector lacking E1, e2a, E3, and all of E4 except open reading frame 3. J Virol 1999; 73:6048–6055.
136. Lusky M, Christ M, Rittner K, et al. In vitro and in vivo biology of recombinant adenovirus vectors with E1, E1/E2A, or E1/E4 deleted. J Virol 1998; 72:2022–2032.
137. Hu H, Serra D, Amalfitano A. Persistence of an (E1-, polymerase-) adenovirus vector despite transduction of a neoantigen into immune-competent mice. Hum Gene Ther 1999; 10:355–364.
138. Schiedner G, Morral N, Parks RJ, et al. Genomic DNA transfer with a high-capacity adenovirus vector results in improved in vivo gene expression and decreased toxicity [published erratum appears in Nat Genet 1998 18(3):298]. Nat Genet 1998; 18: 180–183.
139. Balague C, Zhou J, Dai Y, et al. Sustained high-level expression of full-length human factor VIII and restoration of clotting activity in hemophilic mice using a minimal adenovirus vector. Blood 2000; 95:820–828.
140. Morsy MA, Gu M, Motzel S, et al. An adenoviral vector deleted for all viral coding

sequences results in enhanced safety and extended expression of a leptin transgene. Proc Natl Acad Sci USA 1998; 95:7866–7871.

141. Fisher KJ, Jooss K, Alston J, et al. Recombinant adeno-associated virus for muscle directed gene therapy. Nat Med 1997; 3:306–312.

142. Jooss K, Yang Y, Fisher KJ, Wilson JM. Transduction of dendritic cells by DNA viral vectors directs the immune response to transgene products in muscle fibers. J Virol 1998; 72:4212–4223.

143. Brockstedt DG, Podsakoff GM, Fong L, Kurtzman G, Mueller-Ruchholtz W, Engleman EG. Induction of immunity to antigens expressed by recombinant adeno-associated virus depends on the route of administration. Clin Immunol 1999; 92:67–75.

144. McLachlan G, Stevenson BJ, Davidson DJ, Porteous DJ. Bacterial DNA is implicated in the inflammatory response to delivery of DNA/DOTAP to mouse lungs. Gene Ther 2000; 7:384–392.

145. McMahon JM, Wells KE, Bamfo JE, Cartwright MA, Wells DJ. Inflammatory responses following direct injection of plasmid DNA into skeletal muscle. Gene Ther 1998; 5:1283–1290.

146. Pisetsky DS. The influence of base sequence on the immunostimulatory properties of DNA. Immunol Res 1999; 19:35–46.

147. Krieg AM. Direct immunologic activities of CpG DNA and implications for gene therapy. J Gene Med 1999; 1:56–63.

148. Messina JP, Gilkeson GS, Pisetsky DS. Stimulation of in vitro murine lymphocyte proliferation by bacterial DNA. J Immunol 1991; 147:1759–1764.

149. Krieg AM, Yi AK, Matson S, et al. CpG motifs in bacterial DNA trigger direct B-cell activation. Nature 1995; 374:546–549.

150. Klinman DM, Yi AK, Beaucage SL, Conover J, Krieg AM. CpG motifs present in bacteria DNA rapidly induce lymphocytes to secrete interleukin 6, interleukin 12, and interferon gamma. Proc Natl Acad Sci USA 1996; 93:2879–2883.

151. Ballas ZK, Rasmussen WL, Krieg AM. Induction of NK activity in murine and human cells by CpG motifs in oligodeoxynucleotides and bacterial DNA. J Immunol 1996; 157:1840–1845.

152. Sparwasser T, Koch ES, Vabulas RM, et al. Bacterial DNA and immunostimulatory CpG oligonucleotides trigger maturation and activation of murine dendritic cells. Eur J Immunol 1998; 28:2045–2054.

153. Behboudi S, Chao D, Klenerman P, Austyn J. The effects of DNA containing CpG motif on dendritic cells. Immunology 2000; 99:361–366.

154. Krieg AM, Yi AK, Schorr J, Davis HL. The role of CpG dinucleotides in DNA vaccines. Trends Microbiol 1998; 6:23–27.

155. Yew NS, Zhao H, Wu IH, et al. Reduced inflammatory response to plasmid DNA vectors by elimination and inhibition of immunostimulatory CpG motifs. Mol Ther 2000; 1:255–262.

156. Xiao W, Berta SC, Lu MM, Moscioni AD, Tazelaar J, Wilson JM. Adeno-associated virus as a vector for liver-directed gene therapy. J Virol 1998; 72:10222–10226.

157. Hagstrom JN, Couto LB, Scallan C, et al. Improved muscle-derived expression of

human coagulation factor IX from a skeletal actin/CMV hybrid enhancer/promoter. Blood 2000; 95:2536–2542.

158. Watson GL, Sayles JN, Chen C, et al. Treatment of lysosomal storage disease in MPS VII mice using a recombinant adeno-associated virus. Gene Ther 1998; 5: 1642–1649.

159. Vicat JM, Boisseau S, Jourdes P, et al. Muscle transfection by electroporation with high-voltage and short-pulse currents provides high-level and long-lasting gene expression. Hum Gene Ther 2000; 11:909–916.

160. Koeberl DD, Alexander IE, Halbert CL, Russell DW, Miller AD. Persistent expression of human clotting factor IX from mouse liver after intravenous injection of adeno-associated virus vectors. Proc Natl Acad Sci USA 1997; 94:1426–1431.

3
Expression Systems: Regulated Gene Expression

Jeffrey L. Nordstrom
Valentis, Inc., Burlingame, California, U.S.A.

I. INTRODUCTION

A. Why Drug-Regulated Gene Transfer?

The Orkin-Motulsky report, a critical review of gene therapy published in December 1995 [1], highlighted the need for "improving vectors for gene delivery, enhancing and maintaining high-level expression of genes transferred to somatic cells, achieving tissue-specific and regulated expression of transferred genes, and directing gene transfer to specific cell types." Gene transfer was deemed to be very early in development. Since the report was published, gene transfer methods have made significant improvements, particularly for helper-dependent adenoviral vectors delivered intravenously [2–4] (see Chap. 7), for adenovirus-associated virus (AAV) vectors delivered intravenously or intramuscularly [5-7] (see Chap. 8), and for plasmids delivered intramuscularly with the assistance of in vivo electroporation [8-10] (Chap. 10). These systems have demonstrated the ability to maintain the production of secreted proteins at therapeutic levels for months to years in rodents and larger animals.

The ability to achieve sustained transgene expression by gene transfer methods creates a need for robust, drug-dependent regulation systems, because most proteins of therapeutic value are associated with side effects and toxicities when overproduced. A drug-dependent gene regulation system provides an ability to adjust the time or level of transgene expression, thereby conferring an ability to maximize therapeutic benefit and minimize side effects, to tailor expression to compensate for variations in therapeutic need over time, or to shut down expression if unexpected effects due to the gene product are encountered.

B. Strategies for Drug-Dependent Gene Regulation

Many systems have been developed to provide drug-dependent regulation of transgene expression in mammalian cells. Most are based on two key components. The first component is a chimeric transcription factor, which can be controlled, at the level of its activity or synthesis, by a low molecular weight drug. The transcription factor is usually a fusion protein with domains for drug binding, DNA binding, and transcription activation that are derived from different sources. The second component is the regulated promoter that is linked to the target gene of interest. The regulated promoter usually contains multiple tandem copies of a DNA binding site for the novel transcription factor, and binding of the drug-dependent active form of the transcription factor usually activates target gene transcription.

Current drug-dependent gene-regulation systems fall into three categories, whereby chimeric transcription factors are based on bacterial repressor proteins (such as tetracycline, β-galactoside, or streptogramin-regulated systems) [11–13], nuclear hormone receptors (such as antiprogestin, antiestrogen, ecdysteroid, or glucocorticorticoid-regulated systems) [14–17], or heterodimeric proteins resulting from chemical-induced dimerization (such as rapamycin-regulated systems) [18].

C. Focus

The four systems that have been evaluated most extensively in preclinical gene transfer studies are the tetracycline-repressible, tetracycline-inducible, rapamycin-inducible, and antiprogestin-inducible systems. There are many variants for each of these four systems. This chapter will focus on the systems that have been evaluated in vivo and applied to the regulated production of secreted proteins that circulate in the blood stream. Virus-based as well as plasmid-based gene transfer studies, published as of December 2001, will be discussed.

II. TETRACYCLINE-REPRESSIBLE SYSTEM FOR TRANSGENE REGULATION

A. Description

Gossen and Bujard first described the tetracycline-repressible (Tet-OFF) system in 1992 [11]. It is based on a bacterial repressor/transcriptional activator fusion protein that is allosterically inhibited by tetracycline binding. The system

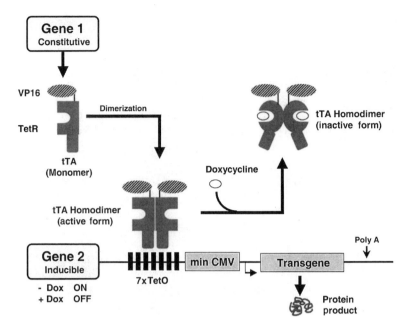

Figure 1   Schematic diagram of the tetracycline-repressible system for transgene regulation. Two genes are required. Gene 1 is a constitutively active gene that codes for the tTA (tetracycline transactivator) protein. Gene 2 is the regulated transgene. The tTA protein is a homodimeric protein that is active in the absence of doxycycline. It binds to TetO (tetracycline operator) sites in the inducible promoter and activates transgene expression. When doxycycline (corepressor) is added, the tTA homodimer is converted to an inactive form, which does not remain bound to TetO sites, thereby causing transgene expression to return to baseline levels. Thus, in the absence of doxycycline, transgene expression is ON; in the presence of doxycycline, transgene expression is OFF.

works via negative control, which means that expression of the target gene is ON in the absence of the small molecule corepressor (tetracycline) but OFF in its presence.

The tetracycline repressible system is schematically depicted in Figure 1. The system requires two genes. One gene encodes the tTA (tetracycline transactivator) fusion protein. The second gene is the target gene, and it contains a promoter with multiple TetO (tetracycline operator)–binding sites for the tTA protein. In the absence of tetracycline, tTA protein has high affinity for TetO sites in the promoter and stimulates transcription of the target gene. Following tetracycline binding, tTA protein undergoes a conformational change, which greatly reduces its affinity for TetO sites and causes transcription of the target gene to be shut down [19].

The structure of the tTA protein is [TetR$_{1-207}$]/S/[VP16$_{352-479}$]. TetR$_{1-207}$ is the full-length repressor protein for the tetracycline operon from bacterial transposon Tn10 (Genbank accession number NP_052930), S is serine, which serves as a linker, and VP16$_{352-479}$ is the activation domain of the herpes simplex virus VP16 protein (Genbank accession number P04486). The tTA protein contains one fusion junction and none of its sequence is of human origin. tTA monomers are 335 amino acids in length and have a molecular weight of 37 kD. A dimerization domain in the repressor segment leads to the formation of homodimers, which comprise the active form of tTA protein. The tTA protein is usually expressed from a gene that is controlled by a strong, constitutively active promoter, such as the cytomegalovirus (CMV) promoter, but tissue-specific or autogenously controlled promoters also have been utilized.

The target gene's promoter consists of seven TetO sites linked to a minimal fragment of the human CMV promoter. The 19-bp TetO element (TCCCTATCAGTGATAGAGA) is an imperfect palindromic sequence. Seven TetO elements are arranged in direct repeat but separated by 23-bp linkers (AAAGTGAAAGTCGAGTTTACCAC) and joined via a 33-bp linker (AAAGTGAAAGTCGAGCTCGGTACCCGGGTCGAG) to a fragment of the human CMV immediate early 1 gene (Genbank accession number X03922) that extends from -51 to $+69$ bp (where $+1$ is defined as the site of initiation of transcription). The TATA box element is located at -29 to -24 bp. Thus, the CMV promoter fragment lacks most sequences upstream of the TATA box, and is therefore considered to be mostly inactive. The overall length of the tetracycline-controllable promoter, from the first TetO site to the presumed site of transcription initiation, is 355 bp.

The tTA protein could potentially influence the expression of endogenous genes if appropriate target sequences are present. When human DNA sequences were analyzed, no perfect sequence matches to the 19-bp TetO sequence were found. Frequencies for partial matches were as follows: one mismatch (0), two mismatches (0), three mismatches (3), four mismatches (60), five mismatches (111), six mismatches (469) and seven mismatches (356). These frequencies were obtained from the Genbank Sequence Database, Release 127.0, December 15, 2001, using BLASTN 2.2.2. The paucity of sequences with complete or nearly complete identity to the TetO sequence suggests that the risk of activation of an endogenous human gene by the tTA protein is low.

Tetracyclines are commonly used broad-spectrum antibiotics that act by interrupting prokaryotic polypeptide chain elongation. Doxycycline (molecular weight [MW] 462.46), a tetracycline derivative, is the preferred corepressor molecule. Doxycycline is lipophilic and is virtually completely absorbed after oral ingestion. It binds to plasma proteins. Doxycycline has an elimination half-life of 18–22 hours in humans and ~3 hrs in mice [20]. Doxycycline hyclate is water soluble. Dosages approved for human use are 100–200 mg (~1.7–3.3 mg/kg body weight [BW]) per day. Doxycycline, like any tetracycline, has an affinity for calcium binding and forms stable calcium complexes in any bone-forming tissue.

In cell culture studies, repression of transgene expression occurs at doxycycline concentrations that range from 0.2 to 20 nM (0.1 to 10 ng/mL), with half-maximal repression occurring at 2 nM [21]. In transiently transfected cells, the dynamic range of gene expression may be as high as 100-fold, and in stable cell lines, the dynamic range of expression may be greater than 1000-fold [11]. In transgenic mice, the dynamic range of transgene expression may be as high as 10^5-fold [22]. In animal studies, doxycycline was dissolved in 5% sucrose and added to drinking water at a concentration of 2 mg/mL [22] or injected intravenously or intraperitoneally (IP) at doses of 25–100 mg/kg BW [23].

B. Plasmid-Based Gene Transfer Studies

One of the first gene transfer studies on regulated transgene expression was performed with a plasmid mixture that was delivered to murine tibialis muscle by direct intramuscular injection [24]. One plasmid contained a gene for the tTA protein: CMV promoter/tTA/SV40(L) poly(A); and the other contained a regulated luciferase gene: 7xTetO/minCMV promoter/Luc/SV40(L) poly(A). Administration of tetracycline (100 mg/kg/day, IP or 2.2 mg/mL in drinking water) for 7 days caused luciferase expression to be repressed by approximately 100-fold. Repression was incomplete, however, since the lowest level of repressed expression was 10-fold greater than that exhibited by the regulated luciferase plasmid alone. The extent of luciferase repression was dependent on the tetracycline dose when it was varied from 0.05 to 2.2 mg/mL in drinking water, and repression was reversible, because luciferase activity returned to nonrepressed levels within 2 days of cessation of oral tetracycline dosing. Similar data on regulated luciferase expression were obtained when plasmids for a modified version of the tetracycline-regulated system also were delivered by direct intramuscular injection [25]. Both studies were limited by the fact that each animal only provided a single time point, and, thus data on transitions in the levels expressed by individual animals were not obtained. However, the studies were notable, because they provided proof-of-concept that plasmid-based expression systems can be delivered to muscle and regulated in a drug-dependent manner.

C. Virus-Based Gene Transfer Studies

Rendahl and coworkers [26] demonstrated the ability to repress erythropoietin (EPO) transgene expression in mice in a doxycycline-dependent manner following intramuscular delivery of two recombinant adeno-associated virus (AAV) vectors. One vector contained a constitutively expressed gene for the tTA protein:

CMV promoter/tTA/murine protamine intron/murine protamine poly(A). The second vector contained the regulated murine EPO (mEPO) gene: 7xTetO/minCMV promoter/β-globin intron/mEPO/bGH poly(A). The vectors were mixed and injected into quadriceps muscles of Balb/c mice. With 10^9 particles of each vector, hematocrit increased from 47 to 65% in 18 weeks and low levels of mEPO protein were detected. No increase in hematocrit occurred with 10^9 particles of the mEPO vector alone. With 10^{10} particles of each vector, hematocrit increased much more rapidly, exceeding 80% by week 8. When slow-release tetracycline pellets (~29 mg/kg/day for 21 days) were implanted at week 4, mEPO expression was repressed by approximately 20-fold and hematocrit decreased from 70 (week 4) to 52% (week 9). By week 12, mEPO expression levels began to rise (due to depletion of the tetracycline pellet) and the hematocrit reached 70% by week 18. The data clearly demonstrated that the tetracycline-repressible system could be used to regulate mEPO expression and hematocrit in mice. However, some shortcomings were apparent. First, delivery of 10^{10} particles of the mEPO vector alone led to a gradual increase in hematocrit, indicating that the 7xTetO/minCMV promoter provided a significant level of basal expression. Thus, to achieve tight regulation of hematocrit, 10^{10} particles of the tTA vector were mixed with 10^9 particles of the mEPO vector. This was effective, but the magnitude of the increase in hematocrit was lower than that observed at the higher vector dose. Second, no attempt was made to block the initial rise in hematocrit by initiating tetracycline dosing at the time of vector delivery. Third, since tetracycline was administered only as single-dose, slow-release pellets, no data were obtained for the kinetics following cessation of tetracycline dosing or for a tetracycline dose response.

D. Summary

The data show that the tetracycline-repressible (Tet-OFF) system is functional in muscle when delivered as a plasmid-based or AAV-based system. Expression was repressed when tetracycline was administered at doses of 30–100 mg/kg/day and the magnitude of repression in vivo was ~100-fold for the plasmid-based system and ~20-fold for the AAV-based system. These values are similar to those observed in transiently transfected cells, but considerably less than those observed when the genes were integrated into chromosomes of stably transfected cells or transgenic mice. The low magnitude of regulation exhibited by the AAV-based system is interesting, especially since AAV vectors are capable of chromosomal integration. The plasmid-based studies were short—only encompassing a two-week period—whereas the AAV-based study demonstrated that repressible transgene expression was retained in murine muscle for at least 3 months, which suggests a lack of immune responses to either the transgene product or to the foreign tTA protein. Control experiments with the AAV–EPO vector alone that led to substantial increases in hematocrit indicate that the 7xTetO/minCMV promoter has a significant level of basal activity.

III. TETRACYCLINE-INDUCIBLE SYSTEM FOR TRANSGENE REGULATION

A. Description

Gossen and coworkers first described the tetracycline inducible (Tet-ON) system in 1995 (27). It is based on the properties of a mutant version of the TetR (tetracycline repressor) protein. Fusion proteins derived from the mutant TetR sequence are allosterically activated, not inhibited, by tetracycline binding. The system works via positive control, which means that expression of the target gene is OFF in the absence of the small molecule inducer (tetracycline) but ON in its presence.

The tetracycline-inducible system is schematically depicted in Figure 2. The system requires two genes. One gene encodes the rtTA (reverse tetracycline transactivator) fusion protein. The promoter for the target gene is identical to the one used for the tetracycline-repressible system and has multiple TetO sites linked to a minimal CMV promoter. In the absence of tetracycline, rtTA protein has low affinity for TetO sites and is inefficient in stimulating target gene transcription. Following tetracycline binding, rtTA protein undergoes a conformational change, which greatly increases its affinity for TetO sites and stimulates transcription of the target gene.

The mutant TetR protein was isolated following random mutagenesis of the Tn10-TetR gene and selection in *Escherichia coli* for tetracycline-dependent repression of gene expression (in contrast, wild-type TetR protein mediates tetracycline-dependent derepression of gene expression in *E. coli*). The mutant TetR protein had four amino acid substitutions (E71K, D95N, L101S, and G102D), and it was fused to the VP16 activation domain to generate the rtTA protein, which has the structure: [TetR$_{1-207}$ (4 point mutations)]/S/[VP16 $_{352-479}$]. Like the tTA protein, the rtTA protein forms homodimers. The monomers are 335 amino acids in length with a molecular weight of 37 kD. No portion of its sequence is of human origin.

Doxycycline is an effective inducer for the system, but tetracycline is an inefficient inducer, yielding transgene expression levels that are less than 1% of that observed with doxycycline [27]. The concentrations of doxycycline that produce allosteric changes in the rtTA protein are approximately 2 logs higher than those required to cause allosteric changes in the tTA protein. In cell culture, the responsive doxycycline range for rtTA is 1–1000 ng/mL, with half-maximal induction at 100 ng/mL; and in stable cell lines, the dynamic range of expression varied from 260- to >5000-fold 27). In transgenic mice, the dynamic range of

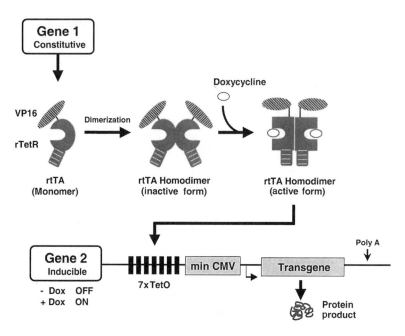

Figure 2 Schematic diagram of the tetracycline-inducible system for transgene regulation. Two genes are required. Gene 1 is a constitutively active gene that codes for the
rtTA (reverse tetracycline transactivator) protein. Gene 2 is the regulated transgene. The
rtTA protein is a homodimeric protein that is inactive in the absence of doxycycline. It
does not bind to TetO (tetracycline operator) sites in the inducible promoter and transgene
expression is low. When doxycycline (inducer) is added, the rtTA homodimer is converted
to an active form, which gains the ability to bind TetO sites, thereby activating transgene
expression. Thus, in the absence of doxycycline, transgene expression is OFF; in the presence of doxycycline, transgene expression is ON.

transgene expression, in response to doxycycline added at 2 mg/mL to drinking
water, ranged from 10^2- to 10^5-fold [22].

B. Plasmid-Based Gene Transfer Studies

Plasmid delivery to muscle by electroporation is efficient and capable of providing high-level, long-term transgene expression at low plasmid doses. In vivo
electroporation can enhance transgene expression by at least 100-fold [8–10]
(Chap. 10). Following the delivery of a mixture of two plasmids to skeletal muscle with electroporation, Rizzuto and coworkers (28) demonstrated induction of
mEPO expression and hematocrit by doxycycline. One of the plasmids used a

muscle-specific promoter to express the rtTAnls, an rtTA protein that was modified to include an N-terminal nuclear localization sequence: Muscle creatine kinase promoter/rtTAnls/bGH poly(A). The target plasmid contained an inducible murine EPO gene: 7xTetO/minCMV promoter/mEPO/SV40 poly(A). Delivery of the inducible EPO plasmid alone caused hematocrit to increase if the plasmid dose exceeded 10 μg, indicating that the 7XTetO/minCMV promoter has a significant level of basal activity. To achieve regulated EPO expression, 10 μg of the inducible EPO plasmid was mixed with 0.5, 2.0, or 10.0 μg of the transactivator (rtTAnls) plasmid. At the two higher doses of transactivator plasmid, an increase in hematocrit occurred in the absence of doxycycline. These data show that the inactive form of the transactivator protein contributes to the basal expression level, probably by low-level binding to TetO sites in the inducible promoter. When 10.0 μg of inducible EPO plasmid and 0.5 μg of transactivator plasmid were delivered to mouse quadriceps muscle with electroporation, and animals were not treated with doxycycline, mEPO expression was undetectable, but a small increase in hematocrit (from 43 to 48%) occurred over a 30-day period. Daily administration of doxycycline (0.4 mg/mL in drinking water) induced mEPO expression and caused hematocrit to increase from 48 to 56% in 10 days, and this level was maintained for an additional 30 days. Reversibility was demonstrated by a group of animals that, having reached hematocrit levels of 58% after 30 days of doxycycline dosing, were withdrawn from drug, which caused hematocrit to decline to 44% (baseline) over a 40-day period. The data demonstrated that the plasmid-based, doxycycline-inducible transgene regulation system remained functional in murine muscle for at least 2 months. However, the leakiness of the system, evidenced by increases in hematocrit in the absence of doxycycline dosing, restricted the amounts of plasmid that could be delivered, which probably limited the magnitude of the desired biological effects.

C. Virus-Based Gene Transfer Studies

Bohl and coworkers (29) utilized a single AAV vector for inducible mEPO expression that contained two genes in converging orientation, 7xTetO/minCMV promoter/mEPO/SV40 bidirectional poly(A)/rtTA/LTR promoter, where EPO transcripts initiate from the inducible promoter and are processed at the simian virus 40 (SV40) late poly(A) site, and rtTA transcripts initiate from the long terminal repeat (LTR) promoter and are processed at the SV40 early poly(A) site. The AAV vector was injected into the tibialis anterior muscle of mice at two doses (2.5 or 5 x 10^{10} genomes). In the absence of doxycycline treatment, mEPO expression was measurable and hematocrit gradually increased from 45 to 54% at the low dose and to 63% at the high dose. These data demonstrate the existence of a significant level of basal expression in the absence of doxycycline dosing. Initiation of daily doxycycline dosing (0.2 mg/mL in drinking water) caused mEPO expression to increase by 13-fold, which caused hematocrit to increase after 6 weeks to >75% for the low- and high-dose groups. Withdrawal of doxycycline dosing caused mEPO expression to decline gradually and incompletely, which caused hematocrit to decline over a 10-week period to 50–57%, levels that were similar to noninduced but higher than naïve animals (45%). Finally, reinitiation of doxycycline dosing at week 19 or 24 caused mEPO to be induced to the same level observed at week 2, which reestablished elevated hematocrits (>75%) within 4–6 weeks. These data demonstrated that the inducible EPO system retained function for at least 6 months after gene transfer. Southern blot analysis revealed the retention at week 29 of approximately 0.04 double-stranded vector genomes per diploid cell genome, and using a polymerase chain reaction (PCR) assay, vector was detected in DNA from injected muscle, but not spleen, liver, lung, or brain. No evidence for histological abnormalities was detected in recipient muscles analyzed at the time of sacrifice, nor were antibodies to rtTA detected, thus suggesting an absence of immune responses directed against the xenogeneic epitopes of the rtTA protein.

D. Summary

The data demonstrate that the doxycycline-inducible system is functional for multiple months in mice following the intramuscular delivery of plasmids (facilitated by electroporation) or AAV vectors. The main shortcoming was a significant level of basal expression that occurred in the absence of doxycycline dosing and was exacerbated by increases in the levels of transactivator (rtTA) protein.

IV. DIMERIZER-INDUCIBLE SYSTEM FOR TRANSGENE REGULATION

A. Description

Rivera and coworkers first described the dimerizer-inducible system in 1996 (18). It is based on the ability of drugs like rapamycin to bind to certain protein moieties and trigger the formation of unique heterodimeric proteins. Thus, rapamycin is a chemical inducer of dimerization. For the transgene-regulation system, each partner of the heterodimer is a fusion protein. One has rapamycin-binding and DNA-binding domains, and the other has rapamycin-binding and transcription-activation domains. The system works via positive control, which means that expression of the target gene is OFF in the absence of small molecule inducer (rapamycin) but ON in its presence.

The dimerizer-inducible system is schematically depicted in Figure 3. The system requires three genes. One gene encodes the ZFHD1-3xFKBP protein, in which a DNA-binding domain (ZFHD1) is fused to three copies of a rapamycin-

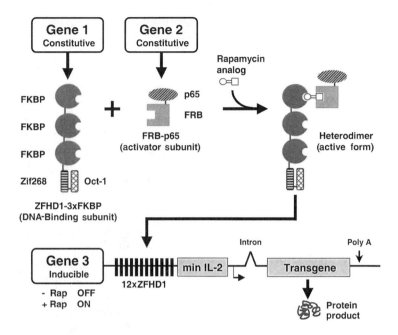

Figure 3 Schematic diagram of the dimerizer-inducible system for transgene regulation. The system requires three genes. Gene 1 is a constitutively active gene that codes for the DNA-binding subunit (ZFHD1-3xFKBP). Gene 2 is a constitutively active gene that codes for the transcriptional activator subunit (FRB-p65). Gene 3 is the inducible transgene. The DNA-binding and activator subunits remain separate and inactive in the absence of rapamycin and transgene expression is low. When rapamycin (inducer) is added, the DNA-binding and activator subunits form an active heterodimeric protein through rapamycin's dimerizing activity. The heterodimer binds to the ZFHD1 sites in the inducible promoter, thereby activating transgene expression. Thus, in the absence of rapamycin, transgene expression is OFF; in the presence of rapamycin, transgene expression is ON.

binding moiety (FKBP). The second gene encodes the FRB-p65 protein, in which a transcriptional activation domain (p65) is fused to a different rapamycin-binding moiety (FRAP). The different rapamycin-binding moieties bind to different portions of the dimerizing agent, rapamycin. The third gene, the inducible target gene, contains a promoter with multiple binding sites for the ZFHD1 DNA-binding domain. In the absence of rapamycin, the DNA-binding and transcriptional-activation activities are on separate proteins that are ineffective in stimulating target gene transcription. The DNA-binding protein might bind to sites in the inducible promoter, but, because it lacks an activation domain, transcription is not induced. Addition of rapamycin leads to the formation of heterodimers that have both DNA-binding and transcriptional-activation activity, which results in the activation of target gene transcription.

The structure of ZFHD1-3xFKBP, the fusion protein with DNA-binding activity, is HA/NLS/hZif268$_{335-392}$/GG/hOct-1$_{378-439}$/SR/hFKBP$_{1-107}$/TR/hFKBP$_{1-107}$/TR/hFKBP$_{1-107}$/TS. HA is a 16-amino acid epitope from *Haemophilus influenzae* hemagglutinin, NLS is a nuclear localization sequence from SV40 large T antigen, and GG, SR, TR, and TS are amino acid sequences that serve as linkers. hZif268$_{335-392}$ is a fragment of the human Zif268 transcription factor (Genbank accession number P18146) that contains two zinc finger DNA-binding domains. hOct-1$_{378-439}$ is the DNA binding portion of the human homeodomain transcription factor Oct-1 (Genbank accession number CAA31767). The entire chimeric DNA-binding entity, hZif268/GG/hOct-1, is called ZFHD1. hFKBP$_{1-107}$ is a nearly complete fragment of the human protein FKBP-12, also called FK506-binding protein 1A (Genbank accession number A35780). The ZFHD1-3xFKBP protein is approximately 474 amino acids in length and has a molecular weight of approximately 53 kD. Excluding the epitope and NLS sequences at the amino terminus, the protein contains four fusion junctions. Approximately 94% of its sequence is of human origin. A constitutively active CMV promoter controls expression of the gene for the ZFHD1-3xFKBP protein. The structure of p65-FRAP, the fusion protein with transcriptional activation activity, is HA/NLS/SR/hFRB$_{2025-2113}$ (T2098L)/TR/hP65$_{361-550}$/TS. HA is a 16–amino acid epitope from *H. influenzae* hemagglutinin, NLS is the SV40 nuclear localization sequence, and SR, TR and TS are amino acid sequences that serve as linkers. hFRB$_{2025-2113}$ (T2098L) is a fragment of human FRAP-1 protein, also called FK506-binding protein 12–rapamycin associated protein 1 (Genbank accession number P42345); T2098L indicates that it contains a threonine to leucine mutation at residue 2098. hP65$_{361-550}$ is a fragment of the activation domain of the p65 subunit of human transcription factor nuclear fragment-κB (NFκB) (Genbank accession number AAA36408). The p65-FRAP protein is approximately 306 amino acids in length and has a molecular weight of approximately 32 kD. Excluding the sequences at the amino terminus, the protein contains one fusion junction. Approximately 90% of its sequence is of human origin. A constitutively active CMV promoter controls the gene for the p65-FRAP protein.

The rapamycin-inducible promoter for the target gene consists of 12 ZFHD1 sites linked to a minimal fragment of the human interleukin-2 (IL-2) promoter. ZFHD1 elements are 12 bp in length (TAATGATGGGCG) and arranged in direct repeats that are separated by 6-bp linkers (GTCGAC, CTCGAG, or TCTAGC). The region with 12 ZFHD1 elements is joined via a 17-bp linker (TCTAGAACGCGAATTCA) to a fragment of the human IL-2 gene (Genbank accession number X00695) that extends from −72 to +45 bp (where +1 is defined as the site of initiation of transcription). The TATA box element is located at -32 to -27 bp. Thus, the human IL-2 promoter fragment lacks most sequences upstream of the TATA box and is therefore mostly inactive. The overall length of the rapamycin-inducible promoter, from the first ZFHD1 element to the presumed site of transcription initiation, is 299 bp.

Heterodimers could potentially influence the expression of endogenous genes if appropriate target sequences are present. When human DNA sequences in Genbank were analyzed, 30 perfect matches to the 12-bp ZFHD1 sequence were found, and 969 sequences exhibited only one mismatch. These frequencies were obtained from the Genbank Sequence Database, Release 127.0, December 15, 2001, using BLASTN 2.2.2. The substantial number of sites with zero or only one mismatch suggests that there may be a risk for activating the transcription of certain endogenous human genes.

Rapamycin is the best-characterized dimerizer that has been used as an inducer. Rapamycin (MW 914.2) is an immunosuppressive agent that inhibits T-lymphocyte activation and proliferation that occurs in response to antigenic and cytokine stimulation. It also inhibits antibody production. Rapamycin binds to the immunophilin, FK binding protein-12, to generate an immunosuppressive complex. Rapamycin is rapidly absorbed with an oral bioavailability of $\sim 18\%$ and is associated with plasma proteins. It has a terminal elimination half-life of ~ 62 hrs in humans and 2–5 hrs in mice [31]. Rapamycin is indicated for the prophylaxis of organ rejection in patients receiving renal transplants. The approved dosage is 2 mg (~ 0.03 mg/kg BW) per day.

In cell culture, transgene induction occurs at rapamycin concentrations that range from 0.5 to 10.0 nM, with half-maximal induction occurring at 2 nM (18). The dynamic range of expression was approximately 1000-fold in transiently transfected cells and as high as 10^4-fold in stable cell lines. The low background is attributed to the lack of an inherent affinity between the two fusion proteins in the absence of rapamycin. In mice that received intramuscular implants of stably transfected cells, induction of transgene expression occurred at intravenous rapamycin doses of 0.1–10.0 mg/kg BW, with half-maximal induction at 1 mg/kg. The dynamic range of transgene expression in vivo was approximately 40-fold [18].

It is anticipated that clinical use of this gene-regulation system will require dimerizers that are nonimmunosuppressive rapamycin analogs (rapalogs). The strategy for obtaining nonimmunosuppressive analogs involves the addition of substituents to rapamycin that block its binding to FKBP or FRAP. However, modification of amino acid residues in the drug-binding domains of the fusion proteins is needed to establish high-affinity binding of the analogs. Nonimmunosuppressive analogs have been synthesized and appropriate modifications in the drug-binding domains have been constructed [30], but their applicability to in vivo gene transfer studies has not been reported. The ability to alter the structures of the low molecular weight inducer and its binding target should provide an extremely powerful tool for customizing and refining the properties of this inducible gene-regulation system.

B. Plasmid-Based Gene Transfer Studies

No gene transfer studies with plasmid-based versions of the dimerizer inducible system have been reported.

C. Virus-Based Gene Transfer Studies

Ye and coworkers [32] utilized two AAV vectors to achieve inducible mEPO expression in mice. One vector (termed TF) had a bicistronic transcription unit that encoded both of the fusion proteins: CMV promoter/HA epitope/SV40nls/FRB-p65/IRES/HA epitope/SV40nls/ZFHD1-3xFKBP/RβG intron/RβG poly(A). The other vector contained the inducible mEPO gene: 12xZFHD1/minIL-2 promoter/chimeric intron/mEPO/SV40(L) 3'UTR fragment/2.7 kb stuffer fragment/hGH 3'UTR fragment/RβG intron fragment/RβG poly(A). The two AAV vectors were mixed in a 1:1 ratio and 4×10^{11} genomes were injected into quadriceps and tibialis anterior muscles of immunocompetent mice. No increases in plasma EPO or hematocrit occurred during periods (30 or 70 days in length) prior to challenge with rapamycin. Intraperitoneal administration of rapamycin induced mEPO expression and caused hematocrit to increase from 45 to 70%. Four or five cycles of EPO induction were achieved within a 5-month period with rapamycin doses that ranged from 0.1 to 10 mg/kg, and there was no diminution of inducible EPO expression over time. Maximum EPO levels were achieved within a few days of rapamycin administration, but the returns to undetectable levels were more gradual, generally requiring several weeks. Following the induction of hematocrit to 70%, hematocrit values remained elevated regardless of extended gaps in the schedule of rapamycin administration, which suggests that significant basal expression occurred following the cessation of rapamycin dosing.

An AAV vector with an inducible rhesus monkey EPO (rmEPO) gene was also constructed [32]. The rmEPO gene was identical to the inducible mEPO gene except it lacked the chimeric intron and the rabbit β-globin intron fragment. A 10:1 (rmEPO:TF1) vector mixture was injected into 10 sites of the vastus lateralis muscles of a 5-kg rhesus monkey. No increase in plasma EPO or hematocrit occurred during the 29-day period between vector injection and initiation of rapamycin dosing. Two cycles of induction of rmEPO expression by single doses of rapamycin were achieved within a 40-day period. Peak EPO levels occurred after ~4 days, which required ~2 weeks to return to baseline levels. Each induction cycle caused hematocrit to rise to 60%, which was followed by a gradual decline to baseline levels. However, inducibility in the monkey was lost by day 90, for treatment with rapamycin failed to induce rmEPO expression or elevate hematocrit. This could be indicative of an immune response directed against the chimeric fusion proteins or a decrease in the level of transcription factors below a critical threshold level.

When an AAV vector for inducible human growth hormone (hGH) expression, 12xZFHD1/minIL-2 promoter/hGH gene with four introns/SV40(L) poly(A), was mixed with TF1 vector in a 1:1 mixture and injected into mice, little induction of hGH was observed [33]. Based on cell culture studies, it appeared that expression of the FRB-p65 protein was limiting. To correct this deficiency, a second FRB-p65 gene, CMV promoter/HA epitope/SV40nls/FRB-p65/RβG intron / RβG poly(A), was inserted into the hGH vector downstream of the inducible hGH gene. Although this AAV construct yielded high basal levels of hGH in vitro, it functioned well in vivo when mixed with the TF1 vector in a 1:1 ratio. In immunocompetent mice, four cycles of undiminished hGH induction were observed in a 10-week period, and in immunodeficient mice, 9 cycles were observed in a 10-month period. The usual induction pattern was followed: peak hGH levels occurred within days of a single dose of rapamycin, but reestablishment of baseline was slow, requiring 10–20 days. Induction of hGH was proportional to rapamycin doses (administered intraperitoneally) that ranged from 0.05 to 0.35 mg/kg and higher doses (up to 1 mg/kg) were not inhibitory. To achieve similar levels of induction, oral rapamycin doses had to be four-fold higher, which is consistent with its oral bioavailability of 15–20%.

An adenoviral (Ad) system for inducible hGH expression was identical to that used for the AAV vector, but an additional gene for FRBp65 was not needed [33]. In immunodeficient mice, hGH expression was induced by more than 100-fold at times during the first 2 months. At 4 and 6 months, the magnitude of induction decreased to 20- and 10-fold, respectively. The maximal level of hGH induction was dependent on viral doses that ranged from 3×10^9 to 10^{11} particles per mouse. Relatively stable levels of hGH in blood were maintained by administration of 0.05 or 0.5 mg/kg doses of rapamycin every other day, which resulted in hGH levels of 1000–2000 or 3000–5000 pg/mL, respectively. In immunocompetent mice, the ad vector was much less effective owing to antiviral immune responses; only one round of induction was observed and its magnitude was small—approximately four-fold.

D. Summary

The dimerizer-inducible system is durable and effective, particularly when incorporated into AAV vectors that are delivered by intramuscular injection. Drug-dependent regulation of EPO expression was observed in mice and primates, and no increases in hematocrit occurred before initiation of rapamycin dosing.

However, once hematocrit levels were induced and drug dosing was terminated for extended periods of time, declines in hematocrit were extremely gradual or nonexistent, suggesting the existence of low-level rapamycin-independent expression. The requirement for two copies of the FRB-p65 gene to achieve regulated hGH expression indicates that the intracellular level of the protein moiety that carries the transcriptional activation domain is critical and rate limiting.

V. ANTIPROGESTIN-INDUCIBLE SYSTEM FOR TRANSGENE REGULATION

A. Description

Wang and coworkers first described the antiprogestin-inducible system in 1994 [14]. It is based on a truncated version of a progesterone receptor's ligand-binding domain (LBD) that completely alters its specificity of activation [34]. The wild-type LBD responds to natural progestins as agonists but to synthetic antiprogestins as antagonists. In contrast, the truncated LBD fails to respond to natural progestins but responds to synthetic antiprogestins like mifepristone as agonists. The antiprogestin-inducible system, also known as the GeneSwitch system,* works via positive control, which means that expression of the target transgene is OFF in the absence of inducer but ON in its presence.

The components of the antiprogestin inducible system and the mechanism by which it works are outlined in Figure 4. The system requires two genes. One gene encodes the regulator protein, in which a nuclear hormone receptor's truncated LBD is fused to a DNA-binding domain (GAL4) and a transcriptional activation domain (p65). The second gene, the inducible target gene, contains a promoter with multiple binding sites for the GAL4 DNA-binding domain. In the absence of antiprogestin, the regulator protein resides in an inactive monomeric form that is ineffective in stimulating target gene transcription. Inactive monomers, like other inactive nuclear hormone receptors, are probably maintained as inert complexes with heat shock and other proteins. Following antiprogestin binding, the regulator protein undergoes a conformational change, which leads to homodimer formation and release from the components of the inert complex. The active form of the regulator protein binds to GAL4 sites in the inducible promoter and activates target gene transcription.

The structure of the regulator protein (version 3.1) is MDSQQPDL/ $yGAL4_{2-93}$/EFPGVDQV/$hPR_{640-914}$/GST/$hP65_{285-551}$. MDSQQPDL is an N-terminal amino acid sequence and EFPGVDQV and GST are amino acid sequences that serve as linkers. $yGAL4_{2-93}$ is the DNA-binding domain from the yeast GAL4 protein (Genbank accession number AAA34626), $hPR_{640-914}$ is a C-terminal truncated version of the LBD of the human progesterone receptor (Genbank accession

---

* GeneSwitch® is a registered trademark of Valentis, Inc., Burlingame, CA.

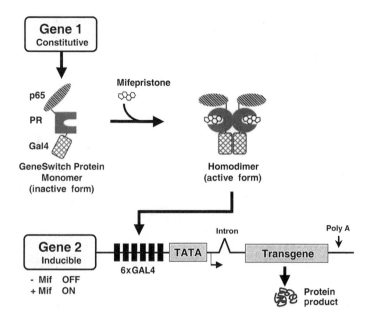

Figure 4   Schematic diagram of the antiprogestin-inducible system for transgene regulation. The system requires two genes. Gene 1 is a constitutively active gene that codes for the GeneSwitch regulator protein. Gene 2 is the inducible transgene. The regulator protein exists as an inactive monomer in the absence of the antiprogestin, mifepristone, which does not bind GAL4 sites efficiently and transgene expression is low. When mifepristone (inducer) is added, the regulator protein is converted to an active homodimeric form that efficiently binds to GAL4 sites in the inducible promoter, thereby activating transgene expression. Thus, in the absence of antiprogestin, transgene expression is OFF; in the presence of antiprogestin, transgene expression is ON.

number AAA60081), and $hP65_{285-551}$ is the activation domain of the p65 subunit of human transcription factor NFκB (Genbank accession number AAA46408). The regulator protein contains two fusion junctions, and 83% of its sequence is of human origin. Monomers of the regulator protein are 652 amino acids in length and have a molecular weight of 73 kD. The gene for the regulator protein is usually controlled by a weakly active tissue-specific promoter, but highly active CMV promoters or autogenously controlled promoters also have been utilized.

The antiprogestin-inducible promoter for the target gene consists of six (or four) GAL4 sites linked to a TATA box. GAL4 elements are 17 bp in length (CGGAGTACTGTCCTCCG) and arranged in direct repeats that are separated by 2-bp linkers (AG). The region with six GAL4 elements is joined via a 12-bp linker (AGTCGACTCTAG) to the TATA box region of an adenoviral E1b promoter (residues 1665–1677, Genbank accession number J01917). The remaining portion of the promoter consists of linker (GGATCTCGAGATATCGGAGCT). This minimal promoter, since it only consists of a TATA box, is highly inactive. The overall length of the antiprogestin-inducible promoter, from the first GAL4 element to the presumed site of transcription initiation, is 170 bp.

The regulator protein could potentially influence the expression of endogenous genes if appropriate target sequences are present. When human DNA sequences were analyzed, no perfect sequence matches to the 17-bp GAL4 sequence were found. Frequencies for partial matches were as follows: one mismatch (1), two mismatches (1), three mismatches (15), four mismatches (122), five mismatches (496), and six mismatches (363). These frequencies were obtained from the Genbank Sequence Database, Release 127.0, December 15, 2001, using BLASTN 2.2.2. These frequency values suggest that there may be some risk for activating endogenous human genes.

Mifepristone is the best-characterized inducer for the GeneSwitch system. Mifepristone (MW 429.6) has antiprogestational activity, which results from competitive interaction with progesterone at progesterone receptor sites. Doses of 1 mg/kg BW or greater antagonize the endometrial and myometrial effects of progesterone in women, and, during pregnancy, the myometrium becomes sensitized to the contraction-inducing activity of prostaglandins. Mifepristone is rapidly absorbed with an oral bioavailability of ~69% and is 98% bound to plasma proteins. Mifepristone has a terminal elimination half-life of 18 hrs in humans, approximately 5 hrs in mice (P. Babij, personal communication) and 1 hr in rats [35]. Mifepristone is indicated for the medical termination of intrauterine pregnancy through 49 days' pregnancy and is administered as a single 600-mg (~10 mg/kg BW) dose, which is followed in 2 days by a 400-µg dose of misoprostol (a prostaglandin). In other clinical trials, designed to assess its antitumor effects, male and female patients with meningioma tolerated long-term chronic dosing of mifepristone (1–3 mg/kg BW administered daily for 12 months or more) [36,37].

In cell culture, transgene induction occurs at mifepristone concentrations that range from 0.01 to 1.0 nM, with half-maximal induction occurring at 0.1 nM [14,38]. The dynamic range of expression was approximately 100-fold in transiently transfected cells and as high as 10^4-fold in stable cell lines. In mice that received intramuscular implants of stably transfected cells, induction of transgene expression occurred at mifepristone doses of 0.1–10 mg/kg, with half-maximal induction at 1 mg/kg. In transgenic mice, induction of transgene expression occurred at mifepristone doses of 0.1–0.5 mg/kg. The dynamic range of transgene expression in transgenic mice was approximately 10,000-fold [39].

B. Plasmid-Based Gene Transfer Studies

Abruzzese and coworkers [38] were the first to deliver plasmids for an inducible transgene-regulation system to animals by intramuscular injection with in vivo electroporation. One plasmid encoded the GeneSwitch regulator protein: CMV promoter/GeneSwitch protein v.3.1/SV40(E) intron/SV40(E) poly(A). The second plasmid contained an inducible secreted alkaline phosphatase (SEAP) gene: 6x GAL4/TATA box promoter/synthetic intron/SEAP/SV40(L) poly(A). The two plasmids were mixed in a 1:1 ratio and 150 µg was injected into tibialis anterior and gastrocnemius muscles of mice. The injected hindlimbs were then subjected to electroporation (four short-duration, square wave pulses) using two stainless steel parallel plate caliper electrodes, which were placed noninvasively in contact with the skin of the leg [40]. In the absence of mifepristone dosing, baseline serum SEAP levels were low. Following administration of a single mifepristone dose, SEAP expression was induced 10- to 20-fold with peak levels occurring at 2–3 days, and baseline levels were reestablished by 7 days. Induced SEAP levels were equal to or greater than that achieved by a plasmid with a constitutively active CMV promoter. SEAP induction occurred at mifepristone doses that ranged from 0.01 to 1.0 mg/kg BW, with half-maximal induction at approximately 0.03 mg/kg BW, a very low dose. Multiple rounds of SEAP induction were observed, but in immunocompetent mice, the magnitude of each successive induction was diminished because of formation of anti-SEAP antibodies. However, in immunodeficient mice, the ability to induce SEAP expression by weekly doses of mifepristone persisted for at least 5 weeks. Although the dynamic range of transgene regulation observed in vivo was low compared to that observed in vitro (100-fold), these studies demonstrated that the antiprogestin-regulated system was functional for more than a month following intramuscular plasmid delivery with electroporation.

Abruzzese and coworkers [41] showed that the dynamic range of in vivo transgene regulation could be improved by an order of magnitude if the constitutive CMV promoter of the plasmid for the GeneSwitch regulator protein was replaced by an autoinducible promoter that consisted of GAL4 sites linked to a minimal herpes simplex virus thymidine kinase (tk) promoter. For these studies, the GeneSwitch plasmid was: 4× GAL/min tk promoter/synthetic intron/GeneSwitch regulator protein v.3.1/hGH poly(A) and the inducible plasmid was: 6× GAL4/TATA box promoter/synthetic intron/transgene (SEAP, EPO, or VEGF)/hGH poly(A). In transiently transfected cells, the autoinducible system was able to regulate the level of transgene expression by 10^2- to 10^4-fold in response to mifepristone administration. When plasmid mixtures were delivered intramuscularly to mice with electroporation, SEAP expression was induced more than 200-fold by mifepristone dosing. Since peak levels were maintained, the greater magnitude of induction was achieved by a reduction in the level of baseline expression.

When the plasmid-based autoinducible system for human VEGF expression was delivered to mice by intramuscular injection and electroporation, VEGF protein was induced to high levels (>1 ng/mg muscle tissue) in response to mifepristone dosing [41]. Peak levels were approximately 25% of that achieved by a CMV–vascular endothelial growth factor (VEGF) plasmid. Three rounds of

VEGF induction were achieved over a 3-week period and no VEGF expression was detected in muscle either in the absence of drug or 1 week following induction. High levels of VEGF expression were associated with severe edema, an undesirable side effect of VEGF expression, but the extent and duration of edema were controlled by the schedule of mifepristone dosing. The results provide an example of how the plasmid-based antiprogestin-inducible system could serve as a gene transfer "safety switch."

The plasmid-based autoinducible system also provided drug-dependent regulation of murine EPO expression and hematocrit in mice [41]. In these studies, the amount of the inducible EPO plasmid, relative to that of the GeneSwitch plasmid, had to be reduced, because the inducible EPO plasmid caused hematocrit to increase when delivered alone at doses greater than 7.5 μg. These data are similar to those obtained by the plasmid-based and virus-based tetracycline systems [26,28,29], and highlight the sensitivity of mice to low-level EPO expression. Following delivery of the plasmid mixture (7.5 μg of inducible murine EPO plasmid plus 19 μg of autoinducible GeneSwitch plasmid), mifepristone-dependent induction of murine EPO expression was observed. The increase in EPO expression was associated with an increase in hematocrit, which was sustained for several weeks. In the absence of mifepristone, no EPO expression or an increase in hematocrit was observed. Thus, regulation of EPO expression and its biological activity was achieved following delivery of the autoinducible system to murine muscle. The major limitation was that induction at later times was less effective than induction early on. Thus, the duration of responsiveness of the autoinducible system in vivo appears to be limited by mechanisms that are not yet understood.

Studies in mice with a constitutively expressed GeneSwitch plasmid, CMV promoter/GeneSwitch regulator protein v.3.1/SV40(E) intron/SV40(E) poly(A), and an inducible murine EPO plasmid, $6 \times$ GAL4/TATA box promoter/synthetic intron/murine EPO/hGH poly(A), demonstrated mifepristone-dependent induction of murine EPO expression at various times during an entire year following plasmid delivery with electroporation (J. Nordstrom, unpublished data). However, hematocrit was not regulated in these animals in a drug-dependent manner, because basal EPO expression was high enough to elevate hematocrit in the absence of mifepristone dosing.

To make EPO expression more strictly dependent on mifepristone dosing, the GeneSwitch and inducible EPO plasmids were modified in three ways, resulting in SK promoter/synthetic intron/GeneSwitch regulator protein v.4.0/hGH poly(A) and $6 \times$ GAL4/TATA box promoter with 30-bp deletion/synthetic intron/EPO/hGH poly(A). First, the SK promoter is a muscle-specific, avian skeletal α-actin promoter that is 1–10% as active as the CMV promoter. In addition to tissue specificity, the SK promoter is expected to decrease the intracellular levels of the GeneSwitch regulator protein. Second, GeneSwitch regulator protein v.4.0

was constructed by truncating the GAL4 domain by 20 amino acids to reduce the length of a coiled-coil structure, which was believed to enhance the formation of homodimers that bind to GAL4 sites in the inducible promoter and partially activate transgene transcription in the absence of mifepristone dosing. Third, a 30-bp deletion in the transcription initiation region of the inducible EPO plasmid reduced the basal level of EPO expression by approximately 10-fold without impairing the ability of the gene to be induced. When these modified plasmids were delivered to muscle with electroporation, EPO expression and hematocrit were regulated in a mifepristone-dependent manner. Increases in hematocrit that occurred in the absence of mifepristone dosing were low and transient, and, importantly, the ability to regulate hematocrit persisted for many months following plasmid delivery (J. Nordstrom, unpublished data). An example of long-term regulation of hematocrit in rats by mifepristone dosing following delivery of the improved plasmid-based EPO/GeneSwitch system is shown in Figure 5.

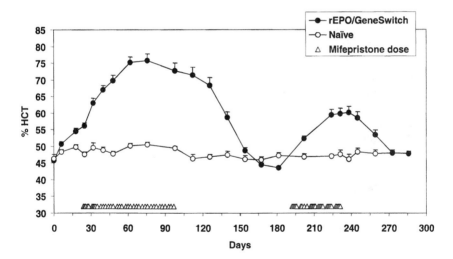

Figure 5 An example of drug-dependent regulation of hematocrit in rats following intramuscular delivery of plasmids for an antiprogestin-inducible system. A 1:1 mixture of plasmids for an inducible rat EPO (rEPO) gene and a GeneSwitch regulator protein gene driven by a muscle-specific promoter was formulated in poly-L-glutamate and delivered to the gastrocnemius muscles of Sprague–Dawley rats (n = 24) by intramuscular injection followed by electroporation with caliper electrodes. Mifepristone was administered by intraperitoneal injection three or five times weekly on days 24–97 (first induction period) or 192–231 (second induction period), as indicated by the open triangles. Hematocrit levels were monitored in animals that received the rEPO/GeneSwitch plasmids (filled circles) and in naïve animals (open circles). Data are presented as mean ± SEM.

C. Virus-Based Gene Transfer Studies

Burcin and coworkers [42] put an antiprogestin-inducible system for human growth hormone (hGH) expression into a single gutless Ad vector. The two genes were arranged as follows: 4× GAL4/TATA box promoter/hGH gene (genomic version with four introns)/hGH poly(A) / TTRB promoter/regulator protein v.3.0/SV40(E) intron / SV40(E) poly(A). The TTRB (transthyretin B) promoter is a liver-specific promoter. GeneSwitch regulator proteins v.3.0 (also called GLp65) and v.3.1 are identical except the former has an extended linker (GSTRYQA) between the progesterone receptor and p65 domains. Ad vector (5 × 10^{12} genomes) was administered to mice by intravenous injection. Induction of hGH protein in serum was observed if mifepristone doses were administered after, but not before, the eighth day after viral delivery. This 8-day lag period might reflect the time necessary for the regulator protein to reach functional steady-state levels in liver cells. hGH expression was undetectable in the absence of mifepristone dosing, and induced levels of hGH expression were high, reaching 2–10 μg/mL, resulting in an induction that was 3–4 logs in magnitude. Peak expression occurred 12–24 hrs after a single mifepristone dose and returned to undetectable levels by 192 hrs. Mifepristone doses that provided induction were 0.1–0.5 mg/kg. hGH expression could be reinduced, and three consistent rounds of induction were observed over a 50-day period. Chronic mifepristone dosing led to sustained, elevated levels of hGH in blood (2–5 μg/mL), which produced substantial weight gains (increases of ~60%) over a 4-week period. Weight gains in animals that received the Ad vector but were dosed only with vehicle (sesame oil) were less than 10% and similar to those of naïve animals.

D. Summary

The antiprogestin-inducible system is effective and durable for many months when incorporated into plasmids that are delivered by intramuscular injection with electroporation or into helper-dependent Ad vectors that are delivered to liver by intravenous injection. Earlier versions of the plasmid-based system exhibited significant levels of basal expression, but modifying the regulator protein, changing to a muscle-specific promoter, and altering the initiation region of the inducible promoter markedly reduced this problem. The autoinducible strategy was also effective in reducing basal expression, but the ability to reinduce expression at later times was limited. For the Ad-based system, where both genes were arranged in a single cassette, the quality of drug-dependent transgene regulation was exceptional and comparable to that observed in transgenic mice. The antiprogestin-inducible system appears to work best when a tissue-specific promoter is used to express the regulator protein.

VI. DISCUSSION

All four systems were effective in providing drug-dependent transgene regulation following delivery by in vivo gene transfer methods. Thus, preclinical proof-of-concept has been achieved, and each system shows promise. However, all of the systems have deficiencies that require attention, and many questions pertaining to the development of drug-regulated gene therapies for humans remain to be answered.

A. Basal Expression

Basal expression is the level of expression that occurs in the absence of drug for an inducible system or in the presence of a saturating amount of drug for a repressible system. An important factor that contributes to basal expression is the inherent activity of the inducible promoters, which consist of minimal promoters linked to multiple copies of specific DNA-binding elements. Even a minimal promoter with just a TATA box element has a low level of activity, which can be measured by quantitative reverse transcriptase–PCR (RT-PCR) assays. Sequence context is important, because cryptic cis-acting elements can influence the activity of minimal promoters regardless of distance or orientation. For example, a typical scan of a 1000-bp sequence reveals hundreds of potential transcription factor binding sites. Most of these sites are expected to be irrelevant, but some could increase basal expression in a cell- or tissue-specific manner. It should never be assumed that a particular inducible promoter would exhibit the same level of basal expression in different vectors or in different contexts or orientations, and the possibility of cryptic transcription factor–binding sites in the coding sequences of different transgenes also needs to be considered.

A second factor that contributes to basal expression is the extent to which regulatory fusion proteins activate inducible promoters in the absence of drug. For the tetracycline- and antiprogestin-dependent systems, this depends on the relative binding affinities of the inactive and active forms of the regulator proteins. For the dimerizer-dependent system, this depends on whether the monomeric subunits have any activity or whether active heterodimers could form in the absence of drug. In all cases, unwanted transcriptional activation is favored as the number of binding sites in the inducible promoter is increased or as intracellular levels of regulator proteins is increased. Accordingly, the strength of the promoter or the quality of the posttranscriptional signals (5' UTR, intron, 3' UTR, poly[A] signal) in the gene for the regulator protein can influence the level of basal transgene expression [43].

A third factor is the gene delivery strategy, for virus-based systems appear to exhibit lower levels of basal expression than plasmid-based systems. Differences in genome sequence, topology, size, structure, interactions with cellular or viral proteins, or subnuclear localization could all be contributing factors. AAV and Ad genomes are probably more effective in forming integrated or noninte-grated chromatin, or chromatinlike, structures that provide an improved state of repression. By contrast, chromatinlike structures do not readily form when non-replicating plasmids are introduced into mammalian cells [44]. Thus, plasmids delivered to terminally differentiated skeletal muscle cells are not expected to form chromatinlike structures and, to date, there is no evidence for plasmid inte-gration into chromosomal DNA. Thus, plasmid-based regulation systems may require more fine tuning to control basal expression.

Studies with regulated EPO expression systems provide the most sensitive measure of biologically relevant basal expression in vivo, for trace levels of EPO protein can trigger substantial increases in hematocrit. All four transgene-regula-tion systems show some evidence of basal EPO expression. Basal expression was high enough to cause unscheduled increases in hematocrit with the AAV-based tetracycline-repressible system, the plasmid-based tetracycline-inducible system, and the plasmid-based antiprogestin-inducible system. However, alterations in the inducible promoter and the regulator protein, combined with a promoter change designed to decrease the intracellular level of the regulator protein, sub-stantially reduced the unscheduled increases in hematocrit exhibited by the plas-mid-based antiprogestin-inducible system. No increases in hematocrit occurred during the weeks following delivery of the AAV-based dimerizer inducible sys-tem, which implies a lack of basal EPO expression. However, this is not a rigor-ous conclusion, since AAV vectors require weeks to mature and reach full capac-ity, and, unfortunately, no data on animals not subjected to rapamycin dosing were reported. Evidence of basal expression is suggested by studies that showed hematocrit levels to be stable or only gradually declining when rapamycin dosing was withdrawn for periods of 40–100 days.

B. Single Versus Multiple Vectors

Single vector systems have advantages, because they simplify and reduce the cost of manufacturing and avoid variations in the ratios of components following in vivo delivery. Indeed, a single Ad vector that carried a cassette with both genes required for the antiprogestin-inducible system worked extremely well, exhibiting low basal and high induced expression. However, the performance of the system was severely compromised when a similar cassette was inserted into a single plasmid (J. Nordstrom, unpublished data). The main problem was high basal expression, and the most likely explanation was that enhancer and promoter ele-ments in the promoter for the regulator protein gene also activated the minimal promoter for the inducible transgene. Thus, plasmid-based systems work best when each gene is carried by a separate plasmid. An advantage of this approach, of course, is that plasmid ratios can be adjusted to optimize performance. This feature was exploited for the plasmid-based tetracycline- and antiprogestin-inducible systems [28,41].

A single AAV vector that carried the two genes for a tetracycline-inducible system also suffered from elevated levels of basal expression [29]. Owing to strict space limitations, the AAV-based dimerizer-inducible system will always require at least two separate vectors even when single vectors carry multiple components. For one embodiment [32], one vector carried the inducible transgene, and the second vector carried a transcription unit for both monomeric subunits by utilizing an internal ribosome entry site (IRES) element. A problem with this approach is that IRES-dependent translation is about 10-fold less efficient than normal cap-dependent translation, and thus the subunits are not produced in equimolar amounts. For another embodiment [33], a gene for one of the subunits was combined into the vector carrying the inducible transgene. When this vector, plus the one that encoded both subunits, was evaluated in vitro, substantial elevations in basal expression were observed. However, when evaluated in vivo, the levels of basal expression were acceptable. These data suggest that the transformations of the AAV genomes that occur following in vivo delivery can have a substantial impact on levels of basal expression.

C. Maximal Levels of Expression

The levels of expression attained by the drug-dependent regulation systems, when maximally turned on, appear to be comparable to those of constitutively active CMV promoter–driven constructs [28,33,38]. Thus, the chimeric regulator proteins employed by the various systems are highly potent transcriptional activators. The most potent system was the Ad-based antiprogestin-inducible system, delivered by intravenous injection, which yielded circulating levels of hGH of 10 μg/ml in mice [42]. This was particularly impressive, since hGH has a serum half-life of approximately 4 mins. In contrast, the Ad-based and AAV-based rapamycin-inducible systems, delivered by intramuscular injection, provided circulating levels of hGH in mice of 5 ng/mL and 1 ng/mL, respectively [33]. No data on hGH expression with a plasmid-based regulation system have been reported.

D. Kinetics

The antiprogestin-inducible system displayed induction and decay kinetics that were rapid. Induction of EPO or hGH expression by plasmid-based or Ad-based systems reached peak levels in mice within 24 hrs of mifepristone dosing, and baseline levels were restored after 72–96 hours [38,42]. Induction of SEAP expression peaked at 2–3 days, but this reflects the long half-life of the SEAP protein in blood [38]. In contrast, the dimerizer-inducible system displayed rapid induction kinetics but slow decay kinetics. Thus, induction of EPO or hGH expression by Ad- or AAV-based systems was maximal in mice or rhesus monkeys within 24 hrs of rapamycin dosing, but the return to baseline required 10–20 days and was sometimes incomplete [32,33]. These data suggest that rapamycin-induced EPO or GH mRNAs are very stable, or that the heterodimeric regulator protein, which is held together by rapamycin, remains in an active state for a prolonged period of time even after free rapamycin has been completely eliminated (the elimination half-life for rapamycin in mice is 2–5 hrs). Detailed kinetics for the AAV- or plasmid-based tetracycline-dependent systems in vivo have not been reported except that induced levels of EPO expression returned to baseline levels within 10 days of cessation of doxycycline dosing [28].

E. Duration

All of the regulation systems remained functional for many months following delivery to animals. AAV-based systems, when delivered to mice by intramuscular injection, showed retention of function for at least 4 months (tetracycline repressible), 7 months (tetracycline inducible), or 10 months (dimerizer inducible) [26,29,33]. Ad-based systems, when delivered to mice by intramuscular or intravenous injection, retained function for at least 2 months (antiprogestin inducible) or 5 months (rapamycin inducible) [33,42]. Plasmid-based systems, when delivered to mice or rats by intramuscular injection with electroporation, remained functional for at least 2 months (tetracycline inducible) or 8 months (antiprogestin inducible) [28] (see Fig. 5). These data demonstrate that the transfected cells tolerated the chimeric regulator proteins, and that robust cellular immune responses directed against these cells were not generated. For the AAV- and Ad-based systems, induced levels of expression were similar at different times following vector delivery, implying that the viral genomes have attained a steady-state condition. This is fortunate, for readministration of viral vectors is difficult owing to immune responses against viral protein moieties. For the plasmid-based systems, induced levels of expression may decline gradually, implying that plasmid genomes are slowly lost over time (see Fig. 5). However, an important feature of plasmid-based systems is an ability to be readministered [28]. Thus, it should be possible to restore regulated transgene expression by delivering plasmids with electroporation to the same or different muscle site.

In rhesus monkeys, the retention of function for an AAV-based rapamycin-inducible system for rhesus EPO expression was erratic, lasting for only a few months in some animals, but exceeding a year in others [32,43]. This could reflect instability of AAV viral genomes or cellular immune responses. Antibodies to the strongly immunogenic HA epitope of the regulator proteins were detected in serum several months after gene delivery, but circulating antibodies are not necessarily indicative of a cellular immune response. As these regulated gene therapy systems move toward human clinical use, it will be important to understand better the limitations presented by nonhuman primates.

F. Ligand

An ideal ligand would be one that regulates transgene expression in a dose-dependent manner, is orally bioavailable, and has no other activities. None of the currently available ligands fits all of these criteria. Doxycycline is a widely used antibiotic, rapamycin is a potent immune suppressor, and mifepristone is an antiprogestin that has abortifacient activity at high doses. However, better ligands can be developed. The binding pockets of the different chimeric regulator proteins are well characterized in structural terms and are known to be malleable. Moreover, the development of cell lines for high throughput screens to identify next-generation ligands is straightforward. Using computer-aided design or directed mutagenesis and selection, one can identify modified LBDs that specifically interact with modified ligands. These types of methods were used to identify the original progesterone receptor mutant that gained the ability to be activated by mifepristone [34], and should be applicable to the identification of related compounds with reduced antiprogestin activity. In addition, these methods have been used to isolate rapamycin analogs (rapalogs) with reduced immunosuppressive activity [30].

Key issues for the currently available ligands are the nature and extent of side effects that occur at the dosages and schedules needed to produce the desired therapeutic effect. For example, intermittent administration of low doses of ligand, which are likely to be well tolerated by patients, may be sufficient to induce EPO expression to elevate and maintain hematocrit levels in anemic patients. However, the applicability of this type of ligand dosing strategy to other regulated transgenes of therapeutic interest would have to be determined on a case by case basis.

G. Immunogenicity and Toxicity

The chimeric structure of the regulator proteins introduces a risk of immune responses, and the potent DNA-binding and transcriptional-activation activities carried by these proteins could interfere with normal intracellular mechanisms and introduce toxicities. With all of the systems to date, however, the durability of drug-controllable expression in rodents and primates suggests that these effects are minor.

The tetracycline-regulator proteins (tTA and rtTA) are composed of bacterial and viral sequences, with one fusion junction, and are not humanized at all. They have been associated with the inhibition of the proliferation of cells in culture and toxicities in transgenic mice [46,47], but no evidence for toxicity has emerged from in vivo gene transfer studies so far. However, the risk for toxicity is increased if strong promoters, like the CMV promoter, are required to express the gene for the tetracycline-regulator protein. The use of a less active, tissue-specific promoter, like that utilized by Rizzuto and coworkers [28], reduces the risk of immunogenicity or toxicity.

For the dimerizer-inducible system, the DNA-binding subunit is 94% humanized in sequence with four fusion junctions, and the transcription activator subunit is 90% humanized with one fusion junction. The large proportion of human sequence in these proteins should minimize the prevalence of foreign epitopes, but novel epitopes might nonetheless arise from the multiplicity of junctions. The DNA-binding subunit should bind to its DNA target with equal affinity either as an inactive monomer or as an active heterodimer, but only the latter should activate transcription. The activator subunit should be without function as a monomer, but it could interact with other transcription factors or related proteins and cause toxicity by squelching or other mechanisms. The intracellular concentration of the activator subunit appears to a critical variable, which makes sense if the DNA-binding subunit is primarily bound to the DNA and the activator subunit is freely diffusible. The need to produce high levels of the activator subunit increases the risk for toxic or immune responses. The elements in the inducible promoter that comprise the binding sites for the regulator protein are only 12 bp in length, which means that multiple exact copies, as well as hundreds of near perfect copies, are found in endogenous human DNA. Thus, there may be a risk of altering the expression of certain endogenous genes, the consequences of which are unknown.

The GeneSwitch regulator protein, composed of yeast and human sequences, is 83% humanized with two fusion junctions. The yeast DNA-binding domain and the two junctions may be sources of foreign epitopes. The inactive monomeric form of the regulator protein, like other inactive nuclear hormone receptors, is probably maintained in inert complexes with heat shock proteins, immunophilins, and other proteins. This may be important, because inert complexes should limit the availability of the regulator protein for squelching or other deleterious interactions, and thereby reduce the chance of toxic responses. The antiprogestin-inducible system appears to function best when tissue-specific promoters are used to control the gene for the regulator protein. Since tissue-specific promoters are relatively inactive, the intracellular levels of regulator protein should be reduced, which further decreases the likelihood of toxicity. Tissue-specific promoters also limit the possibility of expression in nontarget tissues, particularly cells of the immune system, which also decrease the likelihood of immune responses.

VII. SUMMARY

The ability to provide long-term drug-controllable expression of genes for circulating therapeutic proteins is a significant advance to the field of gene therapy. Controlled, continuous production of natural forms of therapeutic proteins should be more convenient than regular, repeated administration of injected recombinant proteins that often have short half-lives. Long-acting controllable gene therapy systems have the potential to be more efficacious by avoiding troughs that drop below therapeutic levels, and to be safer by avoiding peaks that exceed the level required for therapeutic effect. One of the challenges of gene transfer is to achieve predictable and consistent dose-dependent expression of genes following delivery. An advantage of drug-controlled gene transfer systems is that genes can be delivered in a relatively dose-independent manner, because expression levels are mainly controlled by the dose and/or schedule of the inducing drug. Thus, drug-regulated gene systems have potential to provide more consistent and predictable expression than is achievable with constitutively expressed genes.

Achievement of these performance goals will require improvements in the properties of drug-controllable gene transfer systems. The immediate challenge is further improvement of the tightness of regulation, and this will be addressed by alterations in minimal promoter structures, by modifications to the structures of regulator proteins, and by optimizing the transcription units and vectors that carry them. Additional challenges are the characterization of new, safer ligands and understanding the long-term consequences of chimeric protein production in transfected cells. Work in all of these areas is in progress, and substantial improvements have been reported for several systems [48–52]. In vivo gene therapy studies, particularly ones conducted in larger animals, will be needed to evaluate fully the various modifications. The near future promises to be exciting, and one should look forward to the development of drug-controlled systems that will be more fully suitable for human gene transfer studies.

REFERENCES

1. Orkin SH, Motulsky AG. Report and recommendations of the panel to assess the NIH investment in research on gene therapy, 1995; *http://www.nih.gov/news/panelrep.html*.
2. Kochanek S, Clemens PR, Mitani K, Chen HH, Chan S, Caskey CT. A new adenoviral vector: replacement of all viral coding sequences with 28 kb of DNA independently expressing both full-length dystrophin and beta-galactosidase. Proc Natl Acad Sci USA 1996; 93:5731–5736.
3. Morral N, O'Neal W, Rice K, Leland M, Kaplan J, Piedra PA, Zhou H, Parks RJ, Velji R, Aguilar-Cordova E, Wadsworth S, Graham FL, Kochanek S, Carey KD, Beaudet AL. Administration of helper-dependent adenoviral vectors and sequential delivery of different vector serotype for long-term liver-directed gene transfer in baboons. Proc Natl Acad Sci USA 1999; 96:12816–12821.
4. Kim IH, Jozkowicz A, Piedra PA, Oka K, Chan L. Lifetime correction of genetic deficiency in mice with a single injection of helper-dependent adenoviral vector. Proc Natl Acad Sci USA 2001; 6:13282–13287.
5. Kessler PD, Podsakoff GM, Chen X, McQuiston SA, Colosi PC, Matelis LA, Kurtzman GJ, Byrne BJ. Gene delivery to skeletal muscle results in sustained expression and systemic delivery of a therapeutic protein. Proc Natl Acad Sci USA 1996: 93: 14082–14087.

6. Herzog RW, Yang EY, Couto LB, Hagstrom JN, Elwell D, Fields PA, Burton M, Bellinger DA, Read MS, Brinkhous KM, Podsakoff GM, Nichols TC, Kurtzman GJ, High KA. Long-term correction of canine hemophilia B by gene transfer of blood coagulation factor IX mediated by adeno-associated viral vector. Nat Med 1999;5:56–63.

7. Xu L, Daly T, Gao C, Flotte TR, Song S, Byrne BJ, Sands MS, Parker Ponder K. CMV-beta-actin promoter directs higher expression from an adeno-associated viral vector in the liver than the cytomegalovirus or elongation factor 1 alpha promoter and results in therapeutic levels of human factor X in mice. Hum Gene Ther 2001; 12:563–573.

8. Mir LM, Bureau MF, Rangara R, Schwartz B, Scherman D. Long-term, high level in vivo gene expression after electric pulse-mediated gene transfer into skeletal muscle. C R Acad Sci III 1998; 321:893–899.

9. Mathiesen I. Electropermeabilization of skeletal muscle enhances gene transfer in vivo. Gene Ther 1999; 6:508-514.

10. Smith LC, Nordstrom JL. Advances in plasmid gene delivery and expression in skeletal muscle. Curr Opin Mol Ther 2000; 2:150–154.

11. Gossen M, Bujard H. Tight control of gene expression in mammalian cells by tetracycline-responsive promoters. Proc Natl Acad Sci USA 1992; 89:5547–5551.

12. van Sloun PP, Lohman PH, Vrieling H. The design of a new mutation model for active genes: expression of the Escherichia coli lac operon in mammalian cells. Mutat Res 1997; 382:21–33.

13. Fussenegger M, Morris RP, Fux C, Rimann M, von Stockar B, Thompson CJ, Bailey JE. Streptogramin-based gene regulation systems for mammalian cells. Nat Biotechnol 2000; 18:1203–1208.

14. Wang Y, O'Malley BW Jr, Tsai SY, O'Malley BW. A regulatory system for use in gene transfer. Proc Natl Acad Sci USA 1994; 91:8180–8184.

15. Putzer BM, Stiewe T, Crespo F, Esche H. Improved safety through tamoxifen-regulated induction of cytotoxic genes delivered by Ad vectors for cancer gene therapy. Gene Ther 2000; 7:1317–1325.

16. No D, Yao TP, Evans RM. Ecdysone-inducible gene expression in mammalian cells and transgenic mice. Proc Natl Acad Sci USA 1996; 93:3346–3351.

17. Narumi K, Kojima A, Crystal RG. Adenovirus vector-mediated perforin expression driven by a glucocorticoid-inducible promoter inhibits tumor growth in vivo. Am J Respir Cell Mol Biol 1998; 19: 936–941.

18. Rivera VM, Clackson T, Natesan S, Pollock R, Amara JF, Keenan T, Magari SR, Phillips T, Courage NL, Cerasoli F Jr, Holt DA, Gilman M. A humanized system for pharmacologic control of gene expression. Nat Med 1996; 2:1028–1032.

19. Hinrichs W, Kisker C, Duvel M, Muller A, Tovar K, Hillen W, Saenger W. Structure of the Tet repressor-tetracycline complex and regulation of antibiotic resistance. Science 1994; 264:418–420.

20. Bocker R, Estler CJ, Maywald M, Weber D. Comparison of distribution of doxycycline in mice after oral and intravenous application measured by a high-performance liquid chromatographic method. Arzneimittelforschung 1981; 31:2116–2117.

21. Baron U, Schnappinger D, Helbl V, Gossen M, Hillen W, Bujard H. Generation of conditional mutants in higher eukaryotes by switching between the expression of two genes. Proc Natl Acad Sci USA 1999; 96:1013–1018.

22. Kistner A, Gossen M, Zimmermann F, Jerecic J, Ullmer C, Lubbert H, Bujard H. Doxycycline-mediated quantitative and tissue-specific control of gene expression in transgenic mice. Proc Natl Acad Sci USA 1996; 3:10933–10938.

23. Tremblay P, Meiner Z, Galou M, Heinrich C, Petromilli C, Lisse T, Cayetano J, Torchia M, Mobley W, Bujard H, DeArmond SJ, Prusiner SB. Doxycycline control of prion protein transgene expression modulates prion disease in mice. Proc Natl Acad Sci USA 1998; 95:12580–12585.

24. Dhawan J, Rando TA, Elson SL, Bujard H, Blau HM. Tetracycline-regulated gene expression following direct gene transfer into mouse skeletal muscle. Somat Cell Mol Genet 1995; 21233–240.

25. Liang X, Hartikka J, Sukhu L, Manthorpe M, Hobart P. Novel, high expressing and antibiotic-controlled plasmid vectors designed for use in gene therapy. Gene Ther 1996; 3:350–356.

26. Rendahl KG, Leff SE, Otten GR, Spratt SK, Bohl D, Van Roey M, Donahue BA, Cohen LK, Mandel RJ, Danos O, Snyder RO. Regulation of gene expression in vivo following transduction by two separate rAAV vectors. Nat Biotechnol 1998; 16:757–761.

27. Gossen M, Freundlieb S, Bender G, Muller G, Hillen W, Bujard H. Transcriptional activation by tetracyclines in mammalian cells. Science 1995; 268:1766–1769.

28. Rizzuto G, Cappelletti M, Maione D, Savino R, Lazzaro D, Costa P, Mathiesen I, Cortese R, Ciliberto G, Laufer R, La Monica N, Fattori E. Efficient and regulated erythropoietin production by naked DNA injection and muscle electroporation. Proc Natl Acad Sci USA 1999; 96:6417–6422.

29. Bohl D, Salvetti A, Moullier P, Heard JM. Control of erythropoietin delivery by doxycycline in mice after intramuscular injection of adeno-associated vector. Blood 1998; 92:1512–1517.

30. Liberles SD, Diver ST, Austin DJ, Schreiber SL. Inducible gene expression and protein translocation using nontoxic ligands identified by a mammalian three-hybrid screen. Proc Natl Acad Sci USA 1997; 94:7825–7830.

31. Supko JG, Malspeis L. Dose-dependent pharmacokinetics of rapamycin-28-N,N-dimethylglycinate in the mouse. Cancer Chemother Pharmacol 1994; 33:325–330.

32. Ye X, Rivera VM, Zoltick P, Cerasoli F Jr, Schnell MA, Gao G, Hughes JV, Gilman M, Wilson JM. Regulated delivery of therapeutic proteins after in vivo somatic cell gene transfer. Science 1999; 283:88–91.

33. Rivera VM, Ye X, Courage NL, Sachar J, Cerasoli F Jr, Wilson JM, Gilman M. Long-term regulated expression of growth hormone in mice after intramuscular gene transfer. Proc Natl Acad Sci USA 1999; 96:8657–8662.

34. Vegeto E, Allan GF, Schrader WT, Tsai MJ, McDonnell DP, O'Malley BW. The mechanism of RU486 antagonism is dependent on the conformation of the carboxy-terminal tail of the human progesterone receptor. Cell 1992; 69:703–713.

35. Deraedt, R, Bonnat C, et al. In: Balieu EE, Segal, S, eds. The Antiprogestin Steroid RU486 and Human Fertility Control. New York: Plenum Press, 1985:103–122.

36. Grunberg SM, Weiss MH, Spitz IM, Ahmadi J, Sadun A, Russell CA, Lucci L, Stevenson LL. Treatment of unresectable meningiomas with the antiprogesterone agent mifepristone. J Neurosurg 1991; 74:861–866.

37. Spitz IM, Bardin CW. Mifepristone (RU 486)—a modulator of progestin and gluco-corticoid action. N Engl J Med 1993; 329:404–412.

38. Abruzzese RV, Godin D, Burcin M, Mehta V, French M, Li Y, O'Malley BW, Nordstrom JL. Ligand-dependent regulation of plasmid-based transgene expression in vivo. Hum Gene Ther 1999; 10:1499–1507.

39. Wang Y, DeMayo FJ, Tsai SY, O'Malley BW. Ligand-inducible and liver-specific target gene expression in transgenic mice. Nat Biotechnol 1997; 15:239–243.

40. Abruzzese RV, MacLaughlin F, Smith LC, Nordstrom JL. Regulated expression of plasmid-based gene therapies. In: Morgan JR, ed. Molecular Medicine, Vol. 69, Gene Therapy Protocols, 2nd Ed. Totowa, NJ: Humana Press, 2001:109–122.

41. Abruzzese RV, Godin D, Mehta V, Perrard JL, French M, Nelson W, Howell G, Coleman M, O'Malley BW, Nordstrom JL. Ligand-dependent regulation of vascular endothelial growth factor and erythropoietin expression by a plasmid-based autoinducible GeneSwitch system. Mol Ther 2000; 2:276–287.

42. Burcin MM, Schiedner G, Kochanek S, Tsai SY, O'Malley BW. Adenovirus-mediated regulable target gene expression in vivo. Proc Natl Acad Sci USA 1999; 96: 355–360.

43. Nordstrom JL. Expression plasmids for nonviral gene therapy. In: Rolland A, ed. Advanced Gene Delivery: from Concepts to Pharmaceutical Products. Amsterdam: Harwood Academic Publishers, 1999: 15–43.

44. Smith CL, Hager GL. Transcriptional regulation of mammalian genes in vivo. A tale of two templates. J Biol Chem 1997; 272: 27493–27496.

45. Clackson T, Rivera VM, Pollock RM, Zoller KE, Wang X, Courage N, Wardwell S, Zoltick P, Gao G, Wilson JM. Regulated delivery of secreted proteins: preclinical evaluation of systems for drug-controlled transcription and secretion. Mol Ther 2000; 1:S143.

46. Shockett P, Difilippantonio M, Hellman N, Schatz DG. A modified tetracycline-regulated system provides autoregulatory, inducible gene expression in cultured cells and transgenic mice. Proc Natl Acad Sci USA 1995; 92:6522–6526.

47. Gallia GL, Khalili K. Evaluation of an autoregulatory tetracycline regulated system. Oncogene 1998; 16:1879–1884.

48. Rossi FM, Guicherit OM, Spicher A, Kringstein AM, Fatyol K, Blakely BT, Blau HM. Tetracycline-regulatable factors with distinct dimerization domains allow reversible growth inhibition by p16. Nat Genet 1998; 20:389–393

49. Freundlieb S, Schirra-Muller C, Bujard H. A tetracycline controlled activation/repression system with increased potential for gene transfer into mammalian cells. J Gene Med 1999; 1:4–12.

50. Urlinger S, Baron U, Thellmann M, Hasan MT, Bujard H, Hillen W. Exploring the sequence space for tetracycline-dependent transcriptional activators: novel mutations yield expanded range and sensitivity. Proc Natl Acad Sci USA 2000; 97:7963–7968.

51. Pollock R, Issner R, Zoller K, Natesan S, Rivera VM, Clackson T. Delivery of a stringent dimerizer-regulated gene expression system in a single retroviral vector. Proc Natl Acad Sci USA 2000; 97:13221–13226.

52. Mehta V, Abruzzese RV, Bruce B, Florack V, Anscombe I, Fewell JG, Nordstrom JL. Durable, ligand-dependent regulation of erythropoietin transgenes following intramuscular delivery of formulated plasmid to animals. Mol Ther 2001; 3:S397.

4

Mechanisms for Cationic Lipids in Gene Transfer

Lisa S. Uyechi-O'Brien
Onyx Pharmaceuticals, Richmond, California, U.S.A.

Francis C. Szoka, Jr.
University of California at San Francisco, San Francisco, California, U.S.A.

I. CATIONIC LIPIDS IN GENE TRANSFER

For over a decade, cationic lipids have been used for gene transfer both in cell culture and in animals [1,2]. A cationic lipid is a positively charged amphiphile with a hydrophilic head and a hydrophobic tail that self-associates in aqueous solution to form either micelles or bilayer liposomes (Table 1). This results in a multivalent particle which can then interact with the negative charges on a polynucleotide, such as DNA. This complex of cationic lipid and DNA has been given the term *lipoplex* [3].

The goal of cationic lipid–mediated gene transfer is to deliver a plasmid of DNA to a cell such that it becomes transcribed and translated into a desired protein/peptide (Fig. 1). The sequence of events involved in cationic lipid–mediated gene transfer include, but are not exclusive to (Fig. 1): (1) the formation of the cationic lipid into a liposome, providing a multivalent surface charge that is (2) attracted via electrostatics to the negative charges on the DNA phosphate backbone. This mixture of cationic lipid and DNA forms, (3) a small particle, or lipoplex, on the order of 80–400 nm in diameter. If these particles have an excess of positive charge they will, (4) bind to the negative charges on the surface of cells. The principal component through which the lipoplex binds is thought to be the sulfated sugar residues of cell surface proteoglycans. The complexes become (5) internalized through a vesicular pathway, followed by (6) the release of DNA from the lipoplex into the cell cytoplasm. The release of DNA is thought

Table 1 Cationic Lipids

	Lipid/DNA Ratio	Reference
DOTMA	2.67 mol:1 mol (in COS-7 cells)	1
	1.21 mol:1 mol (in CV-1 cells)	1
	0.6 mol:1 mol	10
	0.5 to 2.67 mol:1 mol	51
DC-Chol	1.65 mol:1 mol	10
DOGS	1–3 mol/1 mol	9
	2–3 mol/1 mol	35
GL-67	0.67 mmol/4.8 mmol	6
DMRIE	0.09–0.7 mol/1 mol	5

to occur concomitantly with the release from the cationic lipid. Then some fraction of the released DNA is (7) successfully trafficked by, as yet, an unknown mechanism to the nucleus where it is transcribed and later translated to protein. Each of these uptake and trafficking processes represents a stochastic event in which only a small fraction of DNA proceeds to the next step. These cellular barriers to gene transfer are reviewed in depth by Meyer et al. [4] and will only be discussed as relevant to the known and hypothesized mechanisms of cationic lipid–mediated gene transfer.

As new molecules are synthesized and studied, we gather more information about how these lipids interact with polynucleotides, how they deliver their genomic payload, and how they can induce cell toxicity at higher doses. For most

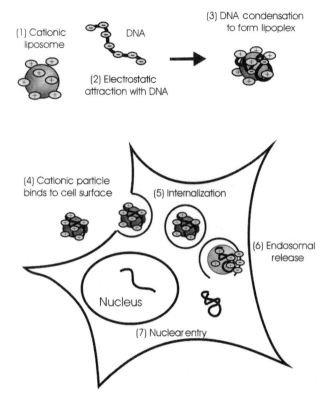

Figure 1 Proposed mechanisms of cationic lipoplex condensation and uptake. (1) A cationic liposome is attracted by electrostatics to (2) the negative charges of DNA forming (3) a lipoplex. (4) Lipoplex binding to the cell surface, (5) internalization, (6) the release of DNA from the lipoplex, and (7) in the nucleus, where RNA is transcribed.

applications, cationic lipid vectors still require higher gene transfer efficiency to render a significant therapeutic effect. Knowledge of the specific mechanisms by which the cationic lipids promote gene transfer would give insight into the limitations and the potential of these vectors and lead to new innovations for improved in vivo delivery. In addition, researchers would be able to make predictions regarding which applications would benefit from the utilization of cationic lipids in gene delivery.

A. Development of Cationic Lipid Formulations

The earliest applications of cationic lipids as gene transfer agents were reported by Felgner et al. [1] where gene expression was demonstrated in vitro using

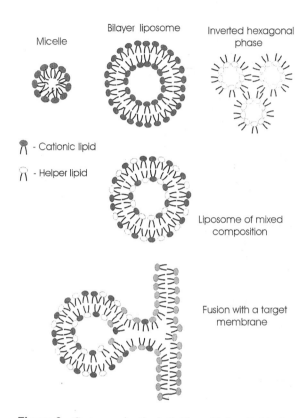

Micelle   Bilayer liposome   Inverted hexagonal phase

ᛉ - Cationic lipid

ᛈ - Helper lipid

Liposome of mixed composition

Fusion with a target membrane

Figure 2   Cartoons of cationic lipids and helper lipids shown as a micelle, single composition and mixed composition liposomes, and in an inverted hexagonal phase. Proposed lipid mixing model of fusion of cationic liposome with a target, i.e., endosomal, membrane. (Adapted from Ref. 24.)

DOTMA (see Fig. 1). Subsequent utilization of the same formulation in animals by Brigham and colleagues showed evidence of gene transfer in the lung [2]. Since then, numerous cationic lipids have been synthesized, varying in head group chemistry, acyl chain length and saturation, and head group to tail linking chemistry. Felgner et al. [5], Lee et al. [6], Byk et al. [7], and Wang et al. [8] are a few of the many studies attempting to correlate chemical structure with transfection activity. Ironically, some of the original cationic lipids including DOTMA, DOGS [9], DC-Chol [10], DOTAP , and lipid #67 [6] remain among the more effective lipids currently used in vitro, in vivo, and in the clinic.

The first indication that a helper or colipid could modulate gene transfer efficiency came from the original Felgner study [1], in which DOTMA formulated with dioleoylphosphatidylethanolamine (DOPE) had greater transfection activity than when formulated with dioleolyphosphatidylcholine (DOPC). DOPE tends to form an inverted hexagonal (Fig. 2) phase and does not promote a stable bilayer by itself at neutral pH. It was proposed that DOPE promoted fusion of the lipid-based complex with the cell membrane better than did DOPC. This difference in activity between DOPE and DOPC was also observed using DC-cholesterol (DC-Chol) formulations [11]. Numerous studies have examined a variety of helper lipid derivatives [12,13] to varying degrees of improvement. One of the commonly used colipids is cholesterol, which has been demonstrated to improve gene transfer in the presence of serum-supplemented culture medium and in vivo [14,15,32]. One hypothesis for this effect is that cholesterol stabilizes the bilayer against interactions with serum components, thus preventing degradation of the lipoplex particle. Although this property may not be necessary for in vitro transfections, it is important in vivo to prevent premature destabilization of complex before it reaches its target tissue (see Sec. III.A).

Despite efforts to converge on a rational approach to selecting a cationic liposomal formulation, no robust structure–activity relationship has been derived. The important features of the cationic liposome appear to be (1) a tertiary or quaternary amine on the cationic lipid head group; (2) a sufficient membrane-destabilizing/fusion characteristic either by the acyl chain lengths or unsaturation within the cationic or colipid; (3) a liposome of uniform size; and for intravenous efficacy, (4) a cholesterol component; and (5) an excess of positive charge. Although subtle formulation and technique differences may affect the overall levels of gene transfer and expression, these are "rules of thumb" that allow the investigator to attain consistent and efficient gene transfer.

II. NAKED DNA MEDIATES GENE TRANSFER

The injection of plasmid DNA alone has been shown to be sufficient for gene expression in the muscle, liver, lung, and brain. The ability for naked DNA to transfect striated muscle is attributed to the structural characteristics of mature myofibers (transverse tubule system, sarcoplasmic reticulum, and multinucleation), facile DNA diffusion through the cytoplasm of a myofiber [16] and a putative receptor for polynucleotide uptake [17,18]. In the liver, several administration techniques mediate naked DNA gene transfer including direct liver injection [19], intraportal injection [20], high volume hydrodynamics injection [21]. Direct instillation of DNA into the lung airway has also been shown to produce low levels of gene transfer [22].

These data indicate that naked DNA is effective in certain applications; however, the incorporation of cationic lipids may be advantageous depending on the property that it provides as a gene delivery system. Some of the advantages of cationic lipids include protection from nuclease degradation, binding and targeting properties, lipid mixing interactions with cell membranes (for intracellular entry of DNA/nucleotide), or adjuvant effects for vaccination and immune response. These mechanisms and the toxicity associated with cationic lipids will be the scope of this discussion in the following sections.

III. MECHANISMS OF CATIONIC LIPID GENE DELIVERY

A. Protection/Stability in Physiological Medium

Cationic lipids protect plasmid DNA from degradation in both cell culture and in the physiological milieu [23,24]. Early biophysical studies showed that DNA becomes encapsulated or encapsidated by the lipid, with a charge neutralizing effect which promotes condensation of the DNA polymer into a compacted particle [25]. Increasing the amount of cationic lipid preserves the circular integrity of plasmid DNA when exposed to either blood serum, cell culture medium, or purified nucleases [24,26,27]. It is thought that the excess lipid covering the DNA creates a steric hindrance, and that nucleases are prevented from binding to DNA in its compacted state. As a result, in vitro gel retardation and DNase I degradation assays are now employed to demonstrate and predict sufficient particle condensation for use in transfection [23,24].

The relevance of protection from DNases depends upon the route of administration, since the plasmid DNA may be exposed to different amounts of degrading enzymes in different compartments. The nuclease activity in mouse serum is reported at 5.0×10^4 U/mL of serum, and is characterized primarily as endonuclease in specificity [28]. Endonucleases are also measured in mouse liver (3.16×10^5 U/mg), lung (5.93×10^4 U/mg), and muscle (4.07×10^4 U/mg), and are most likely present in the extracellular space [28]. The existence of nucleases in serum and tissues influences the half-lives of plasmid DNA in these compartments. Kawabata and colleagues measured the half-life of plasmid DNA at 10 mins in whole blood (29). When administered by instillation into the lung, 50% of the original DNA dose is eliminated by 60 mins with no detectable plasmid DNA by 6 hrs by Southern analysis [22]. In muscle, the reports vary from rapid elimination, that is, 95% within 90 mins [30] to a moderate elimination rate where DNA persists until 3 hrs [17] or 6 hrs [31].

Formulation of a plasmid DNA with a cationic lipid prolongs its half-life in bronchiolar lavage fluid and its in vivo residence time in the airways [32,33]. Similar studies with cationic lipoplexes incubated with isolated blood serum have shown prolonged half-life [26,34]. Indeed, the exposure of cationic lipid/DNA complexes to serum or serum components results in some dissociation of DNA from the lipoplex [24,26,27], allowing degradation by serum nucleases to occur. However, formulations that are more resistant to serum dissociation also tend to have better activity in vivo [26,32]. Presumably the more stable complex will prolong the half-life of the intact plasmid in the physiological milieu, thereby increasing the probability that a transcriptionally active plasmid will be taken up by the cells.

B. Cellular Binding

1. Effect of Charge Ratio and Size/Sedimentation

Part of the gene transfer activity/efficiency is attributed to the particular chemistry of cationic lipid used (e.g., DOTMA, DOGS, DMRIE, DC-Chol) and to an attractive force between the positive charges of the vector and the negative charges on the cell membrane. With all cationic lipoplexes developed to date, increasing the lipid to DNA composition results in increasing levels of cell binding and of gene expression until the high concentration of lipid results in toxicity and cell death. The composition at which transfection becomes detectable at high levels usually depends on the number of amines on the lipid head group and the nature of those amines (e.g., quaternary, tertiary). The most effective and consistent cationic lipid formulations are prepared with a net excess of positive charge [5,11,35,36]. Thus, the amount of binding is positively correlated with increasing cationic charge ratio.

The computed cation to anion ratio can be compared to the actual charge exhibited by the lipoplex using measurements of particle surface charge, or zeta-potential. The zeta-potential usually transitions from negative to positive between the $1:1$ and $2:1$ theoretical net charge. Slight variations in the transition point may occur based on the ionic concentration of the medium, the pH of the medium, and the actual pKa's of the amines [5,37]. Complexes made with an excess of lipid have been shown to be a heterogenous mixture of lipid/DNA particles and free liposomes [38,39]. Such preparations may be purified of free lipid/liposomes by gradient centrifugation, thus obtaining a stable particle which usually exhibits a slight positive charge [38,39]. This suggests that an equilibrium state can be achieved for condensation of the complex. Whereas these separated complexes had reduced gene transfer activity in serum-free conditions, they exhibited less toxicity and demonstrated better activity in serum supplemented medium [39].

In regimens where a negative surface charge is exhibited, size of the complex and its sedimentation appear to be important. As the charge composition approaches unity, aggregates sediment onto cells. Felgner and colleagues [5], in a study using multi-lamellar vesicles and small unilamellar vesicles found that larger complexes were more effective in cell culture [5]. Ross et al. have also reported that a larger sized particle can bring about transfection in the presence of serum [40]. This is reflected in the transfection time required for activity, and levels of expression are, in general, diminished.

2. Role of Heparan Sulfate Proteoglycans

The binding of cationic lipid complexes to cell surface proteoglycans was first characterized by Mislick and Baldeschwieler [41]. These sulfated polysaccharides present a strong negative charge on the surface of the cell and are ideal binding sites for positively charged particles. In particular, heparan sulfate proteoglycans are implicated in many polyamine uptake processes [42]. The susceptibility for cationic lipid–mediated gene transfer can be correlated to the level of proteoglycans expressed on the cell surface, which may vary between different cell culture lines [41,43]. The enzymatic digestion (heparinase I and II and chondroitinase ABC) of cell surface proteoglycans, or the inhibition of proteoglycan sulfation, has been reported to significantly decrease binding of cationic complexes and reduce the level of gene expression [41,42,43]. Complex uptake and expression was also shown to be modulated by stably transfecting a syndecan-1 cell surface proteoglycan on the cell surface [43]. These data provide strong evidence that heparan sulfate proteoglycans are a major binding site for cationic lipid complexes.

In addition to providing surface binding sites for lipoplexes, glycosaminoglycans may also interfere with gene transfer at different steps in the uptake pathway. Externally added proteoglycans have been shown to inhibit cellular binding of positively charged complexes by competing with cell surface receptors for binding of the complex. When added to cell culture medium, heparan sulfate and heparin, but not chondroitin sulfates A, B, or C or hyaluronic acid, were shown to reduce cellular binding of lipoplexes and overall gene expression of the cationic lipid/DNA complexes [41,43]. In addition, heparin and other sulfated glycosaminoglycans (dextran sulfate and heparan sulfate in particular) will cause a dissociation of the lipid and DNA components [24,44,45]. This would lead to increased degradation by nucleases and a reduced cationic charge for binding to cells. This dissociative effect has also been demonstrated to disrupt previously bound lipoplex in the pulmonary vasculature and reduce gene expression in the lung [46].

There is also evidence that soluble, extracellular proteoglycans can be internalized with the cationic complexes and may inhibit expression of gene products [47]. This may explain why hyaluronic acid was observed to abrogate gene expression mediated by DOGS and fractured dendrimer without having a direct effect on the complex stability [45]. This could also be due to coating the positive charges on the lipoplex to provide "shielding" so the lipoplex cannot interact with the cell surface through charge interactions.

It is clear that proteoglycans play multiple roles in the gene transfer process. Further investigation into the mechanisms of transfer and cellular protection will be important to the application of cationic lipids in vivo/gene delivery.

C. Internalization/Endosomal Escape

Although increased binding will usually lead to increased expression in a given formulation, comparative levels of binding between formulations do not always correlate with the levels of gene expression [48–50]. This observation suggests that cationic lipids have a mechanistic effect downstream of binding.

Internalization of the cationic lipid complex is known to occur via a vesicular process; most likely endocytosis [51–54] in vitro and in vivo [55]. Both lipid and DNA components have been visualized by fluorescence microscopy and by electron microscopy within vesicular compartments of the cell [52,53]. However, since a specific receptor for the complex has not been identified, it is difficult to predict the typical trafficking pathway for these vesicles. It is assumed that the endosomes mature to late endosomes and then become fused with lysosomes. With certain cationic lipids, the addition of a pH buffering agent (chloroquine–ammonium chloride), a pH/salt gradient disrupter (monensin, bafilomycin A) or a nuclease inhibitor [56] have been reported to increase gene expression levels. These agents are most active in perturbing the lysosome, and thus may prevent acid-mediated plasmid degradation or prevent acid activation of nucleases. In this way, DNA may persist for longer times within the vesicle. Presumably, this increased in situ time would be sufficient to allow a greater probability of plasmid escape from the vesicle.

In all electron microscopic studies of internalization, both lipid and DNA can be observed in the vesicular compartment. However, DNA is rarely observed free within the cytoplasm, indicating that release of DNA from the endosome is a relatively infrequent process [13,52,53,55]. Because cationic lipoplexes have been observed to interchange or mix their lipid with anionic liposomes, it is reasonable to hypothesize that lipoplex interactions with the cellular membrane are likely. In studies with oligonucleotides, endocytosis is followed by the concentration of the oligo within the nucleus, whereas the lipid is localized in punctate cytoplasmic compartments (44). This suggests that lipid and nucleotide dissociate completely before entering the nucleus.

Fusion of the cationic lipid complex does not appear to occur to a large degree at the cell surface. However, the close proximity of complex and cell membrane within the enclosed volume of the endosome provides a large surface area for contact between the cationic lipid and endosomal membrane [24,44]. Fusion of the membranes may be further enhanced by ion pairing and lipid mixing between the cationic lipid and the anionic lipids within the endosomal membrane [24,44,57].

The capacity of the lipid complexes to intermix with secondary membranes/liposomes has been demonstrated using DNA release data [24], fluorescence dequenching [57], and fluoresence energy transfer [44]. With some cationic lipids, the inclusion of DOPE, which tends to form an inverted hexagonal phase (see Fig. 2), increases in vitro levels of gene transfer compared to the cationic lipid alone [5,10,11]. Wrobel and Collins [54] observed that DOTAP:DOPE complexes exchanged lipids more rapidly than DOTAP:DOPC complexes–correlating with the expression data [1,10,13] Analogs of phosphatidylethanolamine had varied potency on gene expression; again suggesting that the lipid has an effect down-

stream of binding [11,12]. The positive correlation between the propensity to exchange lipids and efficient gene transfer further supports the notion that cationic
lipids facilitate release of DNA from the endosome.

Cationic lipids which have multiple, protonizable amines such as, for example, DOGS and DC-Chol may take advantage of the inherent buffering capacity
of these polyamines. This "proton-sponge" effect which promotes a continual influx of ions and water, causing a disruption of the vesicle [9,58,59]. In the case of
these lipids, the addition of other lysosmotropic agents has little additional benefit.

D. Nuclear Entry

The nuclear envelope is composed of two lipid membrane bilayers (Fig. 3), creating a formidible barrier between the cytoplasmic and nuclear compartments. The
inner and outer membranes are connected at nuclear pore complexes (NPCs)
interspersed throughout the envelope (Fig. 3). Transport in and out of the nucleus
is controlled by these pore complexes, limiting diffusion to molecules less than

60 kD (less than 9 nm in diameter) and actively transports larger macromolecules
with protein carriers. These "karyopherins" are free in the cytoplasm and bind
to nuclear localization sequences on proteins. Once bound to their cargo, the
karyopherins (usually in the form of a heterodimer) bind to the cytoplasmic face
of the nuclear pore complex. Hydrolysis of adenosine triphosphate (ATP),
through a Ran/GAP ATPase, drives the transport of the cargo protein into the
nucleus [4]. Viral gene delivery vectors often have their own nuclear targeting
proteins associated with the genome. This greatly facilitates the rate of nuclear
uptake of DNA and the efficiency of viral gene transfer [4].

Nuclear entry is a tremendous barrier for DNA in the cytoplasm. The signif-
cance of this barrier for delivery of a DNA plasmid (10 MD in size) has been
elegantly demonstrated in a series of nuclear and cytoplasmic microinjection
studies [52,60,61]. Injection into the cytoplasm yields approximately 1000-fold
less expression than injection into the nucleus. One potential solution to this
problem is to use a cytoplasmic expression system, where a RNA–polymerase
is coadministered with the plasmid [62]. Just getting to the nuclear membrane is a
barrier, since diffusion within the cytoplasm is size limited [63], and cytoplasmic
nucleases can degrade the plasmid into unreadable fragments [64].

Cationic lipids are not believed to have a specific role in the nuclear uptake
of plasmid DNA. In fact, nuclear microinjection of cationic lipid/DNA complexes inhibits transcription [16]. However, there may be a protective role for
cationic lipids and polymers within the cytoplasm, preventing degradation or permitting diffusion of the compacted particle. Cationic polymers, on the other hand,
have been suggested to mediate nuclear transport of DNA. This is based on the
observation that polylysine is reminiscent of the nuclear localization sequence
peptide (NLS sequence, PPKKKRKV). In addition, polyethyleneimine has been

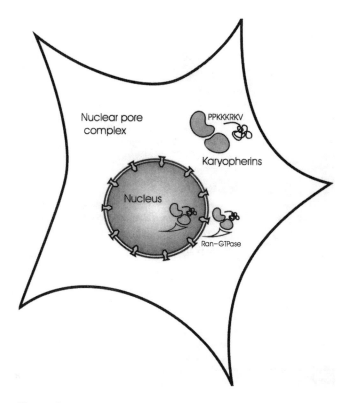

Figure 3 Cartoon of the nucleus with nuclear pore complexes. Active transport of proteins into the nucleus is stimulated by the exposure of a nuclear targeting peptide sequence, such as the SV40 large T-antigen NLS consensus sequence, shown (PPKKKRKV). A heterologous dimer of α and β karyopherins bind the NLS and mediate docking to the nuclear pore complex and to a Ran–GTPase, which mediates translocation of the protein.

reported to have nuclear localization activity that results in nuclear transport of cytoplasmic microinjected polymer/DNA complexes [65]. Conjugation of a NLS peptide to DNA has been shown to increase gene transfer, but only to a modest extent [66,67]. The best enhancement by NLS peptides were seen using a single NLS rather than multiple conjugates [67].

The most likely route of nuclear entry is the incorporation of plasmid during reformation of the nucleus after cell division. This is indirectly supported by evidence that gene transfer occurs most readily in dividing cells [48,52,68] soon after seeding. Direct evidence was reported by correlating gene expression with the localization of a fluorescent oligo:plasmid within doublets of cells [69]. Unfortunately, this would imply that only a subset of the cell population would be transfection competent, and some tissues, such as nervous tissue, would be particularly refractory to gene transfer. The use of nuclear localization signals have shown some improvement in gene transfer efficiency, but other strategies may also be employed. These might include the use of a DNA cis-element which is recognized by a nuclear protein. Alternatively, a transport protein, such as the simian virus 40 (SV40) large T antigen, could be delivered with the plasmid [70]. A convergence of both viral and synthetic systems are likely on this issue.

The cationic lipid–mediated mechanisms of lipoplex binding, internalization, endosome escape, and nuclear entry are believed to be relevant to both cell culture transfections and to gene transfer in animals. However, there are additional complications in animals, such as targeting the appropriate site of action and interactions with biofluids and serum proteins which modulate the nature of the lipoplex. The following sections will discuss the pertinent issues of lipoplex administration intravenously and via the airways (intratracheal administration) and the hypothesized mechanisms of in vivo lipid-mediated gene transfer.

IV. IN VIVO MECHANISMS OF CATIONIC LIPID GENE DELIVERY

A. Lung Anatomical and Physiological Considerations

Most investigations of gene delivery to the lung use one of two routes of administration: through the airways or through systemic administration. Administration via the airways eliminates direct exposure of lipoplex to several organ systems and, in theory, is fairly well contained to the lung. Anatomically, the airways begin at the nasal cavity and continue down through the oropharynx and trachea. Air is channeled into the two bronchi, which progressively branch out into the lung parenchyma, forming the conducting airways of the lungs (Fig. 4). The conducting airways serve to hydrate, warm, and condition the incoming air and to trap airborne particulates and pathogens. This function is provided, in part, by the layer of mucus that coats the upper airways of the lung. Note that in the mouse, there is no significant mucous layer present in the trachea and bronchi [71,72].

The surface epithelia of the trachea and bronchi/bronchioles are composed of ciliated cells, goblet or secretory cells with microvilli, and basal cells (Fig. 5). The underlying submucosal cells, composed of serous and mucus-secreting cells, provide the highly negatively charged and viscous protective layer that lines the airways. These submucosal cells are believed to regulate the Cl^- and Na^+ transport that modulates the water content and viscosity of the aqueous solution layer and mucous. The ciliated cells of the mucosal epithelium beat in synchrony to move continually the mucous layer upward and away from the deep regions of the lung.

The epithelia of the terminal and respiratory bronchioles (Fig. 6) are less

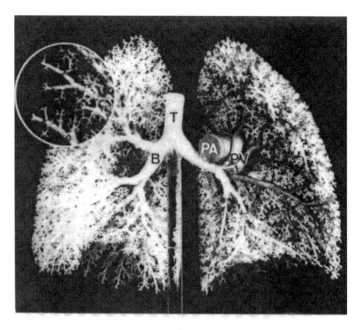

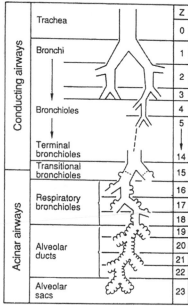

Figure 4 Branching airways cast of the lung airways. Trachea (T), bronchus (B), pulmonary artery (PA), and pulmonary vein (PV). Designation of branching through the airways; Z is the branch generation. (From Ref. 71.)

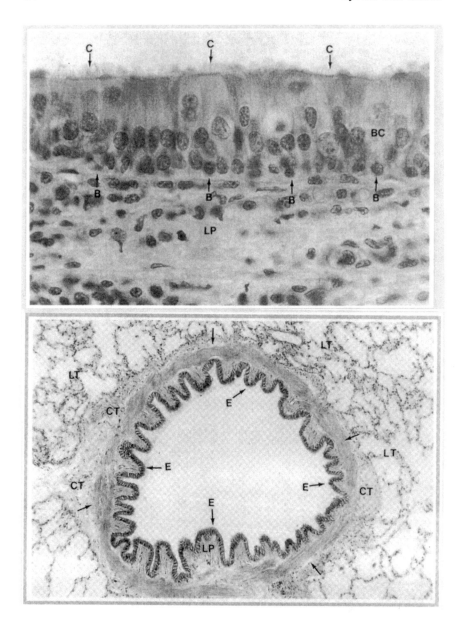

Figure 5 Epithelia of the trachea (top) and bronchioles (bottom). Note the cilia of the columnar cells (C); brush cells (BC), basal cells (B), and lamina propria (LP) of the trachea. Epithelium (E) of the bronchioles is folded; arrows point to the prominent smooth muscle. Connective tissue (CT), lamina propria (LP), and lung tissue (LT) are shown. (From Ref. 107.)

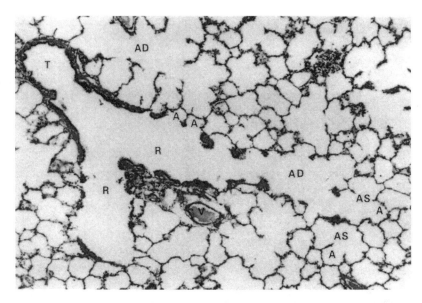

Figure 6 Terminal bronchus (T) splits into respiratory bronchioles (R),which lead into the alveolar ducts (AD). Alveolar sacs (AS) and individual alveoli (A) can be seen. A blood vessel (B) running parallel to the bronchioles can also be seen. (From Ref. 108.)

columnar and are nonciliated, although they continue to generate secretions through both goblet cells and clara cells. Clara cells are cuboidal, nonciliated cells that are more prevalent through the transition region between conducting airways and the gas exchange alveoli. The respiratory bronchioles are the first regions to permit gas exchange with the pulmonary blood vessels, and they fan out to form the delicate alveolar acini (alveolar sacs and individual alveoli). In the alveoli, there is intimate contact between the type I/type II pneumocytes of the epithelia and the endothelium of the pulmonary capillaries. The distance between airway and blood vessel lumen can be as close as 200 nm (Fig. 7). Type II pneumocytes are characterized by the their cuboidal morphology and by their large secretory granules which supply the surfactant for the alveoli gas exchange surface. Type I pneumocytes are flat, squamous cells with protruding nuclei (composing 95% of the alveolar surface but only half the number of alveolar epithelial cells) and are the cells which facilitate gas exchange.

The surfactant layer is composed mostly of phospholipids (primarily saturated and unsaturated phosphatidylcholine) with a small contribution, 5% by weight, by surfactant proteins/apoproteins. It covers the large surface area of the alveolar sacs and reduces surface tension at the air–water interface, so as to reduce the effort required for repeated expansion and collapse of the lung. Surfac-

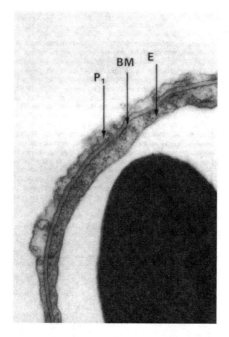

Figure 7 Electron micrograph of the alveolar septum. A type I pneumocyte (P₁) lies adjacent to the basement membrane (BM) and capillary endothelial cell (E). A red blood cell is shown in the capillary lumen. (From Ref. 108.)

tant protein and lipids are secreted from clara cells and type II pneumocytes, some of which form the surfactant layer, hydrophobic tail side out, and the remaining form a lattice of tubular myelin. It is reasonable that these lipid components have the potential to intermix with cationic lipid/DNA complexes and interfere with their delivery to the underlying epithelial cells.

Systemic administration of lipoplex gains access to the lung through two distinct circulatory systems: the pulmonary and bronchial systems. The bronchi of the lung obtain their blood supply from the intercostal arteries off the aorta. They feed the lung tissue through a network of arteries proceeding around the bronchi, penetrating the adventitial layers and submucosal layers. Some of the bronchial capillaries will anastamose with the pulmonary circulatory system, whereas the majority will return to the heart through the vena azygos. The pulmonary vasculature, on the other hand, makes a small circuit to and from the heart and is responsible for replenishing the blood's oxygen supply. The pulmonary capillaries are separated from the lung epithelium by a thin basement membrane, and provide a large surface area for gas exchange to occur ($\sim$50 mm^2/20 g mouse;

140 m^2 in humans [71]. This large epithelial–endothelial contact area provides a large target surface to which the cationic lipoplex can be delivered.

The pulmonary capillaries are approximately 5–6 µg in diameter and are lined with a thin contiguous endothelium. The endothelial cells form a thin monolayer connected by tight junctions and are characterized by their lack of organelles as well as highly vesiculated membranes. These vesicles provide the function of endocytosis into the cell interior and transcytosis to the basal membrane. Nonspecific membrane interactions via electrostatics, may result in fluid phase uptake via pinocytosis, whereas specific adsorption leads to classic endocytosis into clathrin-coated vesicles. Another attribute of the capillary endothelium is the highly anionic glycocalyx that is composed of substantial amounts of heparan sulfate and heparin glycosaminoglycans. The purpose of the glycocalyx is selectively to permit binding of proteins, such as basic fibroblast growth factor (BFGF), and of cells, such as platelets. This also provides an electronegative surface to which the lipoplex can bind and deliver its contents.

B. In Vivo Distribution: Intravenous Administration

The utility of cationic lipid complexes for gene transfer in animals was first reported by Brigham et al. [2] and was based on the hypothesis that electrostatics of the positively charged lipid complex would be an important step in cellular binding. Whereas DNA alone is rapidly cleared by the liver [29], complexes of lipid and polynucleotides exhibit a transient accumulation in the lung. Eighty-five percent of the administered lipoplex dose is retained in the lung after the first 5 mins, followed by a gradual redistribution of both lipid and DNA to the liver [34,46,73,74–76]. This redistribution is presumed to occur because of serum protein disruption of surface-bound complexes.

The pharmacokinetics of the preformed complex appear to be sensitive to the cationic lipid and the formulation used by the investigators. Liu et al. [77] report that increasing the amount of colipid (Tween 80) or reducing the overall formulated complex charge (48:1 vs 6:1) decreases the initial retention in the lung ($t_{1/2}$ <15 mins) and can affect the rate of clearance from the lung. Rapid redistribution of lipoplex from the lung to the liver and spleen was also noted using two commercial preparations (DOTMA:DOPE and DDAB:DOPE) and a detergent dialyzed formulation of DOSPA:DOPE [75,76]. In contrast, the inclusion of cholesterol in the formulation gives prolonged retention (40–50% after 1 hr) in the lung and results in higher levels of gene expression [14,34,74,78].

It was initially proposed that embolism or a first-pass effect was the primary mechanism for retention of cationic complexes in the lung [10,73]. Complexes are known to form aggregates in the presence of serum [55,79], and certain formulations aggregate more readily than others (i.e. DOTAP:Chol over DOTAP alone), correlating with distribution pharmacokinetics [26]. In contrast, systemic administration of cationic liposomes alone does not appear to aggregate sufficiently to cause a first-pass filtration through the lung. Rather the cationic liposomes are rapidly cleared by the liver [34,73,78]. This would suggest that a positively charged lipoplex is subject to different distribution mechanisms than a cationic liposome.

The initial accumulation of cationic lipid/DNA in the lung is responsible for the high level of gene transfer in this organ, and only the first hour (60–90 mins) of complex retention is required for the high levels of gene expression [46]. However, other organs such as the lymph nodes (Peyer's patch in particular), ovaries, and pituitary endothelium also show high levels of uptake in the endothelium, suggesting a more specific mechanism of binding/uptake [55]. Suggestions that this mechanism might be associated with the fortuitous binding of plasma proteins to the lipoplex followed by receptor-mediated uptake of the coated lipoplex have not proven to be correct for complement proteins [78] or for albumin [80].

It has been suggested that the binding and uptake of DNA is mediated by a receptor for DNA in the lung endothelium [77]. If this were the case, then high concentrations of plasmid DNA in the blood should result in gene transfer, and a nonrelevant plasmid should compete for DNA binding sites. Instead, Barron and colleagues found that perfusion of high levels of plasmid without cationic lipid was ineffective in mediating gene transfer to the lung without cationic lipid (81). In addition, perfusion of a noncoding plasmid did not compete with the expression of a previously injected complex [81]. Thus, a complex of both lipid and DNA are required for binding in the lung vasculature, leading to subsequent internalization, as observed by confocal fluorescence microscopic studies [55].

As previously indicated, heparan sulfate proteoglycans (HSPGs) are important to cationic lipid/DNA complex binding in vitro, and are prevalent in the lung endothelium [82]. This was similarly demonstrated in vivo through a reduction in gene expression commensurate with enzymatic degradation (heparinase) of HSPGs in the lung [43]. The investigators observed that the heparinase preinjection was accompanied by a reduction in pulmonary lipoplex retention; presumably due to the loss of binding sites for the positively charged complex. We, however, have found that enzymatic degradation of HSPGs reduces gene expression with no significant difference in lipoplex binding. Other, nonheparan sulfate proteoglycans may be binding lipoplex but are not able to internalize the lipoplex into the appropriate compartment [83]. Thus, heparan sulfate proteoglycans appear to play a significant role in the binding of cationic lipid/DNA complexes in vivo, and further studies should elucidate a potential secondary role in internalization of lipoplex.

The presence of free liposomes in addition to lipoplex has been demonstrated to be important for in vivo gene transfer. The gradient separation of formulated complex from excess lipid results in the loss of activity through the intravenous route [38,84]. Recent work has demonstrated that by simply preinjecting cationic liposomes (DOTAP, DOTMA, and DC-Chol), one can attain high levels of pulmonary gene expression using either naked DNA or low charge–ratio complexes [79,85]. The potentiating effect of the preinjected liposomes persists long after initial binding has become diminished (from 50% to 4% in the lung). Twelve hours after initial administration, only 4% of the cationic liposomes are retained in the lung. Yet, gene expression of a neutral charged complex administered at this time still results in higher expression than the no preinjection controls [85,86]. These results are significant in two respects. First, they indicate that a preformulated complex is not inherently necessary for intravenous administration. Second, there may be a modulating effect of the excess liposomes, perhaps in the plasma or with the endothelial biology, that is still undefined. Saturation of elimination pathways, perhaps in the lung or in the liver, may be partly responsible for the pulmonary expression in the lung. Alternatively, there may be an activation of a cellular pathway that persists over longer periods (12 hrs) and permits high efficiency gene transfer. These questions await resolution.

C. In Vivo Distribution: Intratracheal Administration

Administration of cationic lipid complexes via the airways has a different set of challenges than does intravenous administration. Since the route of administration is fairly well contained, no specific organ targeting/ligand is required. However, the anatomy and physiology of the lung have implications for cationic lipid gene delivery vectors. The positively charged particles may become deposited within the upper airways, being detained by the highly viscous and negatively charged mucous layer. The complex may then be eliminated through the upward movement of the mucous, or it may be dissociated into its lipid and DNA, thus allowing degradation of the plasmid.

There are different techniques used to administer DNA into the lung airways. Nasal instillation of cationic lipid/DNA complexes involves a small volume (20–50 uL for a mouse) of solution to be dropped in through the nostrils. Often this stimulates a gag reflex which redirects the dose to the esophagus, and only a small portion of the administered volume goes into the trachea (Jiang et al. [68] estimates 30% of the dose may be lost). This technique is relatively simple, but can result in inadequate and variable dosing. Administration of cationic lipid / DNA complexes via the nostrils results in reporter gene expression in the alveolar regions of the lung, but there is little detection of gene transfer in the epithelia of the upper airways [68,87].

Intratracheal instillation involves an intubation of the trachea followed by instillation of a small fluid volume (100 uL in mice, one-eighth of the maximum inspired lung volume). With this technique, it is presumed that the cationic complexes will have distribution to all the lung lobes, although some preference for particular lobes may occur [22]. The lung will typically absorb the aqueous vehicle within 5–20 mins. By this method, the gene delivery vector is delivered to the lower airways and has the potential to reach the deep regions of the lung parenchyma. Intratracheal instillation usually results in transfection of the alveolar sacs of mostly type II pneumocytes [88,89], but there have been occasional reports of Clara cells in the bronchioles [32,90] and one report of reporter gene expression primarily in the tracheal epithelia [91]. It is interesting that type II cells, which comprise only 5% of the alveolar surface, appear to be the primary site for gene expression. This may be due to the fact that type II pneumocytes are involved in the internalization and recycling of surfactant, and perhaps the lipoplex is internalized by the same pathway utilized by endogenous surfactant.

The mechanism by which cationic lipids enhance gene delivery by these routes is partly due to the protection of the plasmid from degradation within the airways. Desphande et al. [33] showed that incubation of DNA with bronchial lavage fluid causes a time-dependent degradation which can be retarded when complexed with cationic lipids. Meyer et al. [32] showed that an effective cationic lipid formulation with cholesterol allows DNA to persist in the lung longer than a DOPE formulation. This is presumably because the cholesterol, with a higher crystalline gel phase transition than DOPE, forms a more stable cationic lipid/DNA particle. Also, Glasspool-Malone and Malone [92] reported that coadministration of aurintricarboxylic acid (ATA) with naked DNA could increase plasmid residence time and integrity, and consequently observed a 10- to 65-fold enhancement of gene expression. This again suggests that DNA metabolism in the lung contributes to its removal.

The destabilization of cationic lipid complexes in the airway is fundamentally different from that which takes place within the vasculature. Although anionic proteins are most likely to interact with positively charged complexes within the circulation, the anionic glycosaminoglycans of the mucin layer and the lipids (mostly phosphatidylcholines) of the surfactant layer are likely to be more important. It is known that anionic liposomes readily interact with the cationic lipids of the complex [24,44,57]. However, the extent to which this occurs on the surface of the alveolar sacs can only be inferred.

Increasing lung permeability may be one mechanism to increase gene transfer. This mechanism has not been specifically demonstrated to be associated with cationic lipids, but evidence supports that this mechanism would be important in vivo. Sawa and colleages [93] noted that naked DNA in sterile water or preconditioning the lung with aerosolized water gave higher gene expression than naked DNA in buffered solution. They proposed that administration of water increased the permeability of the apical membrane for a brief time ($\sim$10–20 mins), permitting penetration of the DNA into epithelial cells. Other penetration enhancers have been shown to be effective in increasing the levels of naked DNA transfer in the lung (sodium glycocholate [94], organic solvents [95]). Thus, an increase in cell permeability by cationic lipids may promote greater amounts of DNA uptake into the cell and thereby facilitate gene transfer.

Other pharmaceutical solutions to airway delivery are to aerosolize or nebulize the complexes, which permits delivery in the deep lung. Most of the advancements in aerosolized cationic lipid vectors have focused on reducing the shear degradation of DNA as it flows through the aerosolizing nozzle. Shear degradation reduces the amount of intact, transcriptionally active DNA and can result in a severe drop in transfection activity; by as much as 60 and 70% after cycling 10 mins within the device [96]. When incorporated with cationic lipids, the DNA plasmid is no longer a freely extended polymer but a compacted lipoplex particle which maintains the plasmid integrity through the nebulization/aerosolization process [96,97]. This ultimately leads to more transcriptionally active DNA that is deposited in the lung.

Higher concentrations of DNA can be administered when a polyethylene glycol conjugated lipid is added to the cationic lipid formulation (98)–up to 0.3 mg/mL. The PEG component, with its steric effects, reduces the aggregation seen with high concentrations of the complex. However, this would also shield the positive charge that seems to assist in binding and uptake of the cationic lipid/DNA complex, and thus would seem to preclude gene transfer. A specific ligand may be required for PEG complexes to stimulate binding and uptake of lipoplex, or alternatively this formulation may be useful as a controlled-release system for DNA.

Thus, cationic lipoplexes are able to elicit gene transfer and gene expression in vivo. A net positive charge and additional lipoplex stability provided by a helper lipid, such as cholesterol, appear to be two attributes of lipoplexes which have activity, in vivo. Subtle differences in administration and in formulation also appear to impact the overall levels of gene expression obtained. However, as discussed in the next section, the utilization of cationic lipids in vivo have associated toxicity.

V. TOXICITY ASSOCIATED WITH CATIONIC LIPID GENE DELIVERY

The toxicity associated with cationic lipoplex administration has been observed after both intratracheal and intravenous administrations. This toxicity can be due to the cationic lipid, to the plasmid DNA, or to the physical attributes of the lipoplex. Toxicity associated with the DNA component is due to the presence of unmethylated CpG dinucleotide sequences [99]. Mammalian DNA contains four to five-fold more methylated CpG dinucleotide sequences than the plasmid grown in bacteria. The presence of these unmethylated sequences in the plasmid vector has been demonstrated to induce various immune responses in vivo. Although the sequences alone generate a response greater than their methylated counterparts, the effect is exacerbated when cationic lipids are used [100–102].

Different investigators have observed elevations in cytokine levels after either nasal or intravenous administration of complex [100–103]. The increase in cytokines can be noted as early as 2 hrs after cationic lipid/DNA administration (interleukin-6, IL-6), but most of the production levels peak around 24–48 hrs and usually are diminished within a week (IL-1β, IL-12, tumor necrosis factor (TNF-α), interferon-γ (IFN-γ) [101,103]. Elevations in IL-6 and IL-12 were specifically associated with the unmethylated CpG sequences, whereas TNF-α was associated with the cationic lipid. Only when both TNF-α and either IL-6 or IL-12 levels were high did IFN-γ levels increase (100, 101). Enzymatic modification of the CpG sequences resulted in significant reduction in cytokine production, and elimination of unnecessary CpG sequences may be one strategy to reduce these immunoflammatory effects [101,102].

A common observation resulting from airway instillation of cationic complexes are infiltrates of neutrophils and necrosis of the airway epithelium [103,101,102]. The cellular influx was not observed with plasmid DNA alone (or with unmethylated CpG sequences), and was specifically associated with the cationic lipid. This is highlighted by the observation that individual spermine and colipid components did not cause a similar level of infiltrates [103], suggesting that the amphiphilic nature of the cation was critical to its toxicity. Neutrophilic response could be reduced using antibodies (Mac-1A, LFA-1) or by a dexamethasone treatment, both of which suppressed cytokine production and, significantly, increased the level of gene expression [100,102]. We can attribute this positive result to reduced neutrophilic phagocytosis of the gene delivery complex at early times after administration and to an overall reduction in immunoflammatory response against the heterologous gene-expressing cells. However, these immunoinflammatory events may be beneficial in the treatment of metastatic tumors, where thymocyte activation may assist in the elimination of cancerous cells [89].

Intravenous administration of cationic lipoplex in mice has been associated with thrombocytopenia and leukopenia and the depression of residual serum complement activity [55,78] (L. Barron and F.C. Szoka, Jr., unpublished observations). Activation of complement by various cationic lipids and polymers has been demonstrated to be dependent on the concentration and the valence of the cation [104]. Although increasing doses of cationic lipoplex generally leads to higher gene transfer levels, maximum tolerated doses have been observed at approximately 1500–1800 nmol of cationic lipid (per 20–25 g mouse basis, or approximately 50 mg/kg). The lethal effect of these high doses of cationic lipid is associated with the redistribution of lipoplex to the liver and visible liver toxicity. Cationic lipids based upon carnitine have been synthesized in an attempt to design biodegradable lipids which might circumvent this toxicity in vivo [8]. Intravenous administration of large lipoplex aggregates have the potential to cause embolism upon injection. This manifests itself in rapid death of the mouse, usually due to embolism in the lung and brain and possibly within the liver. These are important limiting factors in the utilization of lipoplex by the intravenous route. Controlling the rate of administration by infusion is one tactic to minimize toxicity associated with cationic lipoplexes (L. Barron and F.C. Szoka, Jr., unpublished observations).

Large lipoplex formulations (>800 nm particles) were also found to be lethal upon instillation in the lungs of mice [38]. Although the instillation protocol is invasive to the animal, the intratracheal administration of cationic lipid complexes is less well tolerated than DNA, anionic liposomes, and equivalent doses of cationic liposomes alone (our unpublished observations). The large cationic particulates are likely to concentrate in the delicate airway regions of the lung and possibly contribute to the necrosis and epithelial denudation noted by Eastman and colleagues [97].

At a molecular level, Farhood et al. [105] noted the inhibitory effect of a series of cationic lipids on protein kinase C (PKC). The inhibition was strongest with cationic lipids containing a quartenery amine, and the increase in PKC inhibition was associated with lower levels of gene expression. This may be a correlate of the effect observed by Reston et al., who showed an increase in gene transfer by using PKC activators [106]. The reason for this effect was not clear; however, it does suggest that the PKC activators, tetradecanoyl-phorbol-12,13-acetate (TPA) and 1,2-dioctanoylglycerol (DiC8) may be useful in increasing the efficacy of cationic lipid gene transfer systems.

These studies illustrate that cationic lipids exhibit acute and chronic toxicity, and more research is required to evaluate adequately the potential benefits and risks of cationic lipid–mediated gene transfer. In addition, mechanistic studies will be required to determine how the gene transfer characteristics of cationic lipids can be separated from their toxic side effects. In this manner, we can devise more effective and less toxic lipoplex formulations.

VI. SUMMARY

A number of gene transfer mechanisms have been attributed to cationic lipid structure and lipoplex formulations in vitro. These include a quaternary amine head group, a fusogenic moeity, and a characteristic binding to cell surface heparan sulfate proteoglycans. The extension of these principles to describe in vivo mechanisms are confounded by the interactions and changes that occur to the lipoplex prior to its reaching the target site. A substantial body of work from different groups has confirmed that binding and retention of cationic lipoplexes occur in the lung after intravenous and intratracheal administration. Although some of the lipoplex gene transfer mechanisms may be shared among formulations and alternate routes of administration, the subsequent mechanisms of internalization and trafficking may account for the variations in efficacy. The application of confocal microscopy combined with physiological perturbations and biochemical studies are just a few strategies one can use to probe the mechanisms which govern in vivo complex uptake and release, as well as characterize cationic lipid–associated toxicity.

ACKNOWLEDGMENTS

This work was supported by NIH DK 46052.

REFERENCES

1. Felgner PL, Gadek TR, Holm M, Roman R, Chan HW, Wenz M, Northrop JP, Ringold GM, Danielsen M. Lipofection: A highly efficient, lipid-mediated DNA-transfection procedure. Proc Natl Acad Sci 1987; 84:7413–7417.
2. Brigham KL, Meyrick B, Christman B, Magnuson M, King G, Berry LCJ. In vivo transfection of murine lungs with a functioning prokaryotic gene using a liposome vehicle. Am J Med Sci 1989; 298:278–281.
3. Felgner PL, Barenholz Y, Behr JP, Cheng SH, Cullis P, Huang L, Jessee JA, Seymour L, Szoka FC, Thierry AR, Wagner E, Wu G. Nomenclature for synthetic gene delivery systems. Hum Gene Ther 1997; 8:511–512.
4. Meyer KEB, Uyechi LS, Szoka FC. Manipulating the intracellular trafficking of nucleic acids. In: Brigham KL, ed. Gene Therapy for Diseases of the Lung. Vol 104. New York: Marcel Dekker, 1997:135–180.
5. Felgner J, Kumar R, Sridhar CN, Wheeler CJ, Tsai YJ, Border R, Ramsey P, Martin M, Felgner PL. Enhanced gene delivery and mechanism studies with a novel series of cationic lipid formulations. J Biol Chem 1994; 269:2550–2561.
6. Lee ER, Marshall J, Siegel CS, Jiang C, Yew NS, Nichols MR, Nietupski JB, Ziegler RJ, Lane MB, Wang KX, Wan NC, Scheule RK, Harris DJ, Smith AE, Cheng SH. Detailed analysis of structures and formulations of cationic lipids for efficient gene transfer to the lung. Hum Gene Ther 1996; 7:1701–1717.
7. Byk G, Dubertret C, Escriou V, Frederic M, Jaslin G, Rangara R, Pitard B, Crouzet J, Wils P, Schwartz B, Scherman D. Synthesis, activity, and structure-activity relationship studies of novel cationic lipids for DNA transfer. J Med Chem 1998; 41: 224–235.
8. Wang J, Guo X, Xu Y, Barron L, Szoka FC. Synthesis and characterization of long chain alkyl acyl carnitine esters. Potentially biodegradable cationic lipids for use in gene delivery. J Med Chem 1998; 41:2207–2215.
9. Behr J-P, Demeneix B, Loeffler J-P, Perez-Mutul J. Efficient gene transfer into mammalian primary endocrine cells with lipopolyamine-coated DNA. Proc Natl Acad of Sci 1989; 86:6082–6986.
10. Gao X, Huang L. A novel cationic liposome reagent for efficient transfection of mammalian cells. Biochem Biophys Res Commun 1991; 179:280–285.

11. Farhood H, Serbina N, Huang L. The role of dioleoyl phosphatidylethanolamine in cationic liposome mediated gene transfer. Biochim Biophys Acta 1995; 1235: 289–295.

12. Hui SW, Langner M, Zhao Y-L, Ross P, Hurley E, Chan K. The role of helper lipids in cationic liposome-mediated gene transfer. Biophys J 1996; 71:590–599.

13. Fasbender A, Marshall J, Moninger TO, Grunst T, Cheng S, Welsh MJ. Effect of co-lipids in enhancing cationic lipid-mediated gene transfer in vitro and in vivo. Gene Ther 1997; 4:716–725.

14. Templeton NS, Lasic DD, Frederik PM, Strey HH, Roberts DD, Pavlakis GN. Improved DNA: liposome complexes for increased systemic delivery and gene expression. Nat Biotechnol 1997; 15:647–652.

15. Crook K, Stevenson BJ, Dubouchet M, Porteous DJ. Inclusion of cholesterol in DOTAP transfection complexes increases the delivery of DNA to cells in vitro in the presence of serum. Gene Ther 1998; 5:137–143.

16. Dowty ME, Williams P, Zhang G, Hagstrom JE, Wolff JA. Plasmid DNA entry into postmitotic nuclei of primary rat myotubes. Proc Natl Acad Sci 1995; 92: 4572–4576.

17. Wolff JA, Dowty ME, Jiao S, Repetto G, Berg RK, Ludtke JJ, Williams P. Expression of naked plasmids by cultured myotubes and entry of plasmids into T tubules and caveolae of mammalian skeletal muscle. J Cell Sci 1992; 103:1249–1259.

18. Dowty ME, Wolff JA. Possible mechanisms of DNA uptake in skeletal muscle. In: Wolff JA, ed. Gene Therapeutics: Methods and Applications of Direct Gene Transfer. Boston: Birkhauser, 1994:82–98.

19. Hickman MA, Malone RW, Kehmann-Bruinsma K, Sih TR, Knoell D, Szoka FC, Walzem R, Carlson DM, Powell JS. Gene expression following direct injection of DNA into liver. Hum Gene Ther 1994; 5:1477–1483.

20. Budker V, Zhang G, Knechtle S, Wolff J. Naked DNA delivered intraportally expresses efficiently in hepatocytes. Gene Ther 1995; 3:593–598.

21. Liu F, Song Y, Liu D. Hydrodynamics-based transfection in animals by systemic administration of plasmid DNA. Gene Ther 1999; 6:1258–1266.

22. Meyer KB, Thompson MM, Levy MY, Barron LG, Szoka FC. Intratracheal gene delivery to the mouse airway: characterization of plasmid DNA expression and pharmacokinetics. Gene Ther 1995; 2:450–460.

23. Van der Woude I, Visser HW, Ter Beest MBA, Wagenaar A, Ruiters MHJ, Engberts JBFN, Hoekstra D. Parameters influencing the introduction of plasmid DNA into cells by the use of synthetic amphiphiles as a carrier system. Biochim Biophys Acta 1995; 1240:34–40.

24. Xu Y, Szoka FC. Mechanism of DNA release from cationic liposome/DNA complexes used in cell transfection. Biochemistry 1996; 35:5616–5623.

25. Gershon H, Ghirlando R, Guttman SB, Minsky A. Mode of formation and structural features of DNA-cationic liposome complexes used for transfection. Biochemistry 1993; 32:7143–7151.

26. Li S, Tseng W-C, Stolz DB, Wu S-P, Watkins SC, Huang L. Dynamic changes in the characteristics of cationic lipidic vectors after exposure to mouse serum: implications for intravenous lipofection. Gene Ther 1999; 6:585–594.

27. Zelphati O, Uyechi LS, Barron LG, Szoka FC. Effect of serum components on

the physico-chemical properties of cationic lipid/oligonucleotide complexes and on their interactions with cells. Biochim Biophys Acta 1998; 1390:119–133.

28. Barry ME, Pinto-Gonzales D, Orson FM, McKenzie GJ, Petry GR, Barry MA. Role of endogenous endonucleases and tissue site in transfection and CpG-mediated immune activation after naked DNA injection. Hum Gene Ther 1999; 10:2461–2480.

29. Kawabata K, Takakura Y, Hashida M. The fate of plasmid DNA after intravenous injection in mice: involvement of scavenger receptors in its hepatic uptake. Pharm Res 1995; 12:825–830.

30. Manthorpe M, Cornefert-Jensen F, Hartikka J, Felgner J, Rundell A, Margalith M, Dwarki V. Gene therapy by intramuscular injection of plasmid DNA: studies on firefly luciferase gene expression in mice. Hum Gen Ther 1993; 4:419–431.

31. Levy MY, Barron LG, Meyer KB, Szoka FC. Characterization of plasmid DNA transfer into mouse skeletal muscle: evaluation of uptake mechanism, expression and secretion of gene products into blood. Gene Ther 1996; 3:201–211.

32. Meyer KEB, Thompson MM, Barron LG, Szoka FC. Cationic liposome formulations containing cholesterol increase plasmid persistence and gene expression in the mouse lung after intratracheal administration. In preparation.

33. Deshpande D, Blezinger P, Pillai R, Duguid J, Freimark B, Rolland A. Target specific optimization of cationic lipid-based systems for pulmonary gene therapy. Pharm Res 1998; 15:1340–1347.

34. Mahato RI, Kawabata K, Nomura T, Takakura Y, Hashida M. Physicochemical and pharmacokinetic characteristics of plasmid DNA/cationic liposome complexes. J Pharm Sci 1995; 84:1267–1271.

35. Barthel F, Remy J-S, Loeffler J-P, Behr J-P. Laboratory methods: Gene transfer optimization with lipospermine-coated DNA. DNA Cell Biology 1993; 12:553–560.

36. Lasic DD, Ruff D, Templeton NS, Belloni P, Alfredson T, Podgornik R. Cationic lipid-based gene delivery systems. In: Rolland A, ed. Advanced Gene Delivery, from Concepts to Pharmaceutical Products. Vol 10. Singapore: Harwood, Publishers, 1999:103–121.

37. Huang L, Viroonchatapan E. Introduction. In: Huang L, Hung M-C, Wagner E, eds. Nonviral Vectors for Gene Therapy San Diego: Academic Press, 1999:3–22.

38. Smith JG, Wedeking T, Vernachio JH, Way H, Niven RW. Characterization and in vivo testing of a heterogeneous cationic lipid-DNA formulation. Pharm Res 1998; 15:1356–1363.

39. Xu Y, Hui S-W, Frederick P, Szoka FC. Physicochemical characterization and purification of cationic lipoplexes. Biophys J 1999; 77:341–353.

40. Ross PC, Hui SW. Lipoplex size is a major determinant of in vitro lipoplex efficiency. Gene Ther 1999; 6:651–659.

41. Mislick KA, Baldeschwieler JD. Evidence for the role of proteoglycans in cation-mediated gene transfer. Proc Natl Acad Sci 1996.

42. Belting M, Persson S, Fransson L-A. Proteoglycan involvement in polyamine uptake. Biochem J 1999; 338:317–323.

43. Mounkes LC. Proteoglycans mediate cationic liposome-DNA complexed-based gene delivery in vitro and in vivo. J Biol Chem 1998; 273:26164–26170.

44. Zelphati O, Szoka FC. Mechanism of oligonucleotide release from cationic liposomes. Proc Natl Acad Sci 1996; 93:11493–11498.

45. Ruponen M, Yla-Herttuala S, Urtti A. Interactions of polymeric and liposomal gene delivery systems with extracellular glycosaminoglycans: physicochemical and transfection studies. Biochim Biophys Acta 1999; 1415:331–341.

46. Barron LG, Gagne L, Szoka FC. Lipoplex-mediated gene delivery to the lung occurs within 60 minutes of intravenous administration. Hum Gene Ther 1999; 10: 1683–1694.

47. Belting M, Petersson P. Intracellular accumulation of secreted proteoglycans inhibits cationic lipid-mediated gene transfer. Co-transfer of glycosaminoglycans to the nucleus. J Biol Chem 1999; 274:19375–19382.

48. Fasbender A, Zabner J, Zeiher BG, Welsh MJ. A low rate of cell proliferation and reduced DNA uptake limit cationic lipid-mediated gene transfer to primary cultures of ciliated human airway epithelia. Gene Ther 1997; 4:1173–1180.

49. Tanswell AK, Staub O, Iles R, Belcastro R, Cabacungan J, Sedlackova L, Steer B, Wen Y, Hu J, O'Brodovich H. Liposome-mediated transfection of fetal lung epithelial cells: DNA degradation and enhanced superoxide toxicity. Am J Physiol 1998; 275:L452–L460.

50. Yang J-P, Huang L. Overcoming the inhibitory effect of serum on lipofection by increasing the charge ratio of cationic liposome to DNA. Gene Ther 1997; 4:950–960.

51. Legendre J-Y, Szoka FC. Delivery of plasmid DNA into mammalian cell lines using pH-sensitive liposomes: comparison with cationic liposomes. Pharma Res 1992; 9:1235–1242.

52. Zabner J, Fasbender AJ, Moninger T, Poellinger KA, Welsh MJ. Cellular and molecular barriers to gene transfer by a cationic lipid. J Biol Chem 1995; 270:18997–19007.

53. Labat-Moleur F, Steffan A-M, Brisson C, Perron H, Feugeas O, Furstenberger P, Oberling F, Brambilla E, Behr J-P. An electron microscopy study into the mechanism of gene transfer with lipopolyamines. Gene Ther 1996; 3:1010–1017.

54. Wrobel I, Collins D. Fusion of cationic liposomes with mammalian cells occurs after endocytosis. Biochim Biophys Acta 1995; 1235:296–304.

55. McLean JW, Fox EA, Baluk P, Bolton PB, Haskell A, Pearlman R, Thurston G, Umemoto EY, McDonald DM. Organ-specific endothelial cell uptake of cationic liposome-DNA complexes in mice. Am J Physiol 1997; 273:H387–404.

56. Coonrod A, Li F-Q, Horwitz M. On the mechanism of DNA transfection: efficient gene transfer without viruses. Gene Ther 1997; 4:1313–1321.

57. Bailey AL, Cullis PR. Membrane fusion with cationic liposomes: effects of target membrane composition. Biochemistry 1997; 36:1628–1634.

58. Haensler J, Szoka FC. Polyamidoamine cascade polymers mediate efficient transfection of cells in culture. Bioconjug Chem 1993; 4:371–379.

59. Tang MX, Szoka FC. Mechanisms of in-vitro transfection mediated by cationic polymer/DNA complexes. In preparation.

60. Capecchi MR. High efficiency transformation by direct microinjection of DNA into cultured mammalian cells. Cell 1980; 22:479–488.

61. Mirzayans R, Aubin RA, Paterson MC. Differential expression and stability of foreign genes introduced into human fibroblasts by nuclear versus cytoplasmic microinjection. Mutat Res 1992; 281:115–122.

62. Gao X, Huang L. Cytoplasmic expression of a reporter gene by co-delivery of T7 RNA polymerase and T7 promoter sequence with cationic liposomes. Nucleic Acids Res 1993; 21:2867–2872.

63. Lukacs G, Haggie P, Seksek O, Lechardeur D, Freedman N, Verkman AS. Size-dependent DNA mobility in cytoplasm and nucleus. J Biological Chem 2000; 275: 1625–1629.

64. Lechardeur D, Sohn K-J, Haardt M, Joshi PB, Monck M, Graham RW, Beatty B, Squire J, O'Brodovich H, Lukacs GL. Metabolic instability of plasmid DNA in the cytosol: a potential barrier to gene transfer. Gene Ther 1999; 6:482–497.

65. Pollard H, Remy J-S, Loussouarn G, Demolombe S, Behr J-P, Escande D. Polyethyleneimine but not cationic lipids promotes transgene delivery to the nucleus in mammalian cells. J Biol Chem 1998; 273:7507–7511.

66. Collas P, Alestrom P. Rapid targeting of plasmid DNA to zebrafish embryo nuclei by the nuclear localization signal of SV40 antigen. Mol Marine Biol Biotechnol 1997; 6:48–58.

67. Zanta MA, Belguise-Valladier P, Behr J-P. Gene delivery: a single nuclear localization signal peptide is sufficient to carry DNA to the cell nucleus. Proc Natl Acad Sci 1999; 96:91–96.

68. Jiang C, O'Conner SP, Fang SL, Wang KX. Efficiency of cationic lipid-mediated transfection of polarized and differentiated airway epithelial cells in vitro and in vivo. Hum Gene Ther 1998; 9:1531–1542.

69. Zelphati O, Liang X, Hobart P, Felgner PL. Gene chemistry: functionally and conformationally intact fluorescent plasmid DNA. Hum Gene Ther 1999; 10:15–24.

70. Dean DA, Dean BS, Muller S, Smith LC. Sequence requirements for plasmid nuclear import. Exp Cell Res 1999; 253:713–722.

71. Crystal RG, West JB, Barnes PJ, Cherniack NS, Weibel ER. The Lung: Scientific Foundations. Vol 1. New York: Raven Press, 1991:1208.

72. Kaplan HM, Brewer NR, Blair WH. Physiology. In: Foster HL, Small JD, Fox JG, eds. The Mouse in Biomedical Research: Normative Biology, Immunology and Husbandry. Vol 3. New York: Academic Press, 1983:248–278.

73. Litzinger DC, Brown JM, Wala I, Kaufman S, Van GY, Farrell CL, Collins D. Fate of cationic liposomes and their complex with oligonucleotide in vivo. Biochim Biophys Acta 1996; 1281:139–149.

74. Liu Y, Mounkes LC, Liggitt HD, Brown CS, Solodin I, Heath TD, Debs RJ. Factors influencing the efficiency of cationic liposome-mediated intravenous gene delivery. Nat Biotechnol 1997; 15:167–173.

75. Hofland HE, Nagy D, Liu J-J, Spratt K, Lee Y-l, Danos O, Sullivan SM. In vivo gene transfer by intravenous administration of stable cationic lipid/DNA complex. Pharm Res 1997; 14:742–749.

76. Mahato RI, Anwer K, Tagliaferri F, Meaney C, Loenard P, Wadhwa MS, Logan M, French M, Rolland A. Biodistribution and gene expression of lipid/plasmid complexes after systemic administration. Hum Gene Ther 1998; 9:2083–2099.

77. Liu F, Qi H, Huang L, Liu D. Factors controlling the efficiency of cationic lipid-mediated transfection in vivo via intravenous administration. Gene Ther 1997; 4: 517–523.

78. Barron LG, Meyer KB, Szoka FC. Effects of complement depletion on the pharma-

cokinetics and gene delivery mediated by a cationic lipid-DNA complexes. Hum Gene Ther 1998; 9:315–323.

79. Li S, Huang L. In vivo gene transfer via intravenous administration of cationic lipid-protamine-DNA (LPD) complexes. Gene Ther 1997; 4:891–900.

80. Barron LG, Szoka FC. The perplexing mechanism of cationic lipid mediated gene delivey. In: Huang L, Hung M-C, Wagner E, eds. Nonviral Vectors for Gene Therapy. San Diego: Academic Press, 1999:229–266.

81. Barron LG, Uyechi LS, Szoka FC. Cationic lipids are essential for gene delivery mediated by intravenous administration of lipoplexes. Gene Ther 1999; 6:1179–1183.

82. Simionescu M, Simionescu N, Silbert JE, Palade GE. Differentiated microdomains on the luminal surface of the capillary endothelium II. Partial Characterization of their anionic sites. J Cell Biol 1981; 90:614–621.

83. Uyechi LS, Gagné L, Thurston G, Szoka FC. Mechanism of Lipoplex gene delivery in mouse lung: binding and internalization of fluorescent lipid and DNA components. Gene Therapy 2001; 8:828–836.

84. Hofland HEJ, Shephard L, Sullivan SM. Formation of stable cationic lipid/DNA complexes for gene transfer. Proc Natl Acad Sci USA 1996; 93:7305–7309.

85. Song Y, Liu F, Liu D. Enhanced gene expression in mouse lung by prolonging the retention time of intravenously injected plasmid DNA. Gene Ther 1998; 5:1531–1537.

86. Song YK, Liu D. Free liposomes enhance the transfection activity of DNA/lipid complexes in vivo by intravenous administration. Biochim Biophys Acta 1998; 1372:141–150.

87. Zabner J, Cheng SH, Meeker D, Launspach J, Balfour R, Perricone MA, Morris JE, Marshall J, Fasbender A, Smith AE, Welsh MJ. Comparison of DNA-lipid complexes and DNA alone for gene transfer to Cystic Fibrosis airway epithelia in vivo. J Clin Invest 1997; 100:1529–1537.

88. Gorman CM, Aikawa M, Fox B, Fox E, Lapuz C, Michaud B, Nguyen B, Nguyen H, Roche E, Sawa T, Wiener-Kronish JP. Efficient in vivo delivery of DNA to pulmonary cells using the novel lipid EDMPC. Gene Ther 1997; 4:983–992.

89. Blezinger P, Freimark BD, Matar M, Wilson E, Singhal A, Min W, Nordstrom JL, Pericle F. Intratracheal administration of Interleukin-12 plasmid-cationic lipid complexes inhibits murine lung metasteses. Hum Gene Ther 1999; 10:723–731.

90. Griesenbach U, Chonn A, Cassady R, Hannam V, Ackerley C, Post M, Tanswell AK, Olek K, O'Brodovich H, Tsui L-C. Comparison between intratracheal and intravenous administration of liposome-DNA complexes for cystic fibrosis lung gene therapy. Gene Ther 1998; 5:181–188.

91. Oudrhiri N, Vigneron J-P, Peuchmaur M, Leclerc T, Lehn J-M, Lehn P. Gene transfer by guanidinium-cholesterol cationic lipids into airway epithelial cells in vitro and in vivo. Proc Natl Acad Sci, USA 1997; 94:1651–1656.

92. Glasspool-Malone J, Malone RW. Marked enhancement of direct respiratory tissue transfection by aurintricarboxylic acid. Hum Gene Ther 1999; 10:1703–1713.

93. Sawa T, Miyazaki H, Pittet J-F, Widdicombe JH, Gropper MA, Hashimoto S, Conrad DJ, Folkesson HG, Debs R, Forsayeth JR, Fox B, Wiener-Kronish J. Intraluminal water increases expression of plasmid DNA in rat lung. Hum Gene Ther 1996; 7:933–941.

94. Freeman DJ, Niven RW. The influence of sodium glycocholate and other additives on the in vivo transfection of plasmid DNA in the lungs. Pharma Res 1996; 13: 202–209.

95. Schugart K, Bischoff R, Rasmussen UB, Ali Hadji D, Perraud F, Accart N, Boussif O, Silvestre N, Cordier Y, Pavirani A, Kolbe HJV. Solvoplex: a new type of synthetic vector for intrapulmonary gene delivery. Hum Gene Ther 1999; 20:2891–2905.

96. Schwartz LA, Johnson JL, Black M, Cheng SH, Hogan ME, Waldrep JC. Delivery of DNA-cationic liposome complexes by small-particle aerosol. Hum Gene Ther 1996; 7:731–741.

97. Eastman SJ, Tousignant JD, Lukason MJ, Murray H, Siegel CS, Constantino P, Harris DJ, Cheng SH, Scheule RK. Optimization of formulations and conditions for the aerosol delivery of functional cationic lipid:DNA complexes. Hum Gene Ther 1997; 8:313–322.

98. Eastman SJ, Lukason MJ, Tousignant JD, Murray H, Lane MD, St. George JA, Akita GY, Cherry M, Cheng SH, Scheule RK. A concentrated and stable aerosol formulation of cationic lipid:DNA complexes giving high-level gene expression in mouse lung. Hum Gene Ther 1997; 8:765–773.

99. Krieg AM. Direct immunologic activities of CpG DNA and implications for gene therapy. J Gene Med 1999; 1:56–63.

100. Tan Y, Li S, Pitt BR, Huang L. The inhibitory role of CpG immunostimulatory motifs in cationic lipid vector-mediated transgene expression in vivo. Hum Gene Ther 1999; 10:2153–2161.

101. Freimark BD, Blezinger HP, Florack VJ, Nordstrom JL, Long SD, Deshpande DS, Nochumson S, Petrak KL. Cationic lipids enhance cytokine and cell influx levels in the lung following administration of plasmid: cationic lipid complexes. J Immunol 1998; 160:4580–4586.

102. Yew NS, Wang KX, Przbylska M, Bagley RG, Stedman M, Marshall J, Scheule RK, Cheng SH. Contribution of plasmid DNA to inflammation in the lung after administration of cationic lipid:pDNA complexes. Hum Gene Ther 1999; 10:223–234.

103. Scheule RK, St. George JA, Badley RG, Marshall J, Kaplan JM, Akita GY, Wang KX, Lee ER, Harris DJ, Jiang C, Yew NS, Smith AE, Cheng SH. Basis of pulmonary toxicity associated with cationic lipid-mediated gene transfer to the mammalian lung. Hum Gene Ther 1997; 8:689–707.

104. Plank C, Mechtler K, Szoka FC, Wagner E. Activation of the complement system by synthetic DNA complexes: a potential barrier for intravenous gene delivery. Hum Gene Ther 1996; 7:1437–1446.

105. Farhood H, Bottega R, Epand RM, Huang L. Effect of cationic cholesterol derivatives on gene transfer and protein kinase C activity. Biochim Biophys Acta 1992; 1111:239–246.

106. Reston JA, Gould-Fogerite S, Mannino R. Potentiation of DNA mediated gene transfer in NIH3T3 cells by activators of protein kinase C. Biochim Biophys Acta 1991; 1088:270–275.

107. Berman I. Color Atlas of Basic Histology. Norwalk, CT: Appleton & Lange, 1993: 332.

108. Wheater PR, Burkitt HG, Daniels VG. Functional Histology: A Text and Colour Atlas. New York: Churchill Livingstone, 1987:348.

5

Polymer-Based Gene Delivery Systems

Lionel Wightman
Boehringer Ingelheim Austria, Vienna, Austria

Ralf Kircheis
Ingeneon AG, Vienna, Austria

Ernst Wagner
Ludwig-Maximilians-University, Munich, Germany

I. INTRODUCTION

Nonviral vectors for gene delivery are receiving increasing attention for application in a broad variety of gene-mediated therapies for humans. Chemical vectors are attractive to the pharmaceutical industry as alternatives to viral vectors because of compound stability and easy chemical modification. Furthermore, the low cost and consistent standard of production (compared to growth of viruses in bioreactors followed by purification), higher biosafety (less immunogenic as compared to viruses such as adenoviruses), and high flexibility once a formulation has been defined make these compounds very attractive [1,2]. The biosafety of nonviral vectors is higher, as compared to viruses, in not eliciting an acute immune reaction; also, they will not recombine with wild-type viruses and will very rarely insert into the host genome (unless integration sequences are incorporated into the delivered plasmid). However, to date, nonviral vectors are generally less efficient in delivering DNA and expressing proteins as compared to their viral counterparts particularly when used in vivo. The efficiency of nonviral mediated gene delivery can be improved when physical methods such as gene gun [3,4], electroporation [5,6] or hydrodynamic techniques [7,8] are used to enhance delivery.

(3) ==> injection in blood circulation ==> to distant tumors

   --> **passing through blood circulation**
   --> **survival in biological fluids**
   --> **passing liver, spleen (filtration/removal of foreign particles)**

(2) ==> **to the target cells**

   --> **passing network of extracellular matrix**
   --> **phagocytes and immune cells**

(1) ==> uptake into the target cell => gene expression

   --> **DNA condensation**
   --> **endocytosis**
   --> **endosamal release**
   --> **nuclear uptake and gene expression**

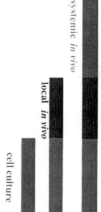

Figure 1   Schematic of the various barriers that polyplexes have to cross in order to achieve gene expression in vitro and in vivo.

To design the optimal nonviral vector for particle-mediated gene delivery, it must be able partially to fulfill a number of predefined biological criteria depending on its specific therapeutic aim. First, the vector must be able to transfer DNA molecules of varying sizes. Second, the vector should be able to transduce dividing and/or nondividing cells, depending on the target cell type. Third, the vector should target a specific cell type. Fourth, the vector should have no or very low toxicity in vivo and avoid stimulating the immune system. Fifth, depending on the therapeutic application, the vector should be able to induce a sustained expression over a defined period of time (optionally integrating into the host genome). And, finally, the vector must not transform the target cell.

Even with all of these conditions for a vector, for an effective nonviral vector which fulfills parts or all of the above criteria, there are numerous barriers for the vector to overcome (Fig. 1). Once the complex is taken up by the cell (applies to in vitro and in vivo gene expression), it has to be released into the cytoplasm, migrate to the nucleus, cross the nuclear membrane, and subsequently release the DNA for gene expression to be initiated. Furthermore, when particles are within tissue (e.g., after local application in vivo), they have to diffuse efficiently between nontarget cells, avoiding interactions with extracellular matrix, and then be taken up by the target cell. Finally, particles that have been applied systemically in vivo have to be inert and not interact with blood components and reticular endothelial system (RES). Also, the complexes have to cross the endothelial cell layer out of the vascular system into the target tissue.

A number of nonviral systems have been described which partially fulfill the above-defined criteria, such as naked DNA, cationic liposome–DNA complexes (lipoplexes), polymer–DNA complexes (polyplexes), and combinations of these [1,2,9–11]. However, no nonviral system (or viral) fulfills all the various criteria described above to date. In this chapter, cationic polymers are reviewed, the various polymers currently used, and how they are modified to achieve greater efficiency and specificity in particle-mediated delivery and gene expression in vitro and, more challengingly, in vivo are discussed.

II. DELIVERY IN VITRO TO TARGET CELL

In vitro, gene delivery into a target cell is essentially a balance between the efficiency of the polycation in initiating particle uptake and gene expression versus the intrinsic toxicity of the compound while disrupting cellular membranes during the transfection procedure. Once the polycation has been condensed with DNA (forming the polyplex; see Sec. II.A), there are a number of cellular barriers that can inhibit particle entry and subsequently reduce transfection efficiency. For example, adsorption onto the cell surface (see Sec. II.A and C) and uptake of the polyplex into primary endosomes (see Sec II.A), followed by release of the polyplex into the cytoplasm and subsequent migration to the nucleus (see Sec. II.B), finally the DNA has to be released from the polycation for transcription to be initiated. This basic sequence of events for gene delivery in vitro is also required for particle-mediated delivery in vivo (locally or systemically applied polycation/DNA complexes) once the polyplex has reached the target cell. However, in vivo, the target cells are usually differentiated and nondividing; furthermore, dividing cells, such as tumor cells in humans, have a longer cell cycle period in comparison to cells cultured in vitro. These added barriers taken together with the other cellular barriers make gene expression in vivo more challenging even when the polyplexes are delivered to the intended site of activity (see Sec. III).

A. Intrinsic Properties of Polycations (DNA Condensation, Complex Uptake, and Endosomal Release)

A number of polymeric molecules have been developed that can interact with the negative phosphate groups of naked DNA condensing DNA into compact particles (complexes), protecting it from degradation, and enhancing the uptake of DNA into the cell resulting in efficient transgene expression. Positively charged cationic polymers, such as polylysine [12,13], polyamidoamine (PAMAM) dendrimers [14], and polyethylenimines (PEIs) [15] (see Sec. II.A.3 for other polycations) are able to interact electrostatically via their protonated amine groups with the phosphate groups of DNA, creating a condensed particle or polyplex. The condensation of the polycation around DNA protects the DNA from nuclease damage [16]. Furthermore, cells more efficiently take up these compact particles via a number of natural processes, such as adsorptive endocytosis, pinocytosis, and phagocytosis. When forming polyplexes, the extent of condensation between the polycation and DNA will depend on the charge ratio [9]. In the case of poly-ethylenimines (where protonation of amine groups depends on the pH), the charge ratio is defined as the N/P ratio of nitrogens (from the polycation) to phosphates (from the DNA). When low amounts of polycation are used to complex DNA, around electroneutrality, often the complexes are large and insoluble forming aggregates; however, at high N/P charge ratios, polyplexes are often soluble, small, and positively charged particles [17]. Polyplexes with a positive surface charge, or zeta (ζ) potential, will interact directly with the cell surface of many types of cells because of the presence of negatively charged cell surface proteo-glycans [18].

A number of physical techniques have been developed to characterize the condensation of polycation/DNA complexes, including: size or shape (electropho-retic mobility in agarose gels, electron or atomic force microscopy, laser light scattering), charge (electrophoretic mobility, zeta (ζ) potential), conformation and condensation (circular dichroism, ethidium bromide exclusion, and centrifuga-tion). These various methods for characterizing the particles are often complex and require specialized equipment that is not accessible to all laboratories. Re-cently, the requirement for the characterization of surface charge and size of com-plexes is becoming obligatory to understand the processes involved in cellular transfection and particle behavior within different biological fluids and solutions.

1. Polylysine-Based Vectors

One of the first polycations characterized and recognized as a potential nonviral vector for DNA condensation was poly-(L)-lysine. Polylysine typically comes in a variety of sizes, and is usually specified as the average number of polylysine molecules within a defined solution rather than a specifically defined number of lysine molecules per polylysine molecule. In generating pharmaceutical products, this heterogeneity represents a disadvantage of this polymer class, as it is difficult to generate polymers that have the same defined size between batches; however, this problem is not limited to polylysine but also to polycations in general. Some researchers have used lysine-rich peptides or oligolysine to circumvent this het-erogeneity problem [19,20].

At physiological pH, the amino group of lysine is positively charged and can ionically interact with polyanions, such as DNA, and has been extensively characterized [21,22]. With all of the polycations, not just polylysine, the proce-dure for complex formation can influence transfection efficiency (e.g., salt con-centration or speed of mixing the cation and anion) [13,17,23–25]. Numerous methods have been developed to generate protocols that consistently and repeat-edly generate predefined polylysine/DNA complexes of defined stability and size, such as flash mixing [26], high-salt conditions followed by dialysis [27,28], or high salt and vigorous agitation [29–31]. These various methods generate polylysine/DNA complexes of between 15 and 30 nm, 50 and 150 nm, and particles that could be filtered through a 0.45-μm filter. However, the size of polylysine used for DNA condensation and its purity also can have a large influence on the sizes of particles generated independent of the mixing [23,32]. Polyplexes formed with poly-D-lysine showed no additional advantage over the (L)-lysine form [26,33].

Polylysine as a compound is not biologically inert; for example, polyplexes have been shown to affect cellular processes during transfection by enhancing pinocytosis and phagocytosis [34,35] and cell division [36]. Also, the activation of proteins and phospholipases has been observed [37,38], which affects membrane permeability [39]. Under certain conditions, polylysine can initiate cell fusion [40] or have a lytic activity on cells [41,42]; however, this fusion activity has not been observed in all cell types [43].

Condensed polylysine/DNA complexes bind nonspecifically to the cell surface, often due to the negative charge of cells electrostatically interacting with the positively charged complex, and is subsequently taken up by absorptive endocytosis. However, the efficiency of uptake and gene expression can be enhanced by the covalent addition of ligands to the polylysine that can specifically target cells and promote cellular uptake; termed receptor-mediated targeting (see Sec. II.C). Although polylysine can initiate DNA uptake into cells, often gene expression is low owing to a number of factors, such as, ineffective endosomal release, degradation in the cytoplasm, limited migration to the nucleus, inefficient nuclear uptake (see Sec. II.B), and finally gene expression (transcription, RNA maturation, translation, and protein processing). All of these factors can decrease transfection efficiency; however, a major disadvantage with polylysine is that it does not have an intrinsic endosomal release activity. Hence, complexes are often not released, or escape, from the endosome/lysosome into the cytoplasm. Furthermore, the addition of a ligand can enhance cellular uptake but not influence endosomal release [44]. Therefore, different strategies have been developed to increase endosomal release by disrupting the endosomal lipid bilayer.

The enhancement of endosomal release of polylysine/DNA complexes has been achieved via a number of strategies. The addition of lysosomotropic agents (e.g., chloroquine) into the cell medium can increase gene expression in a number of cell lines [45]. Chloroquine accumulates in the vesicles and leads to swelling of the endosome by raising the pH and destabilizing it. Furthermore, it may help in the release of the polylysine from the DNA and thus increase gene expression as a result of the release of partially dissociated DNA into the cytoplasm [46,47]. Similarly, the addition of glycerol results in enhanced gene expression in a number of cell lines; probably by the weakening of the endosomal lipid bilayer and allowing polylysine/DNA complex release into the cytoplasm [48].

However, the addition of lysosomotropic agents or glycerol to increase endosomal release of polylysine/DNA complexes is limited to in vitro use only. Another approach developed to enhance endosomal release is the incorporation of viruses that have an intrinsic endosomal disruption mechanism, thus initiating endosomal release of the polylysine complexes. Inactivated adenoviruses have been used where the intrinsic endosomal release of the capsid can be used to enhance polyplex release into the cytoplasm [49–54]. As the activity is from the protein capsid, the viruses can be inactivated by treating them with methoxypso-ralen plus ultraviolet irradiation without affecting the endosomal release activity [55]. Generally, the viral capsid has to be codelivered with the polylysine/DNA complex; and has been achieved, for example, by linking polylysine to the virus [51,53] or a biotin–streptavidin bridge between the virus and polylysine [50]. Other viruses that have been used to achieve endosomal release of polylysine/ DNA complexes are an avian adenovirus (CELO) or human rhinoviruses [56,57]. The inclusion of viruses into polyplexes may enhance endosomal release, but has the drawback of potentially eliciting an immune response in vivo. Therefore, the specific parts of viruses that elicit endosomal release have been incorporated into polylysine/DNA complexes to increase transfection activity [58]. Instead of the whole virus, viral peptide sequences characterized by endosomal activity in-cluded the N-terminal of the hemagglutinin 2 subunit [33], rhinovirus VP-1 [57], or synthetic membrane-destabilizing peptides [19,33,59–62].

2.   Starburst Dendrimers and Polyethylenimine-Based Vectors

Several polycations have been developed efficiently complex DNA and have an intrinsic endosomal release activity, hence removing the need for additional endo-somolytic agents as required for polylysine. The first polycation described with this activity was PAMAM cascade polymer (or Starburst dendrimer) [14,63–66]. These compounds efficiently complex DNA and have endosomal lytic activity as compared to polylysine; however, transfection activity with the dendrimers could be further enhanced by the addition of compounds that improve membrane porosity [67]. The dendrimer's amines (terminal and internal) have a low pKa which prevents DNA degradation within the endosomes/lysosomes; however, for optimum transfection efficiency, high dendrimer:DNA charge ratios (e.g., amine: phosphate $= 6:1$) are required, often causing toxicity to cells in vitro [68]. Par-tially degraded or fractured dendrimers, as compared to intact dendrimers, were found to be more efficient in transfecting cells; probably because the complexes are more soluble and do not aggregate [64,65].

A further polycation, PEI, has been described [15] which can efficiently condense DNA while retaining an improved endosomal activity as compared to dendrimers. PEI can efficiently condense and deliver large DNA up to 2.3 Mb in vitro [69]. PEI/DNA complexes are taken up by the cell and the endosome is acidified, PEI acts as a proton sponge. It is thought that the interaction of PEI amines with protons triggers osmotic swelling of the complex, causing the endosome to rupture and thus release complexes into the cytoplasm [15]. Efficient gene expression has been observed with the PEI/DNA complexes; furthermore, the addition of chloroquine or endosomally active viruses did not improve transfection [15,70,71]. A number of PEI molecules have been described in detail with varying molecular size or structure (branched or linear) [15,72–75]. A problem observed with PEI (and polylysine) is that complexes often precipitate at a low amine:phosphate ratio [17]. This problem has been partially alleviated by the addition of a block or graft polymer, for example, poly(ethylene) oxide onto the PEI backbone [76]; however, often transfection efficiency is reduced through this modification even if solubility is increased.

   PEI can easily be modified by the addition of other chemical groups or ligands (see Sec. II.C), making this group particularly versatile and interesting for further development of nonviral vectors.

3. Other Cationic Polymers

In general, characterization of polycations for DNA condensation and gene delivery have concentrated on polylysine and PEI as the main reagents of this class; however, other synthetic and natural polycations have been described which can complex DNA and promote gene transfer into cells. These include polyornithine [77], polyarginine [78,79], histones [80], high-mobility group protein (HMG1) [81,82], poly(2-(dimethylamino)ethyl)methacrylate (PDMAEMA) [83], polyallylamine derivatives [84], diethylaminoethyl (DEAE)–dextran [85], chitosan [86], poly(N-alkyl-4-vinylpyridinium) salts [87], and histidylated polylysine [88]. Similarly to polylysine and PEI, these other polycations can be modified by the addition of ligands, and, when necessary, endosomal release can be enhanced by the addition of inactivated viruses or peptides.

B. Migration Through the Cytoplasm and Nuclear Entry

Once polyplexes have been taken up by the cell and released from endosomes into the cytosol, complexes have to undergo two processes: (1) migration to the nucleus through the cytoplasm, nuclear trafficking; and (2), the DNA has to free itself from the polycation (disassembly) before it can be transcribed. When these two processes are inefficient, a dramatic decrease in gene expression can be observed, via degradation of DNA, or remaining complexed and the DNA inaccessible to the transcription machinery. Some groups of investigators are developing cytoplasmic expression systems [89,90]; however, the preferred route is still via nuclear targeting and transcription within the nucleus.

   The transport of the complexes through the cytoplasm to the nucleus is poorly understood, and has been shown to be inefficient in polycation-condensed DNA [80]. It has been reported that polylysine and PEI have some intrinsic nuclear targeting activity [91–93]. However, this may be due to the polycations protecting the DNA from cytosol nucleases [94], and that during division the complexes can randomly "migrate" to the nucleus. Brunner et al. [95] demonstrated that the cell cycle could have a dramatic influence on transgenic expression efficiency, where polyplexes (adenovirus-enhanced transferrinfection [AVET], polylysine, and PEI) used to transfect cells in G1 phase of the cell cycle has the lowest efficiency. This was not observed with recombinant adenoviruses. Generally, cells in cell culture are constantly dividing; however, this may not be the case in vivo and could potentially be one of the serious limiting factors for efficient gene expression. The attachment of nuclear localization signal sequences to the polycation or binding sites for transcription factors within the DNA sequence to overcome this barrier has been reported by several groups of investigators [96–99].

The final stage before gene expression is the disassembly of complexes so that the transcription apparatus can access the DNA efficiently for gene expression. Efficient uncoating of DNA is an interesting challenge, as it is diametrically opposite to previous stages of gene delivery, as premature disassembly of complexes would lead to DNA degradation and loss of the gene expression in other parts of the pathway to gene delivery and expression.

C. Receptor-Mediated Gene Delivery

One of the problems with polycations is that they have a nonspecific positive charge, which will interact electrostatically with negatively charged cells. This nonspecific interaction will generally not be a concern in vitro or for locally applied complexes; however, often in vivo specific targeting of complexes to a particular cell type is desired, and these nonspecific interactions will inhibit efficient gene expression (see Sec. IV).

Transfection efficiency can be increased by incorporating ligands covalently onto the polycation backbone, enhancing the binding of polyplexes onto the cell surface, and subsequently uptake; termed receptor-mediated gene transfer. This concept attempts to mimic nature, such as viruses, toxins, and normal host ligands (hormones, growth factors [e.g., insulin, epidermal growth factor (EGF), fibroblast growth factor (FGF)], nutrients (e.g., low-density lipoprotein [LDL], folic acid, transferrin)), which bind to a specific receptor expressed on the cell surface. After binding of the ligand to the receptor, the molecule is internalized into clathrin-coated or clathrin–noncoated vesicles via receptor-mediated endocytosis, phagocytosis, or pinocytosis. Furthermore, many of these receptors are expressed specifically on certain cell types or at a particular stage of the cell's life cycle, making it theoretically possible to target specific cells depending on their type and stage of development or division. The use of a ligand can also just mediate cell attachment (e.g., antibodies specific for a cell surface protein) and not stimulate cellular uptake, hence giving specificity, and the incorporation of a second ligand could mediate efficient uptake of complexes into the cell. The polycations are amenable to this approach, as they have groups such as amines which can be modified covalently via a number of chemical procedures, allowing the attachment of peptides, sugars, proteins, antibodies, and other small moieties (see reviews in Ref. 10).

Wu and coworkers [12,27,100], first described polycation receptor–mediated gene delivery where specific targeting to the asialoglycoprotein receptor, specifically expressed on hepatocytes, was demonstrated by asialoorosomucoid–polylysine/DNA complexes in vitro and in vivo. Since this original demonstration of receptor-mediated gene targeting, other ligands with polylysine or PEI have been demonstrated. These ligands include transferrin [13,71,101], EGF [102], basic FGF (bFGF) [103], insulin [104], integrins [105], folate [106,107], lectins [108,109], surfactant proteins A and B [52,110], and synthetic glycosylated ligands [44,111,112]. The attachment of antibodies has also enhanced the uptake of polyplexes into cells, such as anti-CD3 [71,113,114], anti-CD5 [54], and anti-EGF [115]. An artificial approach to receptor-mediated gene delivery was recently reported demonstrating the potency of the concept, where biotinylated cells were efficiently transfected with streptavidin–PEI/DNA complexes [116].

The attachment of ligands to polylysine backbone has been shown to increase the targeting of these complexes clearly in vitro, for example, transferrin–polylysine/DNA complexes were efficiently taken up by stimulated K562 cells that express the transferrin receptor at high levels on their surface [101]. Furthermore, the transferrin receptor is upregulated in proliferating cells, such as tumor cells, and not just a specific cell type [45,117]. Transferrin–polylysine/DNA–mediated gene delivery was further enhanced by the coupling of inactivated adenoviruses, creating AVET. This system has proven to be versatile in transfecting numerous cell lines [50] and primary human melanoma cell lines in vitro and ex vivo [118]. A potential problem with adenoviruses for application in vivo is that all cells express the adenoviral receptor and the complexes could potentially enter via nonspecific targeting. A number of strategies have been used to block this activity either by modifying the fiber chemically [53], blocking with antibodies (119), or using an avian adenovirus called CELO [56]. Furthermore, these modifications do not alter the endosomal activity of the complexes.

Transferrin has been coupled to PEI (800 kD) and resulted in an up to several hundred-fold increase in transfection efficiency in K562 cells; furthermore, the addition of endosomal active agents was not required to enhance transfection as compared to transferrin–polylysine complexes [71]. In a variety of primary human melanoma cell lines, transferrin–PEI800/DNA polyplexes were found to deliver reporter genes (luciferase) equally to AVET complexes [120].

Antibodies conjugated to polylysine and PEI have also been used to target specifically to a certain cell type, for example, anti-CD3, which is specifically expressed on T lymphocytes or T-cell–derived lymphoid cell lines [71,113]. However, anti-CD3–polylysine/DNA complexes often remained in endosomes, and inactivated adenovirus or membrane-destabilizing peptides have to be incorporated to ensure endosomal release [33,50]. The addition of endosomal release reagents was found not to be required for anti-CD3–PEI800/DNA complexes.

III. LOCAL APPLICATION IN VIVO

In the previous section, entry of polycation/DNA complexes into cells via cell attachment and crossing the various barriers was discussed; these events apply to both in vitro and in vivo gene delivery. However, when applying polycation/DNA complexes to an organism in vivo (locally or systemically), new barriers have to be overcome to ensure efficient and specific gene expression. For locally applied complexes, depending on target site, route of access, or chosen therapy will determine the strategy for complex application. When the target site is easily accessible, local administration such as intradermal, intramuscular, intratracheal, or intratumoral can be used. Similarly, surgery or catheters can deliver the complexes directly to the target site. Other physical methods can be used to increase polyplex gene delivery locally, such as biolistic gold microparticles or gene gun [121], local electroporation [5,6], and continuous application via micropump [122].

Local delivery of genes using nonviral vectors was initially described in the treatment of melanoma using lipoplexes [123]. Polylysine/DNA complexes initially did not demonstrate efficient gene transfer. With the addition of endosomal active agents to the polylysine complexes, such as adenoviruses (generating AVET complexes), efficient gene expression was found in airway epithelium after intratracheal application [124], in nasal epithelium [125], and in tumors after intratumoral application [120,126,127]. Other cationic polymers, such as PEI, have successfully been used for local gene delivery in vivo: mouse brain [15,72,73], kidney [128], lung [129,130], bilary epithelia [131], preimplantation embryos [132], carotid artery [133], and subcutaneous tumors [66,120,122,127, 134].

Several groups of investigators have described local administration of PEI/DNA complexes to subcutaneous tumors. Generally, the low molecular weight PEIs (22 and 25 kD) have a greater transfection competency and lower toxicity as compared to high molecular weight PEI (800 kD) [120,127]. Furthermore, intratumoral delivery of linear PEI (22 kD) into solid or cystic tumor models could transfect 1% of the cells and persisted for 15 days [122]. The addition of ligands, such as transferrin, have also been seen to increase the efficiency of cell transfection with transferrin–PEI (800 kD)/DNA complexes, being 10- to 100-

fold more efficient than naked DNA; similar results were also observed with AVET [120,127].

Once applied, complexes have to diffuse through the solid tumor mass and not remain at the site of injection. The diffusion of complexes will be influenced by solubility and complex size, which can be influenced by the charge ratio and salt conditions [25]. Intratumoral application of complexes often results in a lot of the injected material refluxing through the needle tract and being lost from the intended site of application. This was partially overcome by administering complexes with a micropump [122]. The injection of a large volume of liquid may help in the diffusion of complexes by also weakening the cell-to-cell connections and dispersing the complexes throughout the tissue via a hydrodynamic mechanism, as observed in livers after application of high injection volumes in murine livers [7,135] and muscle [8].

The interaction of transfection complexes with extracellular matrix and necrotic cellular material (e.g., in tumors) can significantly reduce transfection efficiency. Furthermore, some glycosaminoglycans (such as hyaluronic acid and heparin sulfate) can completely inhibit the ability of the complexes to transfect cells, and can relax and release the DNA from complexes, causing degradation of the DNA [136]. Therefore, strategies that shield complexes from these nonspecific interactions have been developed (see Sec. IV). However, compared to lipoplexes, polyplexes are more resistant to inactivation by pulmonary surfactants and retain their ability to condense DNA [137].

The activation of the immune system is often a side effect that is unwanted. Generally, the polycations have a very low level of immunogenicity (e.g., as compared to adenoviruses [138]). The activation of serum complement by polylysine, PEI, and PAMAM has been observed, with highly positive complexes being the most active stimulators [139]. The activation of the complement cascade can be avoided by shielding the positive particles with polyethyleneglycol (PEG; see Sec. IV). Stimulation of the immune system by complexes can also be a result of bacterial DNA contamination and lipopolysaccaride (LPS) or CpG islands within plasmid DNA [140–142]. The avoidance of immune stimulation for certain applications may be desired; however, when generating a vaccination response to an antigen or an antitumoral effect, these nonspecific activities could potentially be beneficial in eliciting an effective treatment.

IV. SYSTEMIC APPLICATION

The delivery of polyplexes to the target organ, or in the case of cancer, metastatic nodules (that cannot be reached via local administration), can only be achieved by applying complexes systemically through the blood circulation system and targeting the specific site. This route of application presents numerous biological and physical obstacles that must be overcome just to reach the intended site and pass out of the blood system into the targeted tissue. Particle size is critical for successful delivery, playing a role in (1), transport of complexes through capillaries, (2), passing out of the blood vessels (extravasation), and (3), passing through the intertissue space as described in Section III. Furthermore, complexes must remain stable and soluble, while not aggregating in the blood, or unraveling, revealing the DNA to degrading enzymes within the blood or intertissue space. Also, complexes should be designed not to interact with components of the blood system, such as plasma, complement, erythrocytes, cells of the RES (e.g., macrophages), extracellular matrix and other nontarget cells.

A. Systemically Delivered Polylysine/DNA Complexes

Wu and colleagues demonstrated that asialoorosomucoid–polylysine complexed with the chloramphenicol acetyltransferase (CAT) gene could specifically target rat livers in vivo after application of the complexes systemically [12,27,100]. Targeted gene expression was further increased by partial hepatectomy after complex administration into the rats [30,143], or via disruption of microtubules with colchicine (144). Subsequently, genes encoding human albumin and LDL protein were delivered specifically to livers using asialoorosomucoid–polylysine complexes in rat and rabbit models [28,145]. Since this first report other researchers replaced the asialoorosomucoid on polylysine with galactose [30] or tri-antennary NAc galactose [111] and demonstrated specific targeting to the liver. The generation of small polyplexes (15–30 nm) under specific conditions was considered to play a primary role in the targeting of these complexes (30). When the fate of delivered DNA was assessed, the majority of DNA was found in the endosome and only a small amount was detected in the nucleus [143], suggesting that endosomal release is a major problem with polylysine delivered complexes in vivo (see Sec. II.A.1). The incorporation of endosomal active peptides into the particles to enhance DNA release from the endosome has been considered. Recently, hepatocyte-targeted gene delivery was described after systemic application of polyornithine polyplexes conjugated with galactose and the hemagglutinin fusogenic peptide [62]. Furthermore, gene expression could be interfered with when galactose–bovine serum albumin was used to compete with the liver-targeted polyplexes. Characterization of radioactive DNA administered revealed that 60% was detected in the liver, with most DNA being seen in parenchymal cells; however, often there is a discrepancy between liver accumulation of DNA and gene expression (see Sec. IV.D). Gene expression of systemically delivered asialoorosomucoid–polylysine complexes could be greatly enhanced after the inclusion of endosomal active peptides; furthermore, nuclear uptake was enhanced in complexes containing the peptide [146].

Recently, optimized AVET complexes have been shown specifically to direct gene expression to tumors (127). Murine neuroblastoma (neuro2a) tumors grown subcutaneously preferentially expressed the reporter gene luciferase, as compared to other organs, when AVET complexes were systemically administered intravenously via the tail vein. Systemic delivery of anti-pIgR Fc-polylysine/DNA complexes resulted in gene expression in the airway epithelium of the lung [31]. However, a humoral immune response against the ligand reduced the efficiency of gene delivery after multiple applications [138].

B. Systemically Delivered PEI/DNA Complexes

Systemically delivered linear low molecular weight PEI(22)/DNA complexes resulted in a high gene expression in the lung, with lower activity gene expression being observed in other organs (spleen, heart, liver, and kidneys) [74,129]. These complexes rapidly crossed the endothelial cells of the lung capillary bed, and gene expression was mainly found in alveolar cells [75,147,148]. However, branched PEIs (25 and 800 kD) were less efficient and often toxic at high polycation nitrogen:DNA phosphate ratios as compared to the linear PEI [74,127]. Surprisingly, systemic delivery of PEI(22)/DNA complexes in mice bearing lung tumors were found to be less efficient at expressing transgenes as compared to normal lungs [122]. This type of complex is often not specific or targeted to a particular cell type even when a ligand is covalently attached [127]. These nonspecific interactions are via the positive surface charge of complexes interacting with components of the blood. These interactions cause the particles to be taken out of circulation; for example, by nonspecifically binding to erythrocytes or being taken up by the RES. The interaction of the polycation with blood proteins, such as albumin, can cause the complex to disassemble, exposing the DNA to degrading enzymes via a similar mechanism observed with polylysine [149]. Hence, these colloidal and physical parameters are critical to efficient gene delivery in vivo [25,150,151]. Among other possible reasons, the toxicity that is associated with PEI polyplexes is thought to arise from the positively charged particles interacting with erythrocytes, causing them to aggregate together and then lodge in the lungs, causing an embolism [25,151].

C. Shielding of Polycation Complexes

The shielding of the surface positive charge of polyplexes has recently been shown to reduce aggregation, nonspecific interactions with serum components, and the lung endothelium. This protection from nonspecific interactions has been achieved via a number of methods.

    1.   *Polyethylene glycol (PEG) shielding of polyplexes.* The chemical linkage of PEG can shield the positive charge of polycation/DNA com-

plexes and reduces the positive surface charge (ζ potential) to near neutral charge [151]. Shielded PEI/DNA complexes when applied in vivo have reduced interaction with erythrocytes and also reduced toxicity; furthermore, they have increased circulation time before being cleared from the blood stream as compared to unshielded complexes [127,151]. PEG–polylysine/DNA–shielded complexes were also found to have decreased ζ potential and reduced toxicity; furthermore, they had an increased solubility as compared to unshielded polylysine/ DNA complexes [152,153]. Copolymers of anionic peptides have been linked to PEG which stabilize cationic complexes by inhibiting aggregation and reducing complement activation [154].

2. *Poly-N-(2-hydroxypropyl)methacrylamide (pHPMA) shielding.* Polylysine grafted with pHPMA via a peptide linker has increased solubility with lower toxicity and a negative surface charge; furthermore, the pHPMA shield reduced the interaction of complexes with blood components (155).

3. *Poloxamer shielding.* PEI/DNA complexes shielded with the polyether pluronic 123 showed reduced lung gene expression with an increased liver expression over nonshielded PEI complexes [76]. Local application of DNA shielded with poloxamers increased gene expression in muscle; furthermore, the complexes diffused throughout the tissue more efficiently [156].

4. *Ligand density.* Ligands that are covalently attached to the polycation can be used as shielding agent. Previously, transferrin–PEI(800) complexes were found to have a reduced ζ potential compared to ligand-free PEI(800) [127]. When the ligand density was increased within the complex, the ζ potential was reduced to near neutral and the nonspecific interactions with cells (such as erythrocytes) (Fig. 2) was inhibited; furthermore, the toxicity of the complexes was reduced when applied in vivo [157]. These initial reports of shielding, combinations of the above methods have helped to generate particles that have reduced nonspecific interactions while gaining circulation time and increased solubility.

D. Targeted Gene Expression with Shielded Polyplexes

The development of systemically applied shielded polycation/DNA complexes that can target specific tissue has been achieved with varying success. An example of systemically applied polycation/DNA complexes is transferrin–PEI shielded with either PEG [127] or by increased ligand density [157]. Vectors were developed that systemically targeted a subcutaneous tumor (murine neuroblastoma [neuro2A]) after application of complexes in the tail vein of syngenic A/J mice.

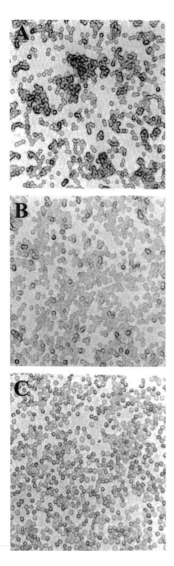

Figure 2 Interaction of shielded and nonshielded polyplexes with erythrocytes. Washed fresh murine erythrocytes were incubated with (A) PEI/DNA complexes; (B) transferrin-shielded PEI/DNA complexes, and (C) control.

Application of PEGylated transferrin–PEI/DNA and high ligand density transferrin–PEI/DNA complexes were found preferentially to target tumors [157,158]. Furthermore, luciferase gene expression was reduced in other organs (liver, lung, heart, spleen, and kidney) as compared to nonshielded complexes. The targeting of PEGylated–transferrin–PEI polyplexes was the first demonstration of tumor-specific targeted gene expression after systemic application. Furthermore, this targeted expression is particularly interesting, as subcutaneous tumors are not directly joined to the main blood supply but via the peripheral blood supply. The expression observed in these tumors is probably via two mechanisms. First, passive targeting due to the shielding of the complexes from nonspecific interactions increases the circulation time, allowing the complexes to reach the tumor. Second, active targeting via receptor-mediated uptake of the transferrin receptor on the tumor cell interacts with transferrin on the polyplex. However, the targeting observed is only efficient when tumors are well vascularized with a higher permeability as compared to normal vasculature (our unpublished observations). The size of fenestration within human tumors will be critical for the future development of nonviral particles, as this will put certain size constraints on particle sizes that can be administered.

When gene expression was characterized within tumors, often the pattern of gene expression was in small foci distributed throughout the tumor with these foci being close to immature blood vessels [127,157,158]. The distribution of the stabilized particles was analyzed by Southern blotting for DNA distribution in the various organs. Particles that were shielded by increasing the density of transferrin on the polyplex surface were mainly observed in the liver followed by the tumor. However, particularly in the liver, the majority of DNA was degraded, with almost no gene expression being observed in the liver. Histochemical analysis of the liver revealed that most of the DNA was taken up by the Kupffer cells within the liver. In contrast, Southern blotting of tumors revealed that only ~0.1% of the administered DNA was in tumors, but histochemical analysis demonstrated that the DNA was associated mainly with tumor cells and not normal host tissue. The potential of the liver to filter out particles will be a challenge for researchers to overcome; ensuring that the majority of material administered reaches the targeted organ or tumor and is not removed on the first pass through the liver.

V. SUMMARY

A therapeutically applicable nonviral vector should be able to incorporate a number of functions that enables specific delivery, uptake, and expression of a desired gene. The incorporation of ligands, endosomal release enhancers, nuclear localization signals and shielding agents protecting from nonspecific interactions and enhancing gene expression as described in this chapter can also be combined with transcriptional and translational regulators to ensure that the gene of interest is only expressed at the desired target site. Some of the parameters that have been characterized in tumor targeting after systemic application will be important for other nononcogenic applications in the specific targeting of an organ while blocking the particles from interacting with blood components or RES.

To date, no nonviral system has been generated which would be effective in the clinic; however, recent advances in tumor targeting, liver targeting with polycations, and observations made with other nonviral systems indicate that with time better and more effective vectors will be developed.

REFERENCES

1. Ledley FD. Nonviral gene therapy: the promise of genes as pharmaceutical products. Hum Gene Ther 1995; 6(9):1129–1144.
2. Lollo CP, Banaszczyk MG, Chiou HC. Obstacles and advances in non-viral gene delivery. Curr Opin Mol Ther 2000; 2(2):136–142.
3. Choi AH, Basu M, Rae MN, McNeal MM, Ward RL. Particle-bombardment–mediated DNA vaccination with rotavirus VP4 or VP7 induces high levels of serum rotavirus IgG but fails to protect mice against challenge. Virology 1998; 250(1): 230–240.
4. Kuriyama S, Mitoro A, Tsujinoue H, Nakatani T, Yoshiji H, Tsujimoto T, et al. Particle-mediated gene transfer into murine livers using a newly developed gene gun. Gene Ther 2000; 7(13):1132–1136.
5. Heller R, Jaroszeski M, Atkin A, Moradpour D, Gilbert R, Wands J, et al. In vivo gene electroinjection and expression in rat liver. FEBS Lett 1996; 389(3):225–228.
6. Mathiesen I. Electropermeabilization of skeletal muscle enhances gene transfer in vivo. Gene Ther 1999; 6(4):508–514.
7. Liu F, Song Y, Liu D. Hydrodynamics-based transfection in animals by systemic administration of plasmid DNA. Gene Ther 1999; 6(7):1258–1266.
8. Budker V, Zhang G, Danko I, Williams P, Wolff J. The efficient expression of intravascularly delivered DNA in rat muscle. Gene Ther 1998; 5(2):272–276.
9. Felgner PL, Barenholz Y, Behr JP, Cheng SH, Cullis P, Huang L, et al. Nomenclature for synthetic gene delivery systems. Hum Gene Ther 1997; 8(5):511–512.
10. Zauner W, Ogris M, Wagner E. Polylysine-based transfection systems utilizing receptor-mediated delivery. Adv Drug Del Rev 1998; 30:97–113.
11. Huang CH, Hung MC, Wagner E. Nonviral Vectors for Gene Therapy. San Diego: Academic Press, 1999.
12. Wu GY, Wu CH. Receptor-mediated in vitro gene transformation by a soluble DNA carrier system. J Biol Chem 1987; 262(10):4429–4432.
13. Wagner E, Zenke M, Cotten M, Beug H, Birnstiel ML. Transferrin-polycation conjugates as carriers for DNA uptake into cells. Proc Natl Acad Sci USA 1990; 87(9): 3410–3414.

14. Haensler J, Szoka FC Jr. Polyamidoamine cascade polymers mediate efficient transfection of cells in culture. Bioconjug Chem 1993; 4(5):372–379.

15. Boussif O, Lezoualch F, Zanta MA, Mergny MD, Scherman D, Demeneix B, et al. A versatile vector for gene and oligonucleotide transfer into cells in culture and in vivo: polyethylenimine. Proc Natl Acad Sci USA 1995; 92(16):7297–7301.

16. Dash PR, Toncheva V, Schacht EH, Seymour LW. Synthetic polymers for vectoral delivery of DNA: characterization of polymer–DNA complexes by photon correlation spectroscopy and stability to nuclease degradation and disruption by polyanions in vitro. J Controlled Release 1997; 48:269–276.

17. Kabanov AV, Kabanov VA. DNA complexes with polycations for the delivery of genetic material into cells. Bioconjug Chem 1995; 6(1):7–20.

18. Mislick KA, Baldeschwieler JD. Evidence for the role of proteoglycans in cation-mediated gene transfer. Proc Natl Acad Sci USA 1996; 93(22):12349–12354.

19. Gottschalk S, Sparrow JT, Hauer J, Mims MP, Leland FE, Woo SL, et al. A novel DNA-peptide complex for efficient gene transfer and expression in mammalian cells. Gene Ther 1996; 3(5):48–57.

20. McKenzie DL, Collard WT, Rice KG. Comparative gene transfer efficiency of low molecular weight polylysine DNA-condensing peptides. J Pept Res 1999; 54(4): 311–318.

21. Laemmli UK. Characterization of DNA condensates induced by poly(ethylene oxide) and polylysine. Proc Natl Acad Sci USA 1975; 72(11):4288–4292.

22. Baeza I, Gariglio P, Rangel LM, Chavez P, Cervantes L, Arguello C, et al. Electron microscopy and biochemical properties of polyamine-compacted DNA. Biochemistry 1987; 26(20):6387–6392.

23. McKee TD, DeRome ME, Wu GY, Findeis MA. Preparation of asialoorosomucoid-polylysine conjugates. Bioconjug Chem 1994; 5(4):306–311.

24. Perales JC, Ferkol T, Molas M, Hanson RW. An evaluation of receptor-mediated gene transfer using synthetic DNA-ligand complexes. Eur J Biochem 1994; 226(2): 255–266.

25. Ogris M, Steinlein P, Kursa M, Mechtler K, Kircheis R, Wagner E. The size of DNA/transferrin-PEI complexes is an important factor for gene expression in cultured cells. Gene Ther 1998; 5(10):1425–1433.

26. Wagner E, Cotten M, Foisner R, Birnstiel ML. Transferrin-polycation-DNA complexes: the effect of polycations on the structure of the complex and DNA delivery to cells. Proc Natl Acad Sci USA 1991; 88(10):4255–4259.

27. Wu CH, Wilson JM, Wu GY. Targeting genes: delivery and persistent expression of a foreign gene driven by mammalian regulatory elements in vivo. J Biol Chem 1989; 264(29):16985–16987.

28. Wu GY, Wilson JM, Shalaby F, Grossman M, Shafritz DA, Wu CH. Receptor-mediated gene delivery in vivo. Partial correction of genetic analbuminemia in Nagase rats. J Biol Chem 1991; 266(22):14338–14342.

29. Ferkol T, Perales JC, Mularo F, Hanson RW. Receptor-mediated gene transfer into macrophages. Proc Natl Acad Sci USA 1996; 93(1):101–105.

30. Perales JC, Ferkol T, Beegen H, Ratnoff OD, Hanson RW. Gene transfer in vivo: sustained expression and regulation of genes introduced into the liver by receptor-targeted uptake. Proc Natl Acad Sci USA 1994; 91(9):4086–4090.

31. Ferkol T, Perales JC, Eckman E, Kaetzel CS, Hanson RW, Davis PB. Gene transfer into the airway epithelium of animals by targeting the polymeric immunoglobulin receptor. J Clin Invest 1995; 95(2):493–502.

32. Wolfert MA, Seymour LW. Atomic force microscopic analysis of the influence of the molecular weight of poly(L)lysine on the size of polyelectrolyte complexes formed with DNA. Gene Ther 1996; 3(3):269–273.

33. Plank C, Oberhauser B, Mechtler K, Koch C, Wagner E. The influence of endosome-disruptive peptides on gene transfer using synthetic virus-like gene transfer systems. J Biol Chem 1994; 269(17):12918–12924.

34. Shen WC, Ryser HJ. Conjugation of poly-L-lysine to albumin and horseradish peroxidase: a novel method of enhancing the cellular uptake of proteins. Proc Natl Acad Sci USA 1978; 75(4):1872–1876.

35. Pruzanski W, Saito S. The influence of natural and synthetic cationic substances on phagocytic activity of human polymorphonuclear cells. An alternative pathway of phagocytic enhancement. Exp Cell Res 1978; 117(1):1–13.

36. Kundahl E, Richman R, Flickinger RA. The effect of added H1 histone and polylysine on DNA synthesis and cell division of cultured mammalian cells. J Cell Physiol 1981; 108(3):291–298.

37. Shier WT, Dubourdieu DJ, Durkin JP. Polycations as prostaglandin synthesis inducers. Stimulation of arachidonic acid release and prostaglandin synthesis in cultured fibroblasts by poly(L-lysine) and other synthetic polycations. Biochim Biophys Acta 1984; 793(2):238–250.

38. Kyriakis JM, Avruch J. pp54 microtubule-associated protein 2 kinase. A novel serine/ threonine protein kinase regulated by phosphorylation and stimulated by poly-L-lysine. J Biol Chem 1990; 265(28):17355–17363.

39. Kornguth SE, Stahmann MA. Effect of polylysine on the leakage and retension of compounds by Ehrlich ascites tumor cells. Cancer Res 1961; 21:907–912.

40. Arnold LJ, Jr., Dagan A, Gutheil J, Kaplan NO. Antineoplastic activity of poly(L-lysine) with some ascites tumor cells. Proc Natl Acad Sci USA 1979; 76(7):3246–3250.

41. Gad AE, Bental M, Elyashiv G, Weinberg H, Nir S. Promotion and inhibition of vesicle fusion by polylysine. Biochemistry 1985; 24(22):6277–6282.

42. Elferink JG. Cytolytic effect of polylysine on rabbit polymorphonuclear leukocytes. Inflammation 1985; 9(3):321–331.

43. Wagner E, Plank C, Zatloukal K, Cotten M, Birnstiel ML. Influenza virus hemagglutinin HA-2 N-terminal fusogenic peptides augment gene transfer by transferrin-polylysine- DNA complexes: toward a synthetic virus-like gene-transfer vehicle. Proc Natl Acad Sci USA 1992; 89(17):7934–7938.

44. Plank C, Zatloukal K, Cotten M, Mechtler K, Wagner E. Gene transfer into hepatocytes using asialoglycoprotein receptor mediated endocytosis of DNA complexed with an artificial tetra-antennary galactose ligand. Bioconjug Chem 1992; 3(6): 533–539.

45. Wagner E, Curiel D, Cotten M. Delivery of drugs, proteins and genes into cells using transferrin as a ligand for receptor mediated endocytosis. Adv Drug Del Rev 1994; 14:113–135.

46. Seglen PO. Inhibitors of lysosomal function. Methods Enzymol 1983; 96:737–764.

47. Erbacher P, Roche AC, Monsigny M, Midoux P. Putative role of chloroquine in gene transfer into a human hepatoma cell line by DNA/lactosylated polylysine complexes. Exp Cell Res 1996; 225(1):186–194.
48. Zauner W, Kichler A, Schmidt W, Mechtler K, Wagner E. Glycerol and polylysine synergize in their ability to rupture vesicular membranes: a mechanism for increased transferrin- polylysine-mediated gene transfer. Exp Cell Res 1997; 232(1):137–145.
49. Curiel DT, Agarwal S, Wagner E, Cotten M. Adenovirus enhancement of transferrin-polylysine-mediated gene delivery. Proc Natl Acad Sci USA 1991; 88(19): 8850–8854.
50. Wagner E, Zatloukal K, Cotten M, Kirlappos H, Mechtler K, Curiel DT et al. Coupling of adenovirus to transferrin–polylysine/DNA complexes greatly enhances receptor-mediated gene delivery and expression of transfected genes. Proc Natl Acad Sci USA 1992; 89(13):6099–6103.
51. Cristiano RJ, Smith LC, Kay MA, Brinkley BR, Woo SL. Hepatic gene therapy: efficient gene delivery and expression in primary hepatocytes utilizing a conjugated adenovirus- DNA complex. Proc Natl Acad Sci USA 1993; 90(24):11548–11552.
52. Baatz JE, Bruno MD, Ciraolo PJ, Glasser SW, Stripp BR, Smyth KL, et al. Utilization of modified surfactant-associated protein B for delivery of DNA to airway cells in culture. Proc Natl Acad Sci USA 1994; 91(7):2547–2551.
53. Wu GY, Zhan P, Sze LL, Rosenberg AR, Wu CH. Incorporation of adenovirus into a ligand-based DNA carrier system results in retention of original receptor specificity and enhances targeted gene expression. J Biol Chem 1994; 269(15): 11542–11546.
54. Merwin JR, Carmichael EP, Noell GS, DeRome ME, Thomas WL, Robert N, et al. CD5-mediated specific delivery of DNA to T lymphocytes: compartmentalization augmented by adenovirus. J Immunol Methods 1995; 186(2):257–266.
55. Cotten M, Saltik M, Kursa M, Wagner E, Maass G, Birnstiel ML. Psoralen treatment of adenovirus particles eliminates virus replication and transcription while maintaining the endosomolytic activity of the virus capsid. Virology 1994; 205(1): 254–261.
56. Cotten M, Wagner E, Zatloukal K, Birnstiel ML. Chicken adenovirus (CELO virus) particles augment receptor- mediated DNA delivery to mammalian cells and yield exceptional levels of stable transformants. J Virol 1993; 67(7):3777–3785.
57. Zauner W, Blaas D, Kuechler E, Wagner E. Rhinovirus-mediated endosomal release of transfection complexes. J Virol 1995; 69(2):1085–1092.
58. Wagner E. Application of membrane-active peptides for nonviral gene delivery. Adv Drug Del Rev 1999; 38(3):279–289.
59. Parente RA, Nir S, Szoka FC Jr. Mechanism of leakage of phospholipid vesicle contents induced by the peptide GALA. Biochemistry 1990; 29(37):8720–8728.
60. Midoux P, Kichler A, Boutin V, Maurizot JC, Monsigny M. Membrane permeabilization and efficient gene transfer by a peptide containing several histidines. Bioconjug Chem 1998; 9(2):260–267.
61. Kichler A, Freulon I, Boutin V, Mayer R, Monsigny M, Midoux P. Glycofection in the presence of anionic fusogenic peptides: a study of the parameters affecting the peptide-mediated enhancement of the transfection efficiency. J Gene Med 1999; 1(2):134–143.

62. Nishikawa M, Yamauchi M, Morimoto K, Ishida E, Takakura Y, Hashida M. Hepatocyte-targeted in vivo gene expression by intravenous injection of plasmid DNA complexed with synthetic multi-functional gene delivery system. Gene Ther 2000; 7(7):548–555.

63. Kukowska Latallo JF, Bielinska AU, Johnson J, Spindler R, Tomalia DA, Baker JR Jr. Efficient transfer of genetic material into mammalian cells using Starburst polyamidoamine dendrimers. Proc Natl Acad Sci USA 1996; 93(10):4897–4902.

64. Tang MX, Redemann CT, Szoka FC Jr. In vitro gene delivery by degraded polyamidoamine dendrimers. Bioconjug Chem 1996; 7(6):703–714.

65. Tang MX, Szoka FC. The influence of polymer structure on the interactions of cationic polymers with DNA and morphology of the resulting complexes. Gene Ther 1997; 4(8):823–832.

66. Maruyama TH, Harada Y, Matsumura T, Satoh E, Cui F, Iwai M, et al. Effective suicide gene therapy in vivo by EBV-based plasmid vector coupled with polyamidoamine dendrimer. Gene Ther 2000; 7(1):53–60.

67. Kukowska Latallo JF, Chen C, Eichman J, Bielinska AU, Baker JR Jr. Enhancement of dendrimer-mediated transfection using synthetic lung surfactant exosurf neonatal in vitro. Biochem Biophys Res Commun 1999; 264(1):253–261.

68. Brazeau GA, Attia S, Poxon S, Hughes JA. In vitro myotoxicity of selected cationic macromolecules used in non-viral gene delivery. Pharm Res 1998; 15(5):680–684.

69. Baker A, Cotten M. Delivery of bacterial artificial chromosomes into mammalian cells with psoralen-inactivated adenovirus carrier. Nucleic Acids Res 1997; 25(10): 1950–1956.

70. Boussif O, Zanta MA, Behr JP. Optimized galenics improve in vitro gene transfer with cationic molecules up to 1000-fold. Gene Ther 1996; 3(12):1074–1080.

71. Kircheis R, Kichler A, Wallner G, Kursa M, Ogris M, Felzmann T, et al. Coupling of cell-binding ligands to polyethylenimine for targeted gene delivery. Gene Ther 1997; 4(5):409–418.

72. Abdallah B, Hassan A, Benoist C, Goula D, Behr JP, Demeneix BA. A powerful nonviral vector for in vivo gene transfer into the adult mammalian brain: polyethylenimine. Hum Gene Ther 1996; 7(16):1947–1954.

73. Goula D, Remy JS, Erbacher P, Wasowicz M, Levi G, Abdallah B, et al. Size, diffusibility and transfection performance of linear PEI/DNA complexes in the mouse central nervous system. Gene Ther 1998; 5(5):712–717.

74. Goula D, Benoist C, Mantero S, Merlo G, Levi G, Demeneix BA. Polyethylenimine-based intravenous delivery of transgenes to mouse lung. Gene Ther 1998; 5(9):1291–1295.

75. Bragonzi A, Boletta A, Biffi A, Muggia A, Sersale G, Cheng SH, et al. Comparison between cationic polymers and lipids in mediating systemic gene delivery to the lungs. Gene Ther 1999; 6(12):1995–2004.

76. Nguyen HK, Lemieux P, Vinogradov SV, Gebhart CL, Guérin N, Paradis G et al. Evaluation of polyether-polyethyleneimine graft copolymers as gene transfer agents. Gene Ther 2000; 7(2):126–138.

77. Dong Y, Skoultchi AI, Pollard JW. Efficient DNA transfection of quiescent mammalian cells using poly-L-ornithine. Nucleic Acids Res 1993; 21(3):771–772.

78. Pouton CW, Lucas P, Thomas BJ, Uduehi AN, Milroy DA, Moss SH. Polycation-

DNA complexes for gene delivery: a comparison of the biopharmaceutical properties of cationic polypeptides and cationic lipids. J Controlled Release 1998; 53(1–3):289–299.

79. Emi N, Kidoaki S, Yoshikawa K, Saito H. Gene transfer mediated by polyarginine requires a formation of big carrier-complex of DNA aggregate. Biochem Biophys Res Commun 1997; 231(2):421–424.

80. Fritz JD, Herweijer H, Zhang G, Wolff JA. Gene transfer into mammalian cells using histone-condensed plasmid DNA. Hum Gene Ther 1996; 7(12):1395–1404.

81. Böttger M, Vogel F, Platzer M, Kiessling U, Grade K, Strauss M. Condensation of vector DNA by the chromosomal protein HMG1 results in efficient transfection. Biochim Biophys Acta 1988; 950(2):221–228.

82. Zaitsev SV, Haberland A, Otto A, Vorobev VI, Haller H, Böttger M. H1 and HMG17 extracted from calf thymus nuclei are efficient DNA carriers in gene transfer. Gene Ther 1997; 4(6):586–592.

83. van de Wetering P, Cherng JY, Talsma H, Crommelin DJ, Hennink WE. 2-(Dimethylamino)ethyl methacrylate based (co)polymers as gene transfer agents. J Controlled Release 1998; 53(1–3):145–153.

84. Boussif O, Delair T, Brua C, Veron L, Pavirani A, Kolbe HV. Synthesis of poly-allylamine derivatives and their use as gene transfer vectors in vitro. Bioconjug Chem 1999; 10(5):877–883.

85. Lopata MA, Cleveland DW, Sollner WB. High level transient expression of a chloramphenicol acetyl transferase gene by DEAE-dextran mediated DNA transfection coupled with a dimethyl sulfoxide or glycerol shock treatment. Nucleic Acids Res 1984; 12(14):5707–5717.

86. Erbacher P, Zou S, Bettinger T, Steffan AM, Remy JS. Chitosan-based vector/DNA complexes for gene delivery: biophysical characteristics and transfection ability. Pharm Res 1998; 15(9):1332–1339.

87. Kabanov AV, Astafieva IV, Maksimova IV, Lukanidin EM, Georgiev GP, Kabanov VA. Efficient transformation of mammalian cells using DNA interpolyelectrolyte complexes with carbon chain polycations. Bioconjug Chem 1993; 4(6):448–454.

88. Midoux P, Monsigny M. Efficient gene transfer by histidylated polylysine/pDNA complexes. Bioconjug Chem 1999; 10(3):406–411.

89. Brisson M, Tseng WC, Almonte C, Watkins S, Huang L. Subcellular trafficking of the cytoplasmic expression system. Hum Gene Ther 1999; 10(16):2601–2613.

90. Schirrmacher V, Förg P, Dalemans W, Chlichlia K, Zeng Y, Fournier P, et al. Intrapinna anti-tumor vaccination with self-replicating infectious RNA or with DNA encoding a model tumor antigen and a cytokine. Gene Ther 2000; 7(13):1137–1147.

91. Page RL, Butler SP, Subramanian A, Gwazdauskas FC, Johnson JL, Velander WH. Transgenesis in mice by cytoplasmic injection of polylysine/ DNA mixtures. Transgenic Res 1995; 4(6):353–360.

92. Pollard H, Remy JS, Loussouarn G, Demolombe S, Behr JP, Escande D. Polyethylenimine but not cationic lipids promotes transgene delivery to the nucleus in mammalian cells. J Biol Chem 1998; 273(13):7507–7511.

93. Godbey WT, Wu KK, Mikos AG. Tracking the intracellular path of poly(ethylenimine)/DNA complexes for gene delivery. Proc Natl Acad Sci USA 1999; 96(9):5177–5181.

94. Lechardeur D, Sohn KJ, Haardt M, Joshi PB, Monck M, Graham RW, et al. Metabolic instability of plasmid DNA in the cytosol: a potential barrier to gene transfer. Gene Ther 1999; 6(4):482–497.

95. Brunner S, Sauer T, Carotta S, Cotten M, Saltik M, Wagner E. Cell cycle dependence of gene transfer by lipoplex, polyplex and recombinant adenovirus. Gene Ther 2000; 7(5):401–407.

96. Sebestyén MG, Ludtke JJ, Bassik MC, Zhang G, Budker V, Lukhtanov EA, et al. DNA vector chemistry: the covalent attachment of signal peptides to plasmid DNA. Nat Biotechnol 1998; 16(1):80–85.

97. Zanta MA, Belguise VP, Behr JP. Gene delivery: a single nuclear localization signal peptide is sufficient to carry DNA to the cell nucleus. Proc Natl Acad Sci USA 1999; 96(1):91–96.

98. Brandén LJ, Mohamed AJ, Smith CI. A peptide nucleic acid-nuclear localization signal fusion that mediates nuclear transport of DNA. Nat Biotechnol 1999; 17(8):784–787.

99. Ludtke JJ, Zhang G, Sebestyén MG, Wolff JA. A nuclear localization signal can enhance both the nuclear transport and expression of 1 kb DNA. J Cell Sci 1999; 112(12):2033–2041.

100. Wu GY, Wu CH. Receptor-mediated gene delivery and expression in vivo. J Biol Chem 1988; 263(29):14621–14624.

101. Cotten M, Längle RF, Kirlappos H, Wagner E, Mechtler K, Zenke M, et al. Transferrin-polycation-mediated introduction of DNA into human leukemic cells: stimulation by agents that affect the survival of transfected DNA or modulate transferrin receptor levels. Proc Natl Acad Sci USA 1990; 87(11):4033–4037.

102. Cristiano RJ, Roth JA. Epidermal growth factor mediated DNA delivery into lung cancer cells via the epidermal growth factor receptor. Cancer Gene Ther 1996; 3(1):4–10.

103. Sosnowski BA, Gonzalez AM, Chandler LA, Buechler YJ, Pierce GF, Baird A. Targeting DNA to cells with basic fibroblast growth factor (FGF2). J Biol Chem 1996; 271(52):33647–33653.

104. Huckett B, Ariatti M, Hawtrey AO. Evidence for targeted gene transfer by receptor-mediated endocytosis. Stable expression following insulin-directed entry of NEO into HepG2 cells. Biochem Pharmacol 1990; 40(2):253–263.

105. Erbacher P, Remy JS, Behr JP. Gene transfer with synthetic virus-like particles via the integrin-mediated endocytosis pathway. Gene Ther 1999; 6(1):138–145.

106. Gottschalk S, Cristiano RJ, Smith LC, Woo SL. Folate receptor mediated DNA delivery into tumor cells: potosomal disruption results in enhanced gene expression. Gene Ther 1994; 1(3):185–191.

107. Mislick KA, Baldeschwieler JD, Kayyem JF, Meade TJ. Transfection of folate-polylysine DNA complexes: evidence for lysosomal delivery. Bioconjug Chem 1995; 6(5):512–515.

108. Batra RK, Wang JF, Wagner E, Garver RIJ, Curiel DT. Receptor-mediated gene delivery employing lectin-binding specificity. Gene Ther 1994; 1(4):255–260.

109. Yin W, Cheng PW. Lectin conjugate-directed gene transfer to airway epithelial cells. Biochem Biophys Res Commun 1994; 205(1):826–833.

110. Ross GF, Morris RE, Ciraolo G, Huelsman K, Bruno M, Whitsett JA, et al. Surfac-

tant protein A-polylysine conjugates for delivery of DNA to airway cells in culture. Hum Gene Ther 1995; 6(1):31–40.

111. Merwin JR, Noell GS, Thomas WL, Chiou HC, DeRome ME, McKee TD, et al. Targeted delivery of DNA using YEE(GalNAcAH)3, a synthetic glycopeptide ligand for the asialoglycoprotein receptor. Bioconjug Chem 1994; 5(6):612–620.

112. Chen J, Stickles RJ, Daichendt KA. Galactosylated histone-mediated gene transfer and expression. Hum Gene Ther 1994; 5(4):429–435.

113. Buschle M, Cotten M, Kirlappos H, Mechtler K, Schaffner G, Zauner W, et al. Receptor-mediated gene transfer into human T lymphocytes via binding of DNA/CD3 antibody particles to the CD3 T cell receptor complex. Hum Gene Ther 1995; 6(6):753–761.

114. Poncet P, Panczak A, Goupy C, Gustafsson K, Blanpied C, Chavanel G, et al. Antifection: an antibody-mediated method to introduce genes into lymphoid cells in vitro and in vivo. Gene Ther 1996; 3(8):731–738.

115. Chen J, Gamou S, Takayanagi A, Shimizu N. A novel gene delivery system using EGF receptor-mediated endocytosis. FEBS Lett 1994; 338(2):167–169.

116. Wojda U, Goldsmith P, Miller JL. Surface membrane biotinylation efficiently mediates the endocytosis of avidin bioconjugates into nucleated cells. Bioconjug Chem 1999; 10(6):1044–1050.

117. Cotten M, Wagner E, Birnstiel ML. Receptor-mediated transport of DNA into eukaryotic cells. Methods Enzymol 1993; 217:618–644.

118. Schreiber S, Kämpgen E, Wagner E, Pirkhammer D, Trcka J, Korschan H, et al. Immunotherapy of metastatic malignant melanoma by a vaccine consisting of autologous interleukin 2-transfected cancer cells: outcome of a phase I study. Hum Gene Ther 1999; 10(6):983–993.

119. Michael SI, Huang CH, Römer MU, Wagner E, Hu PC, Curiel DT. Binding-incompetent adenovirus facilitates molecular conjugate- mediated gene transfer by the receptor-mediated endocytosis pathway. J Biol Chem 1993; 268(10):6866–6869.

120. Wightman L, Patzelt E, Wagner E, Kircheis R. Development of transferrin-polycation/DNA based vectors for gene delivery to melanoma cells. J Drug Target 1999; 7(4):293–303.

121. Yang NS, Burkholder J, Roberts B, Martinell B, McCabe D. In vivo and in vitro gene transfer to mammalian somatic cells by particle bombardment. Proc Natl Acad Sci USA 1990; 87(24):9568–9572.

122. Coll JL, Chollet P, Brambilla E, Desplanques D, Behr JP, Favrot M. In vivo delivery to tumors of DNA complexed with linear polyethylenimine. Hum Gene Ther 1999; 10(10):1659–1666.

123. Nabel GJ, Nabel EG, Yang ZY, Fox BA, Plautz GE, Gao X et al. Direct gene transfer with DNA-liposome complexes in melanoma: expression, biologic activity, and lack of toxicity in humans. Proc Natl Acad Sci USA 1993; 90(23):11307–11311.

124. Gao L, Wagner E, Cotten M, Agarwal S, Harris C, Romer M, et al. Direct in vivo gene transfer to airway epithelium employing adenovirus-polylysine-DNA complexes. Hum Gene Ther 1993; 4(1):17–24.

125. Fasbender A, Zabner J, Chillón M, Moninger TO, Puga AP, Davidson BL, et al. Complexes of adenovirus with polycationic polymers and cationic lipids increase

the efficiency of gene transfer in vitro and in vivo. J Biol Chem 1997; 272(10): 6479–6489.

126. Nguyen DM, Spitz FR, Yen N, Cristiano RJ, Roth JA. Gene therapy for lung cancer: enhancement of tumor suppression by a combination of sequential systemic cisplatin and adenovirus-mediated p53 gene transfer. J Thorac Cardiovasc Surg 1996; 112(5):1372–1376.

127. Kircheis R, Schüller S, Brunner S, Ogris M, Heider KH, Zauner W, et al. Polycation-based DNA complexes for tumor-targeted gene delivery in vivo. J Gene Med 1999; 1(2):111–120.

128. Boletta A, Benigni A, Lutz J, Remuzzi G, Soria MR, Monaco L. Nonviral gene delivery to the rat kidney with polyethylenimine. Hum Gene Ther 1997; 8(10): 1243–1251.

129. Ferrari S, Moro E, Pettenazzo A, Behr JP, Zacchello F, Scarpa M. ExGen 500 is an efficient vector for gene delivery to lung epithelial cells in vitro and in vivo. Gene Ther 1997; 4(10):1100–1106.

130. Ferrari S, Pettenazzo A, Garbati N, Zacchello F, Behr JP, Scarpa M. Polyethylenimine shows properties of interest for cystic fibrosis gene therapy. Biochim Biophys Acta 1999; 1447(2–3):219–225.

131. McKay T, MacVinish L, Carpenter B, Themis M, Jezzard S, Goldin R, et al. Selective in vivo transfection of murine biliary epithelia using polycation-enhanced adenovirus. Gene Ther 2000; 7(8):644–652.

132. Ivanova MM, Rosenkranz AA, Smirnova OA, Nikitin VA, Sobolev AS, Landa V, et al. Receptor-mediated transport of foreign DNA into preimplantation mammalian embryos. Mol Reprod Dev 1999; 54(2):112–120.

133. Turunen MP, Hiltunen MO, Ruponen M, Virkamäki L, Szoka FCJ, Urtti A, et al. Efficient adventitial gene delivery to rabbit carotid artery with cationic polymer-plasmid complexes. Gene Ther 1999; 6(1):6–11.

134. Mendiratta SK, Quezada A, Matar M, Wang J, Hebel HL, Long S, et al. Intratumoral delivery of IL-12 gene by polyvinyl polymeric vector system to murine renal and colon carcinoma results in potent antitumor immunity. Gene Ther 1999; 6(5): 833–839.

135. Zhang G, Budker V, Wolff JA. High levels of foreign gene expression in hepatocytes after tail vein injections of naked plasmid DNA. Hum Gene Ther 1999; 10(10):1735–1737.

136. Ruponen M, Ylä-Herttuala S, Urtti A. Interactions of polymeric and liposomal gene delivery systems with extracellular glycosaminoglycans: physicochemical and transfection studies. Biochim Biophys Acta 1999; 1415(2):331–341.

137. Ernst N, Ulrichskötter S, Schmalix WA, Rädler J, Galneder R, Mayer E, et al. Interaction of liposomal and polycationic transfection complexes with pulmonary surfactant. J Gene Med 1999; 1(5):331–340.

138. Ferkol T, Pellicena PA, Eckman E, Perales JC, Trzaska T, Tosi M, et al. Immunologic responses to gene transfer into mice via the polymeric immunoglobulin receptor. Gene Ther 1996; 3(8):669–678.

139. Plank C, Mechtler K, Szoka FCJ, Wagner E. Activation of the complement system by synthetic DNA complexes: a potential barrier for intravenous gene delivery. Hum Gene Ther 1996; 7(12):1437–1446.

140. Hartmann G, Krieg AM. CpG DNA and LPS induce distinct patterns of activation in human monocytes. Gene Ther 1999; 6(5):893–903.

141. Hartmann G, Weiner GJ, Krieg AM. CpG DNA: a potent signal for growth, activation, and maturation of human dendritic cells. Proc Natl Acad Sci 1999; 96(16): 9305–9310.

142. McLachlan G, Stevenson BJ, Davidson DJ, Porteous DJ. Bacterial DNA is implicated in the inflammatory response to delivery of DNA/DOTAP to mouse lungs. Gene Ther 2000; 7(5):384–392.

143. Chowdhury NR, Wu CH, Wu GY, Yerneni PC, Bommineni VR, Chowdhury JR. Fate of DNA targeted to the liver by asialoglycoprotein receptor–mediated endocytosis in vivo. Prolonged persistence in cytoplasmic vesicles after partial hepatectomy. J Biol Chem 1993; 268(15):11265–11271.

144. Chowdhury NR, Hays RM, Bommineni VR, Franki N, Chowdhury JR, Wu CH, et al. Microtubular disruption prolongs the expression of human bilirubin-uridinediphosphoglucuronate-glucuronosyltransferase-1 gene transferred into Gunn rat livers. J Biol Chem 1996; 271(4):2341–2346.

145. Wilson JM, Grossman M, Wu CH, Chowdhury NR, Wu GY, Chowdhury JR. Hepatocyte-directed gene transfer in vivo leads to transient improvement of hypercholesterolemia in low density lipoprotein receptor-deficient rabbits. J Biol Chem 1992; 267(2):963–967.

146. Schuster MJ, Wu GY, Walton CM, Wu CH. Multicomponent DNA carrier with a vesicular stomatitis virus G-peptide greatly enhances liver-targeted gene expression in mice. Bioconjug Chem 1999; 10(6):1075–1083.

147. Goula D, Becker N, Lemkine GF, Normandie P, Rodrigues J, Mantero S et al. Rapid crossing of the pulmonary endothelial barrier by polyethylenimine/DNA complexes. Gene Ther 2000; 7(6):499–504.

148. Zou SM, Erbacher P, Remy JS, Behr JP. Systemic linear polyethylenimine (L-PEI)-mediated gene delivery in the mouse. J Gene Med 2000; 2(2):128–134.

149. Dash PR, Read ML, Barrett LB, Wolfert MA, Seymour LW. Factors affecting blood clearance and in vivo distribution of polyelectrolyte complexes for gene delivery. Gene Ther 1999; 6(4):643–650.

150. Erbacher P, Bettinger T, Belguise VP, Zou S, Coll JL, Behr JP, et al. Transfection and physical properties of various saccharide, poly(ethylene glycol), and antibody-derivatized polyethylenimines (PEI). J Gene Med 1999; 1(3):210–222.

151. Ogris M, Brunner S, Schüller S, Kircheis R, Wagner E. PEGylated DNA/transferrin-PEI complexes: reduced interaction with blood components, extended circulation in blood and potential for systemic gene delivery. Gene Ther 1999; 6(4): 595–605.

152. Wolfert MA, Schacht EH, Toncheva V, Ulbrich K, Nazarova O, Seymour LW. Characterization of vectors for gene therapy formed by self-assembly of DNA with synthetic block co-polymers. Hum Gene Ther 1996; 7(17):2123–2133.

153. Toncheva V, Wolfert MA, Dash PR, Oupicky D, Ulbrich K, Seymour LW, et al. Novel vectors for gene delivery formed by self-assembly of DNA with poly(L-lysine) grafted with hydrophilic polymers. Biochim Biophys Acta 1998; 1380(3): 354–368.

154. Finsinger D, Remy JS, Erbacher P, Koch C, Plank C. Protective copolymers for

nonviral gene vectors: synthesis, vector characterization and application in gene delivery. Gene Ther 2000; 7(11):1183–1192.

155. Dash PR, Read ML, Fisher KD, Howard KA, Wolfert M, Oupicky D, et al. Decreased binding to proteins and cells of polymeric gene delivery vectors surface modified with a multivalent hydrophilic polymer and retargeting through attachment of transferrin. J Biol Chem 2000; 275(6):3793–3802.

156. Lemieux P, Guérin N, Paradis G, Proulx R, Chistyakova L, Kabanov AV, et al. A combination of poloxamers increases gene expression of plasmid DNA in skeletal muscle. Gene Ther 2000; 7(11):986–991.

157. Kircheis R, Wightman L, Schreiber A, Robitza B, Roessler V, Kursa M, et al. Polyethylenimine/DNA complexes shielded by transferrin target gene expression to tumors after systemic application. Gene Ther 2001; 8:28–40.

158. Kircheis R, Ostermann E, Kursa M, Wightman L, Wagner E. Tumor-targeted gene delivery: an attractive strategy to use highly active effector molecules in cancer treatment. Gene Ther 2002; 9:731–735.

6
Chimeric Gene Delivery Systems

Yasufumi Kaneda
Osaka University, Osaka, Japan

I. INTRODUCTION

Further development of effective gene transfer vector systems is key to the promotion of human gene therapy [1]. Numerous viral and nonviral (synthetic) methods for gene transfer have been developed [2,3], and generally viral methods are more efficient than nonviral methods for delivery of genes to cells. However, viral vectors present safety issues because of cointroduction of essential genetic elements from the parent viruses, leaky expression of viral genes, immunogenicity, and alterations of host genomic structure. In general, nonviral vectors are less toxic and less immunogenic than viral vectors. However, most nonviral methods are less efficient for gene transfer, especially in vivo. Thus, both viral and nonviral vectors have limitations as well as advantages. Therefore, to develop an in vivo gene transfer vector with high efficiency and low toxicity, the limitations of one type of vector system should be compensated for by introducing the strengths of another.

One example, such as a chimeric viral vector is the pseudotype retroviral vector. Retroviral envelope proteins are converted to vesicular stomatitis virus (VSV) G protein by recombination of the retroviral genome [4]. The pseudotype vector can fuse with a broad range of cells, because the G protein recognizes phosphatidylserine on the cell surface, and the chimeric envelope is strong enough to permit concentration of viral particles by high-speed centrifugation. By mixing the pseudotype retroviral vector with the human immunodeficiency virus (HIV) vector, a new lentiviral vector has been constructed [5]. Unlike oncoretroviruses such as Moloney murine leukemia virus, which is a popular retroviral vector, HIV-1 can cross the nuclear envelope in nondividing cells and integrate its genome into the genome of the host cells [6]. However, HIV-1 infects only lymphocytes and macrophages owing to the tissue tropism of the receptor molecules. In converting the envelope of the recombinant HIV vector to VSV–G protein, the novel lentiviral vector has been shown to infect many tissues and induce long-term gene expression in nondividing neurons [6].

In nonviral gene delivery systems, various modifications have been made to enhance the efficiency of gene delivery. Liposomes can be used to target and introduce macromolecules into cells by binding to antibody-recognizing cell surface molecules. Using cross-linking reagents [7], antibodies have been coupled to liposomes for the introduction of macromolecules into human erythrocytes [8], human glioma cells [9], and lymphoid cells [10]. Tissue-specific ligands can be used to target liposomes to specific cells. For example, the poly-L-lysine conjugates asialoglycoprotein [11] and transferrin [12] can be used to target hepatocytes and cancer cells, respectively.

Another approach is construction of novel hybrid gene transfer vectors combining viral and nonviral vectors. We constructed a fusigenic viral liposome with a fusigenic envelope derived from the hemagglutinating virus of Japan (HVJ; Sendai virus) [13,14]. In this delivery system, DNA-loaded liposomes are fused with ultraviolet (UV)–inactivated HVJ to form the fusigenic virus–liposome, HVJ–liposome, which is 400–500 nm in diameter. The advantage of fusion-mediated delivery is protection of the molecules in endosomes and lysosomes from degradation.

Similar approaches for conferring viral function to liposomes have been reported [15]. Influenza virus A has been solubilized with detergent and mixed with a cationic lipid, dioleoyldimethylamonium chloride (DODAC), at 30% to form cationic virosomes. Plasmid DNA has been complexed with these cationic virosomes and successfully transferred to cultured cells by low pH-dependent membrane fusion. In HIV envelope proteins, gp120 and gp41 are required for fusion of the virus with the target cell membrane [16]. Schreier et al. [16] developed artificial HIV envelopes by inserting gp120 into liposomes containing lipids similar to HIV envelope components. To insert proteins into liposomes, they constructed a fusion gene between gp120 and the glycosylphosphatidylinositol (GPI) signal of decay-accelerating factor (DAF). The purified chimeric gp120DAF protein inserted spontaneously into the liposomal membrane via the GPI anchor, and liposomes converting gp120DAF bound specifically to CD4-expressing Chinese hamster ovary cells. The receptor-bound liposomes were internalized and recycled in those cells. Thus, new "virosome" vectors possess both the efficient delivery of viral molecules and the reduced toxicity of the liposomes.

II. DEVELOPMENT OF HVJ-LIPOSOMES

We have developed gene delivery system based on liposomes that provides efficient delivery and enhanced gene expression [13,14]. Our basic concept is the construction of novel, hybrid-type liposomes with functional molecules inserted into them. Since fusion proteins are known to provide efficient gene delivery, insertion of fusion protein was sought. We utilized fusion protein of HVJ. HVJ (Fig. 1a), also known as Sendai virus, can fuse with cell membranes and also with liposomes [17]. When UV-inactivated HVJ is incubated with DNA-loaded liposomes for 10 min on ice, the particles associate with each other; fusion of the particles occurs, forming a vesicle with a single uniform membrane. The

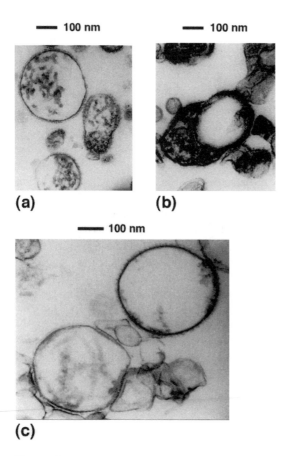

Figure 1 (a) Electron microscopic view showing HVJ, which is approximately 300–400 nm in diameter. The featherlike materials on the envelope are fusion glycoproteins. (b) Ten minutes after incubation of unilammelar DNA-loaded liposomes (right) with HVJ (left) on ice, HVJ and liposomes interact and begin to fuse. (c) Thirty minutes after incubation of HVJ and liposomes at 37°C, fusion of HVJ and liposomes is completed, and a single uniform membrane is formed. When we view the membrane carefully, fusion proteins can be seen on the surface. The HVJ–liposomes are approximately 400–500 nm in diameter.

resulting vesicle, HVJ–liposome, contains fusion proteins on the envelope and DNA inside. The HVJ–liposome is approximately 400–500 nm in diameter. These liposomes can encapsulate DNA smaller than 100 kb, and the trapping efficiency of DNA is approximately 20%. RNA, oligodeoxynucleotides (ODN), proteins, and drugs can also be enclosed and delivered to cells.

HVJ–liposomes are useful for in vivo gene transfer. When HVJ–liposomes containing the *LacZ* gene were injected directly into one rat liver lobe, approximately 70% of cells expressed *LacZ* gene activity, and no pathological hepatic changes were observed [18].

One advantage of HVJ–liposomes is allowance for repeated injections. Gene transfer to rat liver cells was not inhibited by repeated injections. After repeated injections, anti-HVJ antibody generated in the rat was not sufficient to neutralize HVJ–liposomes. Cytotoxic T cells recognizing HVJ were not detected in the rat transfected repeatedly with HVJ–liposomes [18].

III. RECONSTITUTED FUSION LIPOSOMES FOR GENE TRANSFER IN VITRO AND IN VIVO

We have been constructing HVJ–liposomes using inactivated whole HVJ virion. Instead of whole virion, isolated fusion proteins can be used for HVJ–liposomes. HVJ fusion proteins were purified by applying the detergent-lysed HVJ to ion-exchange column chromatography. The 52-kD (F1) and 72-kD (HN) proteins were the dominant proteins eluted in the flow-through fractions [19]. The fusion proteins were mixed with NP-40 solubilized lipid mixture, and the liposomes were prepared by dialysis [20]. The liposomes contained F1 and HN. However, we were unable to trap DNA in these liposomes by dialysis. Fusion particles containing DNA were constructed by incubating empty fusion particles with DNA-loaded liposomes prepared by a vortexing-sonication method [19]. A schema of the construction is shown in Figure 2. These reconstituted fusion liposomes were as effective as conventional HVJ–liposomes with whole virion in terms of delivery of both FITC-ODN and the luciferase gene to cultured cells. LacZ gene was also transferred directly to mouse skeletal muscle in vivo using the reconstituted fusion particles.

IV. IMPROVEMENT OF THE CURRENT VECTOR SYSTEM

The HVJ–liposome gene delivery system has several advantages, but improvement is needed before use in humans. To increase the efficiency of gene delivery, we investigated the lipid components of liposomes [21]. Our conclusion was threefold: The most efficient gene expression occurred with a phosphatidylcho-

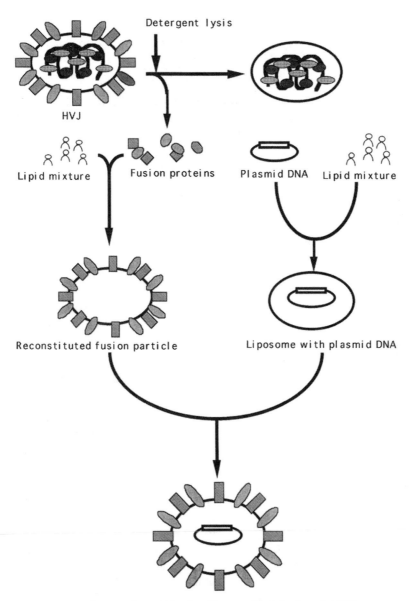

Figure 2   Preparation of reconstituted fusion liposomes. Fusion proteins (HN and F) of
HVJ are isolated by detergent lysis of HVJ particles. The proteins are inserted into lipo-
somes by dialysis to form the reconstituted fusion particles. Liposomes containing plasmid
DNA or FITC–ODN are constructed and fused with the reconstituted fusion particles.
The reconstituted fusion liposomes with plasmid DNA or FITC-ODN are then used as
gene transfer vehicles in vitro and in vivo.

line, phosphatidylethanolamine, and sphingomyelin molar ratio of $1:1:1$; anionic HVJ–liposomes should be prepared using phospatidylserine (PS) as the anionic lipid; and the ratio of phospholipids to cholesterol should be $1:1$. Accordingly, we developed new anionic liposomes called HVJ–AVE liposomes; that is, HVJ–artificial viral envelope liposomes. The lipid components of AVE liposomes are very similar to the HIV envelope and mimic the red blood cell membrane [22]. HVJ–AVE liposomes have yielded gene expression in liver and muscle 5–10 times higher than that observed with conventional HVJ–liposomes [21]. HVJ–AVE liposomes were shown to be very effective for gene delivery to isolated rat heart via the coronary artery. *LacZ* gene expression was observed in the entire heart, whereas expression was not observed with empty HVJ–AVE liposomes [23]. Gene expression efficiency of conventional HVJ–liposomes was approximately 30%.

Another improvement was construction of cationic-type HVJ–liposomes using cationic lipids. Of the cationic lipids, positively charged DC-cholesterol (DC-Chol) [24] has been the most efficient for gene transfer. For luciferase expression, HVJ–cationic DC liposomes were 100 times more efficient than were conventional HVJ–anionic liposomes. Although it has been very difficult to transfer genes to bone marrow and spleen cells using conventional HVJ–liposomes, HVJ–cationic DC-Chol liposomes have been shown to be effective for gene transfer to both types of cells. However, when introduced into mouse muscle or liver, total luciferase expression after transfection with HVJ–cationic liposomes was shown to be 10–150 times lower than that with conventional anionic HVJ–liposomes [21], which were less efficient for in vitro transfection. Furthermore, AVE liposomes were modified further to create AVE+DC-Chol10 (contains 10% phosphatidyl serine and 10% DC-Chol), AVE+DC-Chol20 (containing 10% PS and 20% DC-Chol), and AVE–PS (containing neither PS nor DC-Chol) liposomes. We examined in vivo gene transfection efficiency with these liposomes after conjugation with the HVJ envelope. AVE yielded the highest luciferase expression in liver. AVE–PS and AVE+DC-Chol10 liposomes, which have a net neutral charge, showed intermediate luciferase activities. AVE+DC-Chol20 liposomes, which have an excessive amount of cationic lipid, yielded luciferase activities similar to those of HVJ–DC-Chol liposomes. However, we recently found HVJ–cationic liposomes to be more effective in some cases for in vivo gene transfer. High expression of the *LacZ* gene was obtained in restricted regions of chick embryos after injection of HVJ–cationic liposomes [25], whereas HVJ–anionic liposomes were ineffective. In addition, when HVJ–cationic liposomes containing the *LacZ* gene were administered to rat lung with a jet nebulizer, more efficient gene expression in the epithelium of the trachea and bronchus was observed compared to that found with HVJ–anionic liposomes [26]. HVJ–cationic liposomes were also very effective for antisense (AS)–ODN transfer to a restricted region of rat brain (Saji M, Kabayashi S, Ohno K; personal communication), whereas AS-ODN was broadly distributed in the brain by HVJ–anionic liposomes.

Cationic HVJ–liposomes   Anionic HVJ–liposomes

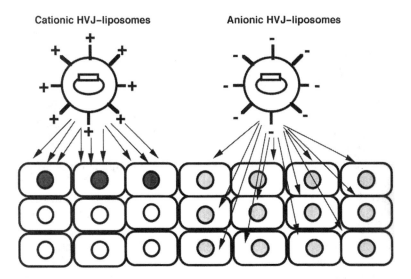

Figure 3   Schema of gene transfer to organs by HVJ–anionic liposomes (right) or HVJ–cationic liposomes (left). It is thought that HVJ-anionic liposomes can better penetrate tissues and distribute to broader areas than cationic liposomes. These HVJ–anionic liposomes gradually fuse with many cells to transfer DNA. However, the level of gene expression per cell is much lower than that observed with HVJ–cationic liposomes. HVJ–cationic liposomes cannot penetrate tissues and remain localized at the injection sites. Therefore, gene expression is detected only at the surface or restricted areas of tissues. Gene expression per cell is much higher with HVJ-cationic liposomes than that with HVJ–anionic liposomes.

Therefore, we conclude that anionic HVJ–liposomes should be used for gene transfer to broad areas of tissues, and that HVJ–cationic liposomes should be used for gene transfer to restricted regions or surface areas (Fig. 3). HVJ–anionic liposomes and HVJ–cationic liposomes can complement each other, and each liposome should be used for proper targeting. Thus, it is difficult to optimize in vivo transfection strategies with in vitro experiments.

V.   IMPROVEMENTS OF GENE EXPRESSION

A.   Sustained Gene Expression In Vitro and In Vivo by the EBV Replicon Vector Coupled with HVJ-Liposomes

In the development of chimeric systems, modification of envelope components has been the primary focus. However, in a broader sense, the chimeric system can be applied to the gene-expression apparatus. One major limitation of in vivo liposome-based gene delivery is transient gene expression. The reason for the short-term gene expression is not fully understood. One possibility is degradation of plasmid DNA in the nucleus [27], and another is silencing of target gene expression in target cells [28].

There are two different approaches for long-term expression of transgenes in vivo. One is insertion of the transgene into the host genome, but insertion must be site specific, and the efficiency is still extremely low. The other is stable retention of the transgene extrachromosomally. For stable retention, a latent viral infection apparatus system can be utilized.

Epstein–Barr virus (EBV) can induce latent infection in human lymphoid cells [29]. The latent viral DNA replication origin sequence (oriP) and EBV nuclear antigen-1 (EBNA-1) protein are necessary for latent infection by the virus [30]. These two components induce autonomous replication of the plasmid in primate cells and episomal retention of the plasmid in the nucleus [31–33]. For long-term gene expression *in vivo*, we constructed an EBV replicon–based plasmid, pEB, that contains oriP and EBNA-1 [34].

Luciferase gene expression was observed in cultured human HEK293 cells for at least 10 days, whereas expression was reduced to very low levels after transfection with a conventional plasmid lacking the EBV sequence. Approximately 20% of the episomal DNA in HEK293 cells transfected with an EBV replicon vector replicated autonomously. In rodent BHK-21 cells, approximately 100–150 copies of the EBV replicon vector were retained in the nucleus on day 9, and then decreased between days 9 and 14. Without autonomous replication in rodent cells, EBV replicon vector can be retained in the nucleus. Nuclear retention of the EBV genome results from the association of the oriP sequence with the nuclear matrix [35], and oriP-dependent transcription can be activated by binding of EBNA-1 with oriP [36]. The disadvantage of the EBV replicon vector is the unequal delivery of the plasmid to daughter cells after cell division due to the lack of centromere sequence. Therefore, the EBV replicon vector appears to be more effective for long-term gene expression in nondividing tissue cells.

The EBV replicon vector was transfected into mouse liver and skeletal muscle using HVJ–AVE liposomes. With the EBV sequence, luciferase expression was observed for at least 35 days in liver and 96 days in muscle, whereas without the EBV sequence significant expression was not observed by day 14 in liver and by day 28 in muscle [14,34]. With HVJ–AVE liposomes carrying the EBV replicon vector, gene expression lasted more than 9 weeks in rat kidney, whereas it disappeared by 2 weeks with conventional plasmid (Isaka Y, Imai E, Tsujie M, unpublished data).

B. Enhancement of Transgene Expression by EBNA-1

It has been suggested that EBNA-1 acts as a transcriptional enhancer at the oriP site. If oriP-binding capacity for EBNA-1 is not saturated, more EBNA-1 can bind and enhance expression of sequences linked to oriP. We simultaneously introduced an EBNA-1 expression vector and a plasmid harboring the oriP sequence into cultured cells and mouse organs using HVJ–liposomes [37]. When a plasmid, pCMV–EBNA-1, which gives high expression of EBNA-1, was co-introduced with an oriP-harboring plasmid, poriP–CMV–luciferase, luciferase gene expression was at most 30 times greater than that without EBNA-1. This enhancement was regulated at the transcriptional level and was dependent upon both the oriP sequence and the amount of EBNA-1. We then cointroduced poriP–CMV–human proinsulin (oriP–HIN) gene with pCMV–EBNA-1 into mouse skeletal muscle or liver with HVJ–liposomes. The mean level of human proinsulin in mouse serum was 5–10 times greater than the level without pCMV–EBNA-1 (Fig. 4a). Gene expression was sustained for more than 4 weeks by the cotransfer into muscle; it disappeared by 2 weeks without EBNA-1. When the plasmid were cointroduced into mouse liver, human proinsulin was not detected at 15 days. With secondary transfer of pCMV–EBNA-1 alone into the liver, the human proinsulin level was elevated again and detected for 15 additional days, whereas no reactivation occurred with pcDNA-3 transfer (Fig. 4b). These results indicate that only the oriP of EBV is sufficient for retention of a plasmid in cells,

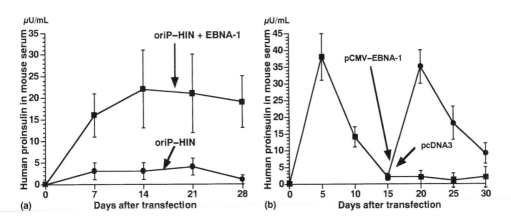

Figure 4 Human proinsulin levels in mouse sera. (a) Approximately 3.8 μg of poriP–CMV–HIN (oriP-HIN) was cotransfected with 12 μg of pCMV–EBNA-1 (EBNA-1) (■) (n = 4) or 8.6 μg of pcDNA3 (●) (n = 4) by HVJ–anionic liposomes into the quadriceps. At 1, 2, 3, and 4 weeks after transfer, the human proinsulin level (μU/ml) was measured by ELISA. (b) Approximately 7.6 μg of poriP–CMV–HIN was transferred directly with 24 μg of pCMV–EBNA-1 (●) (n = 6) to one lobe of the liver. On day 15 after transfer, 24 μg of pCMV–EBNA-1 alone (●) (n = 3) or 17.2 μg of pcDNA3 alone (■) (n = 3) was transferred to one lobe of the liver by HVJ-anionic liposomes. The mean value and standard deviation of samples from four mice are indicated to each time point. A concentration of 1 μU/mL corresponds to 40.67 pg of human proinsulin/mL of mouse serum.

and EBNA-1 enhances transgene expression that is dependent upon oriP. The data also demonstrate that expression of oriP-harboring plasmid DNA tends to be shut off in cells, and EBNA-1 can reactivate expression. Finally, oriP-harboring plasmid bound to EBNA-1 can escape from silencing transgene expression in cells. These findings will be helpful for analyzing the mechanism of regulation of transcription at the chromatin level.

SUMMARY

Numerous novel gene therapy vectors can be produced by combining both viral and non-viral components to solve difficulties encountered in human gene therapy. The possibilities for chimeric systems appear to be unlimited.

REFERENCES

1. Marshall E. Gene therapy's growing pains. Science 1995; 269:1052–1055.
2. Mulligan RC. The basic science of gene therapy. Science 1993; 260:926–932.
3. Ledley F D. Non-viral gene therapy: the promise of genes as pharmaceutical products. Hum Gene Ther 1995; 6:1129–1144.
4. Burns JC, Friedmann T, Driever W, Burrascano M, Yee JK. Vesicular stomatitis virus G glycoprotein pseudotyped retroviral vectors: concentration to very high titer and efficient gene transfer into mammalian and nonmammalian cells. Proc Natl Acad Sci USA 1993; 90:8033–8037.
5. Naldini L, Blomer U, Gallay P, Ory D, Mulligan R, Gage F, Verma I, Trono D. *In vivo* gene delivery and stable transduction of nondividing cells by a lentivirus vector. Science 1996; 272:263–267.
6. Weinberg JB, Matthews TJ, Cullen BR, Malim MH. Productive human immunodeficiency virus type 1 (HIV-1) infection of non-proliferating human monocytes. J Exp Med 1991; 174:1477–1482.
7. Martin FJ, Papahadjopoulos D. Irreversible coupling of immunoglobulin fragments to preformed vesicles. J Biol Chem 1982; 257:286–288.
8. Leserman LD, Barbet J, Kourilsky, F. Targeting to cells of fluorescent liposomes covalently coupled with monoclonal antibody or protein A. Nature 1980; 288:602–604.
9. Mizuno M, Yoshida J, Sugita K, Inoue I, Seo H, Hayashi Y, Koshizaka T, Yagi K. Growth inhibition of glioma cells tranfected with human beta-interferon gene by liposomes coupled with a monoclonal antibody. Cancer Res 1990; 50:7826–7829.
10. Machy P, Lewis F, McMillan L, Jonak, ZL. Gene transfer from targeted liposomes to specific lymphoid cells by electroporation. Proc Natl Acad Sci USA 1988; 85: 8027–8031.
11. Wu GY, Wu CH. Receptor-mediated gene delivery and expression in vivo. J Biol Chem 1988; 263:14621–14624.

12. Zenke M, Steinlein P, Wagner E, Cotten M, Beug H, Birnstiel, ML. Receptor-mediated endocytosis of transferrin-polycation conjugates: an efficient way to introduce DNA into hematopoietic cells. Proc Natl Acad Sci USA 1990; 87:3655–3659.
13. Kaneda Y. Fusigenic Sendai-virus liposomes: a novel hybrid type liposome for gene therapy. Biogenic Amines 1998; 14:553–572.
14. Kaneda Y, Saeki Y, Morishita R. Gene therapy using HVJ-liposomes; the best of both worlds. Mol Med Today 1999; 5:298–303.
15. Schoen P, Chonn A, Cullis PR, Wilschut J, Scherrer P. Gene transfer mediated by fusion protein hemagglutinin reconstituted in cationic lipid vesicles. Gene Ther 1999; 6: 823–832.
16. Schreier H, Mora P, Caras IW. Targeting liposomes to cells expressing CD4 using glycosylphosphatidylinositol-anchored gp120. Influence of liposome composition on intracellular trafficking. J Biol Chem 1994; 269:9090–9098.
17. Okada Y. Sendai-virus induced cell fusion. In: Duzgunes N, ed. Methods in Enzymology. Vol 221. San Diego: Academic Press, 1993: 18–41.
18. Hirano T, Fujimoto J, Ueki T, Yamamoto H, Iwasaki T, Morishita R, Sawa Y, Kaneda Y, Takahashi H, Okamoto E. Persistent gene expression in rat liver in vivo by repetitive transfections using HVJ-liposome. Gene Ther 1998; 5:459–464.
19. Suzuki K, Nakashima H, Sawa Y, Morishita R, Matsuda H, Kaneda Y. Reconstituted fusion liposomes for gene transfer in vitro and in vivo. Gene Ther Reg 2000; 1:65–77.
20. Uchida T, Kim, J, Yamaizumi M, Miyake Y, Okada Y. Reconstitution of lipid vesicles associated with HVJ (Sendai virus) spikes: purification and some properties of vesicles containing nontoxic fragment A of diphtheria toxin. J Cell Biol 1979; 80: 10–20.
21. Saeki Y, Matsumoto N, Nakano Y, Mori M, Awai K, Kaneda Y. Development and characterization of cationic liposomes conjugated with HVJ (Sendai virus): reciprocal effect of cationic lipid for in vitro and in vivo gene transfer. Hum Gene Ther 1997; 8:1965–1972.
22. Chander R, Schreier H. Artificial viral envelopes containing recombinant human immunodeficiency virus (HIV) gp160. Life Sci 1992; 50:481–489.
23. Sawa Y, Kaneda Y, H-Z Bai, Suzuki K, Fujimoto J, Morishita R, Matsuda H. Efficient transfer of oligonucleotides and plasmid DNA into the whole heart through the coronary artery. Gene Ther 1998; 5:1472–1480.
24. Goyal K, Huang L. Gene therapy using DC-Chol liposomes. J Liposome Res 1995; 5:49–60.
25. Yamada G, Nakamura S, Haraguchi R, Sakai M, Terashi T, Sakisaka S, Toyoda T, Ogino Y, Hatanaka H, Kaneda Y. An efficient liposome-mediated gene transfer into the branchial arch, neural tube and the heart of chick embryos: a strategy to elucidate organogenesis. Cell Mol Biol 1997; 43:1165–1169.
26. Yonemitsu Y, Kaneda Y, Muraishi A, Yoshizumi T, Sugimachi K, Sueishi K. HVJ (Sendai virus)-cationic liposomes: a novel and potentially effective liposome-mediated gene transfer technique to the delivery to the airway epithelium. Gene Ther 1997; 4:631–638.
27. Kaneda Y, Iwai K, Uchida T. Introduction and expression of the human insulin gene in adult rat liver, J Biol Chem 1989; 264:12126–12129.

28. Yamano T, Ura K, Morishita R, Nakajima H, Monden M, Kaneda Y. Amplification of transgene expression in vitro and in vivo using a novel inhibitor of histone deacetylase. Mol Ther 2000; 1:425–431.

29. Lindahl T, Adams A, Bjursell G, Bomkamm GW, Kaschka-Dierich C, Jehn U. Covalently closed circular duplex DNA of Epstein-Barr virus in a human lymphoid cell line. J Mol Biol 1976; 102:511–530.

30. Lupton S, Levine A J. Mapping genetic elements of Epstein-Barr virus that facilitate extrachromosomal persistence of Epstein-Barr virus derived plasmids in human cells, Mol Cell Biol 1985; 5:2533–2542.

31. Yates JL, Warren N, Sugden B. Stable replication of plasmids derived from Epstein-Barr virus in various mammalian cells. Nature 1985; 313:812–815.

32. Kelleher ZT, Fu H, Elizabeth L, Wendelburg, B, Gulino, S, Vos J-M. Epstein–Barr-based episomal chromosomes shuttle 100 kb of self-replicating circular human DNA in mouse cells. Nat Biotechnol 1988; 16:762–768.

33. Wendelburg BJ, Vos J-M. An enhanced EBNA-1 variant with reduced IR3 domain for long-term episomal maintenance and transgene expression of oriP-based plasmids in human cells. Gene Ther 1998; 5:1389–1399.

34. Saeki Y, Wataya-Kaneda M, Tanaka K, Kaneda Y. Sustained transgene expression in vitro and in vivo using an Epstein-Barr virus replicon vector system combined with HVJ-liposomes. Gene Ther 1998; 5:1031–1037.

35. Jankelevich S, Kolman JL, Bodnar JW, Miller G. A nuclear matrix attachment region organizes the Epstein-Barr viral plasmid in Raji cells into a single DNA domain. EMBO J 1992; 11:1165–1176.

36. Wysokenski DA, Yates JL. Multiple EBNA-1 binding sites are required to form an EBNA-1-dependent enhancer and to activate a minimal replication origin within oriP of Epstein-Barr virus. J Virol 1989; 63:2657–2666.

37. Kaneda Y, Saeki Y, Nakabayashi M, Zhou W-Z, Wataya-Kaneda M, Morishita, R. Enhancement of transgene expression by cotransfection of oriP plasmid with EBNA-1 expression vector. Hum Gene Ther 2000; 11:471–479.

7

Adenoviral Vectors for Gene Delivery

Jonathan L. Bramson
McMaster University, Hamilton, Ontario, Canada

Robin J. Parks
Ottawa Hospital Research Institute, Ottawa, Ontario, Canada

I. INTRODUCTION

One of the major hurdles to successful gene therapy of genetic and acquired disease is the ability to introduce a foreign gene efficiently into the tissue of interest and, in the case of genetic diseases, achieve long-term transgene expression. Adenoviruses (Ads) have many attractive features which have made them a popular vehicle for gene transfer, including [1] well-defined biology, [2] the capacity to accommodate foreign DNA inserts up to 36 kb, and [3] the ability to infect a wide variety of cell types, tissues, and species in a cell cycle–independent fashion. Perhaps the most important quality of Ads is the fact that they are relatively safe, and are not associated with severe disease in immunocompetent individuals.

This describes recent advances in the clinical application of first-generation Ads and the development of vectors with further deletions of viral sequences (next-generation), including the new class of Ad-based vectors termed fully deleted (gutted) adenoviral vectors (fdAd). The fdAd vectors are deleted of all Ad protein coding sequences, which, in part, has circumvented the problem of short-term transgene expression previously associated with more traditional Ad vectors. As will be shown, Ad vectors do hold great promise as delivery vehicles for gene therapies, but the vector design is critical to successful application.

II. BIOLOGY OF ADENOVIRUSES

Adenoviruses were first discovered in the early 1950s as novel viral agents associated with respiratory ailments in human patients [1,2], and their name derives from the original source of tissue from which the prototype member was isolated. Since that time over 100 family members have been identified and characterized from a wide variety of mammalian and avian species. All Ads have the same general structural characteristics: an icosohedral, nonenveloped capsid (~70–100 nm in diameter) surrounding a nucleoprotein core containing a linear double-stranded genome (~30–40 kb). Of the human Ads, serotypes 2 (Ad2) and 5 (Ad5), both of subclass C, are the most extensively characterized (reviewed in Ref. 3). Their genomes have been sequenced, and are ~95% identical at the nucleotide level, with a similar arrangement of transcriptional units. Ad5 is approximately 36 kb, and encodes genes that are divided into early and late viral functions, depending on whether they are expressed before or after DNA replication (Fig. 1). In general, the early transcription units (E1a, E1b, E3, and E4) encode proteins required for transactivating other viral regions or modifying the host cellular or immunological environment. E2 encodes proteins directly involved in viral DNA replication. The late transcription units, L1–L5, are expressed from a common major late promoter and are generated from alternative

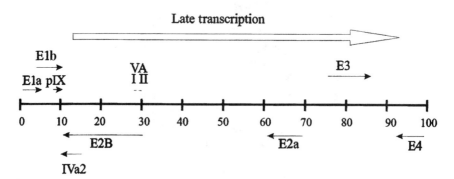

Figure 1 Simplified transcription map of human adenovirus serotype 5 (Ad5). The Ad5 genome, shown as 100 map units (~36 kb), is divided into four early transcription units, E1–E4, that are expressed before DNA replication and five late transcription units, L1–L5 (not shown), which are expressed after DNA replication and are alternative splice products of a common late transcript. Four smaller transcipts are also produced: pIX, IVa2, and VA RNAs I and II. Not shown are the viral inverted terminal repeats, approximately 100 bp at each end of the viral genome, and packaging signal, located from nucleotides 190 to 380 at the left end of the genome (according to the standard Ad map), which are cis-acting elements involved in Ad DNA replication and packaging, respectively.

splicing of a single transcript. The late transcripts generally encode virion structural proteins. Four other small transcripts are also produced: pIX (encoding a minor structural protein), IVa2 (activator of the major late promoter), and VA RNA I and II (block activation of the interferon response).

Viral infection initiates with the Ad fiber protein binding to a specific receptor on the cell surface [4], which acts as a common receptor for both Coxsackie B virus and Ad, followed by a secondary interaction between Ad penton and $\alpha_v\beta_3$ and $\alpha_v\beta_5$ integrins [5]. The efficiency with which Ad binds to and enters cells is directly related to the level of the primary and secondary receptors found on the cell surface [6,7]. Ad is internalized by endocytosis, triggered by the penton–integrin interaction, and escapes from the early endosome prior to formation of the lysosome [8,9]. The virion translocates to the nucleus along the microtubular network [9], during which time there is a sequential disassembly of the Ad virion and, as a final step, Ad hexon remains at the nuclear membrane while the DNA is released into the nucleus [10]. Viral DNA replication and assembly of progeny virions occur within the nucleus of infected cells, and the entire life cycle takes about 24 hr with an output of approximately 10^4 virions per cell. In humans, Ads are not associated with any neoplastic disease, and only cause relatively mild, self-limiting illness in immunocompetent individuals, primarily respiratory illnesses, keratoconjunctivitis, or gastroenteritis (depending on the serotype). For a more comprehensive discussion of Ad biology, the reader is referred to an excellent review by Shenk [3].

III. FIRST-GENERATION Ad VECTORS

The early versions of Ad vectors had modest mutations: deletion of the E1 and/ or the E3 regions. The E3 sequences, which are dispensable for viral propagation in vitro, can be removed and replaced with foreign DNA, resulting in a replication-competent virus capable of producing high levels of the protein encoded by the foreign sequences. Since Ad is a human pathogen, replication competent virus would be expected to elicit a strong inflammatory/immune response, which could greatly limit its value for effective gene transfer, in addition to the risk that replication-competent virus may produce disease in human hosts. As such, gene therapists have focused their attention on replication-deficient ads (RDAs). RDAs can be generated by deletion of the E1 region, which encodes the necessary elements to initiate viral replication. Removal of E1 allows for the insertion of approximately 4.7 kb of foreign DNA which can be further increased to approximately 8 kb by deleting the E3 sequences as well [11,12]. Ad vectors lacking E1 must be propagated on special cell lines that provide the functions of the E1 proteins in trans. From this point forward, we will refer to viruses with deletions of E1 function as "first-generation" adenoviral vectors. The presence or absence of E3 will not be considered, because although these proteins may provide some immunosuppressive qualities, the relative importance of this region remains controversial [13–16]. It was initially thought that, in the absence of replication, these vectors would be safe and weakly immunogenic, providing high-efficiency, long-term gene expression. However, it has been discovered through preclinical and clinical studies that replication is only one of several viral features which must be inactivated to produce a vector for gene transfer in vivo.

A. Construction and Propagation of First-Generation Viruses

Although the full-length Ad genome can be captured in a plasmid form which yields infectious virus following transfection into permissive cell lines, the large size of these plasmids (~40 kb) makes standard cloning procedures difficult [17]. Several methods for viral construction have been developed which benefit from the natural recombination pathways present in prokaryotic and eukaryotic cells. One of the earliest systems for creating first-generation Ad employed the recombination pathway in mammalian cells following the transfection of two DNAs: (1) a plasmid containing the left-hand portion of the virus (including the packaging signal and 5′ITR) with an appropriate expression cassette replacing the E1 sequences and (2) the right-hand portion of the viral genome (either purified viral DNA digested with Xba I/ClaI to remove the packaging signal or a plasmid containing a circularized genome)[18]. Recombination between homologous Ad sequences on the two plasmids leads to the formation of recombinant viral genomes. Unfortunately, this approach was not found to be highly efficient. Recently, this system has been improved 100-fold by using a site-specific recombinase (Cre) to mediate recombination between the two plasmids instead of the natural homologous recombination pathways in mammalian cells (Fig. 2) [19]

Alternatively, a method has been developed using homologous recombination in bacteria which is more efficient than the mammalian system [20]. In this case, bacteria are transformed with two plasmids:(1) a shuttle plasmid similar to the one used for rescue in mammalian cells and (2) a circularized Ad genome with a deletion of the entire 5′region, including the 5′ITR, packaging signal and E1 region. The advantage of this design is that recombinants are identified by a conversion of antibiotic resistance from ampicillin (parental circularized Ad) to kanamycin (recombinant circularized Ad). By screening recombinants in *Escherichia coli* rather than mammalian cells, the length of time required to identify the appropriate virus is greatly reduced. Once the correct viral genome is identified, the plasmid can be isolated, linearized (Ad ITRs must be present at the termini of a linear molecule for efficient DNA replication), and transfected into an E1-complementing cell line, where it will generate recombinant Ad. A recent enhancement to this approaches utilizes a cosmid-based vector [21]. Rather than

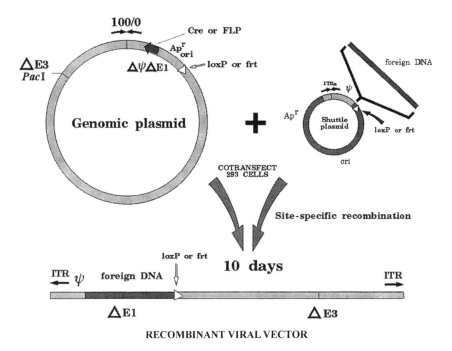

Figure 2 Schematic outline of the AdMax system. The gene of interest is cloned into a shuttle plasmid containing the Ad5 packaging signal and both 5′ and 3′ ITRs. The shuttle plasmid is cotransfected with a genomic plasmid which lacks a packaging signal and contains an expression cassette encoding the Cre recombinase. Cre mediates recombination between loxP sites contained in both the shuttle plasmid and genomic plasmid facilitating the generation of viral genomes. (Courtesy of F. Graham and modified from Ref. 19.)

depending on homologous recombination, the shuttle plasmid is ligated directly to the cosmid prior to packaging in λ particles. The shuttle plasmid provides both the left cosmid arm (necessary for packaging into phage particles) and the lacZ α-peptide allowing identification of appropriate recombinant through classic "blue/white" selection in bacteria. Again, plasmids containing the appropriate recombinants can be verified through restriction analysis, isolated, linearized and transfected into E1-complementing mammalian cells to produce recombinant adenovirus.

Regardless of the method used for generating the recombinant first-generation vector, the absence of E1 in the viral genome must be complemented to propagate the virus. Originally, researchers used a cell line called 293 that was derived from human embryonic kidney cells transformed with the left-hand portion of Ad5 (including the E1 region) [22]. Unfortunately, it has been demonstrated that, over time first-generation vectors propagated on 293 cells can recombine with the E1 sequences contained within the cell line, thus transferring E1 to the vector and yielding replication-competent adenoviruses (RCAs) [23]. As discussed previously, RCA is undesirable for human applications, and alternative cell lines have been created for propagating Ads which prevent the formation of RCA. Cell lines such as PER.C6, GH329, and AE25 have been generated using heterologous promoter systems minimizing the amount of adenoviral DNA required to provide the coding sequences of E1 proteins [24–26]. No RCAs have been detected when these new cell lines are paired with viruses that have matching deletions (i.e., no overlapping sequences between the viruses and the cell line).

B. Applications of First-Generation Viruses

1. Genetic Disease

Owing to the natural tropism of Ad for the lung epithelium, initial studies focused on the use of this virus for lung gene therapy. Delivery of the *CFTR* gene, mutated in patients suffering from cystic fibrosis (CF) seemed an obvious candidate. Although animal studies provided some promising results, it became clear that the recombinant virus could still elicit a strong inflammatory response despite being replication deficient [27]. In clinical studies, initial trials focused on delivering the virus directly to the nasal epithelium and measuring gene transfer, viral shedding, and evidence of chloride transport [28,29]. Although gene transfer was measurable in most patients, the efficiency of gene transfer and correction of chloride transport was low. Nonetheless, Ad delivery was well tolerated. In studies where the virus was administered to the bronchial epithelium, adverse effects were noted in a patient receiving the highest dose of virus (2×10^9 pfu) [30–32]. This quantity of virus is not particularly high, considering that most murine models employ between 10^8 and 10^9 pfu of virus. Thus, it appeared further modifications of the virus were necessary to achieve results in humans comparable to the results in animals. Although the poor outcome of the clinical trials was disappointing with respect to the likelihood of curing CF using first-generation vectors, these studies provided enormous lessons in the fundamentals of practical gene therapy.

Another application for Ad in gene-replacement strategies was as a biological source of circulating proteins such as factor IX or erythropoeitin. To this end, two tissue targets were chosen for infection: the liver and the skeletal muscle. In murine models, Ad vectors transduce the liver with high efficiency, and this organ represents an excellent source of plasma proteins given its access to the systemic circulation. As with the lung, gene delivery to the liver using first-generation vectors resulted in toxicity and immune responses, which limited the duration of gene expression [33,34]. The death of a subject in a clinical trial for gene replacement in patients suffering from ornithine transcarbamylase deficiency has reiterated our limited understanding of Ad vector biology in humans [35]. However, it should be noted that this patient received a remarkably high dose of vector (10^{13} vector particles) administered directly into the liver through the portal vein. These results should not provoke investigators to abandon Ad gene therapy, but they do indicate that further refinement of the technology is required, and great caution should be used when administering vector directly into the liver.

Gene transfer to the muscle successfully yielded transient production of serum factors such as human erythropoeitin [36]. Interestingly, it was discovered that if the murine erythropoeitin cDNA was used instead of the human cDNA in a murine model, the duration of gene expression was greatly extended, whereas expression of foreign genes such as β-galactosidase by first-generation Ad was transient, which is, similar to the results with the human erythropoeitin, [37,38]. These results established another important fundamental of gene therapy: Gene expression could be limited by immunity against the transgene in addition to host responses against the virus. Furthermore, first-generation vectors may be of value for muscle gene therapies provided the gene they were carrying was fully homologous with the host.

2. Cancer Gene Therapy

Although the short-term gene expression provided by first-generation vectors does not seem sufficient for gene-replacement strategies, short-term expression is ideal for applications which only require a brief exposure to gene product. Cytokine immunotherapy of cancer has been limited by the extremely toxic side effects of systemic cytokine exposure. Ad vectors could be used to deliver cytokine genes directly to the tumor site producing high levels of protein within the tumor microenvironment and limiting exposure to the circulatory system. In pre-clinical models, Ads expressing interleukin (IL)-2, IL-4, IL-6, IL-7, and IL-12 and interferon(IFN)–α, IFN-β, and IFN-γ have demonstrated activity against established tumors [39–50]. Clinical trials employing direct intratumoral injection of Ad–IL-2 demonstrated gene transfer in the majority of tumors despite the existence of antibodies against Ad [51]. Most importantly, the therapy was well tolerated. Unfortunately, these studies are at the early stage, so it is difficult to assess whether the treatments provided a clinical benefit.

Ad vectors have also been used for suicide gene therapy with the herpes simplex virus–TK/ganciclovir (HSV-TK/GCV) or E. coli cytosine deaminase/5-fluorocytosine (CdA/5-FC) combinations. In theory, by delivering Ad directly to the tumor, conversion of the prodrugs to the toxic form should occur only within the site of the tumor creating a nontoxic form of chemotherapy. The Ad/suicide gene therapy approach has proven to be effective in murine models of established tumors [52]. Tumor regression may be dependent upon an effective immune response, and the strong immunogenicity of Ad may enhance that effect [53–56]. However, Ad can disseminate from the tumor into peripheral tissues, such as the liver, eliciting dose-limiting toxicities [57–59]. Intralesional administration of AdHSV-TK was well tolerated in patients but, as in previous studies, a number of individuals displayed adverse effects, including transient hepatotoxicity [60,61]. Further modifications of the vector to limit gene expression using tumor-specific promoters or by limiting the tropism of the virus through targeting (see below) should reduce toxic side effects and improve the specificity of this approach.

The genetic alterations in tumors lead to the activation of cellular oncogenes and inactivation of tumor suppressor genes. Tumor-suppressor genes, like Rb or p53, can mediate a dominant growth inhibitory function and represent interesting candidates for cancer gene therapy, because expression of these proteins would be expected to attenuate growth of tumor cells but not affect normal tissues. Although it might be expected that long-term gene expression is essential for the success of tumor-suppressor therapy, transient expression of wild-type p53 in a tumor cell will lead to apoptosis, so long-term expression may not be critical [62]. Several clinical studies have evaluated the utility of Ad–p53 therapy in non–small cell lung cancer and squamous cell carcinoma of the head and neck [63–65]. Clinically, the Ad vector was delivered either via bronchoscope or percutaneously using a computed tomographic (CT) scan as a guide. This treatment was generally well tolerated, although many patients exhibit transient fever. Recombinant vector was detectable in sputum and urine; howevever, viral shedding was not associated with fever or liver damage, so it is likely that this is merely excess virus from the initial instillation rather than from replication of vector. Most interestingly, objective responses were observed in a number of instances suggesting therapeutic effect [63]. Of course, these are early-stage trials and further evaluation is necessary. Cyclin inhibitors have also been used to suppress cell cycle progress in cancer gene therapies employing Ads [66–68]. Alternatively, investigators have employed proapoptotic proteins such as bax to promote cell death in tumors [69,70]. Again, the success of strategies involving the expression of proteins that may be toxic to nontumor cells will depend upon limiting gene expression to the tumor.

A novel approach to cancer gene therapy using first-generation Ad is the development of oncolytic variants [71,72]. The E1 gene products are essential for binding the cellular tumor-suppresor genes p53 and Rb to permit viral replication. Deletion of E1a or E1b attenuates viral replication in normal cells with functional Rb and p53, respectively. However, in cancer cells where p53 and Rb are often mutated during the course of oncogenesis, these viruses are capable of productively replicating and ultimately lysing the cell. Thus, investigators are currently evaluating the benefit of treating cancer with oncolytic, first-generation Ads which selectively replicate in tumor cells but not in normal cells. One such virus, Onyx-015, has progressed to clinical trial [73–75]. Onyx-015 lacks the E1b proteins and, therefore, is unable to replicate in cells that express wild-type p53. In theory, Onyx-015 is expected to be replication deficient in normal cells and replication competent in tumor cells with p53 mutations. However, the actual mechanism of action for this virus may be more complex [76]. Treatment of patients with head and neck cancer using Onyx-015 was well tolerated but flulike symptoms were observed in a number of patients [73]. No objective responses were observed in this early trial, but the investigators were able to demonstrate viral replication in 4 of 22 biopsy samples and each replication-positive biopsy also had mutated p53. In preclinical studies, investigators have increased the potency of Onyx-015 by inserting coding sequences for the HSV-TK gene, creating a bivalent therapeutic [77,78]. However, there will be greater concern for tight regulation of HSV-TK expression in this system, because increased viral dissemination can be expected using a conditionally replicating virus rather than a fully replication-deficient vector.

3. Other Applications

The experience in cancer gene therapy using antiproliferative strategies has been extended into another hyperproliferative disease—restenosis. Using balloon-catheter technology, a methodology has been developed to deliver Ad vectors concomitant with angioplasty. In this setting, the vascular smooth muscle cells, which may multiply following angioplasty and reduce blood flow through the vessel lumen (restenosis), are transduced with either a prodrug-converting enzyme (HSV-TK) or a suppressor of cell growth (Rb) resulting in growth inhibition and diminution of restenosis [79–85]. Ads have also been used for a genetic approach for the treatment in coronary artery disease. Viruses expressing vascular endothelial growth factor (VEGF), a potent angiogenic agent, were delivered to the heart by direct injection into the myocardium as a means of increasing blood flow [86]. This approach will be particularly useful for individuals with multiple obstructions who are not candidates for coronary bypass. Clinically, no significant adverse effects were observed in patients treated by this approach and there was evidence of clinical efficacy (improved oxygenation and reduced angina)[87,88].

Another exciting opportunity for Ads is preoperative modification of transplant tissue. Using genes encoding immunosuppressive factors, it may be possible to delay tissue rejection and enhance graft survival. Ads can efficiently transduce tissues ex vivo under conditions compatible with transplantation in preclinical models [89–93]. Most work has focused on the fusion molecule CTLA4-Ig, which prevents T-cell activation by blocking the B7 family of costimulatory molecules on antigen-presenting cells [94–97]. Other molecules that have also been used include complement receptor 1, cellular IL-10, viral IL-10, viral IL-10, and the p40 subunit of IL-12 [98–104]. Reperfusion injury, which can result from resuming blood flow within the ischemic transplanted organ, is another factor limiting graft survival and delivery of the genes encoding the Cu/Zn superoxide dismutase or heat shock protein 70 (HSP70) via first-generation Ad vectors have been shown to reduce ischemia–reperfusion damage [105–107].

C. Limitations of First-Generation Vectors

As mentioned previously, first-generation vectors can still induce substantial inflammation despite being replication deficient. Inactivated virus has been shown to be capable of initiating the production of chemokines and proinflammatory molecules from transduced cells [108–110]. Therefore, inflammation appears to be a consequence of viral treatment independent of the expression of viral genes, suggesting that Ad vectors will produce limiting inflammation regardless of genetic manipulation. However, in vivo studies have indicated that by deleting all vector sequences (as will be discussed below) or by including inhibitors of inflammation (IκB), it is possible to reduce the inflammatory response and extend gene expression [111].

In animal models, inflammation induced by the virus does not appear to be limiting to the duration of transgene expression, because gene expression can be extended by several months using T-cell–deficient animals; implicating acquired immunity as the major limiting factor in long-term gene expression [27,112–117]. Both viral proteins and therapeutic proteins were found to be targets for immune attack. Surprisingly, despite lack of E1, viral proteins are expressed on first-generation vectors at levels sufficient to elicit a T-cell response [34,118–120]. Further manipulation of the viral backbone should succeed in attenuating expression of all viral proteins. Unfortunately, since gene-replacement strategies rely on introducing a functional gene to replace a defective one, the therapeutic gene will often be recognized as foreign by the host. A recent report has suggested that by sequestering gene expression to the tissue of interest (i.e., using tissue-specific promoters), the immune response could be bypassed [121]. Thus, although first-generation Ads were less successful than anticipated for long-term gene expression in vivo, for many applications, particularly in cancer, first-generation vectors have proven to be quite valuable.

IV. SECOND-GENERATION Ad VECTORS

To improve the utility of Ad vectors for gene therapy applications, researchers have further modified the virus by introducing additional mutations or deletion in the E2 or E4 regions [122–132], both of which are required for normal viral replication, generating second-generation Ad vectors. These vectors, by necessity, must be propagated in cell lines which complement both E1 and the second missing function, and they are generated using similar methods as described for first-generation Ads. Second-generation Ads have produced mixed results ranging from no improved function [123,130] to significantly improved long-term transgene expression and reduced immunogenicity and toxicity [133–135] compared to E1-deleted vectors. One interesting observation arising from analysis of transgene expression from E4-deleted vectors is the influence that this region can have on the persistence of expression of transgenes controlled by viral promoters contained in these vectors (e.g., cytomegalovirus immediate-early promoter). Vectors deleted of most or all of E4 showed reduced transgene expression over time, which was not accompanied by a loss of vector DNA [136,137]. Subsequently, it was determined that the Ad E4ORF3 was able to prevent viral promoter downregulation, which can occur over time in transduced cells [138], and this appears to be a generalized phenomenon for viral, but not cellular, promoters contained in Ad vectors. Inclusion of an E4ORF3 expression cassette in plasmid constructs also resulted in an improved duration of transgene expression in vivo [139].

The utility of second-generation Ads is exemplified by the work of Amalfitano and coworkers [140], who showed that administration of an E1/E2B (Ad DNA polymerase)–deleted vector encoding human acid-α-glucosidase (GAA) to GAA knockout mice resulted in systemic correction of musclar glycogen storage disease. In these experiments, a single retro-orbital delivery of the vector resulted in efficient uptake of the virus by hepatic cells. GAA proenzyme produced from the hepatic protein factories was secreted into the serum and taken up by skeletal and cardiac muscle resulting in decreased glycogen accumulation in the muscle. Although treated mice were only characterized out to 12 days postvector administration, and no long-term data have been presented, this approach may have wide implications for treatment of musclar disease. Studies using a similar vector encoding β-galactosidase (β-gal) showed expression of the protein for at least 2 months in mice [133], suggesting that DNA polymerase–deleted Ad vectors may provide the duration of expression, even of potentially immunogenic transgenes, necessary for the correction of many genetic diseases. Although second-generation Ads are easier to generate than fully deleted Ad vectors (see below), they have not gained widespread use because of their inconsistent performance and marginal increase in cloning capacity compared with first-generation Ads.

V. FULLY DELETED Ad VECTORS

Perhaps the ultimate in Ad vector attenuation results from the deletion of all Ad protein coding sequences, giving rise to fully deleted Ad vectors (fdAd). The only Ad sequences that need be retained in the fdAd are approximately 500 bp of cis-acting DNA elements, including the viral inverted terminal repeats (ITRs) located at each end of the genome that are necessary for viral DNA replication, and the viral packaging signal (Ψ), which is the only element necessary for packaging of the DNA into virions. Since vectors of this type are not able to replicate, the current methods for producing fdAd involve coreplication of the fdAd in the presence of a second helper virus which provides all replicative functions in trans. fdAds retain many of the advantages of first-generation Ad vectors, including high transduction efficiency of mammalian cells, but also have the added advantages of increased cloning capacity (up to ~37 kb) increased safety, and the potential for reduced immune responses due to the elimination of all viral coding sequences.

The prospect of using fdAds for gene therapy was first proposed by Mitani et al. [141], who produced a vector that was deleted of approximately 7.3 kb of essential Ad coding sequences and encoded a β-gal/neomycin fusion reporter gene. The vector DNA and wild-type (i.e., E1+) Ad2 helper DNA were cotransfected into 293 cells, and the resulting blue plaques, presumably containing both the β-gal vector and Ad2 helper virus, were isolated and amplified by serial passage on 293 cells. The resulting vector preparations contained an excess of helper virus (~200-fold over vector), but could transduce β-gal into recipient COS cells. Strategies similar to those employed by Mitani et al. were used to generate fully deleted Ad vectors encoding cDNAs for CFTR [142], dystrophin [143,144], or the γ-subunit of cyclic guanosine monophosphate (GMP) phosphodiesterase for rescue of photoreceptor degeneration [145], although helper functions were provided by a first-generation Ad5 vector. The fdAd were partially purified from the helper virus by centrifugation, and the helper content of the final purified vector stocks ranged from approximately a 1000-fold excess [142] to 3% [145] of the vector titer. Interestingly, fdAd vectors that were constructed significantly below the size of the wild-type Ad genome tended to be unstable and underwent DNA rearrangement, primarily multimerization, increasing the final size of the gene transfer vector to approximately that of wild-type Ad [142,143]. Nevertheless, all of the fdAds described above were able to transduce cells and express transgenes.

A. Propagation of fdAds

To reduce the quantity of helper virus present in the final vector preparations, Kochanek et al. [146] used SV5, a derivative of a first-generation Ad5 helper virus that was deleted for 91 bp of the Ad packaging signal, including three of the five elements believed to be essential for Ad DNA packaging [147]. This mutation resulted in a 90-fold reduction in the packaging efficiency of the helper virus compared to wild-type Ad. Thus, during serial amplification of an fdAd coinfected with SV5, the fdAd, which contains a wild-type packaging signal, was preferentially packaged, resulting in an improvement in the purity of the fdAd (~1% helper virus contamination). In this manner, a fdAd was produced that contained the full-length human dystrophin cDNA, under the regulation of the murine creatine kinase promoter, and human cytomegalovirus (HCMV)–driven β-gal expression cassette, designated AdDYSβ-gal. The β-gal cassette was included to monitor easily vector titer and transduction efficiency. This vector could express both β-gal and dystrophin in muscle, and resulted in partial, albeit transient, phenotypic correction of *mdx* mice, which carry a natural mutation in the dystrophin gene [148]. Intramuscular administration of AdDYSβ-gal resulted in inflammation in immunocompetent mice, and histological examination of transduced tissues showed an infiltration by CD4$^+$ and CD8$^+$ lymphocytes [149]. Transgene expression in wild-type mice was transient (<42 days); however, in β-gal–tolerized mice (i.e., *lac*Z transgenic animals), transgene expression was substantially prolonged (>82 days), suggesting that immune responses to β-gal may have been the reason for a loss of transgene expression. AdDYSβ-gal was also used in an ex vivo approach to transduce myoblasts which were subsequently transferred into *mdx* mice, resulting in expression of dystrophin in the recipient mouse muscle [150]. However, as observed after direct injection of vector into naive animals, the myoblasts were eventually eliminated because of the induction of cellular immune responses to the foreign dystrophin protein. Nevertheless, the results of Clemens and coworkers [148] clearly indicated that fdAds could be used in gene therapy applications.

Recombination systems have been employed to prevent packaging of the helper virus, or to remove large regions of coding sequence from first-generation vectors [151–153]. Lieber et al. [152] engineered a first-generation vector where one *loxP* site, a target for the bacteriophage P1 Cre recombinase, was located immediately downstream from the transgene cassette in the E1 region and a second site was inserted within the E3 region. In the presence of Cre, the *loxP* sites are recombined, excising the Ad coding sequences located between the two *loxP* sites, and leaving a genome of approximately 9 kb. In this system, the unrecombined vector itself acts as the complementing helper virus and the E4 transcriptional unit is retained in the final vector. Unfortunately, these vectors were unstable in vivo and provided only limited transgene expression after intravenous injection. Since the vectors developed by Lieber et al. still retain Ad coding sequences and have similar cloning capacity for foreign DNA as first-generation vectors, this strategy to generate fdAd has not gained widespread use.

A different Cre/*loxP*–based system for the generation of fdAd involves the use of a first-generation helper virus where the packaging signal is flanked by *loxP* recognition sites [151,153]. This virus is easily propagated in normal 293 cells; however, upon infection of a 293-derived cell line that stably expresses the Cre recombinase [154], the packaging signal is excised rendering the helper virus genome unpackageable (Fig. 3). The helper virus DNA retains the ability

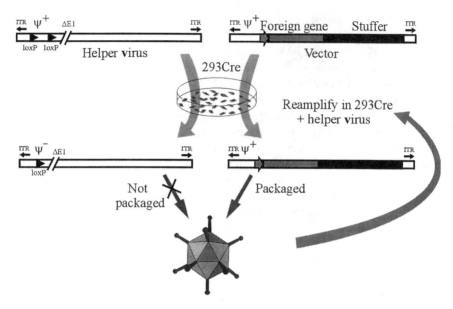

Figure 3 The Cre/*lox*P system for generating helper-dependent Ad vectors. The helper virus is a first-generation vector (i.e., E1-deleted) that contains a packaging signal (Ψ) flanked by *lox*P recombination target sites. Upon infection of a 293-derived cell line that stably expresses the bacteriophage P1 Cre recombinase, the packaging signal of the helper virus is excised, rendering the helper virus DNA unpackagable into virions. The helper virus can still replicate and provides all of the trans-acting factors required for propagation of a second vector deleted of most, if not all, Ad protein coding sequences, but containing the appropriate cis-acting elements, mainly the Ad inverted terminal repeats (ITRs) and packaging signal. The titer of the helper-dependent vector is increased by serial passage through helper virus–infected 293Cre cells (Adapted from Ref. 153.)

to replicate and express all of the functions required in trans for propagating a fdAd. The fdAd vector titer is increased by serial passage in 293Cre cells and, since the helper virus is not packaged in the Cre cells, a constant amount of new helper virus must be added at each passage. This system facilitates the generation of high-titer fdAd preparations with substantially reduced quantities of contaminating helper virus, typically ranging from 0.1 to 0.01% of the fdAd titer, and is the method currently employed to produce most fdAds for gene therapy studies.

B.   Size Constraints and the Importance of Stuffer DNA

Using the Cre/*lox*P fdAd system, it was determined that Ad virions have a lower limit for efficient DNA packaging of approximately 75% of the wild-type genome length [155]. Vectors constructed below 75% of the Ad genome length had undergone DNA rearrangements, resulting in a final vector size of greater than ~27 kb, whereas larger vectors were unaltered. The requirement for a minimum fdAd vector size of 28 kb implies that, in many cases, additional stuffer DNA must be included to prevent rearrangement. However, the nature of the stuffer DNA can have important effects on vector performance [156]. For example, a fdAd which contained 22 kb of bacteriophage lambda DNA expressed poorly in vivo following intravenous delivery to mice, and elicited the formation of cytotoxic T lymphocytes (CTLs) to peptides produced from the lambda DNA. In contrast, a similar fdAd engineered with a fragment from the human hypoxanthine–guanine phosphoribosyltransferase gene (HPRT) as stuffer showed significantly improved expression and did not elicit HPRT-specific CTL. These observations suggest that deletion of all Ad coding sequences, in itself, is not sufficient to permit long-term gene expression, and indicate that the nature of the stuffer segment must be carefully considered. The immediate interpretation of these results is that only eukaryotic DNA should be used as stuffer. It is not clear if there are other DNA sequence elements or physical characteristics of eukaryotic stuffer DNAs which contribute to their effectiveness.

C. In Vitro and In Vivo Studies

Schiedner et al. [157] used the Cre/*lox*P system to prepare an fdAd which encoded a genomic copy of the human α_1-antitrypsin gene (fdAd-AAT), including the endogenous promoter. In these experiments, AAT was used more as a reporter gene rather than as a potential therapeutic. This vector showed tissue-specific gene expression in cell culture and high-level, stable transgene expression in C57B1/6J mice for greater than 10 months following intravenous administration. Similar results were observed in C3H/HeJ mice [121], indicating that long-term gene expression was not a mouse strain–specific phenomenon. Moreover, histopathological examination revealed that the liver morphology of animals treated with the fdAd was essentially normal even at very high vector doses [158], whereas treatment with a first-generation Ad vector resulted in significant acute and chronic liver injury. Long-term expression of AAT (>1 year) was also observed in two of three baboons treated with the fdAd-AAT vector by intravenous delivery [159]. Blood chemistry for all the baboons was normal over the duration of the experiment. Interestingly, the baboon that only displayed short-term AAT expression had developed antibodies to the human AAT protein, suggesting that immunity to the therapeutic transgene may have contributed to the loss of measurable serum AAT. Similar results were observed after intravenous delivery of a fdAd encoding mouse leptin to obese mice (ob/ob, leptin knockout mice), although, in this model, the effectiveness of the treatment was eventually compromised because of the induction of immune responses to the foreign leptin protein in the knockout mice [160].

Using a different design of helper virus for fdAd vector propagation [161],

Balague et al. [l62] produced a fdAd vector containing a human factor VIII (hFVIII) cDNA, the defective gene in hemophilia A, under the regulation of the human albumin promoter. Expression of hfVIII protein was detected for almost 1 year after a single intravenous administration of the vector to C57BL/6J mice, and the quantity of protein expressed was within physiological range (100–800 ng/ml of serum). Interestingly, hFVIII expression declined sharply 14 days following inoculation of Balb/c mice and within 3–8 weeks in 13 of 16 hemophilic mice. There appeared to be an inverse correlation between the duration of hFVIII expression and formation of anti-hFVIII antibodies in the mice, with the strongest and earliest response being observed in the Balb/c strain. All mice developed antibodies to Ad. When assayed 6 days postinjection by tail clipping experiments, hemophilic mice showed improved blood clotting times of 30 min compared to continuous bleeding in untreated mice and approxintmely 10 min in nonhemophilic control mice. Following intravenous delivery, vector could be detected by a polymerase chain reaction (PCR)–based assay in spleen, lung, kidney, and heart, with the majority of vector DNA being in the liver; however, no vector DNA was detected in gonadal tissue. Histopathological examination of tissues revealed no significant changes compared to untreated animals, irrespective of the vector dose, suggesting that, once again, fdAd vectors are nontoxic.

FdAd have also been investigated for the treatment of acquired disease. Using an approach similar to the cancer immunotherapies we described previously a vector encoding murine INFα2 (mIFNα2) under the regulation of the liver-specific transthyretin promoter was employed to treat mouse hepatitis virus type 3 and concanavalin-induced hepatitis [163]. Successful results were obtained at vector doses which did not give rise to elevated serum mIFNα2, thereby eliminating side effects associated normal systemically delivered protein therapy.

It is important to note that the best results observed after gene delivery using fdAd vectors were in the absence of immune responses to the transgene, where protein expression persisted for greater than 1 year in some studies. Should humoral or cellular immune responses form against the foreign transgene product, protein expression will not persist. This suggests that one very active area of research in gene therapy in the coming years should be methods to prevent or circumvent immune responses to foreign proteins.

D. Readministration of fdAds

Although the fdAd DNA can persist within nondividing or slowly cycling cells for long periods of time, the episomal nature of Ads (and fdAds) may mean that the vector DNA will inevitably be lost from the cell as a consequence of normal cell division. Thus, although fdAds provide greater longevity of transgene expression than observed for first-generation Ads, there may be a requirement for repeat vector administration in order to boost transgene expression levels. Unfortunately, in most cases Ad (and fdAd), administration to immunocompetent individuals results in the formation of anti-Ad neutralizing antibodies which present a significant barrier to vector readministration [164–166]. However, Ads come in several different serotypes. For example, Ad2 and Ad5 are from different serotypes, so neutralizing antibodies to one virus, Ad2, will not prevent infection by the second virus, Ad5. A key feature of the fdAd system is that the vector serotype is defined by the helper virus, and thus a series of genetically identical fdAds of different serotypes can be generated simply by changing the helper used for vector rescue. By alternating the serotype of the helper virus, it was possible to use a serotype 5 fdAd to deliver a transgene effectively to the liver of mice previously exposed to a fdAd rescued with a type 2 helper virus [167]. Interestingly, sequential administration of first-generation Ads of different serotypes has been associated with a rapid decrease in transgene expression within a few days after administration of the second vector, presumably as a result of cross-reacting cellular immune responses elicited after the first injection [168]. This phenomenon was not observed during serotype switching with fdAds, suggesting that either fdAds, do not elicit such destructive processes or, alternatively, the infected cells are poor targets.

Experiments by Maione et al. [169] showed that a fdAd–murine erythropoietin (mEPO) construct produced 100-fold higher levels of expression compared to a first-generation Ad containing an identical expression cassette. As a result, only small quantities of fdAd–mEPO were required for a favorable effect, and such low quantities of vector did not result in the induction of anti-Ad neutralizing antibodies in the mice. Thus, readministration of the same vector was achieved without the need of immunosuppression or serotype switching, and illustrates that although serotype switching can be accomplished easily using fdAd, it may not be necessary in some cases.

E. Recent Advances in Vector Design

Recently, efforts have been made to increase the effectiveness of fdAd amplification and expression by optimizing the helper virus and fdAd vector design. In order to reduce the chances of recombination between homologous elements in the viral packaging signal of the helper virus and the fdAd, which would result in a loss of the left-most *loxP* site flanking the packaging signal and, therefore, loss of selection in the Cre/*lox*P system, Sandig et al. [170] reconstructed the packaging signal of the helper virus, reducing the length of homologous regions to as little as 22 bp. Second, in addition to identifying new DNA elements which can effectively act as stuffer in fdAd, vector propagation could be improved by including in the fdAd the E4 promoter element located at the right end of the Ad genome immediately adjacent to the left ITR. This effect was attributed to enhanced packaging of the E4-modified fdAd DNA compared to vectors lacking the promoter, and suggests that both the left and right ends of Ad are involved in packaging the viral DNA into virions.

The large cloning capacity of fdAds means that additional features can be accommodated, such as expression cassettes which can be regulated or integrate into the genome. Regulated gene expression can be accomplished either by the use of tissue-specific promoters, such that the transgene is only turned on in a particular tissue type [157], or by using promoters which are active only after the addition of an exogenous compound, allowing on/off regulation in all transduced cells. The combination of regulated gene expression and long-term vector persistence in vivo represents a powerful tool for gene delivery and expression, as elegantly illustrated by Burcin et al. [171]. A fdAd was generated to express the chimeric transactivator, GLp65, which consisted of a mutated progesterone receptor–ligand-binding domain fused to GAL4 DNA-binding domain and part of the activation domain of the human p65 protein (a component of NF-κB complex) under the regulation of a liver-specific promoter. The fdAd also contained the cDNA for human growth hormone under the control of a minimal promoter linked to GAL4-binding sites. When mice were inoculated with this fdAd by tail vein injection, high levels of serum growth hormone were detected following treatment with mifepristone, an inducer of the GLp65 transactivator, whereas no expression was detected in the absence of the inducer. Moreover, the transgene could be repeatedly induced (i.e., fully off to fully on) over a period of several weeks. These data illustrate an approach that should have wide applications in animal studies and gene therapy.

To permit stable insertion into the mammalian genome of an Ad-delivered expression cassette, Recchia et al. [172] produced an Ad/adeno-associated virus (AAV) hybrid which efficiently transduced cells and led to integration of a transgene cassette. AAV is a parvovirus capable of integrating into the host chromosome, and very encouraging results have been obtained using this virus as a gene delivery vector both in animals and humans [173,174]. Unfortunately, AAV has only a limited capacity for foreign DNA (~4.7 kb). The Ad/AAV hybrid system is an attempt to combine the large cloning capacity of fdAds and the integrative capacity of AAV. The hybrid vector system contained two components. The first component was a fdAd that contained the coding sequence for the AAV REP78 protein, responsible for AAV replication and site-specific integration, under the regulation of a liver-specific promoter. A second vector contained an AAV-ITR-flanked reporter gene. Coinfection of human liver cells in culture with the two vectors resulted in the excision and amplification of the AAV ITR-flanked transgene cassette and, in some cells, integration of the cassette into the cellular genome. Some of the integration events occurred in a site-specific manner into the AAVS1 site located on chromosome 19, the normal site for integration of wild-type AAV. Once integrated, the expression cassette could be passed to all daughter cells during cell division, thus circumventing the normal episomal nature of Ads.

VI. CELL TARGETING

Altering the tropism of Ad can be advantageous in a number of settings. If the primary or secondary receptor for Ad is not present on the target cell type, or is present only at low levels, modifying the virus to contain a ligand which can bind the target cell should enhance cell transduction. Alternatively, viral infection can be limited to specific cell types by abrogating elements on the Ad virion responsible for binding to its cognate receptor and adding a ligand specific to the target tissue.

One way to retarget the Ad virion is through the use of bispecific antibodies in which one end binds the Ad fiber and the other recognizes a cell surface receptor. This approach was used successfully to improve Ad transduction of endothelial and smooth muscle cells by using a bispecific antibody which bound α_v, integrins, which are prevalent on these tissues [6]. Unfortunately, the methodology required to generate Ad/antibody conjugates leads to formulation variability.

A second approach to Ad retargeting is genetic modification of Ad fiber. Ad knob, the terminal region of fiber, has been crystallized and the structure determined [175,176], and an extensive mutational analysis of amino acid residues responsible for the Ad fiber knob/CAR interaction has been performed [177,178]. Studies have shown that short ligands can be added to Ad fiber either at the carboxy-terminus of the protein [179] or within the variable HI loop of knob [180]. In many cases, these additions do not alter Ad binding to CAR. Genetic modification of Ad fiber has led to improved transduction of a variety of tissues and cell types, including mature muscle [181], ovarian tumors [182], and macrophage, endothelial, smooth muscle, and T cells [179]. More importantly, the thorough analysis of knob/CAR interactions has lead to the identification of residues which, when mutated, abrogate CAR binding, leading to true Ad receptor redirection [177,178]. The combination of Ad cell targeting redirection and the use of tissue-specific promoters will undoubtedly improve the utility of Ads in gene therapy.

VII. SUMMARY

The popularity of first-generation Ad vectors arose from their ease of genetic manipulation, high transduction efficiency of many different cell types and tissues, and their relative safety compared to other vector systems. Unfortunately, these characteristics came at a price, mainly transient transgene expression due to the induction of strong immune responses in the host and vector toxicity. Whereas strong immunogenicity can be advantageous for vaccine design and therapeutic strategies for cancer and infectious diseases, it is detrimental for gene-replacement approaches. Fully deleted Ad vectors have retained the high transduction efficiency of Ads in addition to gaining other desirable qualities, including large cloning capacity (~36 kb) and reduced vector-associated toxicity, inflammation, and cell-mediated immune responses. An unexpected advantage of fdAd vectors is that, apparently, they can persist as an episome in nondividing cells for long periods of time, giving rise to sustained transgenic expression for greater than 1 year. This may be an advantage over integrating vectors (e.g., AAV, retrovirus, and lentivirus), since integration can lead to inadvertent gene inactivation or activation depending on the site of integration in the host chromosome. Taken together, these characteristics suggest that fdAd vectors may prove to be quite useful for applications which are not amenable to first-generation Ads.

As with all fields of science, vectorology is continually evolving. The development of fdAds from first-generation vectors was necessitated because of the shortcomings of those vectors. For now, fdAds appear to be one of the best choices for gene therapy applications directed toward genetic disease, where the goal is long-term gene expression, with minimal immunological insult. Almost certainly, fdAds will continue to be improved upon. Indeed, the ultimate gene delivery vehicle will likely be a fusion of the best features of the current spectrum of vectors. Over the next few years, we will witness many innovations in vectorology which, as in the past, should continue to make gene therapy an exciting field of research for both basic and clinical scientists.

ACKNOWLEDGMENTS

Funding for this work was provided by the Ottawa Hospital Research Institute, the Medical Research Council of Canada (MRC), the Ontario Research and Development Challenge Fund, and the Hamilton Health Sciences Corporation. R.J.P. is an MRC scholar.

REFERENCES

1. Hilleman MR, Werner JH. Recovery of new agents from patients with acute respiratory illness. Proc Soc Exp Biol Med 1954; 85:183–188.
2. Rowe WP, Huebner RJ, Gilmore LK, Parrott RH, Ward TG. Isolation of a cytopathogenic agent from human adenoids undergoing spontaneous degeneration in tissue culture. Proc Soc Exp Biol Med 1953; 84:570–573.

3. Shenk T. 1996. Adenoviridae: the viruses and their replication. In: Fields BN, Knipe DM, Howley PM, eds. Fields Virology. Philadelphia: Lippincott-Raven, 1996: 2111–2148.

4. Bergelson JM, Cunningham JA, Droguett G, Kurt-Jones EA, Krithivas A, Hong JS, Horwitz MS, Crowell RL, Finberg RW. Isolation of a common receptor for Coxsackie B viruses and adenoviruses 2 and 5. Science 1997; 275:1320–1323.

5. Wickham TJ, Mathias P, Cheresh DA, Nemerow GR. Integrins alpha v beta 3 and alpha v beta 5 promote adenovirus internalization but not virus attachment. Cell 1993; 73:309–319.

6. Wickham TJ, Segal DM, Roelvink PW, Carrion ME, Lizonova A, Lee GM, Kovesdi I. Targeted adenovirus gene transfer to endothelial and smooth muscle cells by using bispecific antibodies. J Virol 1996; 70:6831–6838.

7. Goldman M, Su Q, Wilson JM. Gradient of RGD-dependent entry of adenoviral vector in nasal and intrapulmonary epithelia: implications for gene therapy of cystic fibrosis. Gene Ther 1996; 3:811–818.

8. Mellman I. The importance of being acidic: the role of acidification in intracellular membrane traffic. J Exp Biol 1992; 172:39–45.

9. Leopold PL, Ferris B, Grinberg I, Worgall S, Hackett NR, Crystal RG. Fluorescent virions: dynamic tracking of the pathway of adenoviral gene transfer vectors in living cells. Hum Gene Ther 1998; 9:367–378.

10. Greber UF, Willetts M, Webster P, Helenius A. Stepwise dismantling of adenovirus 2 during entry into cells. Cell 1993; 75:477–486.

11. Bett AJ, Prevec L, Graham FL. Packaging capacity and stability of human adenovirus type 5 vectors. J Virol 1993; 67:5911–5921.

12. Bett AJ, Haddara W, Prevec L, Graham FL. An efficient and flexible system for construction of adenovirus vectors with insertions or deletions in early regions 1 and 3. Proc Natl Acad Sci USA 1994; 91:8802–8806.

13. Ilan Y, Droguett G, Chowdhury NR, Li Y, Sengupta K, Thummala NR, Davidson A, Chowdhury JR, Horwitz MS. Insertion of the adenoviral E3 region into a recombinant viral vector prevents antiviral humoral and cellular immune responses and permits long-term gene expression. Proc Natl Acad Sci USA 1997; 94:2587–2592.

14. Poller W, Schneider-Rasp S, Liebert U, Merklein F, Thalheimer P, Haack A, Schwaab R, Schmitt C, Brackmann HH. Stabilization of transgene expression by incorporation of E3 region genes into an adenoviral factor IX vector and by transient anti-CD4 treatment of the host. Gene Ther 1996; 3:521–530.

15. Schowalter DB, Tubb JC, Liu M, Wilson CB, Kay MA. Heterologous expression of adenovirus E3-gp19K in an E1a-deleted adenovirus vector inhibits MHC I expression in vitro, but does not prolong transgene expression in vivo. Gene Ther 197; 4:351–360.

16. Sparer TE, Tripp RA, Dillehay DL, Hermiston TW, Wold WS, Gooding LR. The role of human adenovirus early region 3 proteins (gp19K, 10.4K, 14.5K, and 14.7K) in a murine pneumonia model. J Virol 1996; 70:2431–2439.

17. Graham FL. Covalently closed circles of human adenovirus DNA are infectious. EMBO J 1984; 3:2917–2922.

18. Graham FL, Prevec L. Methods for construction of adenovirus vectors. Mol Biotechnol 1995; 3:207–220.

19. Ng P, Parks RJ, Cummings DT, Evelegh CM, Graham FL. An enhanced system for construction of adenoviral vectors by the two-plasmid rescue method. Hum Gene Ther 2000; 11:693–699.

20. He TC, Zhou S, da Costa LT, Yu J, Kinzler KW, Vogelstein B. A simplified system for generating recombinant adenoviruses. Proc Natl Acad Sci USA 1998; 95:2509–2514.

21. Danthinne X and Werth E. New tools for the generation of E1- and/or E3-substituted adenoviral vectors. Gene Ther 2000; 7:80–87.

22. Graham FL, Smiley J, Russell WC, Nairn R. Characteristics of a human cell line transformed by DNA from human adenovirus type 5. J Gen Virol 1977; 36:59–74.

23. Lochmuller H, Jani A, Huard J, Prescott S, Simoneau M, Massie B, Karpati G, Acsadi G. Emergence of early region 1–containing replication-competent adenovirus in stocks of replication-defective adenovirus recombinants (delta E1 + delta E3) during multiple passages in 293 cells. Hum Gene Ther 1994; 5:1485–1491.

24. Fallaux FJ, Bout A, van der Velde I, van den Wollenberg DJ, Hehir KM, Keegan J, Auger C, Cramer SJ, van Ormondt H, van der Eb AJ, Valerio D, Hoeben RC. New helper cells and matched early region 1–deleted adenovirus vectors prevent generation of replication-competent adenoviruses. Hum Gene Ther 1998; 9:1909–1917.

25. Gao GP, Engdahl RK, Wilson JM. A cell line for high-yield production of E1-deleted adenovirus vectors without the emergence of replication-competent virus. Hum Gene Ther 2000; 11:213–219.

26. Bruder JT, Appiah A, Kirkman WM III, Chen P, Tian J, Reddy D, Brough DE, Lizonova A, Kovesdi I. Improved production of adenovirus vectors expressing apoptotic transgenes. Hum Gene Ther 2000: 11:139–149.

27. Yang Y, Li, Q, Ertl HC Wilson JM. Cellular and humoral immune responses to viral antigens create barriers to lung-directed gene therapy with recombinant adenoviruses. J Virol 1995; 69:2004–2015.

28. Hay JG, McElvaney NG, Herena J, Crystal RG. Modification of nasal epithelial potential differences of individuals with cystic fibrosis consequent to local administration of a normal CFTR cDNA adenovirus gene transfer vector. Hum Gene Ther 1995; 6:1487–1496.

29. Zabner J, Couture LA, Gregory RJ, Graham SM, Smith AE, Welsh MJ. Adenovirus-mediated gene transfer transiently corrects the chloride transport defect in nasal epithelia of patients with cystic fibrosis. Cell 1993; 75:207–216.

30. Crystal RG, McElvaney NG, Rosenfeld MA, Chu CS, Mastrangeli A, Hay JG, Brody SL, Jaffe HA, Eissa NT, Danel C. Administration of an adenovirus containing the human CFTR cDNA to the respiratory tract of individuals with cystic fibrosis [see comments]. Nat Genet 1994; 8:42–51.

31. Bellon G, Michel-Calemard L, Thouvenot D, Jagneaux V, Poitevin F, Malcus C, Accart N, Layani MP, Aymard M, Bernon H, Bienvenu J, Courtney M, Doring G, Gilly B, Gilly R, Lamy D, Levrey H, Morel Y, Paulin C, Perraud F, Rodillon L, Sene C, So S, Touraine-Moulin F, Pavirani A, et al. Aerosol administration of a recombinant adenovirus expressing CFTR to cystic fibrosis patients: a phase I clinical trial. Hum Gene Ther 1997; 8:15–25.

32. Harvey BG, Leopold, PL, Hackett NR, Grasso TM, Williams PM, Tucker AL, Kaner RJ, Ferris B, Gonda I, Sweeney TD, Ramalingam R, Kovesdi I, Shak S, Crystal RG. Airway epithelial CFTR mRNA expression in cystic fibrosis patients after repetitive administration of a recombinant adenovirus [see comments]. J Clin Invest 1999; 104:1245–1255.

33. Sullivan DE, Dash S, Du H, Hiramatsu N, Aydin F, Kolls J, Blanchard J, Baskin G, Gerber MA. Liver-directed gene transfer in non-human primates. Hum Gene Ther 1997; 8:1195–1206.

34. Yang Y, Nunes FA, Berencsi K, Furth EE, Gonczol E, Wilson JM. Cellular immunity to viral antigens limits E1-deleted adenoviruses for gene therapy. Proc Natl Acad Sci USA 1994; 91:4407–4411.

35. Balter M. Gene therapy on trial [news]. Science 2000; 288:951–957.

36. Tripathy SK, Goldwasser E, Lu MM, Barr E, Leiden JM. Stable delivery of physiologic levels of recombinant erythropoietin to the systemic circulation by intramuscular injection of replication-defective adenovirus. Proc Natl Acad Sci USA 1994; 91:11557–11561.

37. Tripathy SK, Black HB, Goldwasser E, Leiden JM. Immune responses to transgene-encoded proteins limit the stability of gene expression after injection of replication-defective adenovirus vectors. Nat Med 1996; 2:545–550.

38. Jooss K, Yang Y, Fisher KJ, Wilson JM. Transduction of dendritic cells by DNA viral vectors directs the immune response to transgene products in muscle fibers. J Virol 1998; 72:4212–4223.

39. Addison CL, Braciak T, Ralston R, Muller WJ, Gauldie J, Graham, FL. Intratumoral injection of an adenovirus expressing interleukin 2 induces regression and immunity in a murine breast cancer model. Proc Natl Acad Sci USA 1995; 92: 8522–8526.

40. Cordier L, Duffour MT, Sabourin JC, Lee MG, Cabannes J, Ragot T, Perricaudet M, Haddada H. Complete recovery of mice from a pre-established tumor by direct intratumoral delivery of an adenovirus vector harboring the murine IL-2 gene. Gene Ther 1995; 2:16–21.

41. Yoshikawa K, Kajiwara K, Ideguchi M, Uchida T, Ito H. Immune gene therapy of experimental mouse brain tumor with adenovirus-mediated gene transfer of murine interleukin-4. Cancer Immunol Immunother 2000; 49:23–33.

42. Felzmann T, Ramsey WJ, Blaese RM. Characterization of the antitumor immune response generated by treatment of murine tumors with recombinant adenoviruses expressing HSVtk, IL-2, IL-6 or B7-1. Gene Ther 1997; 4:1322–1329.

43. Toloza EM, Hunt K, Miller AR, McBride W, Lau R, Swisher S, Rhoades K, Arthur J, Choi J, Chen L, Chang P, Chen A, Glaspy J, Economou, JS. Transduction of murine and human tumors using recombinant adenovirus vectors. Ann Surg Oncol 1997; 4:70–79.

44. Bramson JL, Hitt M, Addison CL, Muller WJ, Gauldie J, Graham FL. Direct intratumoral injection of an adenovirus expressing interleukin-12 induces regression and long-lasting immunity that is associated with highly localized expression of interleukin-12. Hum Gene Ther 1996; 7:1995–2002.

45. Caruso M, Pham-Nguyen K, Kwong YL, Xu B, Kosai KI, Finegold M, Woo SL,

Chen, SH. Adenovirus-mediated interleukin-12 gene therapy for metastatic colon carcinoma. Proc Natl Acad Sci USA 1996; 93:11302–11306.

46. Osaki T, Hashimoto W, Gambotto A, Okamura H, Robbins PD, Kurimoto M, Lotze MT, Tahara H. Potent antitumor effects mediated by local expression of the mature form of the interferon-gamma inducing factor, interleukin-18 (IL-18). Gene Ther 1999; 6:808–815.

47. Ahmed CM, Sugarman BJ, Johnson DE, Bookstein RE, Saha DP, Nagabhushan TL, Wills KN. In vivo tumor suppression by adenovirus-mediated interferon alpha2b gene delivery. Hum Gene Ther 1999; 1;0:77–84.

48. Lu W, Fidler IJ, Dong, Z. Eradication of primary murine fibrosarcomas and induction of systemic immunity by adenovirus-mediated interferon beta gene therapy. Cancer Res 199; 59:5202–5208.

49. Qin XQ, Tao N, Dergay A, Moy P, Fawell S, Davis A, Wilson JM, Barsoum, J. Interferon-beta gene therapy inhibits tumor formation and causes regression of established tumors in immune-deficient mice. Proc Natl Acad Sci USA 1998; 95: 14411–14416.

50. Fathallah-Shaykh HM, Zhao LJ, Kafrouni AI, Smith GM, Forman, J. Gene transfer of IFN-gamma into established brain tumors represses growth by antiangiogenesis. J Immunol 2000; 164:217–222.

51. Stewart AK, Lassam NJ, Graham FL, Gauldie J, Addison CL, Bailey DJ, Dessureault S, Dube ID, Gallenger S, Krajden M, Rotstein LE, Quirt IC, Moen R. A phase I study of adenovirus mediated gene transfer of interleukin 2 cDNA into metastatic breast cancer or melanoma. Hum Gene Ther 1997; 8:1403–1414.

52. Springer CJ, Niculescu-Duvaz I. Prodrug-activating systems in suicide gene therapy. J Clin Invest 2000; 105:1161–1167.

53. Hall SJ, Sanford MA, Atkinson G, Chen SH. Induction of potent antitumor natural killer cell activity by herpes simplex virus-thymidine kinase and ganciclovir therapy in an orthotopic mouse model of prostate cancer. Cancer Res 1998; 58:3221–3225.

54. Morris JC. Enzyme/prodrug-based tumor vaccination: all politics (and immunity are local [editorial; comment]. J Natl Cancer Inst 1999; 91:1986–1989.

55. Consalvo M, Mullen CA, Modesti A, Musiani P, Allione A, Cavallo F, Giovarelli M, Forni G. 5-Fluorocytosine-induced eradication of murine adenocarcinomas engineered to express the cytosine deaminase suicide gene requires host immune competence and leaves an efficient memory. J Immunol 1995; 154:5302–5312.

56. Kuriyama S, Kikukawa M, Masui K, Okuda H, Nakatani T, Sakamoto T, Yoshiji H, Fukui H, Ikenaka K, Mullen CA, Tsujii T. Cytosine deaminase/5-fluorocytosine gene therapy can induce efficient anti-tumor effects and protective immunity in immunocompetent mice but not in athymic nude mice. Int J Cancer 1999; 81:592–597.

57. van der Eb MM, Cramer SJ, Vergouwe Y, Schagen FH, van Krieken JH, van de Eb AJ, Rinkes IH, van de Velde CJ, Hoeben RC. Severe hepatic dysfunction after adenovirus-mediated transfer of the herpes simplex virus thymidine kinase gene and ganciclovir administration. Gene Ther 1998; 5:451–458.

58. Bramson, JL, Hitt M, Gauldie J, Graham FL. Pre-existing immunity to adenovirus does not prevent tumor regression following intratumoral administration of a vector expressing IL-12 but inhibits virus dissemination. Gene Ther 1997; 4:1069–1076.

59. Brand K, Loser P, Arnold W, Bartels T, Strauss M. Tumor cell-specific transgene expression prevents liver toxicity of the adeno-HSVtk/GCV approach. Gene Ther 1998; 5:1363–71.

60. Sterman DH, Treat J, Litzky LA, Amin KM, Coonrod L, Molnar-Kimber K, Recio, A, Knox L, Wilson JM, Albelda SM, Kaiser LR. Adenovirus-mediated herpes simplex virus thymidine kinase/ganciclovir gene therapy in patients with localized malignancy: results of a phase I clinical trial in malignant mesothelioma. Hum Gene Ther 1998; 9:1083–1092.

61. Herman JR, Adler HL, Aguilar-Cordova E, Rojas-Martinez A, Woo S, Timme TL, Wheeler TM, Thompson TC, Scardino PT. In situ gene therapy for adenocarcinoma of the prostate: a phase I clinical trial. Hum Gene Ther 1999; 10:1239–1249.

62. May P, May E. Twenty years of p53 research: structural and functional aspects of the p53 protein [published erratum appears in Oncogene 2000 Mar 23;19(13): 1734]. Oncogene 1999; 18:7621–7636.

63. Clayman GL, el-Naggar AK, Lippman SM, Henderson YC, Frederick M, Merritt JA, Zumstein LA, Timmons TM, Liu TJ, Ginsberg L, Roth JA, Hong WK, Bruso P, Goepfert H. Adenovirus-mediated p53 gene transfer in patients with advanced recurrent head and neck squamous cell carcinoma. J Clin Oncol 1998; 16:2221–2232.

64. Schuler M, Rochlitz C, Horowitz JA, Schlegel J, Perruchoud AP, Kommoss F, Bolliger CT, Kauczor HU, Dalquen P, Fritz MA, Swanson S, Herrmann R, Huber, C. A phase I study of adenovirus-mediated wild-type p53 gene transfer in patients with advanced non-small cell lung cancer. Hum Gene Ther 1998; 9:2075–2082.

65. Swisher SG, Roth JA, Nemunaitis J, Lawrence DD, Kemp BL, Carrasco CH, Connors DG, El-Naggar AK, Fossella F, Glisson BS, Hong WK, Khuri FR, Kurie JM, Lee JJ, Lee JS, Mack M, Merritt JA, Nguyen DM, Nesbitt JC, Perez-Soler R, Pisters KM, Putnam JB Jr, Richli WR, Savin M, Waugh MK, et al. Adenovirus-mediated p53 gene transfer in advanced non-small-cell lung cancer. J Natl Cancer Inst 1999; 91:763–771.

66. Tsao YP, Huang SJ, Chang JL, Hsieh J, Pong RC, Chen SL. Adenovirus-mediated p21((WAF1/SDII/CIP1)) gene transfer induces apoptosis of human cervical cancer cell lines. J Virol 1999; 73:4983–4990.

67. Schreiber M, Muller WJ, Singh G, Graham FL. Comparison of the effectiveness of adenovirus vectors expressing cyclin kinase inhibitors p16INK4A, p18INK4C, p19INK4D, p21(WAF1/CIP1) and p27KIP1 in inducing cell cycle arrest, apoptosis and inhibition of tumorigenicity. Oncogene 1999; 18:1663–1676.

68. Kobayashi S, Shirasawa H, Sashiyama H, Kawahira H, Kaneko K, Asano T, Ochiai T. P16INK4a expression adenovirus vector to suppress pancreas cancer cell proliferation. Clin Cancer Res 1999; 5:4182–4185.

69. Tai YT, Strobel T, Kufe D, Cannistra SA. In vivo cytotoxicity of ovarian cancer cells through tumor-selective expression of the BAX gene. Cancer Res 1999; 59: 2121–2126.

70. Kagawa S, Pearson SA, Ji L, Xu K, McDonnell TJ, Swisher SG, Roth JA, Fang, B. A binary adenoviral vector system for expressing high levels of the proapoptotic gene bax. Gene Ther 2000; 7:75–79.

71. Heise C, Sampson-Johannes A, Williams A, McCormick F, Von Hoff DD, Kirn

DH. ONYX-015, an E1B gene-attenuated adenovirus, causes tumor-specific cytolysis and antitumoral efficacy that can be augmented by standard chemotherapeutic agents [see comments]. Nat Med 1997; 3:639–645.

72. Fueyo J, Gomez-Manzano C, Alemany R, Lee PS, McDonnell TJ, Mitlianga P, Shi YX, Levin VA, Yung WK, Kyritsis AP. A mutant oncolytic adenovirus targeting the Rb pathway produces anti-glioma effect in vivo. Oncogene 2000; 19: 2–12.

73. Ganly I, Kirn D, Eckhardt SG, Rodriguez GI, Soutar DS, Otto R, Robertson AG, Park O, Gulley ML, Heise C, Von Hoff DD, Kaye SB. A phase I study of Onyx-015, an E1B attenuated adenovirus, administered intratumorally to patients with recurrent head and neck cancer. Clin Cancer Res 2000; 6:798–806.

74. Onyx plans phase III trial of ONYX-015 for head & neck cancer [news]. Oncologist 1999; 4:432.

75. Kirn D, Hermiston T, McCormick F. ONYX-015: clinical data are encouraging [letter; comment]. Nat Med 1998; 4:1341–1342.

76. Rothmann T, Hengstermann A, Whitaker NJ, Scheffner M, zur Hausen H. Replication of ONYX-015, a potential anticancer adenovirus, is independent of p53 status in tumor cells: J Virol 1998; 72:9470–9478.

77. Wildner O, Blaese RM, Morris JC. Therapy of colon cancer with oncolytic adenovirus is enhanced by the addition of herpes simplex virus-thymidine kinase. Cancer Res 1999; 59:410–413.

78. Wildner O, Morris JC Vahanian NN, Ford H Jr, Ramsey WJ, Blaese RM. Adenoviral vectors capable of replication improve the efficacy of HSVtk/GCV suicide gene therapy of cancer. Gene Ther 1999; 6:57–62.

79. Ohno T, Gordon D, San H, Pompili VJ, Imperiale MJ, Nabel GJ, Nabel EG. Gene therapy for vascular smooth muscle cell proliferation after arterial injury [see comments]. Science 1994; 265:781–784.

80. Macejak DG, Lin H, Webb S, Chase J, Jensen K, Jarvis TC, Leiden, JM, Couture L. Adenovirus-mediated expression of a ribozyme to c-myb mRNA inhibits smooth muscle cell proliferation and neointima formation in vivo. J Virol 1999; 73:7745–7751.

81. Guzman RJ, Hirschowitz EA, Brody SL, Crystal RG, Epstein SE, Finkel T. In vivo suppression of injury-induced vascular smooth muscle cell accumulation using adenovirus-mediated transfer of the herpes simplex virus thymidine kinase gene. Proc Natl Acad Sci USA 1994; 91:10732–10736.

82. Claudio PP, Fratta L, Farina F, Howard CM, Stassi G, Numata S, Pacilio C, Davis A, Lavitrano M, Volpe M, Wilson JM, Trimarco B, Giordano A, Condorelli G. Adenoviral RB2/p130 gene transfer inhibits smooth muscle cell proliferation and prevents restenosis after angioplasty. Circ Res 1999; 85:1032–1039.

83. Chang MW, Ohno T, Gordon D, Lu MM, Nabel GJ, Nabel EG, Leiden JM. Adenovirus-mediated transfer of the herpes simplex virus thymidine kinase gene inhibits vascular smooth muscle cell proliferation and neointima formation following balloon angioplasty of the rat carotid artery. Mol Med 1995; 1:172–181.

84. Chang MW, Barr E, Seltzer J, Jiang YQ, Nabel GJ, Nabel EG, Parmacek MS, Leiden J M. Cytostatic gene therapy for vascular proliferative disorders with a con-

stitutively active form of the retinoblastoma gene product. Science 1995; 267:518–522.

85. Chang MW, Barr E, Lu MM, Barton K, Leiden JM. Adenovirus-mediated over-expression of the cyclin/cyclin-dependent kinase inhibitor, p21 inhibits vascular smooth muscle cell proliferation and neointima formation in the rat carotid artery model of balloon angioplasty. J Clin Invest 1995; 96:2260–2268.

86. Muhlhauser J, Merrill MJ, Pili R, Maeda H, Bacic M, Bewig B, Passaniti A, Edwards NA, Crystal RG, Capogrossi MC. VEGF165 expressed by a replication-deficient recombinant adenovirus vector induces angiogenesis in vivo. Circ Res 1995; 77:1077–1086.

87. Rosengart TK, Lee LY, Patel SR, Sanborn TA, Parikh M, Bergman GW, Hachamovitch R, Szulc M, Kligfield PD, Okin PM, Hahn RT, Devereux RB, Post MR, Hackett N, Foster T, Grasso TM, Lesser ML, Isom OW, Crystal RG. Angiogenesis gene therapy phase I assessment of direct intramyocardial administration of an adenovirus vector expressing VEGF121 cDNA to individuals with clinically significant severe coronary artery disease. Circulation 1999; 100:468–474.

88. Rosengart TK, Lee LY, Patel SR, Kligfield PD, Okin PM, Hackett NR, Isom OW, Crystal RG. Six-month assessment of a phase I trial of angiogenic gene therapy for the treatment of coronary artery disease using direct intramyocardial administration of an adenovirus vector expressing the VEGF121 cDNA. Ann Surg 1999; 230:466–470; discussion 470–472.

89. Chapelier A, Danel C, Mazmanian M, Bacha EA, Sellak H, Gilbert MA, Herve P, Lemarchand P. Gene therapy in lung transplantation: feasibility of ex vivo adenovirus mediated gene transfer to the graft. Hum Gene Ther 1996; 7:1837–1845.

90. Chia SH, Geller DA, Kibbe MR, Watkins SC, Fung JJ, Starzl TE, Murase N. Adenovirus mediated gene transfer to liver grafts: an improved method to maximize infectivity. Transplantation 1998; 66:1545–1551.

91. Csete ME, Benhamou PY, Drazan KE, Wu L, McIntee DF, Afra R, Mullen Y, Busuttil RW, Shaked A. Efficient gene transfer to pancreatic islets mediated by adenoviral vectors. Transplantation 1995; 59:263–268.

92. Drazan KE, Csete ME, Da Shen X, Bullington D, Cottle G, Busuttil RW, Shaked A. Hepatic function is preserved following liver-directed, adenovirus-mediated gene transfer. J Surg Res 1995; 59:299–304.

93. Wang J, Ma Y, Knechtle SJ. Adenovirus-mediated gene transfer into rat cardiac allografts. Comparison of direct injection and perfusion. Transplantation 1996; 61:1726–1729.

94. Guillot C, Mathieu P, Coathalem H, Le Mauff B, Castro MG, Tesson L, Usal C, Laumonier T, Brouard S, Soulillou JP, Lowenstein, PR, Cuturi MC, Anegon I. Tolerance to cardiac allografts via local and systemic mechanisms after adenovirus-mediated CTLA4Ig expression. J Immunol 2000; 164:5258–5268.

95. Kita Y, Li XK, Ohba M, Funeshima N, Enosawa S, Tamura A, Suzuki K, Amemiya H, Hayashi S, Kazui T, Suzuki S. Prolonged cardiac allograft survival in rats systemically injected adenoviral vectors containing CTLA4Ig-gene. Transplantation 1999; 68:758–766.

96. Lu L, Gambotto A, Lee WC, Qian S, Bonham CA, Robbins PD, Thomson AW.

Adenoviral delivery of CTLA4Ig into myeloid dendritic cells promotes their in vitro tolerogenicity and survival in allogeneic recipients. Gene Ther 1999; 6:554–563.

97. Feng S, Quickel RR, Hollister-Lock J, McLeod M, Bonner-Weir S, Mulligan RC, Weir GC. Prolonged xenograft survival of islets infected with small doses of adenovirus expressing CTLA4Ig. Transplantation 1999; 67:1607–1613.

98. David A, Chetritt J, Guillot C, Tesson L, Heslan JM, Cuturi MC, Soulillou JP, Anegon I. Interleukin-10 produced by recombinant adenovirus prolongs survival of cardiac allografts in rats. Gene Ther 2000; 7:505–510.

99. Shinozaki K, Yahata H, Tanji H, Sakaguchi T, Ito H, Dohi K. Allograft transduction of IL-10 prolongs survival following orthotopic liver transplantation. Gene Ther 1999; 6:816–822.

100. Wang CK, Zuo XJ, Carpenter D, Jordan S, Nicolaidou E, Toyoda M, Czer LS, Wang H, Trento A. Prolongation of cardiac allograft survival with intracoronary viral interleukin-10 gene transfer. Transplant Proc 1999; 31:951–952.

101. Hammel JM, Elfeki SK, Kobayashi N, Ito M, Cai J, Fearon DT, Graham FL, Fox IJ. Transplanted hepatocytes infected with a complement receptor type 1 (CR1)–containing recombinant adenovirus are resistant to hyperacute rejection. Transplant Proc 1999; 31:939.

102. Yasuda H, Nagata M, Arisawa K, Yoshida R, Fujihira K, Okamoto N, Moriyama H, Miki M, Saito I, Hamada H, Yokono K, Kasuga M. Local expression of immunoregulatory IL-12p40 gene prolonged syngeneic islet graft survival in diabetic NOD mice. J Clin Invest 1998; 102:1807–1814.

103. Qin L, Ding Y, Pahud DR, Robson ND, Shaked A, Bromberg JS. Adenovirus-mediated gene transfer of viral interleukin-10 inhibits the immune response to both alloantigen and adenoviral antigen. Hum Gene Ther 1997; 8:1365–1374.

104. Drazan KE, Wu L, Olthoff KM, Jurim O, Busuttil RW, Shaked A. Transduction of hepatic allografts achieves local levels of viral IL-10 which suppress alloreactivity in vitro. J Surg Res 1995; 59:219–223.

105. Lehmann TG, Wheeler MD, Schoonhoven R, Bunzendahl H, Samulski RJ, Thurman RG. Delivery of Cu/Zn-superoxide dismutase genes with a viral vector minimizes liver injury and improves survival after liver transplantation in the rat. Transplantation 2000; 69:1051–1057.

106. Hiratsuka M, Mora BN, Yano M, Mohanakumar T, Patterson GA. Gene transfer of heat shock protein 70 protects lung grafts from ischemia-reperfusion injury. Ann Thorac Surg 1999; 67:1421–1427.

107. Danel C, Erzurum SC, Prayssac P, Eissa NT, Crystal RG, Herve P, Baudet B, Mazmanian M, Lemarchand P. Gene therapy for oxidant injury-related diseases: adenovirus-mediated transfer of superoxide dismutase and catalase cDNAs protects against hyperoxia but not against ischemia-reperfusion lung injury. Hum Gene Ther 1998; 9:1487–1496.

108. Bruder JT, Kovesdi I. Adenovirus infection stimulates the Raf/MAPK signaling pathway and induces interleukin-8 expression. J Virol 1997; 71:398–404.

109. Muruve DA, Barnes MJ, Stillman IE, Libermann TA. Adenoviral gene therapy leads to rapid induction of multiple chemokines and acute neutrophil-dependent hepatic injury in vivo. Hum Gene Ther 1999; 10:965–976.

110. Otake K, Ennist DL, Harrod K, Trapnell BC. Nonspecific inflammation inhibits adenovirus-mediated pulmonary gene transfer and expression independent of specific acquired immune responses. Hum Gene Ther 1998; 9:2207–2222.

111. Lieber A, He CY, Meuse L, Himeda C, Wilson C, Kay MA. Inhibition of NF-kappaB activation in combination with bc1–2 expression allows for persistence of first generation adenovirus vectors in the mouse liver. J Virol 1998; 72:9267–9277.

112. Barr D, Tubb J, Ferguson D, Scaria A, Lieber A, Wilson C, Perkins J, Kay MA. Strain related variations in adenovirally mediated transgene expression from mouse hepatocytes in vivo: comparisons between immunocompetent and immunodeficient inbred strains. Gene Ther 1995; 2:151–155.

113. Michou AI, Santoro L, Christ M, Julliard V, Pavirani A, Mehtali M. Adenovirus-mediated gene transfer: influence of transgene, mouse strain and type of immune response on persistence of transgene expression. Gene Ther 1997; 4:473–82.

114. Quinones MJ, Leor J, Kloner RA, Ito M, Patterson M, Witke WF, Kedes L. Avoidance of immune response prolongs expression of genes delivered to the adult rat myocardium by replication-defective adenovirus. Circulation 1996; 94:1394–1401.

115. Kaplan JM, Armentano D, Sparer TE, Wynn SG, Peterson PA, Wadsworth SC, Couture KK, Pennington SE, St. George JA, Gooding LR, Smith AE. 1997. Characterization of factors involved in modulating persistence of transgene expression from recombinant adenovirus in the mouse lung. Hum Gene Ther 1997; 8:45–56.

116. Hermens WT, Verhaagen J. Adenoviral vector-mediated gene expression in the nervous system of immunocompetent Wistar and T cell–deficient nude rats: preferential survival of transduced astroglial cells in nude rats. Hum Gene Ther 1997; 8:1049–1063.

117. Zsengeller ZK, Wert SE, Hull WM, Hu X, Yei S, Trapnell BC, Whitsett, JA. Persistence of replication-deficient adenovirus-mediated gene transfer in lungs of immune-deficient (nu/nu) mice. Hum Gene Ther 1995; 6:457–467.

118. Brough DE, Hsu C, Kulesa VA, Lee GM, Cantolupo LJ, Lizonova A, Kovesdi I. 1997. Activation of transgene expression by early region 4 is responsible for a high level of persistent transgene expression from adenovirus vectors in vivo. J Virol 1997; 71:9206–9213.

119. Engelhardt JF, Yang Y, Stratford-Perricaudet LD, Allen ED, Kozarsky K, Perricaudet M, Yankaskas JR, Wilson JM. Direct gene transfer of human CFTR into human bronchial epithelia of xenografts with E1-deleted adenoviruses. Nat Genet 1993; 4:27–34.

120. Armentano D, Zabner J, Sacks C, Sookdeo CC, Smith MP, St. George JA, Wadsworth SC, Smith AE, Gregory RJ. Effect of the E4 region on the persistence of transgene expression from adenovirus vectors. J Virol 1997; 71:2408–2416.

121. Pastore L, Morral N, Zhou H, Garcia R, Parks RJ, Kochanek S, Graham FL, Lee B, Beaudet AL. Use of a liver-specific promoter reduces immune response to the transgene in adenoviral vectors. Hum Gene Ther 1999; 10:1773–1781.

122. Zhou H, O'Neal W, Morral N, Beaudet AL. Development of a complementing cell line and a system for construction of adenovirus vectors with E1 and E2a deleted. J Virol 1996; 70:7030–7038.

123. Fang B, Wang H, Gordon G, Bellinger DA, Read MS, Brinkhous KM, Woo SL, Eisensmith RC. Lack of persistence of E1-recombinant adenoviral vectors con-

taining a temperature-sensitive E2A mutation in immunocompetent mice and hemophilia B dogs. Gene Ther 1996; 3:217–222.

124. Gorziglia MI, Kadan MJ, Yei S, Lim J, Lee GM, Luthra R, Trapnell BC. Elimination of both E1 and E2 from adenovirus vectors further improves prospects for in vivo human gene therapy. J Virol 1996; 70:4173–4178.

125. Amalfitano A, Chamberlain JS. Isolation and characterization of packaging cell lines that coexpress the adenovirus E1, DNA polymerase, and preterminal proteins: implications for gene therapy. Gene Ther 1997; 4:258–263.

126. Brough DE, Lizonova A, Hsu C, Kulesa VA, Kovesdi I. A gene transfer vector-cell line system for complete functional complementation of adenovirus early regions E1 and E4. J Virol 1996; 70:6497–6501.

127. Krougliak V, Graham FL. Development of cell lines capable of complementing E1, E4, and protein IX defective adenovirus type 5 mutants. Hum Gene Ther 1995; 6:1575–1586.

128. Yeh P, Dedieu JF, Orsini C, Vigne E, Denefle P, Perricaudet M. Efficient dual transcomplementation of adenovirus E1 and E4 regions from a 293-derived cell line expressing a minimal E4 functional unit. J Virol 1996; 70:559–565.

129. Gao GP, Yang Y, Wilson JM. Biology of adenovirus vectors with E1 and E4 deletions for liver-directed gene therapy. J Virol 1996; 70:8934–8943.

130. Lusky M, Christ M, Rittner K, Dieterle A, Dreyer D, Mourot B, Schultz H, Stoeckel F, Pavirani A, Mehtali M. In vitro and in vivo biology of recombinant adenovirus vectors with E1, E1/E2A, or E1/E4 deleted. J Virol 1998; 72:2022–2032.

131. Wang Q, Jia XC, Finer MH. A packaging cell line for propagation of recombinant adenovirus vectors containing two lethal gene-region deletions. Gene Ther 1995; 2:775–783.

132. Armentano D, Sookdeo CC, Hehir KM, Gregory RJ, St. Prince GA, Wadsworth SC, Smith AE. Characterization of an adenovirus gene transfer vector containing an E4 deletion. Hum Gene Ther 1995; 6:1343–1353.

133. Hu H, Serra D Amalfitano A. Persistence of an [E1-, polymerase-] adenovirus vector despite transduction of a neoantigen into immune-competent mice. Hum Gene Ther 1999; 10:355–64.

134. Dedieu JF, Vigne E, Torrent C, Jullien C, Mahfouz I, Caillaud JM, Aubailly N, Orsini C, Guillaume JM, Opolon P, Delaere P, Perricaudet M, Yeh P. Long-term gene delivery into the livers of immunocompetent mice with E1/E4-defective adenoviruses. J Virol 1997; 71:4626–4637.

135. Wang Q, Greenburg G, Bunch D, Farson D, Finer MH. Persistent transgene expression in mouse liver following in vivo gene transfer with a delta E1/delta E4 adenovirus vector. Gene Ther 1997; 4:393–400.

136. Brough DE, Hsu C, Kulesa VA, Lee GM, Cantolupo LJ, Lizonova A, Kovesdi I. Activation of transgene expression by early region 4 is responsible for a high level of persistent transgene expression from adenovirus vectors in vivo. J Virol 1997; 71:9206–9213.

137. Armentano D, Zabner J, Sacks C, Sookdeo CC, Smith MP, Wadsworth SC, Smith AE, Gregory RJ. Effect of the E4 region on the persistence of transgene expression from adenovirus vectors. J Virol 1997; 71:2408–2416.

138. Armentano D, Smith MP, Sookdeo CC, Zabner J, Perricone MA, St George JA,

Wadsworth SC, Gregory RJ. E4ORF3 requirement for achieving long-term transgene expression from the cytomegalovirus promoter in adenovirus vectors. J Virol 1999; 73:7031–7034.

139. Yew NS, Marshall J, Przybylska M, Wysokenski DM, Ziegler RJ, Rafter PW, Li C, Armentano D, Cheng SH. Increased duration of transgene expression in the lung with plasmid DNA vectors harboring adenovirus E4 open reading frame 3. Hum Gene Ther 1999; 10:1833–1843.

140. Amalfitano A, McVie-Wylie AJ, Hu H, Dawson TL, Raben N, Plotz P, Chen YT. Systemic correction of the muscle disorder glycogen storage disease type II after hepatic targeting of a modified adenovirus vector encoding human acid-alpha-glucosidase. Proc Natl Acad Sci USA 1999; 96:8861–8866.

141. Mitani K, Graham FL, Caskey CT, Kochanek S. Rescue, propagation, and partial purification of a helper virus-dependent adenovirus vector. Proc Natl Acad Sci USA 1995; 92:3854–3858.

142. Fisher KJ, Choi H, Burda J, Chen SJ, Wilson JM. Recombinant adenovirus deleted of all viral genes for gene therapy of cystic fibrosis. Virology 1996; 217:11–22.

143. Haecker SE, Stedman HH, Balice-Gordon RJ, Smith DB, Greelish JP, Mitchell MA, Wells A, Sweeney HL, Wilson JM. In vivo expression of full-length human dystrophin from adenoviral vectors deleted of all viral genes. Hum Gene Ther 1996; 7:1907–1914.

144. Kumar-Singh R, Chamberlain JS. Encapsidated adenovirus minichromosomes allow delivery and expression of a 14 kb dystrophin cDNA to muscle cells. Hum Mol Genet 1996; 5:913–921.

145. Kumar-Singh R, Farber DB. Encapsidated adenovirus mini-chromosome-mediated delivery of genes to the retina: application to the rescue of photoreceptor degeneration. Hum Mol Genet 1998; 7:1893–1900.

146. Kochanek S, Clemens PR, Mitani K, Chen HH, Chan S, Caskey CT. A new adenoviral vector: replacement of all viral coding sequences with 28 kb of DNA independently expressing both full-length dystrophin and beta-galactosidase. Proc Natl Acad Sci USA 1996; 93:5731–5736.

147. Grable M, Hearing P. Adenovirus type 5 packaging domain is composed of a repeated element that is functionally redundant. J Virol 1990; 64:2047–2056.

148. Clemens PR, Kochanek S, Sunada Y, Chan S, Chen HH, Campbell KP, Caskey CT. In vivo muscle gene transfer of full-length dystrophin with an adenoviral vector that lacks all viral genes. Gene Ther 1996; 3:965–972.

149. Chen HH, Mack LM, Kelly R, Ontell M, Kochanek S, Clemens PR. Persistence in muscle of an adenoviral vector that lacks all viral genes. Proc Natl Acad Sci USA 1997; 94:1645–1650.

150. Floyd SS Jr, Clemens PR, Ontell MR, Kochanek S, Day CS, Yang J, Hauschka SD, Balkir L, Morgan J, Moreland MS, Feero GW, Epperly M, Huard J. Ex vivo gene transfer using adenovirus-mediated full-length dystrophin delivery to dystrophic muscles. Gene Ther 1998; 5:19–30.

151. Hardy S, Kitamura M, Harris-Stansil T, Dai Y, Phipps ML. Construction of adenovirus vectors through Cre-lox recombination. J Virol 1997; 71:1842–1849.

152. Lieber A, He CY, Kirillova I, Kay MA. Recombinant adenoviruses with large deletions generated by Cre-mediated excision exhibit different biological properties

compared with first-generation vectors in vitro and in vivo. J Virol 1996; 70:8944–8960.

153. Parks RJ, Chen L, Anton M, Sankar U, Rudnicki MA, Graham FL. A helper-dependent adenovirus vector system: removal of helper virus by Cre-mediated excision of the viral packaging signal. Proc Natl Acad Sci USA 1996; 93:13565–13570.

154. Chen L, Anton M, Graham FL. Production and characterization of human 293 cell lines expressing the site-specific recombinase Cre. Somat Cell Mol Genet 1996; 22:477–488.

155. Parks RJ, Graham FL. A helper-dependent system for adenovirus vector production helps define a lower limit for efficient DNA packaging. J Virol 1997; 71:3293–3298.

156. Parks RJ, Bramson JL, Wan Y, Addison CL, Graham FL. Effects of stuffer DNA on transgene expression from helper-dependent adenovirus vectors. J Virol 1999; 73:8027–8034.

157. Schiedner G, Morral N, Parks RJ, Wu Y, Koopmans SC, Langston C, Graham FL, Beaudet AL, Kochanek S. Genomic DNA transfer with a high-capacity adenovirus vector results in improved in vivo gene expression and decreased toxicity. Nat Genet 1998; 18:180–183.

158. Morral N, Parks RJ, Zhou H, Langston C, Schiedner G, Quinones J, Graham FL, Kochanek S, Beaudet AL. High doses of a helper-dependent adenoviral vector yield supraphysiological levels of alpha 1-antitrypsin with negligible toxicity. Hum Gene Ther 1998; 9:2709–2716.

159. Morral N, O'Neal W, Rice K, Leland M, Kaplan J, Piedra PA, Zhou H, Parks RJ, Velji R, Aguilar-Cordova E, Wadsworth S, Graham FL, Kochanek S, Carey KD, Beaudet AL. Administration of helper-dependent adenoviral vectors and sequential delivery of different vector serotype for long-term liver-directed gene transfer in baboons. Proc Natl Acad Sci USA 1999; 6:12816–12821.

160. Morsy MA, Gu M, Motzel S, Zhao J, Lin J, Su Q, Allen H, Franlin L, Parks RJ, Graham FL, Kochanek S, Bett AJ, Caskey CT. An adenoviral vector deleted for all viral coding sequences results in enhanced safety and extended expression of a leptin transgene. Proc Natl Acad Sci USA 1998; 95:7866–7871.

161. Alemany R, Dai Y, Lou YC, Sethi E, Prokopenko E, Josephs SF, Zhang WW. Complementation of helper-dependent adenoviral vectors: size effects and titer fluctuations. J Virol Methods 1997; 68:147–159.

162. Balague C, Zhou J, Dai Y, Alemany R, Josephs SF, Andreason G, Hariharan M, Sethi E, Prokopenko E, Jan H, Lou YC, Hubert-Leslie D, Ruiz L, Zhang WW. Sustained high-level expression of full-length human factor VIII and restoration of clotting activity in hemophilic mice using a minimal adenovirus vector. Blood 2000; 95:820–828.

163. Aurisicchio L, Delmastro P, Salucci V, Paz OG, Rovere P, Ciliberto G, La Monica N, Palombo F. Liver-specific alpha 2 interferon gene expression results in protection from induced hepatitis. J Virol 2000; 74:4816–4823.

164. Dong JY, Wang D, Van GF, Pascual DW, Frizzell RA. Systematic analysis of repeated gene delivery into animal lungs with a recombinant adenovirus vector. Hum Gene Ther 1996; 7:319–331.

165. St George J, Pennington SE, Kaplan JM, Peterson PA, Kleine LJ, Smith AE, Wads-

worth SC. Biological response of nonhuman primates to long-term repeated lung exposure to Ad2/CFTR-2. Gene Ther 1996; 3:103–116.

166. Kaplan JM, St George J, Pennington SE, Keyes LD, Johnson RP, Wadsworth SC, Smith AE. Humoral and cellular immune responses of nonhuman primates to long-term repeated lung exposure to Ad2/CFTR-2. Gene Ther 1996; 3:117–127.

167. Parks RJ, Evelegh CM, Graham FL. Use of helper-dependent adenoviral vectors of alternative serotypes permits repeat vector administration. Gene Ther 1999; 6: 1565–1573.

168. Mack CA, Song WR, Carpenter H, Wickham TJ, Kovesdi I, Harvey BG, Magovern CJ, Isom OW, Rosengart T, Falck-Pedersen E, Hackett NR, Crystal RG, Mastrangeli A. Circumvention of anti-adenovirus neutralizing immunity by administration of an adenoviral vector of an alternate serotype. Hum Gene Ther 1997; 8:99–109.

169. Maione D, Wiznerowicz M, Delmastro P, Cortese R, Ciliberto G, La Monica N, Savino R. Prolonged expression and effective readministration of erythropoietin delivered with a fully deleted adenoviral vector. Hum Gene Ther 2000; 11:859–868.

170. Sandig V, Youil R, Bett AJ, Franlin LL, Oshima M, Maione D, Wang F, Metzker ML, Savino R, Caskey CT. Optimization of the helper-dependent adenovirus system for production and potency in vivo. Proc Natl Acad Sci USA 2000; 97: 1002–7.

171. Burcin MM, Schiedner G, Kochanek S, Tsai SY, O'Malley BW. Adenovirus-mediated regulable target gene expression in vivo. Proc Natl Acad Sci USA 1999; 96: 355–360.

172. Recchia A, Parks RJ, Lamartina S, Toniatti C, Pieroni L, Palombo F, Ciliberto G, Graham FL, Cortese R, La Monica N, Colloca S. Site-specific integration mediated by a hybrid adenovirus/adeno-associated virus vector. Proc Natl Acad Sci USA 1999; 96:2615–2620.

173. Snyder RO, Miao C, Meuse L, Tubb J, Donahue BA, Lin HF, Stafford DW, Patel S, Thompson AR, Nichols T, Read MS, Bellinger DA, Brinkhous KM, Kay MA. Correction of hemophilia B in canine and murine models using recombinant adeno-associated viral vectors. Nat Med 1999; 5:64–70.

174. Kay MA, Manno CS, Ragni MV, Larson PJ, Couto LB, McClelland A, Glader B, Chew AJ, Tai SJ, Herzog RW, Arruda V, Johnson F, Scallan C, Skarsgard E, Flake AW, High KA. Evidence for gene transfer and expression of factor IX in haemophilia B patients treated with an AAV vector [see comments]. Nat Genet 2000; 24: 257–261.

175. Xia D, Henry LJ, Gerard RD, Deisenhofer J. Crystal structure of the receptor-binding domain of adenovirus type 5 fiber protein at 1.7 A resolution. Structure 1994; 2:1259–1270.

176. Xia D, Henry L, Gerard RD, Deisenhofer J. Structure of the receptor binding domain of adenovirus type 5 fiber protein. Curr Top Microbiol Immunol 1995; 199: 39–46.

177. Roelvink PW, Mi Lee G, Einfeld DA, Kovesdi I, Wickham TJ. Identification of a conserved receptor-binding site on the fiber proteins of CAR-recognizing adenoviridae. Science 1999; 286:1568–1571.

178. Kirby I, Davison E, Beavil AJ, Soh CP, Wickham TJ, Roelvink PW, Kovesdi I, Sutton BJ, Santis G. Identification of contact residues and definition of the CAR-binding site of adenovirus type 5 fiber protein. J Virol 2000; 74:2804–2813.

179. Wickham TJ, Tzeng E, Shears LL, Roelvink PW, Li Y, Lee GM, Brough DE, Lizonova A, Kovesdi I. Increased in vitro and in vivo gene transfer by adenovirus vectors containing chimeric fiber proteins. J Virol 1997; 71:8221–8229.

180. Krasnykh V, Dmitriev I, Mikheeva G, Miller CR, Belousova N, Curiel DT. Characterization of an adenovirus vector containing a heterologous peptide epitope in the HI loop of the fiber knob. J Virol 1998; 72:1844–1852.

181. Bouri K, Feero WG, Myerburg MM, Wickham TJ, Kovesdi I, Hoffman EP, Clemens PR. Polylysine modification of adenoviral fiber protein enhances muscle cell transduction. Hum Gene Ther 1999; 10:1633–1640.

182. Vanderkwaak TJ, Wang M, Gomez-Navarro J, Rancourt C, Dmitriev I, Krasnykh V, Barnes M, Siegal GP, Alvarez R, Curiel DT. An advanced generation of adenoviral vectors selectively enhances gene transfer for ovarian cancer gene therapy approaches. Gynecol Oncol 1999; 74:227–234.

8

Gene Delivery Technology: Adeno-Associated Virus

Barrie J. Carter
Targeted Genetics Corporation, Seattle, Washington, U.S.A.

I. INTRODUCTION

Adeno-associated viral (AAV) vectors have a number of advantageous properties as gene delivery vehicles. The parental virus does not cause disease. AAV vectors are the smallest and most chemically defined particulate gene delivery system, and they potentially could be classified as well-characterized biologics for therapeutic applications. AAV vectors contain no viral genes that could elicit undesirable cellular immune responses and appear not to induce inflammatory responses. The primary host response that might impact the use of AAV vectors is a neutralizing antibody response. The vectors readily transduce dividing or nondividing cells and can persist essentially for the lifetime of the cell. Thus, AAV vectors can mediate impressive long-term gene expression when administered in vivo. Consequently, these vectors may be well suited for application where the vector need be delivered only infrequently and where any potential host antibody response to the AAV capsid protein may be less inhibitory. One limitation for AAV vectors is the limited payload capacity of about 4.5 kb per particle.

The lack of good producer systems that could generate high-titer vectors was an early obstacle to development of AAV vectors, but there have been significant advances in upstream production of AAV vectors as well as advances in downstream purification. Clinical development of AAV vectors has progressed significantly, and studies of AAV vectors in clinical trials in cystic fibrosis patients have been extended [1,2]. Other AAV vectors have now entered clinical trials for hemophilia B and limb girdle muscular dystrophy [3,4].

The basic elements of AAV molecular biology, its life cycle, and some of the interesting and seminal early studies on AAV were recently reviewed [5]. Other references provide general reviews of AAV [6,7] and extensive summaries on earlier development of AAV vectors [8–13]. This chapter will summarize improvements in vector production and additional studies on the persistent gene expression with AAV vectors as well as studies relating to host cell responses. It also will describe recent studies on the biology of AAV vectors including uptake into cells, trafficking to the cell nucleus, and the mechanism of genome persistence. These studies suggest possible ways to modify the biological targeting of AAV vectors, to enhance transduction efficiency, and to overcome the packaging limitation.

Most of the early studies on AAV used AAV serotype 2 but genomes of additional AVV serotypes have been cloned and sequenced [14]. The biological properties of individual serotypes may include some differences in the interactions with cellular receptors [15]. Other studies are now providing information on the structure of the AAV capsid and how its interaction with the cell may be modified. Thus, together with additional studies on cellular trafficking pathways, it may be possible to modify the targeting of AAV vectors as well as to enhance their transduction efficiency [16].

Studies on persistence of vector genomes indicate that this involves formation of polymeric structures or concatemers. Concatemers also can be formed between two different vector genomes introduced into the same cell. This provides a way partly to circumvent the packaging limit of AAV by dividing a gene expression cassette between two AAV vectors ("dual vectors") and allowing recombination in the cell to generate the intact expression cassette.

II. ADENO-ASSOCIATED VIRUS

A. Biology

AAV is a small, DNA-containing defective parvovirus that grows only in cells in which certain functions are provided by a coinfecting-helper virus, which is generally an adenovirus or a herpesviruses [6,7]. AAV has not been associated with the cause of any disease but has been isolated from humans [17]. AAV has both a broad cell and tissue specificity, but there may be some limitations to this specificity or some differences in efficiency of transduction of different tissues and organs. These differences may reflect the receptor used for entry into cells as well as a second set of parameters that relate to the nature of the helper function provided by the helper viruses [18].

B. Molecular Biology and Replication

AAV is a nonenveloped particle about 20 nm in diameter comprising a protein coat, containing the three capsid proteins VP1, VP2, and VP3, which encloses a linear single-stranded DNA (ssDNA) genome having a molecular weight of 1.5×10^6. Each particle contains only one single-stranded genome, but strands of either complementary sense, "plus" or "minus" strands, are packaged into individual particles. Several serotypes of AAV have been distinguished [17]. AAV1, AAV2, AAV3, and AAV4 have extensive DNA homology and significant serological overlap, but AAV5 is somewhat less related and a recently characterized AAV 6 may be a recombinant between AAV1 and AAV2 (see below).

The AAV DNA genome is 4.7 kb long with one copy of the 145 nucleotide long inverted terminal repeat (ITR) at each end and a unique sequence region of 4.4 kb. The unique region contains two main open reading frames for the *rep* and *cap* genes in the left and right half of the genome, respectively. The ITR sequences are required in cis to provide functional origins of replication (ori) as well as signals for encapsidation, integration into the cell genome, and rescue from either host cell chromosomes or recombinant plasmids.

In a productive infection of cells in the presence of a helper virus such as adenovirus [9,19], the infecting parental AAV single-stranded genome is converted to a parental duplex replicating form (RF) by a self-priming mechanism. This takes advantage of the ability of the (ITR) to form a hairpin structure in a process that occurs in the absence of a helper virus, but may be enhanced by the helper virus. The parental RF molecule is then amplified to form a large pool of progeny RF molecules in a process which requires both the adenoviral helper functions and the AAV *rep* gene products, Rep78 and Rep68. AAV RF genomes are a mixture of head-to-head or tail-to-tail multimers or concatemers and are precursors to progeny ssDNA genomes that are packaged into preformed empty AAV capsids.

The conversion of the incoming ssDNA genome to a double-stranded molecule is an important event required for efficient function of AAV as a gene delivery vehicle, because it provides the template for transcription and gene expression. This process is termed single-stranded conversion or metabolic activation [20], and the rate at which it occurs depends largely on the physiological state of the cell, but the process may be accelerated by treatment of the cell with genotoxic agents or by certain functions of the helper virus. In some cases, this process may limit the transduction efficiency or alter the rate at which transduction can occur (see below).

The kinetics of AAV replication and assembly have been investigated [19,21]. When human HeLa or 293 cells are simultaneously infected with AAV and adenovirus, there are three phases of the growth cycle. In the first 8–10 hrs, the cell becomes permissive for AAV replication as a result of expression of a subset of the adenoviral genes including *E1*, *E2A*, *E4*, and the *VA RNA* genes. During this period, the infecting AAV genome is converted to the initial parental duplex RF DNA by self-priming from the terminal, base-paired 3' hydroxyl group provided by the ability of the ITR to form a self-paired hairpin structure. This initial generation of a duplex genome also provides a template for transcription and expression of AAV proteins. This is followed by a 10-hr period of Rep and Cap protein synthesis and accumulation of a RF DNA to a constant level. Finally, in the following 5–10 hrs, there is an assembly phase in which the bulk of mature progeny AAV particles accumulate. These studies showed that the AAV growth cycle is highly coordinated with respect to expression of rep and cap proteins and the relationship between replication and assembly [21,22]. Any vector production process which mimics this must necessarily provide the rep and cap functions by complementation and may therefore decrease the efficiency of this highly regulated process.

The cloning of infectious AAV genomes in bacterial plasmids facilitated a molecular genetic analysis of AAV as well as the development of AAV vectors [23,24]. These studies showed that the *rep* and *cap* genes are required in trans to provide functions for replication and encapsidation of viral genomes, respectively, and that the ITR is required in cis [6,9,32]. The *rep* gene is expressed as family of four proteins, Rep78, Rep68, Rep 52, and Rep40, that comprise a common internal region sequence but differ in their amino- and carboxyl-terminal regions. The *cap* gene encodes the proteins VP1, VP2, and VP3 that share a common overlapping sequence but VP1 and VP2 contain an additional amino-terminal sequence. All three proteins are required for capsid production. Genetic studies, together with additional biochemical studies, show that Rep68 and Rep78 are required for replication, that VP2 and VP3 are required to form the capsid, and that Rep52 and Rep40 appear to act in concert with VP1 to encapsidate the DNA and stabilize the particles [9,11].

C. Latency

Infection of cells by AAV in the absence of helper functions may result in persistence of AAV as a latent provirus integrated into the host cell genome. This provided an important demonstration of a way in which AAV can survive in a cell if conditions are not permissive for replication. When such cells containing an AAV provirus cells are superinfected with a helper virus such as adenovirus, the integrated AAV genome is rescued and replicated to yield a burst of infectious progeny AAV particles. In cultured cells, AAV exhibits a high preference for integration at a specific region, the AAVS1 site, on human chromosome 19 [25,26].

Analysis of cells carrying latent AAV showed that they contained a relatively low number of AAV genomes that were integrated into the host cell chromosome mostly as tandem, head-to-tail repeats. Early studies of cells stably transduced with AAV vectors also showed that most stable copies in the cell existed as tandem repeats with a head-to-tail conformation [27]. The tandem repeats could not have been formed by the normal AAV replication process that generates head-to-head or tail-to-tail concatemers. This was an important clue to the mechanism of persistence of AAV that presaged subsequent studies with AAV vectors (see below).

Analysis of chromosomal flanking sequences showed that for wild-type AAV a significant proportion of these integration events occurred in a defined region [25,26]. When wild-type AAV infects human cell lines in culture, up to 50–70% of these integration events occur at a region know as the AAVS1 site on the q arm of chromosome 19. The efficiency and specificity of this process is mediated by the AAV *rep* gene and the cellular DNA sequences at AAVS1, and therefore *rep*-deleted AAV vectors do not retain specificity for integration into the chromosome 19 region [28–31].

D. Permissivity

The precise mechanism of the helper function provided by adenovirus or other helper viruses has not been clearly defined. These helper functions may be complex but relatively indirect and probably affect cellular physiology rather than provide viral proteins that specifically interact with the AAV replication system. Studies with adenovirus [18] have clearly shown that only a limited set of adenoviral genes are required and these comprise the early genes *E1A*, *E1B*, *E2A*, and *E4orf6* and *VA RNA*.

Most of the experimental evidence indicates that the helper viral genes do not appear to provide enzymatic functions required for AAV DNA replication and that these functions are provided by AAV rep protein and the cellular DNA replication apparatus [18]. This is also consistent with the observations that in certain cells lines, particularly if they are transformed with an oncogene, helper-independent replication of AAV DNA can occur if the cells are also treated with genotoxic reagents such as ultraviolet (UV) or x-irradiation or with hydroxyurea [32–34]. In these circumstances, a small proportion of the cells can be rendered permissive for AAV replication but the level of replication production of infectious AAV was very low.

III. AAV VECTORS

A. Design

The general principles of AAV vector construction [9,11,12] are based upon modifying the molecular clones by substituting the AAV coding sequence with foreign DNA to generate a vector plasmid. In the vector, only the cis-acting ITR sequences must be retained intact. The vector plasmid is introduced into producer cells which are also rendered permissive by an appropriate helper virus such as adenovirus. In order to achieve replication and encapsidation of the vector genome into AAV particles, the vector plasmid must be complemented for the trans-acting AAV *rep* and *cap* functions that were deleted in construction of the vector plasmid. AAV vector particles can be purified and concentrated from lysates of such producer cells.

The AAV capsid has three important effects for AAV vectors. There is a limit of about 5 kb of DNA that can be packaged in an AAV vector particle. This places constraints on expression of very large cDNAs and may also constrain the ability to include regulatory control sequences in the vector. The capsid also interacts with the AAV receptor and coreceptors, and thus mediates cell entry. The capsid may also induce humoral immune responses that could limit delivery of AAV vectors for some applications.

Except for the limitation on packaging size and the requirements for ITRs, there are no obvious limitations on the design of gene cassettes in AAV vectors. The ITR can act as a transcription promoter [35,36] but does not interfere with other promoters. Tissue-specifc promoters appear to retain specificity [37,38], and a number of other regulated expression systems have now been used successfully in AAV vectors (see below). Introns function in AAV vectors and may enhance expression, and more than one promoter and gene cassette can be inserted in the same vector. Importantly, transcription from AAV does not seem to be susceptible to in vivo silencing as shown by expression for over a year after intramuscular delivery in rodents [39–41].

B. Production

It is noteworthy that the production of AAV is highly efficient, and AAV has one of the largest burst sizes of any virus. Thus, following infection of cells with AAV and adenovirus as helper, the burst size of AAV may be well in excess of 100,000 particles per cell [11]. This is important in developing vector production systems, because it implies that a high yield of AAV vector particles per cell theoretically is attainable. Attaining high specific productivity is of crucial importance in developing scaled-up production, because the ability to obtain maximum yields ideally requires both high specific productivity (yield of particles per cell) or large biomass (total number of cells). Maximizing the specific productivity may avoid unnecessary increases in biomass.

The cytostatic properties of the AAV rep protein presented an obstacle to generation of stable packaging cell lines for producing AAV vectors [42]. Consequently, AAV vector production initially was based on transfection of a vector plasmid and a second plasmid, to provide complementing *rep* and *cap* functions, into adenovirus-infected cells, usually the transformed human 293 cell line. The original vector production systems yielded a mixture of AAV vector particles and adenoviral particles and exhibited relatively low specific productivity [9,11]. Furthermore, recombination between the vector plasmid and complementing plasmids generates wild-type AAV (wtAAV), pseudo–wild-type or replication-competent AAV (rcAAV) or other recombinant AAV species [43,44].

There have now been improvements in upstream production of AAV vectors by DNA transfection-based methods, as well as development of new cell-based AAV production systems that do not require DNA transfection. Also, transfection systems in which adenoviral infection is replaced by DNA transfection with the relevant adenoviral genes are now available. Both the transfection and cell-based approaches can give much higher productivity—in excess of 10^4 vector particles per cell. However, the cell-based systems are probably more amenable to scale-up for commercial production than are DNA transfection systems. Furthermore, significant improvements in downstream of AAAV vectors has been made by using chromatographic procedures.

1. Complementation Systems

Several approaches have been taken to enhance the upstream production of AAV vectors. First, various modifications have been made to the complementing rep–cap cassette in an attempt to enhance specific productivity and to decrease production of rcAAV. Although one study indicated that expression of rep and cap proteins may be limiting [45], other reports [46,47] suggested that production of cap proteins were limiting owing to downregulation of cap by increased production of rep. This is surprising, because in the AAV replication cycle rep protein increases expression of cap. Perhaps these apparent differences reflect the nonphysiological regulation of rep and cap expression that may occur in complementation systems as was alluded to above. Nevertheless, a packaging plasmid which has the Rep78/69 expression downregulated by changing the initiation codon AUG to ACG was reported to give higher cap expression and higher yields of vector particles [47].

The only adenoviral genes required for full helper function for AAV are *E1*, *E2A*, *E4*, and *VA*, and transfection of the latter three genes into cells which contain the *E1* genes, such as human 293 cells, can provide full permissivity for AAV [5]. Thus, infectious adenovirus (Ad) can be replaced as the helper with a plasmid containing only the adenoviral *E2A*, *E4*, and *VA* genes which, together with the *E1A* genes supplied by 293 cells, provide a complete helper function in the absence of adenoviral production [48,49]. The three plasmid transfection system was simplified further by combining all three adenoviral genes (*E1A*, *E4*, *VA*) and the *rep–cap* genes into a single plasmid [50]. In general, all of these approaches increased vector productivity compared to ealier systems such as pAAV/Ad [51], and productivities of at least 10^4 particles per cell have been reported. Nevertheless, these approaches still require DNA transfection and may be unwieldy for production scale-up.

An alternate approach to AAV vector production is to generate cell lines that contain the *rep-* and *cap-* complementing systems, or the vector genome, or both. In order to avoid DNA transfection, the cells must still be infected by a helper virus, adenovirus, but this can be removed readily as a result of advances in downstream purification processes (see below). Rescue of vector from a producer line having the vector stably integrated was demonstrated by transfecting a *rep–cap* helper plasmid and infecting with adenovius [52]. Also, stable cell lines containing a *rep* gene capable of generating functional *rep* protein were constructed by Yang et al. [42] by replacing the P_5 promoter with a heterologous transcription promoter.

Clark et al. [53,54] generated cell lines containing the *rep* and *cap* gene cassettes but deleted for AAV ITRs. Infection of these cells with adenovirus activates *rep* and *cap* gene expression. Furthermore, the vector can be stably incorporated into the packaging cells to yield AAV vector producer cell lines which need only be infected with adenovirus to generate vector [53]. Producer cell lines provide a scalable AAV vector production system that does not require manufacture of DNA.

A modification of the packaging cell line method is to use a similar cell line containing a *rep–cap* gene cassette that is infected with an Ad/AAV hybrid virus. The Ad/AAV hybrid is an *E1* gene-deleted adenovirus containing the AAV ITR vector cassette [55,56]. After infection of cells containing the *rep–cap* genes, the AAV–ITR cassette is excised from the Ad/AAV and amplified and packaged into AAV particles. This allows the same packaging cell line to be used for production of different AAV vectors simply by changing the Ad/AAV hybrid virus but requires coinfection with adenovirus to provide the E1 gene function.

Another packaging cell system was described in which the packaging cell contains both a *rep–cap* cassette and the AAV ITR vector cassette with both cassettes attached to an semian virus 40 (SV40) replication origin [57]. The cells also carry an SV40 T-antigen gene that is under control of the *tet*-regulated system such that addition of doxycycline induces T antigen that in turn results in amplification of the *rep–cap* and the vector cassettes. Subsequent infection of the cells with adenovirus renders them permissive for vector production. Another modification of the cell-based approach utilizes an HSV/AAV hybrid virus in which the AAV *rep–cap* genes are inserted into the HSV genome. This HSV/AAV *rep–cap* virus can generate AAV vector when infected into cell lines along with a transfected AAV vector plasmid or into cell lines carrying an AAV vector provirus [58,59].

It is worth noting that the hybrid Ad/AAV viruses can also be used as delivery vehicles for AAV vectors [60–62], but this might suffer some disadvantages such as induction of innate immune responses characteristic of the adenoviral capsid interaction with cells. Similarly, an HSV/AAV hybrid virus, having an AAV ITR vector cassette inserted in the herpes simplex virus (HSV) genome was infectious for neural cells, and gene expression from the AAV vector was maintained over 2 weeks even though the HSV genome sequences were lost from the vector [63].

2. Pseudo–Wild-Type or Replication-Competent AAV

The earliest AAV vectors [69,11] were produced by cotransfection with helper plasmids that had overlapping homology with the vector and this generated vector particles contaminated with wild-type AAV due to homologous recombination. Reduction of the overlapping AAV sequence homology between the vector and helper plasmids such as in the widely used pAAV/Ad packaging system of Samulski et al. [43,51,52] reduced, but did not eliminate, generation of wild-type AAV. Flotte et al. [52] described a combination of vector plasmid and packaging plasmid in which the AAV region containing the P_5 promoter was not present in either plasmid. This prevented generation of wtAAV but some pseudo–wild-type AAV was generated at lower frequency by nonhomologous recombination [43]. This nonhomologous recombination may be decreased in a packaging system (split-gene packaging) carrying *rep* and *cap* genes in separate cassettes [43] that would require three or four recombination events to generate rcAAV. An alternate approach to decreasing pseudo–wild-type or recombinant AAV is to insert a large intron within the *rep* gene in the helper plasmid so that any recombinants would tend to be too large to package in AAV particles [64].

It is likely that all vector production systems may have a propensity to generate pseudo–wild-type AAV or other recombinants of AAV at some, albeit low, frequency, because it not possible to eliminate nonhomologous recombination in DNA. This may be more likely in DNA transfection systems, especially in view of the very large genome numbers that are normally introduced into transfected cells. Thus, packaging systems in which transfection is avoided may help to reduce the frequency of such recombination. To detect such pseudo–wild-type or recombinant species in AAV vector, an assay that employs two cycles of amplification by replication and then a sensitive readout such as hybridization or polymerase chain reaction (PCR) (rather than rep or cap immunoassays) is likely to be preferred. Further, the availability of cells line providing *rep*- and *cap*-complementing functions allows evaluation of such recombinant species without having to add wtAAV, and this will facilitate detection of species that are only very poorly replication competent (see below).

3. Purification

Historically, AAV was purified by proteolytic digestion of cell lysates in the presence of detergents followed by banding in CsCl gradients to concentrate and purify the particles and separate adenoviral particles. Significant progress has been made in the downstream processing of AAV vectors and this has led to much higher quality and purity. This is critically importart for preclinical studies and clinical trials. The reliance upon the original CsCl centrifugation techniques is being abandoned, because it is a cumbersome procedure that does not provide high purity, it may inactivate some AAV vector, and it is difficult to envision its use for commercial production.

Several groups have employed nonionic iodixonal gradients as an initial bulk-recovery method [65,66]. A variety of chromatogaphic methods including ion exchange and antibody and heparin affinity resins have been used in both conventional and high-performance liquid chromatography (HPLC) formats have been employed [50,67–69]. Chromatographic procedures in general are more acceptable for the biopharmaceutical manufacturing as therapeutic applications are developed for AAV vectors.

4. Assay

AAV vectors are usually characterized in terms of the total number of particles and the infectivity titer as determined in a replication assay. This in principle allows determination of the ration of the particle to infectivity titer of a vector preparation and could allow for universal comparison of vectors. Universal assays are being discussed by various groups in the field, and eventually some universal assays may be adopted. This will still not provide a measurement of the particular biological efficiency or transducing titer of individual vector preparations, because this will obviously depend on the particular gene that is expressed from the vector. Further, measurement of transducing titers on cell lines in vitro will not necessarily be predictive for biological activity in a particular target cell in vivo.

At the present time, the particle number is usually measured in assays that determine the number of vector genomes present after DNAse treatment of a vector preparation. The number of DNA genomes is determined in slot-blot DNA hybridization assays or more recently by TaqMan real-time DNA PCR [68]. It is assumed that each particle contains one vector genome, and the number of DNase-resistant particles (DRP) is calculated. The infectivity can be measured in either an infectious center assay or an endpoint assay in cells that are also infected with adenovirus and wtAAV [50]. This three-hit assay is difficult to perform reproducibly and robustly. More recently, the availability of cells expressing *rep* and or *cap* has allowed determination of infectivity in a two-hit system in such cells by infecting with the vector and the helper adenovirus [70,71]. These infectivity assays can be configured in a high-throughput mode which may be particularly useful for process development applications [72]. More recently, a rapid assay using analytical HPLC has been used which can give a particle assay readout on a sample in about 20 min [68].

IV. GENERAL PROPERTIES OF AAV VECTORS

AAV vectors have been extensively studied in cells grown in vitro [12,13] and can transduce many types of cells and cell lines. However, the efficiency of this transduction depends on whether the cells are stable, transformed cell lines or primary cultures and also whether the cells are stationary or dividing [73–75]. Two contradictory factors affect transduction in vitro. One is metabolic activation or single-stranded conversion which tends to occur more quickly in dividing cells [76,77]. A second factor is the AAV genome persistence. By contrast, in cells that are rapidly dividing in culture, AAV vectors may give very efficient transient transduction but may be rapidly diluted out as cell division proceeds.

Transduction in cultures of nondividing cells may depend upon the longer time required for conversion of single strands to duplexes. This process presumably requires cellular DNA polymerase and can be enhanced by adenoviral infection or by procedures that stimulate DNA synthesis including DNA-damaging agents such UV, gamma irradiation, genotoxic agents hydroxyurea, topoisomerase inhibitor drugs such as topoisomerase inhibitor, or the adenovirus E4 Orf6 gene [76–78]. However, these treatments are not required for in vivo applications, because in nondividing cells, there is time for generation of double-stranded templates to occur. Further, many in vivo gene therapy targets comprise nondividing or slowly turning over differentiated cells that are excellent targets for AAV transduction and long-term expression (see below). As with many other gene delivery vectors, in vitro assays are useful to study design and confirm biological activity of AAV vectors but are not predictable for behavior in vivo [41]. After confirming biological activity in vitro, most studies with AAV vectors are now focused on direct in vivo delivery.

AAV vectors have proven to be remarkably efficient for long-term gene expression in vivo. Most of these in vivo studies have addressed differentiated target cells that are either slowly proliferating or postmitotic and nondividing, and these types of cells may be the best targets for AAV gene therapy. AAV vectors expressed the human cystic fibrosis transmembrane conductance regulator (CFTR)cDNA in the airway epithelium of rabbits and rhesus macaques for at least 6 months [79,80]. Direct stereotactic injection of AAV vectors expressing reporter genes or tyrosine hydroxylase into the rodent or nonhuman primate brain can result in gene expression for at least 3–6 months [81,81]. Prolonged expression of reporter genes for up to 18 months was achieved in immunocompetent mice by direct intramuscular injection of AAV vectors [39,41]. Delivery of AAV vectors expressing a clotting factor IX cDNA in mice by tail vein or portal vein injection lead to factor IX expression for up to a year [83,84]. Prolonged persistence of a therapeutic level of human erythropoietin (huEpo) was achieved by intramuscular injection of an AAV huEpo vector [40]. Subcutaneous injection into the eye can mediate robust, persistent expression in the retinal pigment cells [38].

V. VECTOR METABOLISM

Some early studies of AAV vector led to the suggestion that the major factor limiting the efficiency of transduction by AAV vectors was the process of converting the single-stranded genome to a duplex molecule. However, as already noted above, this does not appear to be the most limiting factor in vivo. More recent studies are beginning to reveal a complex pathway of events that impact upon the function of AAV vectors. Thus, delivery of the AAV vector genome to the cell nucleus may be influenced by the availability of specific receptors and coreceptors for AAV binding and entry and by cellular trafficking including a potential diversion of AAV particles into a ubiquitin-mediated proteosome degradation pathway.

After the genome is successfully delivered into the nucleus, a novel succession of events results in the vector becoming circularized and ligated into larger concatameric molecules that appear to involve intramolecular and intermolecular recombination. Most of these concatameric molecules appear to be maintained for prolonged periods, perhaps for the lifetime of the cell, in the nucleus as episomal molecules, and very few if any integrate into the host genome.

A. Cellular Binding and Trafficking

AAV appears to have a broad host range and different AAV serotypes infect in vitro in many human cells and a variety of simian and rodent cells. This implied that any cellular receptor for AAV might be relatively common or there might be several different receptors. Recent experiments have shown that AAV2 particles can bind to cells via heparin sulfate proteoglycans (HSPGs) in a manner such that binding and subsequent infection are competed in a dose-dependent fashion by soluble heparin and that cell lines lacking HSPG were significantly impaired for AAV binding and infection [85]. Thus, heparin appears to be a cellular receptor for AAV2. Additional studies identified two coreceptors that are required for efficient internalization of AAV2. One coreceptor is $\alpha_v\beta_5$ integrin [86] and the other is the human fibroblast growth factor receptor 1 (FGFR1) [87].

The $\alpha_v\beta_5$ integrin coreceptor is preferentially located in airway epithelial cells in the more distal cells of the conducting airway, which maybe significant for use of an AAV gene therapy for cystic fibrosis, since the distal airway is the region of the lung most impacted by the disease [88]. FGFR is widely expressed but is particularly abundant in skeletal muscle, as well as neuroblasts and glioblasts in the brain, and these two organs appear to be excellent targets for AAV transduction.

The identification of multiple coreceptors suggests that AAV2 has alternate pathways for cell entry. In support of this concept, AAV2 transduced polarized airway epithelial cells more efficiently via the basolateral surface than the apical surface that correlated with the relative abundance of HSPG [89]. UV irradiation decreased basolateral transduction associated with decreased basolateral HSPG, whereas the same treatment increased apical transduction but did not increase apical HSPG. Rather apical endocytosis of AAV appeared to be enhanced through an independent mechanism [89].

There are five naturally occurring AAV serotypes, 1–5, and another, AAV6, that may be a laboratory recombinant of serotypes 1 and 2 [16]. All have been cloned and sequenced and are being examined as vectors [23,24,90–94]. AAV serotypes 1, 2, and 3 all bind heparin sulfate but serotypes 4 and 5 do not [16,91], so there may be differences in receptor utilization or cellular trafficking. For instance, in a comparison of AAV2, 4, and 5 vectors in rodent brain, type 5 gave the highest transduction efficiency [95]. But comparison of human clotting factor IX in immune-deficient mice using types 1, 2, 3, 4, and 5 vectors suggested that the type 1 vector was most efficient [96]. However, there are significant uncertainties in such comparisons, because precise assays for physical characterization and infectivity of AAV vector particles are not well developed for all serotypes. Further studies may allow a more optimal matching of vector serotype to particular target tissues, and other studies suggest the possibility of altering the cellular binding by modifying the sequence of the capsid by mutagenesis or insertion of additional targeting ligands [97–102].

After binding to the cellular receptor, AAV 2 is internalized by an endocytotic mechanism that for heparin binding appears to be mainly via clathrin-coated pits [103,104], although some clathrin-independent uptake may also occur [103,105]. Additional studies using fluorescent Cy3-conjugated AAV2 vector particles showed that endocytosis can be mediated by an $\alpha_v\beta_5$ integrin/Rac1– dependent mechanism, and that subsequent trafficking to the nucleus requires activation of PI3K pathways as well as functional microtubules and microfilaments [106]. Additional evidence suggests that the AAV particles may be inhibited for entry into the nucleus by being diverted into a ubiquitin-dependent degradation pathway [107]. Treatment of cells with proteosomal inhibitors or ubiquitin ligase inhibitors can significantly increase the transduction efficiency; presumably by allowing the particles to escape this degradation and deliver more genomes to the cell nucleus [105,107].

B. DNA Metabolism

Conversion of AAV vector ssDNA to a duplex genome could be enhanced by a variety of treatments that damage DNA in some in vitro experiments, but as discussed above, this has not proven to be necessary for in vivo applications. Studies in animals show that the conversion of ssDNA to duplex transcription clearly occurs and that robust and therapeutic levels of gene expression can be obtained. In some of these studies and particularly in muscle, expression rises over the first several weeks and then is maintained at a constant level [39–41]. This reflects both the slow process of conversion to duplex genomes and then to concatemeric structures. In rhesus macaques, an AAV CFTR vector was present 3 months after instillation in the lung and appeared to be a dimeric concatemer [108]. In muscle or liver after several weeks, high molecular weight, head-to-tail concatemers are formed [83,109].

The precise mechanism by which the physiological state of the host cell effects the metabolism of AAV DNA including from ssDNA to duplex molecules and then to episomal concatemers or integrated models is not yet well understood. There is good evidence that the single-stranded genome is converted to either linear duplexes or circular duplex molecules [110]. The linear duplexes may be precursors of either the head-to-head and tail-to-tail RF concatemers as seen in normal AAV replication or the head-to-tail concatemers. Interestingly, the adenoviral *E4orf6* gene promotes circularization and head-to-tail concatemers, whereas the adenoviral *E2A* gene promotes RF concatemers [111].

The initial conversion of the vector genome to a duplex molecule might occur by replication and self-priming from the ITR using cellular DNA polymerases, but recently an argument to support a mechanism involving annealing of complementary strands has been advanced [112]. This non-replicative, strand-association model, if it is true, requires the interesting prediction that AAV vectors should demonstrate two-hit kinetics for transduction. The circularization appears to occur by a joining mechanism that is not simply blunt-end joining, because the structure of the monomer circular molecules indicates that they have a "double-D", structure containing one ITR sequence bracketed on either side by a d sequence [110,113]. Blunt-end joining would be expected to abut two copies of the ITR. Generation of the circular concatemers may involve replication, perhaps by rolling circle replication, but this has not been directly demonstrated [112,114]. However, intermolecular recombination has been shown to be involved directly [112,114]. The episomal circular concatemers appear to be the predominant persistent form but over time there appears to be a low level of integration into the host genome.

The evidence that DNA repair and recombination are directly involved in circularization or concatamerization of AAV vector genomes is also supported by recent insights into possible biochemical mechanisms of their formation [115,116]. In fibroblasts from a patient with ataxia-telangiectasia (ATM), there was greatly enhanced formation of AAV vector circular forms and enhanced integration of the head-to-tail concatemers as proviral genomes [115]. The *ATM* gene is a PI-3 kinase that regulates the p53-dependent cell cycle checkpoint and apoptotic pathways, and in these ATM cells the DNA repair systems that normally can be activated by UV irradiation appear to be already activated maximally. Consequently, AAV vectors in these cells yield a high level of transduction and this is not activated further by UV irradiation [115] in contrast to the observations in normal cells [76–78].

A second study recently compared the molecular fate of an AAV vector expressing α1-antitrypsin in rodent muscle using a normal or an immune-deficient strain [116]. The immune-deficient mouse had a mutation in the catalytic subunit (DNA-PKcs) of the DNA-dependent protein kinase (DNA-PK) that plays a role on repair of double-stranded DNA (dsDNA) breaks and in V(D)J recombination by nonhomologous endjoining. In both types of mice, the vector expressed high levels of α1-antitrypsin for over a year. However, in the normal mouse, the vector persisted as circular episomes, but in the immune-deficient mouse, only linear episomes were observed. This implicates the DNA-PK in formation of circular vector genomes.

C. Dual Vectors

The ability of AAV vectors to form concatemers and the evidence that intermolecular recombination is directly involved in their generation has been used to advantage to extend the capacity of AAV vectors beyond the 4.5 payload limit in the dual vector system. It is now possible to divide a gene expression cassette that is up to 10 kb in size between two AAV vectors and following coinfection with both vectors take advantage of the intermolecular recombination to generate the intact expression cassette. This process has been demonstrated in in vitro and in vivo experiments and shows remarkable efficiency [117–120].

Both cis-activation [117] and trans-splicing [118] modes of dual vectors have been described. In cis-activation, one vector carries the transcription promoter or a high-efficiency enhancer of transcription and the other vector carries the gene sequence to be expressed. Recombination after infection places the enhancer in cis with the expression cassette and increases transcription. In the trans splicing mode, one vector carries the transcription promoter and the 5' part of the gene and the other vector carries the 3' part of the gene. By judicious arrangement of splice donor and acceptor sites, the appropriate mRNA can be derived from the readthrough transcript that uses the heteroconcatameric template. Interestingly, the intervening ITR sequence does not appear to inhibit readthrough transcription or RNA splicing [117,118]. It does appear, however, that in these dual vector constructs, some transcription may be seen directly from the ITR either because of its own promoter activity or because of enhancement from the cis-acting elements provided by the second vector [117,118].

Cis-activation was demonstrated using a vector containing a superenhancer element comprising parts of the SV40 and transcription promoters together with a second vector containing the *luciferase* gene. This dual combination gave robust expression of *luciferase* in mouse skeletal muscle at levels that compared well with those from administration of a single vector expressing the same gene [117]. In another study, two vectors containing a LacZ reporter cDNA or an enhancer promoter cassette from human elongation factor EIFα, respectively, were shown to yield reporter gene expression in livers of mice after dual administration by portal vein injection.

Trans-splicing was demonstrated using one vector that contained a transcription promoter and the 5' part of the erythropoietin (epo) genome locus and a second vector that contained the 3' region of the epo locus [118]. Dual injection of the two vectors into mouse skeletal muscle provided therapeutic levels of epo sufficient to protect the mice from adenine-induced anemia. Similarly, another group used two vector carrying the cytomegalovirus (CMV) promoter with the 5' half of the *LacZ* gene or the 3' half of the *LacZ* gene, respectively [120]. These vectors also gave robust expression of intact *LacZ* after dual injection into mouse skeletal muscle.

The dual vector appears to be very efficient and clearly offers a way almost to double the payload capacity of AAV vectors that should extend the utility of this vector system to larger genes. It will require careful analysis to determine what additional safety issues or risks this approach might present for clinical use. For instance, for some applications, such as FVIII protein, it may be deleterious to express a partial protein. Nonetheless, as this system is better characterized, it may well prove to have clinical utility.

D. Host Responses and Toxicity

Host immune responses, including innate immune responses, cellular immunity, or humoral antibody responses, may hinder the use of some gene therapy vectors. However, administration of AAV vectors has not been reported to induce innate immune responses or proinflammatory cytokines. Also, AAV vectors are replication defective and contain no viral genes, so cellular immune responses against the viral components should not be evoked readily. In all of the in vivo studies of AAV vectors in rodents, rabbits, or rhesus macaques, there is no evidence of cellular immune responses to viral components.

AAV vectors may be used mainly for clinical applications requiring only infrequent delivery. But potential humoral immune responses against the viral capsid either preexisting in the human population or induced by vector administration must be considered [5,16]. There is some preliminary information on the likelihood for generation of humoral neutralizing antibody responses to the AAV capsid protein, but more extensive studies will be required to assess whether induction of neutralizing antibody responses will pose any limitations to AAV vectors.

Reinfection of humans by AAV is not prevented by serum neutralizing antibodies [5,16]. However, induction of anti-AAV capsid antibody responses after vector administration may reduce the efficiency of transduction upon re-administration [39,121,122]. This depends upon the route of administration [123–125] and perhaps may depend upon the quality of the vector preparations. In one study [122] in which two AAV vectors expressing as reporter genes either the bacterial β-galactosidase or human alkaline phosphatase were successively administered to lungs of rabbits, expression from the second administered vector was impaired, this was ascribed to a neutralizing antibody response [126]. Similar studies in mice also implied that neutralizing antibodies impaired readministration of AAV vectors, but that this could be partially or completely overcome by transient immunosuppression with anti-CD40 ligand antibodies or soluble CTLA4-immunoglobulin at the time of the initial vector administration [126, 127]. However, interpretation of such studies maybe complicated by the expression of the foreign reporter proteins that could represent confounding variables. Furthermore, other studies showed that immune responses were greatly reduced following airway administration of AAV [123] and that vector transduction can be seen after at least three repeated administrations to the lung of rabbits [128]. Thus, up to three successive administrations of AAV vectors to rabbit lungs over a 20-week period did not prevent gene expression from the third delivery of vector [128].

Whether neutralizing antibody responses to AAV vector capsids will impact applications of AAV vectors remains to be determined. This will most likely require studies in humans to determine if the various animal models such as rodents or rabbits are predictive for the immune response to AAV vectors in humans and whether such immune responses will pose any limitations to their therapeutic application. For uses involving relatively infrequent administration of AAV vectors, transient immune blockade [121–122] may not be an attractive option for therapeutic use of AAV or any other gene delivery vectors. An alternate possibility might be to use vectors that have capsids of different serotypes for subsequent administrations.

Immune responses to the transgene expressed by an AAV vector vary and may depend upon the route of delivery. Both major histocompatibility (MHC) class II–restricted antibody responses and MHC class I cytotoxic T lymphocytes (CTLs) have been reported, but this may vary with the route of administration [125]. In some studies, such as intramuscular delivery in mice, there was no immune response to an expressed foreign reporter gene such as bacterial β-galactosidase, and it was suggested that AAV may be a poor adjuvant or may not readily infect professional antigen-presenting cells in muscle [129,130]. However, an AAV vector expressing the herpes simplex virus type 2 gB protein was delivered intramuscularly into mice and elicited both MHC class I–restricted CTL responses against the gB protein and anti-gB antibodies [131]. Also, following intramuscular delivery of an AAV human factor IX [132] an antibody response, but not a CTL response, was seen. The rules governing a cellular immune response to foreign transgenes following AAV vector delivery remain to be elucidated. However, since there are examples of elicitation of cellular and humoral immune responses mounted against foreign antigens expressed from AAV vectors, then AAV may have utility as a viral vaccination vector.

There have been few, if any, documented reports or indications of toxicity mediated by AAV vectors. The toxicity of AAV vectors has been most extensively tested by delivery of AAV CFTR vector particles directly to the lung in rabbits and nonhuman primates. In rabbits, the vector persisted and expressed for at least 6 months but no short-term or long-term toxicity was observed, and there was no indication of T-cell infiltration or inflammatory responses [79]. Similarly, in rhesus macaques, AAV CFTR vector particles were delivered directly to one lobe of a lung and also persisted and expressed for at least 6 months and no toxicities were detected [80] (see below). Studies in rhesus macaques were also performed to determine if the AAV–CFTR vector could be shed or rescued from a treated individual [108]. AAV–CFTR particles were delivered to the lower right lobe of the lung, and a high dose of adenovirus and wtAAV particles were administered to the nose of the animals. These studies indicated that the vector was not readily rescued and suggested that the probability of vector shedding and transmission to others is likely to be low.

VI. AAV VECTOR APPLICATIONS

AAV vectors were first introduced into clinical trials for treatment of cystic fibrosis (CF) by administration to the airway. On the basis of extensive preclinical studies performed in rabbits and nonhuman primates, an AAV vector expressing the CFTR cDNA has been introduced into clinical trials in CF patients. These are the first clinical trials with an AAV vector [1,2]. Subsequently, AAV vectors have now been entered into clinical trials for treatment of two additional diseases, hemophilia B and limb girdle muscular dystrophy, by intramuscular injection.

A. Cystic Fibrosis

CF is a lethal autosomal recessive disease that is caused by a mutation in the *CFTR* (cystic fibrosis transmembrane conductance regulator) gene. The CFTR protein is a chloride ion channel generally expressed in epithelial cells, and a defect in this protein leads to complex biochemical changes in several organs including lung and often the exocrine pancreas. In the lung, there is decreased mucociliary clearance, increased bacterial colonization, and a chronic neutrophil-dominated inflammatory response which leads to progressive destruction of tissue in the conducting airways. The usual cause of morbidity and eventual mortality in CF patients is progressive loss of lung function. Thus, the goal of a gene therapy for CF is to deliver the CFTR cDNA into the epithelial cells in the conducting airways of the lung.

CF is an attractive target for gene therapy for two reasons. The *CFTR* gene is normally expressed in airway epithelial cells at a very low level of about 1 mRNA molecule per cell. The epithelial cells in the luminal surface of airways can be readily targeted either by direct instillation of vectors or by aerosol inhalation. The earliest in vivo experiments reported with AAV vectors were those to instill an AAV–CFTR vector into rabbit lung using a fiberoptic bronchoscope [79]. Because the size of CFTR cDNA is 4400 nucleotides long, which is at the packaging limit of an AAV vector, the AAV–CFTR vector used the ITR as the transcription promoter [35]. This vector could correct the CF defect in CF epithelial cells in vitro as measured by chloride efflux and electrophysiological patch-clamp assays [35,133].

Extensive preclinical studies of the toxicity of AAV vectors were performed by delivery of AAV–CFTR vector particles directly to the lung in rabbits and nonhuman primates. In either species [79,80], vector persisted and expressed for at least 6 months, but no short-term or long-term toxicity was observed and there was no indication of T-cell infiltration or inflammatory responses. In rhesus macaques [80], no toxicities were observed by several measures including pulmonary function testing, radiology, blood gas analysis, cell counts, and differential in bronchoalveolar lavage or by gross morphological or histopathological examination of organ tissues. There was minimal spread of the vector to organs outside of the lung, and additional studies in rhesus macaques [108] indicated that the vector was not readily rescued and suggested that the probability of vector shedding and transmission to others is likely to be low.

An AAV–CFTR vector has been introduced into a series of clinical trials in CF patients including direct instillation of the vector into the lung and nasal epithelium [1], instillation into the maxillary sinus [2], and by aerosol inhalation to the whole lung. Although the lung is the target for a gene therapy for CF, the maxillary sinuses of CF patients exhibit chronic inflammation and bacterial colonization that is reflective of some aspects of the CF disease in the lung.

In CF patients who have undergone surgical bilateral anstrostomy, the maxillary sinus is directly accessible for instillation of vectors and for electrophysiological measurements and biopsy. In an initial trial, 15 sinuses of CF patients were treated with increasing doses of the AAV–CFTR vector [2]. As measured by PCR analysis of biopsies, instillation of the vector resulted in dose-dependent gene transfer, and the vector genomes persisted in the sinus for at least 70 days after instillation [2]. There were also alterations in the sinus in an electrophysiological measure of the transmembrane epithelial potential difference. In epithelial cell surfaces of CF patients, the transmembrane potential is hyperpolarized compared to normal subjects because of the absence of a functional CFTR chloride channel. Analysis of the transmembrane potential difference across the epithelial surface of the treated sinuses indicated that at the higher doses of vector there was some reversal of the electrophysiological defect. This constitutes presumptive evidence of expression of the CFTR protein from the delivered vector. Administration of the vector was well tolerated and there was little, or no, cytopathic or host immune response.

An additional phase II dose-escalation safety study of vector administration to the whole lung by inhaled aerosol has been completed, and based on this, a phase II randomized, placebo-controlled, double-blinded study of whole lung aerosol administration was initiated late in 2000. AAV–CFTR vector has now been administered to over 60 CF patients in these early clinical trials and has been safe and well tolerated.

B. Hemophilia

Hemophilia B is a severe X-linked recessive disease that results from mutations in the blood coagulation factor IX(FIX) gene, and the absence of functional FIX leads to severe bleeding diathesis. Patients can be treated by protein-replacement therapy. However, the protein has a short half-life and is mainly used therapeutically for acute hemorrhagic episodes. A persistent level of about 5% of normal levels of FIX is believed likely to decrease the risk of spontaneous bleeding into joints and soft tissues. This has stimulated interest in gene therapy approaches to maintain a constitutive, and thus the prophylactic level of FIX protein and AAV vectors may be well suited for this. Also, the FIX cDNA is much smaller than CFTR cDNA and easily fits into an AAV vector.

AAV vectors containing the FIX cDNA and administered one time by injection to muscle or by intravenous delivery to the liver have shown prolonged expression of FIX protein at therapeutically relevant levels in several animal species including hemophilia B mice and dogs [134–141]. Delivery into immunodeficient or immunocompetent mice by portal vein injection into liver resulted in prolonged expression of FIX for up to 36 weeks at levels of 250–2000 ng/mL in serum, which is equivalent to about 20% of the normal human level [83]. Similar delivery of human FIX in hemophiliac dogs resulted in expression at about 1% of normal canine levels, an absence of inhibitors, and sustained partial correction of whole blood clotting time (WBCT) for at least 8 months [140]. In another study by Herzog et al., 1334 intramuscular injections of an AAV human FIX vector in mice led to prolonged expression in immunodeficient mice of about 350 ng/mL for at least 6 months, but in immunocompetent mice, there was generation of inhibitory antibody. These same investigators [137] subsequently showed that in hemophiliac dogs, intramuscular injection of high doses of the AAV–canine FIX achieved expression for over 17 months and demonstrated a stable, dose-dependent partial correction of the WBCT, and at the highest dose, there was a partial correction of activated partial thromboplastin time (APT). Finally, in another study [141], intramuscular injection of an AAV human FIX vector also lead to a transient reduction of WBCT in the first week, but this was rapidly lost as the animals developed anti-FIX antibody.

These and other preclinical studies with AAV–FIX vector have all shown a good safety profile. A clinical trial of an AAV–FIX vector administered by intramuscular injection in hemophiliac patients has now begun. An initial progress summary on three treated patients indicated that the vector was well tolerated [3]. Also, low levels (<1 %) of FIX activity could be measured, and thus far there has been no antibody response to the FIX protein.

C. Muscular Dystrophy

The efficiency of AAV vectors in skeletal muscle may be useful for treatment of muscular–skeletal diseases. Limb girdle muscular dystrophies (LGMDs) are a group of genetically and phenotypically heterogeneous neuromuscular diseases. One type of recessive LGMD is caused by mutations in one of the four muscle sarcoglycan genes (α, β, γ, δ). A phase I clinical trial in LGMD patients was recently initiated to evaluate intramuscular instillation of an AAV vector expressing the α, β, γ, or δ sarcoglycan gene, respectively, depending upon the patient genotype [4]. This trial was based on studies in two animal models. The Bio14.6 cardiomyopathic hamster is a naturally occurring LGMD model due to a deletion in the δ-sarcoglycan gene. Administration of AAV δ-sarcoglycan vectors to these animals showed very efficient correction for a whole muscle [142,143]. Mice that were null mutants for γ-sarcoglycan exhibit severe muscle pathology, but this could be partly corrected by intramuscular injections of an AAV vector expressing the γ-sarcoglycan gene from a muscle-specific promoter [144].

VII. SUMMARY

This chapter has summarized the increasingly rapid progress that is now being made in developing therapeutic applications of AAV vectors. As more groups extend investigations to additional in vivo models, the potential utility of AAV vectors as therapeutic gene delivery vehicles is gaining widespread interest. The development of more sophisticated production systems for AAV vectors has enhanced both the quantity and quality of vectors that can be produced. It is likely that in the next few years there will be a significant increase in testing of AAV vectors for additional therapeutic applications.

REFERENCES

1. Flotte TR, Carter BJ, Conrad CK, Guggino WB, Reynolds TC, Rosenstein BJ, Taylor G, Walden S, Wetzel R: A phase I study of an adeno-associated virus-CFTR

gene vector in adult CF patients with mild lung disease. Hum Gene Ther 1996; 7: 1145–1159.

2. Wagner JA, Messner AH, Moran ML, Daifuku R, Kouyama K, Desch JK, Manley S, Norbash AM, Conrad CK, Friborg A, Reynolds T, Guggino WB, Moss RB, Carter BJ, Wine JJ, Flotte TR, Gardner P. Safety and biological efficacy of an adeno-associated virus vector–cystic fibrosis transmembrane regulator (AAV-CFTR) in the cystic fibrosis maxillary sinus. Laryngoscope 1999; 109:266–274.

3. Kay MA, Manno CS, Rogni MV, Larson PJ, Couto LB, McClelland A, Glader B, Chen AJ, Tai SJ, Herzog RW, Arruda V, Johnson F, Scallen C, Skarsgard E, Flake AW, High KA: Evidence for gene transfer and expression of factor IX in hemophilia B patients treated with an AAV vector. Nat Genet 2000; 24:257–261.

4. Stedman H, Mendell J, Wilson JM, Finke R, Kleckner A-L: Phase I clinical trial utilizing gene therapy for limb girdle muscular dystrophy with α-, β-, γ-, or δ-sarcoglycan gene delivered with intramuscular instillations of adeno-associated vectors. Hum Gene Thera 2000; 11:777–790.

5. Carter BJ. Adeno-associated virus and adeno-associated virus vectors for gene delivery. In: Templeton NS and Lasic DD, eds. Gene Therapy: Therapeutic Mechanisms and Strategies. New York: Marcel Dekker, 2000, pp 41–49.

6. Carter BJ. The growth cycle of adeno-associated virus. In: Tjissen P, ed. Handbook of Parvoviruses. Vol 1. Boca Raton, FL: CRC Press, 1989; pp 155–168.

7. Berns KI. AAV Replication. In: Fields BN, Knipe DM, eds. Virology. New York: Raven Press, 1990, pp 1743–1764.

8. Carter BJ. Parvoviruses as vectors. In: Tjissen T, ed. Handbook of Parvoviruses. Vol 2. Boca Raton, FL: CRC Press, 1989, pp 247–284.

9. Muzyczka N. Use of Adeno-associated virus as a generalized transduction vector in mammalian cells. Curr Topics Microbiol Immunol 1992; 158:97–129.

10. Berns KI, Giraud C. Adeno-associated virus (AAV) vectors in gene therapy. Curr Topics Microbiol Immunol 1996, pp 1–171.

11. Carter BJ. Adeno-associated virus vectors. Curr Opin Biotechnol 1992; 3:533–539.

12. Flotte TR, Carter BJ. Adeno-associated virus vectors for gene therapy. Gene Ther 1995; 2:357–362.

13. Hallek M, Girod A, Braun-Flaco M, Clemens-Martin W, Bogedin C, Horer M. Recombinant adeno-associated virus vectors. Curr Res Mol Therapeut 1998; 1: 417–430.

14. Rabinowitz JE, Samulski R. Adeno-associated virus expression systems for gene transfer. Curr Opin in Biotechnol 1998; 9:470–475.

15. Summerford C, Samulski RJ. Adeno-associated viral vectors for gene therapy. Biogenic Amines 1998; 14:451–475.

16. Rabinowitz JE, Samulski RJ: Building a better vector: Manipulation of AAV virions Virology 2000; 278:301–308.

17. Blacklow NR. Adeno-associated viruses of humans. In: Pattison JR, ed. Parvoviruses and Human Disease. Boca Raton, FL: CRC Press, 1988, pp 165–174.

18. Carter BJ: Adeno-associated virus helper functions. In: (Tjissen P, ed.) Handbook of Parvoviruses. Vol 1. Boca Raton, FL: CRC Press, 1990, pp 255–282.

19. Carter BJ, Mendelson E, Trempe, JP:AAV DNA replication, integration and genet-

ics. In: Tjissen P, ed. Handbook of Parvoviruses. Vol 1. Boca Raton, FL: CRC Press, 1989, pp 169–226.

20. Carter BJ. The promise of adeno-associated virus vectors. Nat Biotechnol 1996; 14:1725–1726.

21. Redemann B, Mendelson E, Carter BJ. Adeno-associated virus rep protein synthesis during productive infection. J Virol 1998; 63:873–882.

22. Myers MW and Carter BJ. 1980. Adeno-associated virus assembly. Virology 1980; 102:71–82.

23. Samulski RJ, Berns KI, Tan N, Muzyczka N. Cloning of infectious adeno-associated virus into pBR322: rescue of intact virus from the recombinant plasmid in human cells. Proc Natl Acad Sci USA 1982; 79:2077–2081.

24. Laughlin CA, Tratschin JD, Coon H, Carter BJ. Cloning of infectious adeno-associated virus genomes in bacterial plasmids. Gene 1983; 23:65–73.

25. Kotin RM, Siniscalco M, Samulski RJ, Zhu XD, Hunter L, Laughlin CA, McLaughlin S, Muzyczka N, Rocchi M, Berns KI. Site-specific integration by adeno- associated virus. Proc Natl Acad Sci USA 1990; 87:2211–2215.

26. Samulski RJ, Zhu X, Xiao S, Brook JD, Housman DE, Epstein N, Hunter LA. Targeted integration of adeno-associated virus (AAV) into human chromosome 19. EMBO J 1991; 10:3941.

27. McLauglin SK, Collis P, Hermonat PL, Muzyczka N. Adeno-associated virus general transduction vectors: analysis of proviral structure. J Virol 1988; 62:1963–1973.

28. Weitzman, MD, Kyostio, SRM, Kotin, RM, and Owens, RA. Adeno-associated virus (AAV) Rep proteins mediate complex formation between AAV DNA and its integration site in human DNA. Proc Natl Acad Sci USA 1994; 91:5808–5812.

29. Kearns WG, Afione SA, Fulmer SB, Pang MG, Erikson D, Egan L, Landrum MJ, Flotte TR, Cutting GR. Recombinant adeno-associated virus (AAV-CFTR) vectors do not integrate in a site-specific fashion in an immortalized epithelial cell line. Gene Ther 1996; 3:748–755.

30. Linden RM, Ward P, Giraud C, Winocour E, Berns KI. Site-specific integration by adeno-associated virus. Proc Natl Acad Sci USA 1996; 93:11288–11294.

31. Young SM, McCarty DM, Degtyareva N, Samulski RJ. Role of adeno-associated virus rep protein and human chromosome 19 in site-specific recombination. J Virol 2000; 74:3953–3966.

32. Yakobson B, Koch T, Winocour E. Replication of adeno-associated virus in synchronized cells without the addition of helper virus. J Virol 1987; 61:972–981.

33. Yakobson B, Hrynko TA, Peak, MJ, Winocour, E. Replication of adeno-associated virus in cells irradiated with UV light at 254 nm. J Virol 1988; 63:1023–1030.

34. Bantel-Schaal U. and zur Hausen H. Adeno-associated viruses inhibit SV40 DNA amplification and replication of herpes simplex virus in SV40-transformed hamster cells. Virology 1988; 164:64–74.

35. Flotte TR, Zeitlin PL, Solow, Afione S, Owens RA, Markakis D, Drumm M, Guggino WB, Carter BJ. Expression of the cystic fibrosis transmembrane conductance regulator from a novel adeno-associated virus promoter. J Biol Chem 1993; 268: 3781–3790.

36. Haberman RP, McCown TJ, Samulski RJ. Novel transcriptional regulatory signals

in the adeno-associated virus terminal repeat A/D junction element. J Virol 2000; 74:8732–8739.

37. Peel AL, Zolotukhin S, Schrimsher GW, Muzyczka N, Reier PJ. Efficient transduction of green fluorescent protein in spinal cord neurons using adeno-associated virus vectors containing cell type-specific promoters. Gene Ther 1997; 4:16–24.

38. Flannery JG, Zolotukhin S, Vaquero MI, LaVail MM, Muzyczka N, Hauswirth WW. Efficient photoreceptor-targeted gene expression in vivo by recombinant adeno-associated virus. Proc Natl Acad Sci USA 1997; 94:6916–6921.

39. Xiao X, Li J, Samulski RJ. Efficient long-term gene transfer into muscle tissue of immunocompetent mice by adeno-associated virus vector. J Virol 1996; 70:8098–8108.

40. Kessler PD, Podsakoff GM, Chen X, McQuiston SA, Colosi PC, Matelis LA, Kurtzman GJ, Byrne B. Gene delivery to skeletal muscle results in sustained expression and systemic delivery of a therapeutic protein. Proc Natl Acad Sci USA 1996; 93:14082–14087.

41. Song S, Morgan M, Ellis T, Poirier A, Chesnut K, Wang J, Brantly M, Muzyczka N, Byrne BJ. Atkinson M, Flotte TR. Sustained secretion of human alpha-l-antitrypsin from murine muscle transduced with adeno-associated virus vectors. Proc Natl Acad Sci USA 1998; 95:14384–14388.

42. Yang Q, Chen F, Trempe JP. Characterization of cell lines that inducibly express the adeno-associated virus Rep proteins. J Virol 1994; 68:4847–4856.

43. Allen JA, Debelak DJ, Reynolds TC, Miller AD. Identification and elimination of replication-competent adeno-associated virus (AAV) that can arise by non-homologous recombination during AAV vector production. J Virol 1997; 71:6816–6822.

44. Wang XS, Khuntirat B, Qing K, Ponnazhagan S, Kube DM, Zhou S, Dwarki VJ, Srivastava A. Characterization of wild-type adeno-associated virus type 2-like particles generated during recombinant viral vector production and strategies for their elimination. J Virol 1998; 72:5472–5480.

45. Fan, P-D, Dong, J-Y. Replication of rep-cap genes is essential or the high-efficiency production of recombinant AAV. Hum Gene Ther 1997; 8:87–98.

46. Vincent, K, Piraino, ST, Wadsworth SC. Analysis of recombinant adeno-associated virus packaging and requirements for rep and cap gene poducts. J Virol 1997; 71: 1897–1905.

47. Li, J, Samulski, RJ, Xiao X. Role for highly regulated rep gene expression in adeno-associated virus vector production. J Virol 1997; 71:5236–5243.

48. Xiao, X, Li, J, Samulski, RJ: Production of high-titer recombinant adeno-associated virus vectors in the absence of helper adenovirus. J Virol 1998; 72:2224–2233.

49. Matushita, T, Elliger, S, Elliger, C, Podskaoff, G, Villareal, Kurtzman, GJ, Iwaki, Y, Colosi, P. Adeno-associated virus vectors can be efficiently produced without helper virus. Gene Ther 1998; 5:938–945.

50. Grimm, D, Kern, A, Rittnet, K, Kleinschmidt, JA. Novel tools for production and purification of recombinant adeno-associated virus vectors. Hum Gene Ther 1998; 9:2745–2760.

51. Samulski RJ, Chang LS, Shenk TE. Helper-free stocks of recombinant adeno-associated viruses: normal integration does not require viral gene expression. J Virol 1989; 63:3822–3828.

52. Flotte TR, Barrazza-Ortiz X, Solow R, Afione SA, Carter BJ, Guggino WB. An improved system for packaging recombinant adeno-associated virus vectors capable of in vivo transduction. Gene Ther 1995; 2:39–47.

53. Clark KR, Voulgaropoulou F, Fraley DM, Johnson PR. Cell lines for the production of recombinant adeno-associated virus. Hum Gene Ther 1995; 6:1329–1341.

54. Liu X, Voulgaropoulou F, Chen R, Johnson PR, Clark KR. Selective rep-cap gene amplifications as a mechanism for high-titer recombinant AAV production from stable cell lines. Mol Ther 2000; 2:394–403.

55. Gao, G-P, Qu, G, Faust, LZ, Engdahl, RK, Xiao, W, Hughes JV, Zoltick, PW, Wilson JM. High-titer adeno-associated viral vectors from a rep/cap cell line and hybrid shuttle virus. Hum Gene Ther 1998; 9:2353–2362.

56. Liu XL, Clark KR, Johnson PR. Production of recombinant adeno-associated virus vectors using a packaging cell line and a hybrid recombinant virus. Gene Ther 1999; 6:293–299.

57. Inoue I, Russell DW. Packaging cells based on inducible gene amplification for the production of adeno-associated virus vectors. J Virol 1998, 72:7024–7031.

58. Conway JE, Zolotukhin S, Muzyczka N, Hayward GS, Byrne BJ. Recombinant adeno-associated virus type 2 replication and packaging is entirely supported by a herpes simplex virus type 1 amplicon expressing rep and cap. J Virol 1997; 71: 8780–8789.

59. Conway JE, Rhys CMJ, Zolotukhin I, Zolotukhin S, Muzyczka N, Hayward GS, Byrne BJ. High-titer recombinant adeno-associated virus production utilizing a recombinant herpes simplex virus type 1 vector expressing AAV-2 rep and cap. Gene Ther 1999; 6:86–993.

60. Fisher KJ, Kelley WM, Burda JF, Wilson JM. A novel adenovirus-adeno-associated virus hybrid vector that displays efficient rescue and delivery of the AAV genome. Hum Gene Ther 1996; 7;2079–2087.

61. Lieber A, Steinwarder DS, Carlson CA, Kay MA. Integrating adenovirus–adeno-associated virus hybrid vectors devoid of all genes. J Virol 1999; 73:9314–9324.

62. Sandolon Z, Gnatenko DM, Bahou WF, Hearing P. Adeno-associated (AAV) rep protein enhances the generation of a recombinant mini-adenovirus utilizing an Ad/AAV hybrid virus. J Virol 2000; 74:10381–10389.

63. Johnston, KM, Jacoby D, Pechan, PA, Fraefel, C, Borghesani, P, Schuback, D, Dunn, RJ, Smith, FI, Breakfield, O. HSV/AAV hybrid amplicon vectors extend transgene expression in human glioma cells. Hum Gen Ther 1997; 8:359–370.

64. Cao L, Liu Y, During M, Xiao W. High-titer wild-type free recombinant adeno-associated virus vector production using intron-containing helper plasmids. J Virol 2000; 74:11456–11463.

65. Zolotukhin S, Byrne BJ, Mason E, Zolotukhin I, Potter M, Chesnut K, Summerford C, Samulski RJ, Muzyczka N. Recombinant adeno-associated virus purification using novel methods improves infectious titer and yield. Gene Ther 1999; 6:973–985.

66. Hermans WT, ter Brakke O, Dijkhuisen A, Sonnemans MA, Grimm D, Kleinschmidt JA, Verhaagen J. Purification of recombinant adeno-associated virus by iodixonal gradient ultracentrifugation allows rapid and reproducible preparation of vector stocks for gene transfer in the nervous system. Hum Gene Ther 1999; 10: 1885–1891.

67. Tamayose K, Hirai Y, Shimada T. A new strategy for large-scale preparation of high titer recombinant adeno-associated virus vectors by using packaging cell lines and sulfonated cellulose column chromatography. Hum Gene Ther 1996; 7:507–513.

68. Clark RW, Liu X, McGrath JP, Johnson PR. Highly purified recombinant adeno-associated virus vectors are biologically active and free of detectable helper and wild type viruses. Hum Gene Ther 1999; 10:1031–1039.

69. Gao G, Gu G, Burnhan MS, Huang J, Chirmule N, Joshi B, Yu Q-C, Marsh JA, Conceicao CM, Wilson JM. Purification of recombinant adeno-associated virus vectors by column chromatography and its performance in vivo. Hum Gene Ther 2000; 11:2079–2091.

70. Clark KR, Voulgaropoulou F, Johnson PR. A stable cell line carrying adenovirus inducible rep and cap genes allows fore infectivity titration of adeno-associated virus vectors. Gene Ther 1996; 3:1124–1132.

71. Atkinson EM, Debelak DJ, Hart LA, Reynolds TC. A high-throughput hybridization method for titer determination of viruses ands gene therapy vectors. Nucleic Acids Res 1998; 26; 2821–2823.

72. Debelak DJ, Fisher J. Iuliano S, Sesholtz D, Sloane DL, Atkinson EM. Cation exchange high-performance liquid chromatography of recombinant adeno-associated virus type 2. J Chromatogr 2000; 740:195–202.

73. Russell DW, Alexander IE, Miller AD. Adeno-associated virus vectors preferentially transduce cells in S phase. Proc Natl Acad Sci USA 1994; 91:8915–8919.

74. Flotte TR, Afione SA, Zeitlin PL. Adeno-associated virus vector gene expression occurs in nondividing cells in the absence of vector DNA integration. Am J Respir Cell Mol Biol 1994; 11:517–521.

75. Podsakoff G, Wong KK, Chatterjee S. Efficient gene transfer into nondividing cells by adeno-associated virus based vectors. J Virol 1994; 68:5656–5666.

76. Ferrari FK, Samulski T, Shenk T, Samulski R. Second-strand synthesis is a rate limiting step for efficient transduction by recombinant adeno-associated virus vectors. J Virol 1996; 70:3227–3234.

77. Fisher KJ, Gao GP, Weitzman MD, DeMatteo R, Burda JF, Wilson JM: Transduction with recombinant adeno-associated virus vectors for gene therapy is limited by leading strand synthesis. J Virol 1996; 70:520–532.

78. Russell DW, Alexander IE, Miller AD. DNA synthesis and topoisomerase inhibitors increase transduction by adeno-associated virus vectors. J Virol 1995; 92: 5719–5723.

79. Flotte, TR, Afione, SA, Solow, R, McGrath, SA, Conrad, C, Zeitlin, PL, Guggino, WB Carter, BJ. In vivo delivery of adeno-associated vectors expressing the cystic fibrosis transmembrane conductance regulator to the airway epithelium. Proc Natl Acad Sci USA 1993; 93:10163–10617.

80. Conrad CK, Allen SS, Afione SA, Reynolds TC, Beck SE, Fee-Maki M, Barrazza-Ortiz X, Adams R, Askin FB, Carter BJ, Guggino WB, Flotte TR. Safety of single-dose administration of an adeno-associated virus (AAV-CFTR) vector in the primate lung. Gene Ther 1996; 3:658–668.

81. Kaplitt MG, Leone P, Samulski RJ, Xiao X, Pfaff D, O'Malley KL, During M.

Long-term gene expression and phenotypic correction using adeno-associated virus vectors in the mammalian brain. Nat Genet 1994; 8:148–154.

82. Xiao X, Li J, McCown TJ, Samulski RJ. Gene transfer by adeno-associated virus vectors into the central nervous system. Exp Neurol 1997; 144:113–124.

83. Snyder RO, Spratt SK, Lagarde C, Bohl C, Kaspar B, Sloan B, Cohen LK, Danos O. Efficient and stable adeno-associated virus-mediated transduction of the skeletal muscle of adult immunocompetent mice. Hum Gene Ther 1997; 8:1891–1900.

84. Snyder RO, Miao CH, Patijn GA, Spratt SK, Danos O, Nagy D, Gown AM, Winther B, Meuse L, Cohen LK, Thompson AR, Kay MA. Persistent and therapeutic concentrations of human factor IX in mice after hepatic gene transfer of recombinant AAV vectors. Nat Genet 1997; 116:270–275.

85. Summerford C, Samulski RJ. Membrane-associated heparan sulfate proteoglycan is a receptor for adeno-associated virus type 2 virions. J Virol 1998; 72:1438–1445.

87. Summerford C, Bartlett JS, Samulski RJ. $\alpha_v\beta_5$ integrin: a co-receptor for adeno-associated virus type 2 infection. Nat Med 1999; 5:78–82.

88. Qing K, Mah C, Hansen J, Zhou S, Dwarki V, Srivastava A. Human fibroblast growth factor receptor 1 is a co-receptor for infection by adeno-asociated virus 2. Nat Med 1999; 5:71–77.

89. Goldman M, Su Q, Wilson JM. Gradient of RGD-dependent entry of adenoviral vector in nasal and intrapulmonary epithelia: implications for gene therapy of cystic fibrosis. Gene Ther 1996; 3:811–818.

89. Duan D, Yue Y, Yan Y, McCray PB, Engelhardt JF. Polarity influences the efficiency of recombinant adeno-associated virus infection in differentiated airway epithelia. Human Gene Ther 1998; 9:2761–2776.

90. Miramatsu S, Mizukami H, Young N, Brown KE. Nucleotide sequencing and generation of an infectious clone of adeno-associated virus 3. Virology 1996; 221: 208–217.

91. Chiorini J, Yang L, Liu Y, Safer B, Kotin R. Cloning and characterization of adeno-associated virus type 5. J Virol 1999; 73:1309–1319.

92. Xiao W, Chirmule N. Berta SC, McCullough B, Gao G, Wilson JM. Gene therapy vectors based on adeno-associated virus type 1. J Virol 1999; 73:3994–4003.

93. Chiorini JA, Yang L, Liu Y, Safer B, Kotin RM. Cloning of adeno-associated virus type 4 (AAV4) and generation of recombinant AAV4 particles. J Virol 1997; 71: 6823–6833.

94. Rutledge EA, Halbert CL, Russell DW. Infectious clones of vectors derived from adeno-associated virus (AAV) serotypes other than AAV type 2. J Virol 1998; 72: 309–319.

95. Davidson B, Stein C, Heth J. Martins L, Kotin R, Derksen T, Zabner J, Rhodes A, Chiorini J. Recombinant adeno-associated virus type 2, 4 and 5 vectors: transduction of variant cell types and regions in mammalian central nervous system. Proc Natl Acad Sci USA 2000; 97:3428–3432.

96. Chao H, Liu Y, Rabinowitz J, Li C, Samulski RJ, Walsh CE. Several log increase in therapeutic transgene delivery by distinct adeno-associated viral serotype vectors. Mol Ther 2000; 2:619–623.

97. Rabinowitz JE, Xiao W, Samulski RJ. Insertional mutagenesis of AAV2 capsid and the production of recombinant virus. Virology 1999; 265:274–285.

98. Girod A, Ried M, Wobus M, Lahm H, Leike K, Kleinschmidt J, Delage G, Haller M. genetic capsid modifications allow efficient re-targeting of adeno-associated virus type 2. Nat Med 1999; 5:1052–1056.

99. Bartlett JS. Prospects for the development of targeted adeno-asociated virus (AAV) vectors. Tumor Target 1999; 4:143–149.

100. Moskalenko M, Chen L, van Roey M, Donahue BA, Snyder RO, McArthur JG, Patel SD. Epitope mapping of human adeno-associated virus type 2 neutralizing anti-antibodies: implications for gene therapy and structure. J Virol 2000; 74:1761–1766.

101. Rabinowitz JE, Samulski RJ. Building a better vector: manipulation of AAV virions. Virology 2000; 278:301–308.

102. Wu P, Xiao W, Conlon T, Hughes J, Agbandje-McKenna M, Fjerkol T, Flotte T, Muzyczka N. Analysis of the adeno-asociated type 2 (AAV2) capsid gene and construction of AAV vectors with altered tropism. J Virol 2000; 74:8635–8647.

103. Duan D, Li Q, Kao AW, Yue Y, Pessin JE, Engelhardt JE. Dynamin is required for recombinant adeno-associated virus type 2 infection. J Virol 1999; 73:10371–10376.

104. Bartlett JS, Wilcher R, Samulski RJ. Infectious entry pathway of adeno-associated and adeno-associated virus vectors. J Virol 2000; 74:2777–2785.

105. Douar AM, Poulard K, Stockholm D, Danos O. Intracellular trafficking of adeno-associated virus vectors: routing to the late endosomal compartment and proteosome degradation. J Virol 2000; 75:1824–1833.

106. Sanlioglu S, Benson PK, Yang J, Atkinson EM, Reynolds T, Engelhardt JF. Endocytosis and nuclear trafficking of adeno-associated virus type 2 are controlled by Rac1 and phosphatidylinositol-3 kinase activation. Virology 2000; 74:9184–9196.

107. Duan D, Yue Y, Yan Z, Yamg J, Engelhardt JE. Endosomal processing limits gene transfer to polarized airway epithelia by adeno-associated virus. J Clin Invest 2000; 105:1573–1587.

108. Afione SA, Conrad CK, Kearns WG, Chunduru S, Adams R, Reynolds TC, Guggino WB, Cutting GR, Carter BJ, Flotte TR. In vivo model of adeno-associated virus vector persistence and rescue. J Virol 1996; 70:3235–3241.

109. Miao CH, Snyder RO, Schowalter DB, Patijn GA, Donahue B, Winther B, Kay MA. The kinetics of rAAV integration into the liver. Nat Genet 1998; 19:13–14.

110. Duan D, Sharma P, Yang J, Yue Y, Dudus L, Zhang Y, Fisher KJ, Engelhardt JF. Circular intermediates of recombinant adeno-associated virus have defined structural characteristics responsible for long-term episomal persistence in muscle tissue. J Virol 1998; 72:8568–8577.

111. Duan D, Sharma P, Dudus L, Zhang Y, Sanlioglu S, Yan Z, Yue Y, Ye Y, Lester R, Yang J, Fisher KJ, Engelhardt JF. Formation of adeno-associated virus circular genomes is differentially regulated by adenovirus E4 orf6 and E2A gene expression. J Virol 1999; 73:161–169.

112. Nakai H, Storm TA, Lay MA. Recruitment of single-stranded recombinant adeno-associated vector genomes and intermolecular recombination are responsible for stable transduction of liver in vivo. J Virol 2000; 74:9451–9463.

113. Duan D, Yan Z, Yue Y, Engelhardt JE. Structural analysis of adeno-associated virus transduction circular intermediates. Virology 1999; 261:8–14.

114. Yang J, Zhou W, Zhang Y, Zidon T, Ritchie T, Engelhardt JE. Concatemerization of adeno-associated virus circular genomes occurs through intermolecular recombination. J Virol 1999; 73:9468–9477.

115. Sanlioglu S, Benson P, Engelhardt JE. Loss of ATM function enhances recombinant adeno-associated virus transduction and integration through pathways similar to UV irradiation. Virology 2000; 268;68–78.

116. Song S, Laipis P, Berns KI, Flotte T. Effect of DNA-dependent protein kinase on the molecular fate of the rAAV2 genome in skeletal muscle. Proc Natl Acad Sci USA 2000; 98:4084–4088.

117. Duan D, Yue Y, Yan Z, Engelhardt. A new dual vector approach to enhance recombinant adeno-associated virus -mediated gene expression through intermolecular cis activation. Nat Med 2000; 6:595–598.

118. Yan Z, Zhang Y, Duan D, Engelhardt JE. Trans-splicing vectors expand the utility of adeno-associated virus for gene therapy. Proc Natl Acad Sci USA 2000; 97: 6716–6721.

119. Nakai H, Storm TA, Kay MA. Increasing the size of rAAV-mediated expression cassettes in vivo by intermolecular joining of two complementary vectors. Nat Biotechnol 2000; 18:527–532.

120. Sun L, Li J, Xiao X. Overcoming adeno-associated virus size limitation through viral DNA heterodimerization. Nat Med 2000; 6:599–602.

121. Chirmule N, Xiao W, Truneh A, Schnell M, Hughes JV, Zoltich P. Wilson JM, Humoral immunity to adeno-associated virus type 2 vectors following administration to murine and non-human primate muscle. J Virol 2000; 74:2420–2425.

122. Halbert CL, Standaert TA, Wilson CB, and Miller AD. Successful readministration of adeno-associated virus vectors to the mouse lung requires transient immunosuppression during initial exposure. J Virol 1998; 72:9795–9805.

123. Hernandez YJ, Wang J, Kearns WG, Wiler S, Poirer A, Flotte TR. Latent adeno-associated virus infection elicits humoral but not cell-mediated immune responses in a non-human primate model. J Virol 1999; 73:8549–8558.

124. Xiao W, Chirmule N, Schnell MA, Tazelaar J, Hughes JV, Wilson JM. Route of administration determines induction of T-cell independent humoral responses to adeno-associated virus vectors. Mol Ther 2000; 1:323–329.

125. Brockstedt DG, Podsakoff GM, Fong L, Kurtzman G, Mueller-Ruchholtz W, Engelmann E. Induction of immunity to antigens expressed by recombinant adeno-associated virus depends on the route of administration. Clin Immunol 1999; 2: 67–75.

126. Halbert CL, Standaert TA, Aitken ML, Alexander IE, Russell DW, Miller AD. Transduction by adeno-associated virus vectors in the rabbit airway: efficiency, persistence and readministration. J Virol 1997; 71:5932–5941.

127. Manning WC, Zhou S, Bland MP, Escobedo JA, Dwarki V. Transient immunosuppression allows transgene expression following readministration of adeno-associated virus vectors. Hum Gene Ther 1998; 9:477–485.

128. Beck SE, Jones LA, Chesnut K, Walsh SM, Reynolds TC, Carter BJ, Askin FB, Flotte TR, Guggino WB. Repeated delivery of adeno-associated virus vectors to the rabbit airway J Virol 1999; 73:9446–9455.

129. Zhang Y, Chirmule N, Gao G-P, Wilson JM:.CD40-ligand dependent activation of

cytotoxic T lymphocytes by adeno-associated virus vectors in vivo: role of immature dendritic cells. J Virol 2000; 74:8003–8010.

130. Jooss K, Yang Y, Fisher KJ, Wilson JM. Transduction of dendritic cells by DNA
viral vectors directs the immune response to transgene products in muscle fibers.
J Virol 1998; 72:4212–4223.

131. Manning WC, Paliard X, Zhou S, Bland MP, Lee AY, Hong K, Walker CM, Escobedo JA, Dwarki V. Genetic immunization with adeno-associated virus vectors
expressing herpes simplex type2 glycoproteins B and D. J Virol 1997; 71:7960–
7962.

132. Fields PA, Kowalczyk DW, Arruda VR, Armstrong E, McClelland ML, Hogstrom
JW, Pasi EJ, Erth HCJ, Herzog RW, High KA. Role of vector in activation of T
cell subsets in immune responses against the secreted transgene product Factor IX.
Mol Ther 2000; 1:225–231.

133. Egan M, Flotte T, Zeitlin P, Afione S, Solow R, Carter BJ, Guggino WB: Defective
regulation of outwardly rectifying Cl-channels by protein kinase A corrected by
insertion of CFTR. Nature 1992; 358:581–583.

134. Herzog RW, Hagstrom JN, Jung S-H, Tai SJ, Wilson JM, Fisher KJ, High KA.
Stable gene transfer and expression of human blood coagulation factor IX after
intramuscular injection of recombinant adeno-associated virus. Proc Natl Acad Sci
USA 1997; 94:5804–5809.

135. Nakai H, Herzog RW, Hagstrom J, Walter J, Kung SH, Yang EY, Tsai SJ, Iwaki
Y, Kurtzman G, Fisher KJ, Colosi P, Couto LB, High KA. Adeno-associated viral
vector mediated gene transfer of human blood coagulation factor IX into mouse
liver. Blood 1998; 91:4600–4607.

136. Wang L, Takabe K, Bidlingmaier SM, I11 CR, Verma IM. Sustained correction
of bleeding disorder in hemophilia B mice by gene therapy. Proc Natl Acad Sci
USA 1999; 96:3906–3910.

137. Herzog RW, Yang EY, Couto LB, Hagstrom JN, Elwell D, Fields PA, Burton M,
Bellinger DA, Read MS, Brinkhous KM, Podsakoff GM, Nichols TC, Kurtzman
GJ, High KA. Long-term correction of canine hemophilia by gene transfer of blood
coagulation factor IX mediated by adeno-associated viral vector. Nat Med 1999;
5:56–63.

138. Snyder RO, Miao C, Meuse L, Tubb J, Donahue BA, Lin HF, Stafford DW, Patel
S, Thompson AR, Nichols T, Read MS, Bellinger DA, Brinkhaus KM, and Kay
MA. Correction of hemophilia B in canine and murine models using recombinant
adeno-associated viral vectors. Nat Med 1999; 5:64–70.

139. Wang L, Nichols TC, Read MS, Bellinger DA, Verma IM. Sustained expression
of therapeutic level of factor IX in hemophilia dogs by AAV mediated gene therapy
in liver. Mol Ther 2000; 1:154–158.

140. Snyder RO, Miao C, Meuse L, Tubb J, Donahue BA, Lin H-F, Stafford, DW, Patel
S, Thompson AR, Nichols T, Read MS, Bellinger DA, Brinkhous KM, Kay MA:
Correction of hemophilia B in canine and murine models using recombinant adeno-
associated vectors. Nat Med 1999; 5:64–69.

141. Monahan PE, Samulski RJ, Tazelar J, Xiao X, Nichols TC, Bellinger DA, Read MS,
Walsh CE. Direct intramuscular injection with recombinant AAV vectors results in
sustained expression in a dog model of hemophilia. Gene Ther 1998; 5:40–49.

142. Xiao X, Li J, Tsao Y-P, Dressman D, Hoffman EP, Watchko JF. Full functional rescue of a complete muscle (TA) in dystrophic hamsters by adeno-associated virus vector-directed gene therapy. J Virol 2000; 74:1436–1442.

143. Greelish JD, Su LT, Lankford EB, Burkman JM, Chen H, Konig SK, Mercier IM, Desjardins PR, Mitchell MA, Zheng X-G, Leferovich J, Gao G-P, Balice-Gardon RJ, Wilson JM, Stedman HH. Stable restoration of the sarcoglycan complex in dystrophic muscle perfused with histamine and a recombinant adeno-associated viral vector. Nat Med 1999; 5:439–443.

144. Cordier L, Hack A-A, Scott MO, Barton-Davis ER, Gao G-P, Wilson JM, McNally EM, Sweeney HL. Rescue of skeletal muscles of γ-sarcoglycan-deficient mice with adeno-associated virus-mediated gene transfer. Mol Ther 2000; 1:119–126.

9
Retrovectors Go Forward

Jean-Christophe Pagès
Centre Hospitalier Régional Universitaire de Tours, Tours, France

Olivier Danos
Genethon, Evry, France

I. INTRODUCTION

Almost 20 years ago the description of retroviral vectors opened a new therapeutic direction that greatly improved the emerging concept of molecular medicine [1]. For the first time, it appeared possible to cure diseases through eradication of the triggering agent rather than simply treating symptoms. During the first decade a huge amount of in vitro experiments prompted physicians to be confident in rapidly being able to cure genetically a broad number of inherited as well as acquired diseases [2]. After a short experimental period, the first series of clinical trials drove everybody back to the bench.

It has been known for years that endogenous retroviruses are important components of our genome [3,4]. Originally viewed as fossils, these chromosomal counterparts of replication-competent genetic elements are now reconsidered since it has recently been shown that some endogenous retroviral code for fully functional proteins having cellular functions [5]. Although LINEs and SINEs mobilization have been observed, it is noteworthy that no mobile human endogenous retroviral sequence has been described so far [4]. Nevertheless, through retroviral gene transfer, retroviruses transiently recover their ability to disseminate from cell to cell while importing a new gene within their host. Somatic gene therapy being the only manipulation allowed, vertical spread of the genetic modification is not possible, thus parallel to ancient heritable mobile elements is partial. The molecular basis for generating retroviral vectors include construction of a retrovirally transportable element with no viral coding sequences and the provision, in trans, of structural and enzymatic information essential for particle formation.

The main goal for gene transfer is to introduce the therapeutic genetic information within the nucleus of the target cell. Human anatomy and histological organization of organs, as well as activation of the immune system, are natural barriers limiting in vivo transduction efficiency. The adapted route of administration and controlled immune response are major, yet unsolved, issues in the field of gene transfer. Furthermore, it has been observed that eukcaryotic cells, although being relatively permissive to retroviral transduction in vitro, proved to be resistant when in vivo access to their genome was attempted. Molecular mechanisms underlying this resistance are of several origins. To circumvent these problems, vector design has been extensively revisited. Improvements are dedicated to generate safer vectors leading to long-lasting, specific expression of the transduced genes. One other remarkable feature is the use of various retroviruses, including the human pathogenic lentiviral AIDS agent, as the basis for the generation of vectors [6]. Nevertheless, we cannot assert that retroviral vectors have now reached their maturity. Safety and in vivo efficiency remain the vectorologist's Grail. It is then paradoxical to note that, although the scientific community feels less interested in the potential of oncoretrovirus-based vectors because of the promise of newly developed lentiviral vectors, the first positive clinical trial, has been performed using an "old-fashioned" Moloney murine leukemia virus (MoMLV) vector [7].

In this chapter, we will concentrate on the structure and biology of retroviral vectors. In the first part, describing the wild-type retroviral life cycle, the rationale for the development of retroviral vectors is presented. Other sections contain some of the latest improvements beneficial to retroviral vectors. Viral vector particles are morphologically indistinguishable from their wild-type parental counterparts. Looking at the genomic organization of vectors, major modifications can be observed. This defines trans-acting sequences, essentially derived from the viral coding sequence and required for particle formation within the producing cell, and cis sequences allowing transgene mobilization. Cis sequences are necessary for reverse transcription, packaging, and integration. Of note, despite some subtle genetic and life cycle differences, in vitro lentiviruses behave just as oncoretroviruses. This is highlighted by the "final" lentiviral vector's structure, which strictly parallels that of MoMLV vectors. Considering the exponential increase of published papers in the field, we have chosen to be relevant rather than exhaustive. Scale-up management and industrial developments aimed at improving retroviral production are briefly discussed.

II. INGREDIENTS OF VIRUSES AND CELLS

A. What Makes a Virus Retro

1. Retroviral Components

Retroviruses are enveloped RNA viruses expressed from a double-stranded DNA intermediate integrated into the host cell genome [4]. This defines, respectively, the genomic RNA and the proviral DNA species. Within the particle, the retroviral genome is made of two identical, partially dimerized, capped mRNAs encompassing the proviral sequence with the exception of two short stretches unique in the RNA form. Thus, the retroviral genome is single stranded and of positive polarity. Since the description of retroviral organization during the early 1970s, seven genera and three groups of retroviruses have been described. Based on molecular data, it is also typical to distinguish simple retroviruses from complex ones [4]. Canonical genetic organization of retroviruses has drawn on the genetic structure of the former, with the latter harboring a more sophisticated regulation of their expression through a subset of different retroviral genes. Despite a common general replication pathway, retroviruses are widely divergent in sequences; a single major homology region (MHR) is found [8]. One can nevertheless define constant features present in all retroviruses, which include at the molecular level:

- The presence of at least four open reading frames, *gag, pro, pol,* and *env* organized in that order within viral genomic material and coding for structural and enzymatic proteins.

    gag codes for at least three structural proteins: the matrix (MA), the capsid (CA), and the nucleocapsid (NC). Gag, is translated as a precursor protein, it could be posttranslationally modified (e.g., myristylation). Most retroviruses express other Gag proteins involved in the correct particle shaping (e.g., p6 in HIV).

    pro codes for an aspartic protease essential for viral infectivity through particle maturation.

    pol codes for two enzymes, the reverse transcriptase (RT) and the integrase (IN). Pol is often produced as a fusion protein with the Gag precursor.

    env encodes a membrane-bound glycoprotein responsible for the recognition of the target cell. The viral envelope is composed of two viral *env* gene products, surface glycoprotein (SU), and a membrane-spanning domain (TM). Exposed on the outside of the virion, Env proteins are arranged in trimers, being anchored within the viral envelope originating from the cellular lipid bilayer.

- The presence of functionally conserved noncoding sequences needed for the viral life cycle.

  R, repeated at both ends of the genomic RNA, is essential for reverse transcription and control genomic RNA size, as it contains the poly-adenylation signal.

  U5 is unique in the 5′ region, is guanethidine (GU) rich and improves the recognition of the R polyadenylation signal.

  PBS, primer binding site, contains a complementary sequence for a specific tRNA (tRNA^{pro} for MoMLV viruses) copackaged with the retroviral genome.

  SD, splice donor sequence, allows the formation of subgenomic mRNA species in the producing cells.

  Ψ is a sequence organized in several stem loops recognized by viral proteins. It allows the specific packaging of viral genomic mRNA and partially overlaps the dimerization signal (DIS).

  SA, splice acceptor. Until recently, simple retroviruses were supposed to have a single SA and complex retroviruses to have several. New acceptors have now been described within MoMLV [9].

  ppt, polypurine track, is a purine-rich sequence which is less sensitive to RT Rnase H degradation. The remaining sequence is used to prime second-strand synthesis of the viral genome. Some retroviruses have two ppt (e.g., lentiviruses) [10].

  U3, unique in the 3′ region within the RNA genome, is duplicated in the provirus and contains a promoter and an enhancer recognized by the cellular transcription machinery.

The noncoding regions are essential components of most retroviral vectors whatever the viral group of origin is.

For their transcription, retroviruses rely on the cellular RNA polymerase II machinery. Spliced and unspliced mRNAs give rise to viral proteins and, for the latter, also to genomic RNA. Most retroviruses are released after budding at the cell surface. Depending on their relative abundance, membrane-anchored cellular proteins could be incorporated into the viral envelope [11]. Within viral particles, small cellular molecules are packaged. Among these, a specific tRNA adheres to the PBS through an interaction with a viral protein, namely, the nucleo-capside (NC).

2. Retroviral Machinery

The retroviral life cycle is composed of four steps, which lead to an active integrated provirus. First, the assembled particle has to modify its internal structure. Shortly after budding, the precursor's proteins are cleaved by the viral protease

[12]. This cleavage induces a shift in the viral morphology corresponding to a reorganization of viral proteins. One other important consequence of this process is the release within the particle of the reverse transcriptase and the integrase in their active forms. Mature particles interact with their target cell through the ENV gene product. Viral trimers of SU recognize a defined receptor on the cell surface [13]. This recognition step induces conformational changes unmasking the SU-associated TM which triggers the fusion between the retroviral envelope and the cellular membrane [14,15]. Finally, a viral nucleoprotein complex is delivered into the cytoplasm where the poorly understood uncoating of the viral genome occurs. The third step consists of the conversion of the RNA genome in the double-stranded DNA provirus. It is now well accepted that RT begins within the particle before cell entry [16]. An extreme example of this is the Foamy virus life cycle where the reverse transcription is almost complete within the particle [17]. RT dimers catalyze RNA-dependent DNA synthesis and DNA-dependent DNA synthesis and have an Rnase H activity localized in the carboxy-terminal domain. As does most nucleic acid polymerase, RT needs a primer. The first strand synthesis is primed by PBS-anchored tRNA. The second DNA strand synthesis is primed at the 3′ end Rnase H–resistant ppt. One other remarkable feature of RT is the duplication of two sequences at each end of the genome, which creates a long terminal repeat (LTR). R sequences are found on both sides of the genomic RNA and are conserved within the LTRs. Conversely U5 and U3, which are unique on either side of the genomic RNA, are present in each LTR. This implies that a single RNA form of U5 or U3 is the template for their two LTR counterparts. Of note, the RNA U5 sequence derives from the 5′ U5 of the parental LTR and the RNA U3 from its 3′ counterpart. An obvious consequence is that any modification observed, or introduced, in the 3′ U3 of a provirus will be exported within the viral genome and thus duplicated in both LTRs of the offspring provirus. The last step of the cycle leads to the integration of the provirus into the host genome. Among the proviral forms observed within the nucleus shortly after infection, only the "linear two LTR provirus" is a substrate for integration. Integration is a three-step process. Integrase specifically cleaves and binds the proviral ends at the attachment (att) sites of the LTR. Thereafter, proviral DNA is integrated through a strand-exchange process. At this stage, the provirus is surrounded by two single-stranded repeats, which are repaired by the host machinery [18].

B. Cellular Partners

1. Finding the Way to the Nucleus

After the viral envelope has fused to the cell membrane, the cytoplasmic retroviral core has to reach the nucleus where the genome will integrate. The cellular partners involved in viral cytoplasmic transport are not all known. Also, depending on the virus, intracellular routes might widely differ. Recent data suggest that incoming HIV particles travel within the cytoplasm along the microtubular network [19,20]. Such binding, also observed for adenoviruses, could explain how viruses reach the vicinity of the nucleus. Following the cytoplasmic transport, nuclear entry is an obligatory step. At this stage, RT is completed and viral genomes are packaged in a heterogeneous preintegration complex (PIC). PICs are composed of the proviral genome, viral proteins, and cellular proteins (e.g., HMG1Y for HIV) [21,22]. PICs are too large to pass across the nuclear envelope by simple diffusion. In cells, proteins with a size over 45 kD are transported through the nuclear pore by specialized cargo proteins [23]. Karyophilic proteins share specific motifs (nuclear localizing sequences, NLS) interacting with cargoes. Several types of motifs have been described, using different families of cargo proteins for the translocation process. Nuclear transport is energy dependent and relies on hydrolysis of guanosine triphosphate (GTP). Cargoes recognize pore proteins by defined motifs and shuttle between the cytoplasm and nucleus. MoMLV is unable to replicate in nondividing cells, and Roe et al. have shown that the block resulted from the inability of MoMLV PICs to become translocated across the nuclear membrane [24]. Yet, nuclear translocation is not sufficient for expression. Rous sarcoma viruses are known to integrate their genome into resting cells but remain dependent on mitosis for expression [25]. For avian viruses, PIC translocation could result from the presence of an NLS in the integrase [26,27]. Three similar motifs have been described in HIV integrase, but the exact localization of these sequences remains controversial [28].

Since it has been observed that differentiated quiescent cells such as macrophages can be infected by HIV, finding the pathways which allow viral entry into the nucleus of G0-arrested cells is a major goal in HIV research [29]. Several nuclear determinants have been ascribed to different HIV proteins, none of which being dominant over the others [28–33]. These nucleophilic signals are remarkably diverse. Within the matrix (MA p15), it relies on phosphorylation of a carboxy-terminal tyrosine. Vpr has been proposed to promote docking of the PIC by direct interaction with karyopherin β. Integrase could be karyophilic through three classic NLS motifs. Finally, nuclear localization of HIV has been attributed to the central Flap sequence, a triplex region formed by strand displacement at the level of the central ppt during reverse transcription [34]. Of note, HIV viruses remain, in part, dependent on the cell cycle, since RT is not completed in T cells until they reach the G1b phase of the cycle [35].

2. Production Supply

Once integrated in the cellular genome, the provirus is expressed. Some viruses have evolved a transactivator to improve and control the pattern of their expression. This is the case for Tat-dependent HIV expression, whereas Foamy viruses control their expression and also their latency by a modulated expression of tas [17].

Viral mRNAs are exported into the cytoplasm where they are translated or packaged. The balance between spliced and unspliced mRNA is important, since spliced mRNAs code for essential proteins and unspliced mRNAs either support the expression of viral proteins or are the genetic material packaged into virions. Viral sequences and proteins as well as cellular proteins are involved in controlling splicing. Complex retroviruses featuring several splice acceptor sites use special pathways in order to provide the budding particle with sufficient amounts of full-length RNA species. HIV produces a viral protein (REV) which bridges a secondary structure of the genomic RNA (RRE) to the CRM1-dependent cellular RNA export pathway [36]. For the same purpose, the Mason–Pfizer monkey virus (MPMV) constitutive transport element (CTE) directly binds a cellular export protein [37].

Most retroviruses bud at the cell surface. The Gag and Gag–Pol precursors are targeted at the inner face of the membrane by motifs in the gag sequence. Viral core proteins are produced in two different forms. The gag precursor is obtained by translation of the genomic mRNA. The pol proteins are also encoded by the genomic RNA and are obtained from a gag–pol precursor. Two different mechanisms lead to the formation of this precursor. MoMLV viruses produce the gag–pol fusion by readthrough suppression of the gag stop codon, whereas HIV uses frameshift suppression. The ratio of Gag versus Gag–Pol is critical for correct virion formation. Clusters of mixed Gag/Gag–Pol precursors are assembled at inner face of the cell membrane triggering the budding of the particle. Except for the Foamy virus, the presence of Env is dispensable for budding. Env glycoproteins follow a classic pathway for maturation and glycosylation. During their traffic through the Golgi apparatus, oligomerized Env gene products are cleaved by a luminal furin protease leading to SU and TM. Only cleaved proteins will exhibit fusion abilities essential for viral infectivity [38]. A second cleavage, performed by the viral protease, is necessary to activate fusion of MoMLV and GaLV envelopes [38,39].

C. Basic Recipe for Generating Retroviral Vectors

1. Trans Giving

From the above description, we can outline the minimal requirements for the production of the viral particle. Essential viral proteins are structural (Gag), enzymatic (Pol), and targeting (Env). Ideally, in order to avoid the generation of replication-competent virus by recombination, all these components must be separated on different transcriptional units. As Pol proteins are translated using the Gag mRNA, and as the ratio of Gag versus Gag–Pol is important, all Gag–Pol–producing constructs are grouped on the same transcription unit. As the ratio of Env to other viral proteins is not critical for generating infectious particles, and as Env is translated from a spliced subgenomic viral RNA, expressing Env from a separated unit is easy. Natural envelopes such as ecotropic, xenotropic, and amphotropic for MoMLV or gp120 for HIV were first used to direct cell specificity [40]. Limitation in the choice of Env gene results from incorporation of membrane-anchored proteins by the virus and by the size of the intracytoplasmic tail of the TM. For most viruses, the level of membrane expressed Env protein is the main driving force for incorporation [11]. Thus, MoMLV and HIV vectors are easily pseudotyped using either retroviral envelopes (for example, gibbon ape leukemia virus (GaLV)) or proteins from other enveloped viruses; the G protein from vesicular stomatitis virus (VSV-G), the HA from the influenza virus, and even the envelope from Ebola virus [41–43]. It is therefore possible to choose an Env gene, according to the target cell. Most retroviral receptors have now been cloned and their expression can be tested. Nevertheless, the level of receptor expression does not always correlate with the observed transduction efficiency [44]. To generate a producing cell line, plasmids containing the constructs described above, namely, Gag–Pol and Env, are transfected into the chosen cell [45]. "Empty" particles are released in the supernatant unless a recombinant vector is coexpressed. Within mice producing cell lines, endogenous retroviruses could be packaged [46–48].

For both Gag–Pol and Env constructs, major modifications aimed at improving the original constructs have focused on the viral protein level of expression. All modified constructs include a deletion of the packaging signal and the substitution of the retroviral polyadenylation signal by a heterologous equivalent. The LTR can be replaced by a stronger promoter such as those from the cytomegalovirus (CMV) and the EF-1 promoter [49,50]. Another improvement has come from the introduction of a translational link between the viral genes and a marker of selection. In these constructs, the sequence of the selectable marker is placed 72 bp 3′ to the viral product without any IRES or splice site [49,50]. Thus, translation of the marker depends on reentry of ribosomes, an event occurring more frequently if translation of the viral gene is high. This selection also improves the stability of viral production frequently hampered by DNA methylation of the promoter. Homologous recombination is the major mechanism by which replication-competent viruses arise. In every new construct, one important goal is to minimize the overlap between the trans- and cis-acting sequences. Codon wobbling has been employed to reduce the homology of coding sequence with recombinant vectors [51].

One other crucial element is the nature of the cell on which the packaging system is based. Criteria for the choice of a packaging cell line comprise absence of pathogens, traceability of the parental cell, low rate of homologous recombination, and high production abilities. Several cell lines have been tested for their ability to produce recombinant retroviruses at high titer [50,52]. Three cell lines are now commonly used to support viral production. NIH 3T3 came first and is presently the only one which received approval for viral production in clinical trials. Recently, two human-derived cell lines proved to be more efficient. 293 cells have the advantage of being easily transfected and have been approved for recombinant adenoviral production [53]. TE 671, a human rhabdomyosarcoma–derived cell line, is the latest cell line introduced in the field of retroviral production [50].

2. Cis Required

A recombinant retroviral vector is a stretch of DNA comprising a transgene of interest surrounded by genetic elements allowing its mobilization by the mean of a retroviral particle. From this simple definition, one can define features for an optimal vector design. Cis-acting sequences can then be separated into three types: sequences essential for RT, sequences needed for packaging, and sequences controlling vector transcription (promoter):

- Reverse transcription relies on two primers, PBS-anchored tRNA and RNase H–resistant ppt, and one jumping and transcriptional stop sequence, R. The disposition of these elements is precisely defined and should not be changed. The transgene must use the viral polyadenylation signal, since "internal" polyadenylation will lead to an RNA which cannot be copied by RT.
- The first-generation vectors contained the minimal packaging signal originally defined by deletion experiments. Some investigators suggested that extending the packaging signal up to 400 bp into gag could be beneficial [54]. One important drawback of this is the overlap between vectors and producing sequences, which increases the risk of homologous recombination. Recently, Kim et al. have shown that the gag sequence could be removed with almost no detrimental effect on viral titer [49]. Accordingly, Zhao et al. suggested that the 400-bp gag sequences function as nuclear export signals [55]. When this short gag stretch was moved 3′ or replaced by CTE from the MPMV, vectors showed higher titer. Other export elements from different viruses have been described, especially for lentiviral vectors [55,56].
- To drive the recombinant vector transcription, one can either use the natural U3 sequence or replace it by a strong promoter, CMV or EF-1α [49,57,58]. The latter solution is more popular, since these promoters have higher transcription efficiency. Of note, any 3′ LTR promoter is also able to drive transgene expression in transduced cell.

Transgene size is an obvious limiting factor. The cloning capacity of retroviral vectors depends on the parental virus; 8 kb is a commonly accepted maximal size for a transgene using MoMLV vectors. A recent study has shown that it was possible to package up to 20 kb in MoMLV-derived vector [59]. The efficiency of full-length template transduction was limited by the RT step, especially when the genomic size was over 15 kb.

It also important to bear in mind that introns included within the construct will not be present in the transduced cell. Overcoming this could be achieved by the introduction of an HIV RRE within the intron in a Rev-expressing packaging cell. Cloning the transgene in a reverse orientation, relative to transcription of the recombinant virus, is another solution. Introns within cis-acting retroviral sequences have paradoxical effects. Krall et al. have shown that intron-containing vectors (e.g., MFG) have a higher titer compared to intronless constructs [58]. The reason for this phenomenon is unclear, since splicing should remove the packaging signal. In these constructs, the SA site originates from a 377-bp stretch of MoMLV *pol* gene. Hunting for homologous regions would demand the removal of this sequence. Using a pragmatic approach, Yu et al. have shown that the synthetic SA can advantageously replace the pol SA, and that the usefulness of the SA depends on the transgene [60]. Classically, transgenes are cloned in place of the retroviral coding sequences between the packaging signal and the ppt. An opportunistic use of the RT mechanism consists of "double-copy" vectors where the transgene is introduced in the 3′ U3 of the proviral recombinant vector. In the transduced cell, the transgene is duplicated leading to a higher level of expression [61]. Few reports of double-copy vectors have confirmed this hypothesis. The presence of two adjacent promoters could result in promoter occlusion or competition, leading to poor transcription efficiency. Still, a smart application of the double-copy system has been developed to obtain a "clean" site of transgene insertion [62]. Choulika et al. have introduced in the 3′ U3 of a cre-expressing recombinant retrovirus, a single loxP sequence placed 5′ to a promoter and transgene. After transduction and duplication of the U3, cre recombinase deletes all the sequences between the loxP. Thus, at the site of integration remains a simple copy of the parental 3′ LTR, including the transgene. Having a cre-expressing unit, these vectors have a reduced capacity. Finally, taking advantage of increased RT template switch in the presence of direct repeats, Delviks et al. have shown efficient removal of the packaging signal in the integrated provirus [63]. This feature prevents vector mobilization, improving safety.

3. Viral Factories

Recombinant retroviruses are released into the supernatant of producing cells. Several techniques using either selection of stable producer clones or transient transfection have been developed to obtain high-titer supernatants [50,64–66].

Transient transfection has the advantages of speed and adaptability, whereas stable clone selection offers advantages in the reliability of long-term viral production and in the ease of scale-up. Nevertheless, the generation of a stable high-titer producer–clone remains time consuming. Vanin et al. described a strategy to facilitate rapid generation of high-titer producer lines [67]. It is based on the selection of a highly efficient genomic transcription site where a provirus containing a lox sequence is introduced. The lox sequence allows the introduction of any transgene of interest. The way the acceptor lox constructs were designed is unfortunately not suitable for a wide variety of recombinant viruses; it thus needs to be modified. Alternatively, direct "high-throughput" screening of transfected clones allows the selection of high-titer clones from a large population. To this end, several techniques of viral titer estimation, based on either dot blot screening or competitive polymerase chain reaction (PCR), have been described [68–70]. Direct blotting of crude supernatant has also been reported. Taking advantage of retroviral endogenous first-strand synthesis, it is possible to develop a rapid PCR-based assay for viral titration by the use of the real-time PCR [71]. As time goes by, stable clones frequently show a drop in recombinant viral production. As described above, design of packaging constructs can help to minimize this phenomenon. It is also possible chemically to activate transcription of packaging constructs [72,73]. Using Vanin's approach, surrounding the lox site by insulators should help to keep the production stable. Other industrial techniques aimed at improving viral titers have been described. Culture conditions could ameliorate viral stability before supernatant collection; procedures for viral concentration and purification are also under technical development [74,75]. Bioreactors using fibers or beads are under evaluation. It is important to bear in mind that owing to envelope shedding vectors having a retroviral envelope have a short half-life, and that some producing cells simultaneously release inhibitors and virus [76]. Recently developed cell lines are, to our knowledge, free of this drawback. Finally, viral titration depends on several parameters including the target cell, the viral volume, viral half-life, and time of exposure [76,77]. To compare and to predict accurately the efficiency of different batches of vector, titration has to be standardized. Andreadis and coworkers developed a convenient formula, based on physical parameters, which should be useful in standardization of retroviral titers [77].

Packaging constructs have also been inserted into three recombinant adenoviruses: one driving gag–pol production, one providing Env, and the third carrying the recombinant construct [52,78]. As adenoviral vectors transduce efficiently cells in vitro as well as in vivo, it is possible to achieve retroviral production in vivo after direct injection of the different adenoviruses. This local retroviral production would theoretically lead to highly efficient in vivo transduction. As recombinant adenoviruses are immunogenic, transduced cells will be rapidly cleared out [79]. Furthermore, repeated target cell infection will augment the risk of insertional mutagenesis

[80]. Consequently, this approach should be restricted to cancer therapy, the purpose of which is to eradicate the transduced cells [81]. Furthermore, transient retroviral production using recombinant adenovirus is efficient but suffers from complex and time-consuming procedures required for generating the recombinant adenoviral vector expressing the retroviral vector.

III. ADAPTIVE IMPROVEMENT OF RETROVIRAL VECTORS

A. Envelope Modifications

The advantage conferred in vitro by an envelope with large tropism, one envelope with many applications, appears less attractive for in vivo delivery. Except for hematopoietic cells, most gene therapy targets are part of organs that cannot be easily explanted for in vitro transduction. This implies that the therapeutic virus has to be delivered either intravenously or by direct injection [82,83]. Owing to tight cellular junctions in organs, this latter mode of administration yields poor transfer efficiency [83,84]. Intravenous administration is easy and allows the delivery of a large amount of virus [82,85,86]. On the other hand, when injected, the virus faces the host immune system and has to find a way to target the right cell. Controlling the immune response and being able specifically to target a recombinant virus have been the focus of researchers' attention for years. If immunology of gene delivery has taken advantage of allotransplantation research, retargeting of viruses seems to be at a dead end [42].

In vivo injection of recombinant virus induces host immunocompetent cell stimulation [87,88]. Since recombinant retroviruses do not express any viral gene, the cytotoxic response is preferentially directed toward the transgene-expressing cells. Drugs developed to control organ rejection could be used to control this immune response [89]. Plasma also supports a protective response via complement activation. Complement acts by direct interaction with viral determinants or is activated by preexisting antibodies directed against specific epitopes. Viral susceptibility to serum inactivation depends on both the producing cell and the viral envelope. A major determinant identified, conferring sensitivity to human serum, is the presence on the envelope glycoprotein of galactosyl(α1–3)galactosyl (α-Gal) terminal glycosidic epitopes [90]. Human and Old World monkey cells are deficient for the (α1–3)galactosyltransferase (α-GT) activity. New World monkeys, as do most mammals and bacteria, have an active form of the enzyme. Thus, shortly after birth, bacteria present in the gut strongly stimulate the production of antibodies directed against α-Gal epitopes, with up to 1% constitutively present IgG recognizing this epitope [91]. First-generation packaging cells, NIH 3T3, do contain an α-GT activity. Recombinant viruses produced by these cells are then highly sensitive to human serum inactivation. The outcome of human-based packaging cells has reduced serum sensitivity. It is noteworthy that not all combinations of envelope glycoproteins and packaging cells have similar capacities to produce resistant viruses [92]. Complement-control proteins (CD55 and CD59) can be incorporated into the HIV retroviral envelope [93]. Hence, differential incorporation would explain part of serum inactivation sensitivity. In order to improve complement-modulating protein incorporation into the vector envelope, some investigators have engineered these proteins by fusing them with a retroviral TM. Such modified vectors show moderate serum resistance [94,95]. The use of the VSV-G envelope is an attractive way to pseudotype retroviral vectors. It confers a wide specificity of infection, and it is the more suitable envelope when concentrated virus is desired. On the other hand, compromising its use for in vivo application, VSV-G seems to be one of the more complement-sensitive envelope [96].

An effective in vivo administration should lead to specific expression of the transgene. Vector targeting could be obtained through the transcriptional control of transgene expression or by the specific interaction of vectors with their target cell [42]. Specific particle targeting offers two advantages. First, fewer viruses have to be injected. Second, specific distribution appears to be safer, especially for integrating vectors. Indeed, one mechanism for retroviral transformation involves random integration, leading to oncogene activation or anti-oncogene knockout [4]. In the absence of targeting, increasing the number of transduced cells to maximize the therapeutic effect would increase the probability of these deleterious events.

For retroviruses, receptor-binding determinants are localized in SU [13]. SU not only promotes receptor recognition but post-binding events rely on its conformational change, which unleashes TM fusion capacities [15]. Thus, one simple way to obtain specificity is the use of an existing viral envelope protein with a defined tropism. This approach is possible for a limited number of situations; for example, targeting HIV-infected cells with an HIV envelope [97]. A second strategy is to retarget the viral coat protein, SU [42]. This could be accomplished by modifying the recognition motif of a retroviral glycoprotein envelope or by expressing a defined ligand within the viral envelope [41,42,98–102]. This latter strategy is possible either by engineering the *env* gene, by expressing an anchored ligand on the retroviral envelope, or by addition of bispecific ligands to mature particles [103,104]. Bispecific ligands proved to efficiently induce specificity when using avian retroviral envelopes [105,106]. Bates and collaborators have shown that the presence of ASLV-A soluble receptor triggers SU conformational shift. Activated envelopes could then promote viral infection of cells not expressing the receptor. This approach may become difficult when large amounts of vectors are needed or for in vivo applications where the ligand should bind the virus with high affinity. Dealing with MoMLVs, several attempts aimed at adding ligands to Env have been described. Targeting was performed by N-terminal addition of peptides derived from the ligand-binding domain or using single-chain antibodies. The crystal structure of the ecotropic envelope confirmed that the SU N-terminal position was exposed, a prerequisite for ligand–receptor interaction [107]. Analyzing all these experiments, it appears that in vitro specific cell binding could easily be obtained, but modified virus failed to complete efficient postbinding events. Increasing viral cell surface affinity is not enough for targeting. Hall and collaborators were able to target wounded sites by incorporating a collagen-binding peptide from von Willebrand factor into an Env glycoprotein [108]. The nature of the receptor used for targeting is also important for the process [109]. In a model study, it was possible to redirect an ecotropic envelope glycoprotein to recognize an amphotropic receptor [110]. In this particular case, both the original and the modified envelope had to target a natural viral receptor. Wild-type viruses might have selected receptor proteins with peculiar recycling properties and surface distribution [111]. These observations prompted investigators to design a two-step targeting strategy. Peng and collaborators introduced a protease cleavage site between the additional ligand and an amphotropic envelope [81]. Once the virus is bound to the target cell, a membrane-anchored protease cleaves the recognition peptide from the wild-type SU, which then recognizes its cognate receptor allowing entry of the virus. This approach is adapted to tumor targeting where cancer cells invade their environment by expressing proteases at the cell surface. This approach raises the possibility of specific targeting by defined modifications, and suggests that insertion of random ligands within the SU may be a powerful method to achieve targeting. Bucholz et al. successfully used this strategy to derive protease cleavage sites on the SU [112]. More recently, using DNA shuffling, it has been possible to redirect an Env glycoprotein.

The endothelial bed is the first physical barrier blocking viral access to the target cell. In a model study, intravenously administrated in an undamaged rat leg, adenoviral vectors or adeno-associated virus–derived vectors transduce the endothelium at high efficiency. The targeted striated muscle cells are poorly transduced unless a dramatic vasodilatation is pharmacologically induced. Drugs utilized to assist vasodilatation are vigorously cardiotoxic, preventing wide use of this technique. Targeting should first permit transcytosis, a process essential for several proteins (e.g., interleukin-8 [IL-8]) and viruses [113]. Inhibition of infectivity, resulting from viral envelope modification, could help to develop a strategy for transcytosis. N-terminal addition of a transcytosis component to an amphotropic Env should result in a vector that is unable to infect endothelial cells but is efficiently transported across the endothelium. An obvious drawback of efficient transcytosis would be the enlargement of viral biodistribution territory. This problem could be overcome by selective in vivo administration.

B. Forcing the Quiescence

1. Mitosis: A License for Integration

As discussed above, it is now well accepted that MoMLV-derived vectors do not transduce quiescent cells [24]. Being unable to transduce quiescent cells is not only problematic for in vivo applications but also dramatically affects ex vivo transduction. Hematopoïetic stem cells, still not fully characterized, are a major target in gene therapy. Several protocols using specific culture conditions have been utilized in order to improve transduction efficiency [114]. Hitherto almost all of the protocols were designed to increase the number of cycling cells while transducing with MLV-based vectors. Yet, experiments in mice have shown that stimulated stem cells were less competent at repopulating an animal [115]. This could be deleterious for long-lasting correction in humans. The strategy of exogenous stimulation by cytokines has been successfully used to stimulate hepatocyte transduction ex vivo as well as in vivo [82,85,116]. Murine mammary tumor virus (MMTV) infects circulating B cells, which are quiescent when unstimulated. Within the MMTV genome, an open reading frame, coding a superantigen (sAg), is thought to facilitate viral infection. Incorporated within the viral envelope, the sAg could stimulate a B cell, which becomes permissive to viral infection [117]. Self-activating vectors either coexpress an envelope-anchored cytokine (e.g., hepatocyte growth factor [HGF] or IL-2) together with a wild-type retroviral Env protein, or harbor an Env modified in order to present on its N-terminal extremity a peptide competent for receptor binding and activation. Both systems proved to fulfill the expectancies when assayed in vitro [98,99]. Nevertheless, HGF-exposing viruses failed to improve in vivo transduction of c-met–expressing hepatocytes (T.H. Nguyen, personal communication).

Two approaches have tried to modify the retroviral genome in order to obtain viral transduction capacities on arrested cells. Deminie and Emerman made chimeric MoMLV-expressing HIV_{MA} or HIV_{CA} in place of the wild-type proteins [118,119]. Neither virus was able to infect quiescent cells. More recently, Lieber and coworkers generated a MoMLV-based vector bearing binding sites for an adenoviral protein involved in nuclear import (terminal protein [TP]). This vector was used to transduce arrested TP-expressing cells. Despite nuclear import of the retroviral genome, a very low frequency of integration was observed [120]. Conversely, introducing an NLS into the MA protein of a spleen necrosis virus (SNV), Parveen and collaborators have been able to generate packaging constructs allowing transduction of quiescent cells [121]. This emphasizes life cycle differences between mice oncoretroviruses and avian retroviruses.

2. Unlocking the Nucleus with Lentiviral Vectors: What Makes the Difference?

Structurally, lentiviruses show two important differences compared to simple viruses. HIV-1 genome contains six accessory genes according to the classic triad [122]. The other difference concerns a structural feature of the lentivirus genome: the presence of a second ppt (cppt). During reverse transcription, RT induces a strand displacement of the DNA initiated at this site, leading to the so-called central FLAP [10,34].

The history of lentiviral vectors parallels that of MoMLVs in an accelerated fashion. As for MoMLV-derived vectors, production systems have been designed to use a split viral genome into individual helper plasmid constructs. As some viral proteins are cytotoxic, the first strategy developed for lentiviral production relied on transient transfection. Starting with a trans-complementing construct harboring all viral genes, except the envelope, it rapidly appeared to be possible to delete almost all the accessory genes [123,124]. Indeed, several lentiviral genes (*vif, vpr, vpu,* and *nef*) are dispensable for viral replication in vitro but seem to be crucially important for viral pathogenesis in vivo. The more controversial deletion concerns the gene coding for Vpr. Vpr is a multifunctional protein [122]. It has been implicated in G2 growth arrest, apoptosis, transactivation of various promoters, and importantly in nuclear translocation. Vpr does not contain a canonical NLS, but seems to promote PIC nuclear docking by its preferential interaction with karyopherin β [125–127]. Yet, evidence has been found that only *vif* and *vpu* may be required for optimal transduction of the liver and of resting lymphocytes [83,128].

Until very recently, it appeared to be impossible also to delete rev. This is of special importance for HIV vectors for two reasons. Optimal vector production requires the presence of rev, which interacts with the RRE sequence and positively affects the nuclear export of both the unspliced *gag–pol* mRNAs and the transfer vector genomic RNA [36]. In the absence of Rev, AU-rich *gag* and *gag–pol* mRNAs are targeted for degradation through inhibitory sequences (NIS) present in the coding region. Hence, an additional improvement is brought about by splitting the original viral genome and by expressing Rev from a separate construct [57]. Rev cytotoxicity has hampered its simple constitutive expression in the producing cell. Furthermore, addition of a RRE within the recombinant vector has a positive influence on the viral titer. Several solutions have been proposed for the design of rev-independent vectors. The RRE sequences can be replaced by heterologous viral sequences known to enhance export and/or stability of unspliced transcripts. These are the CTE from the MPV or from the simian retrovirus type 1 or posttranscriptional regulatory elements from human or woodchuck hepatitis B viruses (HPRE and WPRE, respectively) [129–131]. In the recombinant vector, Tat-dependent 5′ LTR could be replaced by the strong heterologous promoter from the human CMV immediate–early promoter or from the Rous sarcoma virus U3 sequence. The self-inactivating (SIN) design described for MLV vectors has been successfully adapted to lentiviral vectors [132]. Crippling the LTR also reduces the odds of replication-competent retrovirus (RCR) generation, and may prevent transcriptional activation of cellular genes adjacent to the provirus. Although this feature is usually advertised as a prime advantage of SIN vectors, it should be noted that the presence of an active internal promoter is associated with equivalent risks of transcriptional readthrough. Tempering readthrough, which is mostly due to a deficient cleavage and polyadenylation of vector transcripts within the 3' LTR, is obtained by replacing retroviral polyadenylation signal by exogenous ones (e.g., β-globin or SV40). Not only does it improve vector safety, because additional viral sequences are removed but it unexpectedly results in increased titers [133]. It is important to note that according to this design, more than 80% of the original genetic information of the virus is absent without notable changes in gene transfer performances. Removing the donor splice site in the recombinant vector has yielded higher titers [134]. Adding the second polypurine tract within the recombinant construct has been shown to enhance lentiviral vector efficiency into both dividing and nondividing cells [34, 135].

A recent improvement of lentiviral vectors resulted from splitting the packaging construct, one expressing *gag–pro* and the other expressing *pol* [136]. This improves safety while allowing a correct viral production. Codon optimization of the gag–pol gene has led to the inactivation of NIS and to an enhanced protein production. It also reduces the sequence homology between the packaging construct and the recombinant vector, which still harbor a short stretch of gag indispensable for packaging [137,138]. Taken together these reports should lead to the generation of stable producing cell lines; two cell lines are currently available [139,140]. Concerning viral targeting, we previously said that lentiviral vectors could be pseudotyped. The most commonly used envelope is the VSV-G, which allows the production of high-titer vector and confers great stability on the particles.

It is important to note that lentiviral vectors based on other lentiviruses have been developed [141–144]. Investigators argue that the rationale for developing nonhuman lentiviral vectors is biosafety and the availability of an animal model. To date, none of the nonhuman-based vectors offers any crucial advantage over the others.

C. Therapeutic Expression

Through integration, retroviral vectors offer the opportunity of long-lasting expression. This is a major advantage in the cure of genetic diseases. Since early work by Jaenish and coworkers, who tried to generate transgenic animals by retroviral transduction, it has been known that retroviruses are subject to transcription silencing. This phenomenon is induced by two mechanisms: methylation-dependent and methylation-independent molecular modifications of the LTR [145,146]. Using transcriptional insulators, it has been possible to minimize this downregulation of transgene expression [147].

Retroviral vector design allows the use of a tissue-specific promoter. This promoter can either be introduced within the 3' LTR in the parental provirus or within the virus 5' to the transgene. In the former case, the promoter is duplicated in the transduced cells. The resulting provirus in the transduced cell has the structure of a retroviral genome and is subject to splicing and mobilization. In the latter case, the parental construct can also have a deletion of the 3′ U3 promoter. In such a vector, self-inactivated (SIN), the transgene is expressed by an internal transcript. Several tissue-specific retroviruses have been successfully generated [116,148,149].

Control of transgene expression could be advantageous when using toxic transgene or regulated hormones. Using the tetracycline system, it has been possible to produce regulated retroviral vector [150].

IV. COLLATERAL DAMAGES

In that retroviral vectors are integrating vectors, there are two biosafety issues. One issue deals with integration as a source of mutation. Considering integration as a random process, it is possible to set a theoretical calculation of the risk [80]. Other risks come from the vector preparation itself. Three questions need to be examined when evaluating the safety of retroviral vector preparations: (1) What is the pathogenic risk of the parental virus? (2) What are the possibilities of generating RCR during vector preparation or use? (3) Is there any detectable toxicity of vector preparations related to the presence of viral proteins or of compounds derived from the production system?

As far as the choice of the vector is concerned, the relevance of developing non–HIV-1 vectors is disputable. Indeed, there is a detailed knowledge of the pathogenicity of HIV-1 in humans, which is not the case for FIV, CAEV, and EIAV. Nevertheless, these viruses are not known to infect humans and cause disease, although some FIV strains can productively infect human cells in culture. For MoMLV, which could replicate in human cells, this question is less relevant, as the virus cannot infect quiescent cells.

Generating RCR is a major concern associated with any virus-derived system in the field of gene transfer. This problem has been examined in detail in the context of MLV-based vectors, which are now routinely produced under Good Manufacturing Practice conditions and validated for clinical use. MoMLV vectors have been administered safely to tens of thousands of experimental animals and to more than 1200 patients. Studies in monkeys involving production with known RCR contaminations have pointed out the potential dangers [151]. Furthermore, careful design of vectors and packaging systems has drastically reduced this risk. The ongoing development of lentiviral vectors will take advantage of MoMLV experience for which many questions have been addressed and robust assays have been built. This includes mobilization of a detectable marker or cell transformation assays [152]. Using last-generation vectors, none of the individual components used for the production of retroviral vectors can produce RCR, and thus any outbreak is due to either adventitious contamination or recombination [60,132]. Recombination events that result in the formation of RCR occurs between the recombinant vector genome which carries all sequences needed in cis and the sequences providing the viral proteins in trans. Such events occur at the DNA level between (1) a mixture of cotransfected plasmids; (2) a transfected plasmid and a chromosomal sequence bearing homologous sequences; and (3) the proviral vector and a chromosomal sequence of the target cell. For retroviruses, recombination is more frequent at the RNA level. During RT, polymerase switches templates, producing one proviral DNA out of two genomic RNA. The rate of this "copy choice" recombination in a single round of replication has been estimated to be about 4% per kilobase [153]. Therefore, biosafety essentially relies on minimizing packaging of nonspecific or helper RNAs. Of note, superinfection, essentially observed with the VSV-G pseudotyping system, of the producer cell is known to augment the rate of recombination. Human endogenous retroviruses (HERVs) represent more than 1% of the genome [4]. All cloned HERV seqences proved to be related to either onco or spuma retroviruses. Thus, the potential of endogenous retroviral sequences as recombination partners, both at the DNA and RNA levels, raise safety concerns. Of note, for lentiviruses in mammalian genomes, no such homologous sequences have been identified so far. A functional equivalent, and putative ancestor of HIV Rev, is encoded by HERV-K sequences, but no sequence homology exists between the two genes [54].

Toxicity could originate from vector preparation. The quality and safety of clinical grade vectors dependent on the design of the production process also require effective purification procedures which reliably separate the vectors from potential contaminants. Impurities to be removed from retroviral vectors preparation include culture medium, chromatographic buffers, host cell proteins, DNA, antibiotics used for cell line maintenance, and induction of packaging functions. Hence, inducible expression systems should be such that vector harvesting is performed in the absence of drug (i.e., the Tet-OFF system is preferred to Tet-ON). Strict requirements exist regarding the source and the traceability of serum. Development and assessment of serum-free culture media is an important goal [155]. Retroviral vector production based on transient transfection is prone to contamination with plasmid DNA. A second group of potential contaminants consists of adventitious agents such as viruses, viruslike particles, bacteria, fungi, or mycoplasmas. Extensive testing of culture is required to provide reasonable assurance that producing cells are free of agents.

Retroviral particles incorporate a variety of components of cellular origin. Among these cellular elements are essential components of the particle, lipids and the PBS-anchored tRNA, or nonspecifically packaged molecules, various RNA, and proteins. Numerous cellular proteins have been detected in lentiviral particles, and their nature or abundance may depend on the producer cell type.

As with other gene transfer vectors, immunogenicity of retroviral vectors, humoral and cellular responses against the vector, the transgene or molecules coinjected with the vector (e.g., calf serum) will have to be tested in animals.

Since MoMLV vectors are poorly efficient in promoting in vivo transduction, most data come from the use of lentiviral vectors. Prolonged expression in muscle, brain, liver, eye, and cochlea of rodents is obtained with no noticeable lymphocyte or macrophage infiltrates [83,132]. Early examination of the injection site has documented the presence of inflammatory cells; since control animals have the same infiltrate, it was attributed to the surgical procedures [57,123,156–158]. Lentiviral vector gene transfer into the monkey nigrostriatal system has been shown to induce minor perivascular cuffing without an apparent inflammatory response [159]. In the liver, after intraportal infusion of a lentiviral vector in Fischer rats, a dose-dependent increase in serum alanine aminotransferase (ALT) and a mortality rate of 74% for over 8.10^8 transducing units were observed [160]. The presence of contaminants triggering a local inflammatory response shortly after injection can dramatically influence the onset of an immune response.

V. FUTURE-GENERATION VECTORS: NOBODY'S PERFECT

Provided that biosafety will remain the main concern, we would like to highlight some crucial points for imagining future-generation vectors. One way to render lentiviral vectors more acceptable is to show how unlikely would be the formation of a replication-competent virus. In order to minimize the homology between vector and packaging constructs, an in vitro–selected RNA with high affinity for the HIV-1 nucleocapsid protein has been used to mediate packaging into HIV-1 virions and could be substituted to the viral sequence in vectors [161]. Studies performed by Pathak's group [63], showing the importance of a direct repeat to promote a reverse transcriptase template switch, command us to go further in limiting cis-acting viral sequences within vectors. It must, therefore, be possible to change the viral R and U5. Finally, modification could be performed not only within the vector but also in the packaging construct. Even though the wobbling approach has reduced homology between these partners, more dramatic changes are envisionable. Accola et al. have shown that it is possible to obtain particles having less viral structural proteins [162]. If such particles prove to be efficient at RNA packaging, this could be useful for the design of new packaging constructs. Finally, one other mythic goal is the targeted integration of the provirus [163]. Having site-specific integration vectors would dramatically increase the biosafety of retroviral gene transfer.

NOTE ADDED IN PROOF

While this chapter was in press, several events have drawn attention to the risk of insertional mutagenesis when using retroviral vectors. Leukemia was reported in mice transplanted with bone-marow cells modified with a recombinant retrovirus expressing a truncated nerve growth factor receptor. It was shown that the retroviral vector was inserted within the *evi 1* gene, a likely primary event in oncogenesis [164]. Sadly, the first successful gene therapy clinical trial on patients with X-linked severe combined immunodeficiency was stopped after a lymphoproliferative syndrome was detected in one of the patients, three years after gene therapy [165,166]. In this case, the vector was inserted into the *lmo-2* gene, a transcription factor associated with T-cell leukemias in humans [167] whose transcription was activated within tumor cells. Finally, a genome-wide analysis of lentiviral insertion has demonstrated a strong bias for gene-rich regions [168].

REFERENCES

1. Mann R, Mulligan RC, Baltimore D. Construction of a retrovirus packaging mutant and its use to produce helper free defective retrovirus. Cell 1983; 33:153–159.
2. Mulligan RC. The basic science of gene therapy. Science 1993; 260:926–932.
3. Coffin J. Genome structure. In: Weiss R, Teich N, Varmus H, Coffin J, eds. RNA Tumor Viruses. Vol. 2. Cold Spring Harbor, NY: Cold Spring Harbor Laboratory Press, 1985:17–74.
4. Coffin MJ, Hughes HS, Varmus HE. Retroviruses. Cold Spring Harbor, NY: Cold Spring Harbor Laboratory Press, 1997.
5. Blond JL, Lavillette D, Cheynet V, et al. An envelope glycoprotein of the human endogenous retrovirus HERV-W is expressed in the human placenta and fuses cells expressing the type D mammalian retrovirus receptor. J Virol 2000; 74:3321–3329.
6. Buchschacher GL Jr, Wong-Staal F. Development of lentiviral vectors for gene therapy for human diseases. Blood 2000; 95:2499–2504.
7. Cavazzana-Calvo M, Hacein-Bey S, de Saint Basile G, et al. Gene therapy of human severe combined immunodeficiency (SCID)-X1 disease. Science 2000; 288:669–672.
8. Wills JW, Craven RC. Form, function, and use of retroviral gag proteins. AIDS 1991; 5:639–654.
9. Dejardin J, Bompard-Marechal G, Audit M, Hope TJ, Sitbon M, Mougel M. A novel subgenomic murine leukemia virus RNA transcript results from alternative splicing. J Virol 2000; 74:3709–3714.
10. Charneau P, Alizon M, Clavel F. A second origin of DNA plus-strand synthesis is required for optimal human immunodeficiency virus replication. J Virol 1992; 66:2814–2820.
11. Suomalainen M, Garoff H. Incorporation of homologous and heterologous proteins into the envelope of Moloney murine leukemia virus. J Virol 1994; 68:4879–4889.
12. Swanstrom R, Wills JW. Synthesis, assembly, and processing of viral proteins. In: Coffin JM, Hughes SH, Varmus HE, eds. Retroviruses. Cold Spring Harbor, NY: Cold Spring Harbor Laboratory Press, 1998:263–335.
13. Hunter E, Swanstrom R. Retrovirus envelope glycoproteins. Curr Top Microbiol Immunol 1990; 157:187–253.
14. Lavillette D, Ruggieri A, Russell SJ, Cosset FL. Activation of a cell entry pathway common to type C mammalian retroviruses by soluble envelope fragments. J Virol 2000; 74:295–304.

15. Lavillette D, Maurice M, Roche C, Russell SJ, Sitbon M, Cosset FL. A proline-rich motif downstream of the receptor binding domain modulates conformation and fusogenicity of murine retroviral envelopes. J Virol 1998; 72:9955–9965.

16. Trono D. Partial reverse transcripts in virions from human immunodeficiency and murine leukemia viruses. J Virol 1992; 66:4893–4900.

17. Linial ML. Foamy viruses are unconventional retroviruses. J Virol 1999; 73:1747–1755.

18. Yoder KE, Bushman FD. Repair of gaps in retroviral DNA integration intermediates. J Virol 2000; 74:11191–11200.

19. Liu B, Dai R, Tian CJ, Dawson L, Gorelick R, Yu XF. Interaction of the human immunodeficiency virus type 1 nucleocapsid with actin. J Virol 1999; 73:2901–2928.

20. Bukrinskaya A, Brichacek B, Mann A, Stevenson M. Establishment of a functional human immunodeficiency virus type 1 (HIV-1) reverse transcription complex involves the cytoskeleton. J Exp Med 1998; 188:2113–2125.

21. Bushman FD. Host proteins in retroviral cDNA integration. Adv Virus Res 1999; 52:301–317.

22. Farnet CM, Bushman FD. HIV-1 cDNA integration: requirement of HMG I(Y) protein for function of preintegration complexes in vitro. Cell 1997; 88:483–492.

23. Nakielny S, Dreyfuss G. Transport of proteins and RNAs in and out of the nucleus. Cell 1999; 99:677–690.

24. Roe T, Reynolds TC, Yu G, Brown PO. Integration of murine leukemia virus DNA depends on mitosis. EMBO J 1993; 12:2099–2108.

25. Humphries EH, Glover C, Reichmann ME. Rous sarcoma virus infection of synchronized cells establishes provirus integration during S-phase DNA synthesis prior to cellular division. Proc Natl Acad Sci USA 1981; 78:2601–2605.

26. Kukolj G, Jones KS, Skalka AM. Subcellular localization of avian sarcoma virus and human immunodeficiency virus type 1 integrases. J Virol 1997; 71:843–847.

27. Kukolj G, Katz RA, Skalka AM. Characterization of the nuclear localization signal in the avian sarcoma virus integrase. Gene 1998; 223:157–163.

28. Gallay P, Hope T, Chin D, Trono D. HIV-1 infection of nondividing cells through the recognition of integrase by the importin/karyopherin pathway. Proc Natl Acad Sci USA 1997; 94:9825–9830.

29. Heinzinger NK, Bukinsky MI, Haggerty SA, et al. The Vpr protein of human immunodeficiency virus type 1 influences nuclear localization of viral nucleic acids in nondividing host cells. Proc Natl Acad Sci USA 1994; 91:7311–7315.

30. Gallay P, Swingler S, Song J, Bushman F, Trono D. HIV nuclear import is governed by the phosphotyrosine-mediated binding of matrix to the core domain of integrase. Cell 1995; 83:569–576.

31. Gallay P, Stitt V, Mundy C, Oettinger M, Trono D. Role of the karyopherin pathway in human immunodeficiency virus type 1 nuclear import. J Virol 1996; 70:1027–1032.

32. Gallay P, Swingler S, Aiken C, Trono D. HIV 1 infection of nondividing cells: C-terminal tyrosine phosphorylation of the viral matrix protein is a key regulator. Cell 1995; 80:379–388.

33. Bukrinsky MI, Haggerty S, Dempsey MP, et al. A nuclear localization signal within

HIV1 matrix protein that govern infection of non-dividing cells. Nature 1993; 365: 666–669.

34. Zennou V, Petit C, Guetard D, Nerhbass U, Montagnier L, Charneau P. HIV-1 genome nuclear import is mediated by a central DNA flap. Cell 2000; l01:173–185.

35. Sutton RE, Reitsma MJ, Uchida N, Brown PO. Transduction of human progenitor hematopoietic stem cells by human immunodeficiency virus type 1–based vectors is cell cycle dependent. J Virol 1999; 73:3649–3660.

36. Pollard VW, Malim MH. The HIV-1 Rev protein. Annu Rev Microbiol 1998; 52: 491–532.

37. Pasquinelli AE, Ernst RK, Lund E, et al. The constitutive transport element (CTE) of Mason-Pfizer monkey virus (MPMV) accesses a cellular mRNA export pathway. EMBO J 1997; 16:7500–7510.

38. Rein A, Mirro J, Haynes JG, Ernst SM, Nagashima K. Function of the cytoplasmic domain of a retroviral transmembrane protein: pl5E-p2E cleavage activates the membrane fusion capability of the murine leukemia virus Env protein. J Virol 1994; 68:1773–1781.

39. Fielding AK, Chapel-Fernandes S, Chadwick MP, Bullough FJ, Cosset FL, Russell SJ. A hyperfusogenic gibbon ape leukemia envelope glycoprotein: targeting of a cytotoxic gene by ligand display. Hum Gene Ther 2000; 11:817–826.

40. Miller AD, Chen F. Retrovirus packaging cells based on 10A1 murine leukemia virus for production of vectors that use multiple receptors for cell entry. J Virol 1996; 70:5564–5571.

41. Hatziioannou T, Delahaye E, Martin F, Russell SJ, Cosset FL. Retroviral display of functional binding domains fused to the amino terminus of influenza hemagglutinin. Hum Gene Ther 1999; 10:1533–1544.

42. Russell SJ, Cosset FL. Modifying the host range properties of retroviral vectors. J Gene Med 1999; 1:300–311.

43. Wool-Lewis RJ, Bates P. Characterization of Ebola virus entry by using pseudo-typed viruses: identification of receptor-deficient cell lines. J Virol 1998; 72:3155–3160.

44. Uckert W, Willimsky G, Pedersen FS, Blankenstein T, Pedersen L. RNA levels of human retrovirus receptors Pit1 and Pit2 do not correlate with infectibility by three retroviral vector pseudotypes. Hum Gene Ther 1998; 9:2619–2627.

45. Danos O, Mulligan RC. Safe and efficient generation of recombinant retroviruses with amphotropic and ecotropic host ranges. Proc Natl Acad Sci USA 1988; 85: 6460–6464.

46. Patience C, Takeuchi Y, Cosset FL, Weiss RA. Packaging of endogenous retroviral sequences in retroviral vectors produced by murine and human packaging cells. J Virol 1998; 72:2671–2676.

47. Chakraborty AK, Zink MA, Hodgson CP. Expression of VL30 vectors in human cells that are targets for gene therapy. Biophys Biochem Res Commun 1995; 209: 677–683.

48. Chakraborty AK, Zink MA, Hodgson CP. Transmission of endogenous VL30 retro-transposons by helper cells used in gene therapy. Cancer Gene Ther 1994; 1:113–118.

49. Kim SH, Yu SS, Park JS, Robbins PD, An CS, Kim S. Construction of retroviral vectors with improved safety, gene expression, and versatility. J Virol 1998; 72: 994–1004.
50. Cosset F-L, Takeuchi Y, Battini JL, Weiss RA, Collins MKL. High-titer packaging cells producing recombinant retroviruses resistant to human complement. J Virol 1995; 69:7430–7436.
51. Morgenstern JP, Land H. Advanced mammalian gene transfer: high titre retroviral vectors with multiple drug selection markers and a complementary helper-free packaging cell line. Nucleic Acids Res 1990; 18:3587–3596.
52. Duisit G, Salvetti A, Moullier P, Cosset FL. Functional characterization of adenoviral/retroviral chimeric vectors and their use for efficient screening of retroviral producer cell lines. Hum Gene Ther 1999; 10:189–200.
53. Lee M, Kremer E, Perricaudet M. Adenoviral vectors. In: Hall C, ed. Molecular and cell biology of human gene therapeutics. Vol. 5. George Dickson, 1995:20–32.
54. Linial ML, Miller AD. Retroviral RNA packaging: sequence requirements and implications. Curr Top Microbiol Immunol 1990; 157:125–152.
55. Zhao Y, Low W, Collins MK. Improved safety and titre of murine leukaemia virus (MLV)-based retroviral vectors. Gene Ther 2000; 7:300–305.
56. Mautino MR, Keiser N, Morgan RA. Improved titers of HIV-based lentiviral vectors using the SRV-l constitutive transport element. Gene Ther 2000; 7:1421–1424.
57. Dull T, Zufferey R, Kelly M, et al. A third-generation lentivirus vector with a conditional packaging system. J Virol 1998; 72:8463–8471.
58. Krall WJ, Skelton DC, Yu XJ, et al. Increased levels of spliced RNA account for augmented expression from MFG retroviral vector in hematopoïetic cells. Gene Ther 1996; 3:37–48.
59. Shin NH, Hartigan-O'Connor D, Pfeiffer JK, Telesnitsky A. Replication of lengthened Moloney murine leukemia virus genomes is impaired at multiple stages. J Virol 2000; 74:2694–2702.
60. Yu SS, Kim JM, Kim S. High efficiency retroviral vectors that contain no viral coding sequences. Gene Ther 2000; 7:797–804.
61. Hantzopoulos PA, Sullenger BA, Ungers G, Gilboa E. Improved gene expression upon transfer of the adenosine deaminase minigene outside the transcriptional unit of a retroviral vector. Proc Natl Acad Sci USA 1989; 86:3519–3523.
62. Choulika A, Guyot V, Nicolas J-F. Transfer of single gene-contaning long terminal repeats into the genome of mammalian cells by a retroviral vector carrying the cre gene and the loxP site. J Virol 1996; 70:1792–1798.
63. Delviks KA, Hu WS, Pathak VK. Psi- vectors: murine leukemia virus-based self-inactivating and self-activating retroviral vectors. J Virol 1997; 71:6218–6224.
64. Soneoka Y, Cannon PM, Ramsdale EE, et al. A transient three-plasmid expression system for the production of high titer retroviral vectors. Nucleic Acids Res 1995; 23:628–633.
65. Yang Y, Vanin EF, Whitt MA, et al. Inducible high level production of infectious murine leukemia retroviral vector particle pseudotyped with vesicular somatitis virus G envelope protein. Hum Gene Ther 1995; 6:1203–1213.

66. Yang S, Delgado R, King SR, et al. Generation of retroviral vector for clinical studies using transient transfection. Hum Gene Ther 1999; 10:123–132.

67. Vanin EF, Cerruti L, Tran N, Grosveld G, Cunningham JM, Jane SM. Development of high-titer retroviral producer cell lines by using Cre- mediated recombination. J Virol 1997; 71:7820–7826.

68. Tafuro S, Zentilin L, Falaschi A, Giacca M. Rapid retrovirus titration using competitive polymerase chain reaction. Gene Ther 1996; 3:679–684.

69. Onodera M, Yachie A, Nelson DM, Welchlin H, Morgan RA, Blaese RM. A simple and reliable method for screening retroviral producer clones without selectable markers. Hum Gene Ther 1997; 8:1189–1194.

70. Murdoch B, Pereira DS, Wu X, Dick JE, Ellis J. A rapid screening procedure for the identification of high-titer retrovirus packaging clones. Gene Ther 1997; 4:744–749.

71. Towers G, Stockholm D, Labrousse-Najburg V, Carlier F, Danos O, Page J-C. One step screening of retroviral producer clones by real time quantitative PCR. J Gene Med 1999; 1:352–359.

72. Pagès JC, Loux N, Farge D, Briand P, Weber A. Activation of Moloney murine leukemia virus LTR enhances the titer of recombinant retrovirus in psi CRIP packaging cells. Gene Ther 1995; 2:547–551.

73. Olsen JC, Sechelski J. Use of sodium butyrate to enhance production of retroviral vectors expressing CFTR cDNA. Hum Gene Ther 1995; 6:1195–1202.

74. Kotani H, Newton PB III, Zhang S, et al. Improved methods of retroviral vector transduction and production for gene therapy. Hum Gene Ther 1994; 5:19–28.

75. Paul RW, Morris D, Hess BW, Dunn J, Overell RW. Increased viral titer through concentration of viral harvests from retroviral packaging lines. Hum Gene Ther 1993; 4:609–615.

76. Andreadis ST, Brott D, Fuller AO, Palsson BO. Moloney murine leukemia virus-derived retroviral vectors decay intracellularly with a half-life in the range of 5.5 to 7.5 hours. J Virol 1997; 71:7541–7548.

77. Andreadis S, Lavery T, Davis HE, Le Doux JM, Yarmush ML, Morgan JR. Toward a more accurate quantitation of the activity of recombinant retroviruses: alternatives to titer and multiplicity of infection. J Virol 2000; 74:3431–3439.

78. Feng M, Jackson WH Jr, Goldman CK, et al. Stable in vivo gene transduction via a novel adenoviral/retroviral chimeric vector. Nat Biotechnol 1997; 15:866–870.

79. Yang Y, Jooss KU, Su Q, Ertl HJC, Wilson JM. Immune response to viral antigen versus transgene product in the elimination of recombinant adenovirus-infected hepatocytes in vivo. Gene Ther 1996; 3:137–144.

80. Cornetta K, Morgan RA, Anderson WF. Safety issues related to retroviral mediated gene transfer in humans. Hum Gene Ther 1991; 2:5–14.

81. Peng KW, Morling FL, Cosset FL, Murphy G, Russell SJ. A gene delivery system activatable by disease-associated matrix metalloproteinases. Hum Gene Ther 1997; 8:729–738.

82. Bosch A, McCray PB Jr, Chang SM, et al. Proliferation induced by keratinocyte growth factor enhances in vivo retroviral-mediated gene transfer to mouse hepatocytes. J Clin Invest 1996; 98:2683–2687.

83. Kafri T, Blomer U, Peterson DA, Gage FH, Verma IM. Sustained expression of

genes delivered directly into liver and muscle by lentiviral vectors. Nat Genet 1997; 17:314–327.

84. Dubensky TW, Campbell BA, Villarreal LP. Direct transfection of viral and plasmid DNA into liver or spleen of mice. Proc Natl Acad Sci USA 1984; 81:7529–7533.

85. Bosch A, McCray PB Jr, Walters KS, et al. Effects of keratinocyte and hepatocyte growth factor in vivo: implications for retrovirus-mediated gene transfer to liver. Hum Gene Ther 1998; 9:1747–1754.

86. Kitten O, Cosset FL, Ferry N. Highly efficient retrovirus-mediated gene transfer into rat hepatocytes in vivo. Hum Gene Ther 1997; 8:1491–1494.

87. Song ES, Lee V, Surh CD, et al. Antigen presentation in retroviral vector-mediated gene transfer in vivo. Proc Natl Acad Sci USA 1997; 94:1943–1948.

88. McCormack JE, Martineau D, DePolo N, et al. Anti-vector immunoglobulin induced by retroviral vectors. Hum Gene Ther 1997; 8:1263–1273.

89. Halbert CL, Standaert TA, Wilson CB, Miller AD. Successful readministration of adeno-associated virus vectors to the mouse lung requires transient immunosuppression during the initial exposure. J Virol 1998; 72:9795–9805.

90. Takeuchi Y, Porter CD, Strahan KM, et al. Sensitization of cells and retroviruses to human serum by (α1-3) galactosyltransferase. Nature 1996; 379:85–88.

91. Galili U. Evolution and pathophysiology of the human natural anti-alpha-galactosyl IgG (anti-Gal) antibody. Springer Semin Immunopathol 1993; 15:155–171.

92. Takeuchi Y, Liong SH, Bieniasz PD, et al. Sensitization of rhabdo-, lenti-, and spumaviruses to human serum by galactosyl(alpha1-3)galactosylation. J Virol 1997; 71:6174–6178.

93. Saifuddin M, Hedayati T, Atkinson JP, Holguin MH, Parker CJ, Spear GT. Human immunodeficiency virus type 1 incorporates both glycosyl phosphatidylinositol-anchored CD55 and CD59 and integral membrane CD46 at levels that protect from complement-mediated destruction. J Gen Virol 1997; 78:1907–1911.

94. Spitzer D, Hauser H, Wirth D. Complement-protected amphotropic retroviruses from murine packaging cells. Hum Gene Ther 1999; 10:1893–1902.

95. Hiasa A, Watanabe M, Okada H, et al. Retroviruses prepared from human DAF expressing murine packaging cells acquire resistance against human serum. Int J Oncol 1999; 14:1091–1096.

96. DePolo NJ, Reed JD, Sheridan PL, et al. VSV-G pseudotyped lentiviral vector particles produced in human cells are inactivated by human serum. Mol Ther 2000; 2:218–222.

97. Jiang A, Fisher H, Pomerantz RJ, Dornburg R. A genetically engineered spleen necrosis virus-derived retroviral vector that displays the HIV type 1 glycoprotein 120 envelope peptide. Hum Gene Ther 1999; 10:2627–2636.

98. Nguyen TH, Pagès JC, Farge D, Briand P, Weber A. Amphotropic retroviral vectors displaying hepatocyte growth factor- envelope fusion proteins improve transduction efficiency of primary hepatocytes. Hum Gene Ther 1998; 9:2469–2479.

99. Maurice M, Mazur S, Bullough FJ, et al. Efficient gene delivery to quiescent interleukin-2 (IL-2)–dependent cells by murine leukemia virus–derived vectors harboring IL-2 chimeric envelope glycoproteins. Blood 1999; 94:401–410.

100. Somia NV, Zoppe M, Verma IM. Generation of targeted retroviral vectors by using single-chain variable fragment: an approach to in vivo gene delivery. Proc Natl Acad Sci USA 1995; 92:7570–7574.

101. Valsesia-Wittmann S, Morling FJ, Nilson BHK, Takeuchi Y, Russell SJ, Cosset F-L. Improvement of retroviral retargeting by using amino acid spacers between an additional binding domain and the N terminus of Moloney murine leukemia virus SU. J Virol 1996; 70:2059–2064.
102. Valsesia-Wittmann S, Drynda A, Deléage G, et al. Modifications in the binding domain of avian retrovirus envelope protein to redirect the host range of retroviral vectors. J Virol 1994; 68:4609–4619.
103. Srinivasakumar N, Chazal N, Helga-Maria C, Prasad S, Hammarskjold ML, Rekosh D. The effect of viral regulatory protein expression on gene delivery by human immunodeficiency virus type 1 vectors produced in stable packaging cell lines. J Virol 1997; 71:5841–5848.
104. Etienne-Julan M, Roux P, Carillo S, Jeanteur P, Piechaczyk M. The efficiency of cell targeting by recombinant retroviruses depends on the nature of the receptor and the composition of the artificial cell–virus linker. J Gen Virol 1992; 73:3251–3255.
105. Boerger AL, Snitkovsky S, Young JA. Retroviral vectors preloaded with a viral receptor-ligand bridge protein are targeted to specific cell types. Proc Natl Acad Sci USA 1999; 96:9867–9872.
106. Snitkovsky S, Niederman TM, Carter BS, Mulligan RC, Young JA. A TVA–single-chain antibody fusion protein mediates specific targeting of a subgroup A avian leukosis virus vector to cells expressing a tumor-specific form of epidermal growth factor receptor. J Virol 2000; 74:9540–9545.
107. Fass D, Davey RA, Hamson CA, Kim PS, Cunningham JM, Berger JM. Structure of a murine leukemia virus receptor-binding glycoprotein at 2.0 angstrom resolution. Science 1997; 277:1662–1666.
108. Hall FL, Liu L, Zhu NL, et al. Molecular engineering of matrix-targeted retroviral vectors incorporating a surveillance function inherent in von Willebrand factor. Hum Gene Ther 2000; 11:983–993.
109. Porter CD, Collins MK, Tailor CS, et al. Comparison of efficiency of infection of human gene therapy target cells via four different retroviral receptors. Hum Gene Ther 1996; 7:913–919.
110. Valsesia-Wittmann S, Morling FJ, Hatziioannou T, Russell SJ, Cosset FL. Receptor co-operation in retrovirus entry: recruitment of an auxiliary entry mechanism after retargeted binding. EMBO J 1997; 16:1214–1223.
111. Rodrigues P, Heard JM. Modulation of phosphate uptake and amphotropic murine leukemia virus entry by posttranslational modifications of PIT-2. J Virol 1999; 73:3789–3799.
112. Buchholz CJ, Peng KW, Morling FJ, Zhang J, Cosset FL, Russell SJ. In vivo selection of protease cleavage sites from retrovirus display libraries. Nat Biotechnol 1998; 16:951–954.
113. Middleton J, Neil S, Wintle J, et al. Transcytosis and surface presentation of IL-8 by venular endothelial cells. Cell 1997; 91:385–395.
114. Halene S, Kohn DB. Gene therapy using hematopoietic stem cells: Sisyphus approaches the crest. Hum Gene Ther 2000; 11:1259–1267.
115. Wognum AW, Visser TP, Peters K, Bierhuizen MF, Wagemaker G. Stimulation of mouse bone marrow cells with kit ligand, FLT3 ligand, and thrombopoietin leads to efficient retrovirus-mediated gene transfer to stem cells, whereas interleukin 3

and interleukin 11 reduce transduction of short- and long-term repopulating cells. Hum Gene Ther 2000; 11:2129–2141.

116. Pagès JC, Andreoletti M, Bennoun M, et al. Efficient retroviral-mediated gene transfer into primary culture of murine and human hepatocytes: expression of the LDL receptor. Hum Gene Ther 1995; 6:21–30.

117. Ardavin C, Luthi F, Andersson M, et al. Retrovirus-induced target cell activation in the early phases of infection: the mouse mammary tumor virus model. J Virol 1997; 71:7295–7299.

118. Deminie CA, Emerman M. Incorporation of human immunodeficiency virus type 1 Gag proteins into murine leukemia virus virions. J Virol 1993; 67:6499–6506.

119. Deminie CA, Emerman M. Functional exchange of an oncoretrovirus and a lentivirus matrix protein. J Virol 1994; 68:4442–4449.

120. Lieber A, Kay MA, Li ZY. Nuclear import of Moloney murine leukemia virus DNA mediated by adenovirus preterminal protein is not sufficient for efficient retroviral transduction in nondividing cells. J Virol 2000; 74:721–734.

121. Parveen Z, Krupetsky A, Engelstadter M, Cichutek K, Pomerantz RJ, Dornburg R. Spleen necrosis virus-derived C-type retroviral vectors for gene transfer to quiescent cells. Nat Biotechnol 2000; 18:623–629.

122. Frankel AD, Young JA. HIV-1: fifteen proteins and an RNA. Annu Rev Biochem 1998; 67:1–25.

123. Naldini L, Blömer U, Gallay P, et al. In vivo gene delivery and stable transduction of non-dividing cells by a lentiviral vector. Science 1996; 272:263–267.

124. Zufferey R, Nagy D, Mandel RJ, Naldini L, Trono D. Multiply attenuated lentiviral vector achieves efficient gene delivery in vivo. Nat Biotechnol 1997; 15:871–875.

125. Fouchier RA, Meyer BE, Simon JH, et al. Interaction of the human immunodeficiency virus type 1 Vpr protein with the nuclear pore complex. J Virol 1998; 72: 6004–6013.

126. Popov S, Rexach M, Ratner L, Blobel G, Bukrinsky M. Viral protein R regulates docking of the HIV-1 preintegration complex to the nuclear pore complex. J Biol Chem 1998; 273:13347–13352.

127. Popov S, Rexach M, Zybarth G, et al. Viral protein R regulates nuclear import of the HIV-1 pre-integration complex. EMBO J 1998; 17:909–917.

128. Chinnasamy D, Chinnasamy N, Enriquez MJ, Otsu M, Morgan RA, Candotti F. Lentiviral-mediated gene transfer into human lymphocytes: role of HIV-1 accessory proteins. Blood 2000; 96:1309–1316.

129. Srinivasakumar N, Schuening FG. A lentivirus packaging system based on alternative RNA transport mechanisms to express helper and gene transfer vector RNAs and its use to study the requirement of accessory proteins for particle formation and gene delivery. J Virol 1999; 73:9589–9598.

130. Gasmi M, Glynn J, Jin MJ, Jolly DJ, Yee JK, Chen ST. Requirements for efficient production and transduction of human immunodeficiency virus type 1-based vectors. J Virol 1999; 73:1828–1834.

131. Zufferey R, Donello JE, Trono D, Hope TJ. Woodchuck hepatitis virus posttranscriptional regulatory element enhances expression of transgenes delivered by retroviral vectors. J Virol 1999; 73:2886–2892.

132. Zufferey R, Dull T, Mandel RJ, et al. Self-inactivating lentivirus vector for safe and efficient in vivo gene delivery. J Virol 1998; 72:9873–9880.

133. Iwakuma T, Cui Y, Chang LJ. Self-inactivating lentiviral vectors with U3 and U5 modifications. Virology 1999; 261:120–132.

134. Cui Y, Iwakuma T, Chang LJ. Contributions of viral splice sites and cis-regulatory elements to lentivirus vector function. J Virol 1999; 73:6171–6176.

135. Follenzi A, Ailles LE, Bakovic S, Geuna M, Naldini L. Gene transfer by lentiviral vectors is limited by nuclear translocation and rescued by HIV-1 pol sequences. Nat Genet 2000; 25:217–222.

136. Wu X, Wakefield JK, Liu H, et al. Development of a novel trans-lentiviral vector that affords predictable safety. Mol Ther 2000; 2:47–55.

137. Kotsopoulou E, Kim VN, Kingsman AJ, Kingsman SM, Mitrophanous KA. A Rev-independent human immunodeficiency virus type 1 (HIV-1)–based vector that exploits a codon-optimized HIV-1 gag-pol gene. J Virol 2000; 74:4839–4852.

138. Wagner R, Graf M, Bieler K, et al. Rev-independent expression of synthetic gag-pol genes of human immunodeficiency virus type 1 and simian immunodeficiency virus: implications for the safety of lentiviral vectors. Hum Gene Ther 2000; 11: 2403–2413.

139. Kafri T, van Praag H, Ouyang L, Gage FH, Verma IM. A packaging cell line for lentivirus vectors. J Virol 1999; 73:576–584.

140. Klages N, Zufferey R, Trono D. A stable system for the high-titer production of multiply attenuated lentiviral vectors. Mol Ther 2000; 2:170–176.

141. Mangeot PE, Negre D, Dubois B, et al. Development of minimal lentivirus vectors derived from simian immunodeficiency virus (SIVmac251) and their use for gene transfer into human dendritic cells. J Virol 2000; 74:8307–8315.

142. Metharom P, Takyar S, Xia HH, et al. Novel bovine lentiviral vectors based on Jembrana disease virus. J Gene Med 2000; 2:176–185.

143. Mitrophanous K, Yoon S, Rohll J, et al. Stable gene transfer to the nervous system using a non-primate lentiviral vector. Gene Ther 1999; 6:1808–1818.

144. Mselli-Lakhal L, Favier C, Da Silva Teixeira MF, et al. Defective RNA packaging is responsible for low transduction efficiency of CAEV-based vectors. Arch Virol 1998; 143:681–695.

145. Pannell D, Osborne CS, Yao S, et al. Retrovirus vector silencing is de novo methylase independent and marked by a repressive histone code. EMBO J 2000; 19:5884–5894.

146. Challita P-M, Kohn DB. Lack of expression from a retroviral vector after transduction of murine hematopoietic stem cells is associated with methylation in vivo. Proc Natl Acad Sci USA 1994; 91:2567–2571.

147. Rivella S, Callegari JA, May C, Tan CW, Sadelain M. The cHS4 insulator increases the probability of retroviral expression at random chromosomal integration sites. J Virol 2000; 74:4679–4687.

148. Couture LA, Mullen CA, Morgan RA. Retroviral vectors containing chimeric promoter/enhancer elements exhibit cell-type–specific gene expression. Hum Gene Ther 1994; 5:667–677.

149. Ferrari G, Salvatori G, Rossi C, Cossu G, Mavilio F. A retroviral vector containing a muscle-specific enhancer drives gene expression only in differentiated muscle fibers. Hum Gene Ther 1995; 6:733–742.

150. Paulus W, Baur I, Boyce FM, Breakefield XO, Reeves SA. Self-contained, tetracycline-regulated retroviral vector system for gene delivery to mammalian cells. J Virol 1996; 70:62–67.

151. Donahue RE, Kessler SW, Bodine D, et al. Helper virus induced T cell lymphoma in non human primates after retroviral mediated gene transfer. J Exp Med 1992; 176:1125–1135.

152. Printz M, Reynolds J, Mento SJ, Jolly D, Kowal K, Sajjadi N. Recombinant retroviral vector interferes with the detection of amphotropic replication competent retrovirus in standard culture assays. Gene Ther 1995; 2:143–150.

153. Hu WS, Temin HM. Genetic consequences of packaging two RNA genomes in one retroviral particle: pseudodiploidy and high rate of genetic recombination. Proc Natl Acad Sci USA 1990; 87:1556–1560.

154. Yang J, Bogerd HP, Peng S, Wiegand H, Truant R, Cullen BR. An ancient family of human endogenous retroviruses encodes a functional homolog of the HIV-1 Rev protein. Proc Natl Acad Sci USA 1999; 96:13404–13408.

155. Reiser J. Production and concentration of pseudotyped HIV-1-based gene transfer vectors. Gene Ther 2000; 7:910–913.

156. Naldini L, Blomer U, Gage FH, Trono D, Verma IM. Efficient transfer, integration, and sustained long-term expression of the transgene in adult rat brains injected with a lentiviral vector. Proc Natl Acad Sci USA 1996; 93:11382–11388.

157. Blomer U, Naldini L, Kafri T, Trono D, Verma IM, Gage FH. Highly efficient and sustained gene transfer in adult neurons with a lentivirus vector. J Virol 1997; 71: 6641–6649.

158. Miyoshi H, Blomer U, Takahashi M, Gage FH, Verma IM. Development of a self-inactivating lentivirus vector. J Virol 1998; 72:8150–8157.

159. Kordower JH, Bloch J, Ma SY, et al. Lentiviral gene transfer to the nonhuman primate brain. Exp Neurol 1999; 160:1–16.

160. Park F, Ohashi K, Chiu W, Naldini L, Kay MA. Efficient lentiviral transduction of liver requires cell cycling in vivo. Nat Genet 2000; 24:49–52.

161. Clever JL, Taplitz RA, Lochrie MA, Polisky B, Parslow TG. A heterologous, high-affinity RNA ligand for human immunodeficiency virus Gag protein has RNA packaging activity. J Virol 2000; 74:541–546.

162. Accola MA, Strack B, Gottlinger HG. Efficient particle production by minimal Gag constructs which retain the carboxy-terminal domain of human immunodeficiency virus type 1 capsid-p2 and a late assembly domain. J Virol 2000; 74:5395–5402.

163. Goulaouic H, Chow SA. Directed integration of viral DNA mediated by fusion proteins consisting of human immunodeficiency virus type 1 integrase and Escherichia coli LexA protein. J Virol 1996; 70:37–46.

164. Li Z, et al. Murine leukemia induced by retroviral gene marking. Science 2002; 296(5567):497.

165. Cavazzana-Calvo M, et al. Gene therapy of human severe combined immunodeficiency (SCID)-X1 disease. Science 2000; 288(5466):669–672.

166. Marshall E. Clinical research. Gene therapy a suspect in leukemia-like disease. Science 2002; 298(5591):34–35.

167. Murre C. Intertwining proteins in thymocyte development and cancer. Nat Immunol 2000; 1(2):97–98.

168. Schroder AR, et al. HIV-1 integration in the human genome favors active genes and local hotspots. Cell 2002; 110(4):521–529.

10

Device-Mediated Gene Delivery: A Review

Fiona MacLaughlin
Valentis, Inc., The Woodlands, Texas, U.S.A.

Alain Rolland
Vical, Inc., San Diego, California, U.S.A.

I. INTRODUCTION

The principle of somatic gene therapy is the replacement or correction of a defective or missing gene by means of administering DNA encoding for a specific protein. The use of DNA plasmid as part of a medicine or vaccine quells some of the problems associated with low bioavailability and expensive manufacturing costs of most protein drugs, which, in some disease states, currently provide the only therapeutic option. For gene therapy to be viable, the major barriers to cellular translocation of the plasmid are primarily the cell and nuclear membranes. The large size and anionic nature of plasmids makes their cellular uptake a relatively inefficient event. A variety of nonviral systems have subsequently been identified and developed to enhance cellular delivery of plasmids. Such systems are designed to address the key limiting events in the transport of a plasmid from the administration site to the nucleus of the target cell. The attributes of such delivery systems can be summarized by the acronym DART, where D reflects the distribution of plasmid, A, the access of plasmid to a target cell, followed by recognition (R) of the plasmid by a receptor or other cell sensor and ultimate translocation (T) into the cytoplasm and nucleus. For a clinically viable product, a delivery pathway must be identified that enables DART, ultimately leading to controlled in situ production of a safe and nontoxic protein within specific cell types. Overall findings have suggested that each therapy and target site require a specifically tailored gene delivery system. For instance, efficient delivery of plasmid to skeletal muscle is inhibited by precondensation with cationic polymers, for example, but can be enhanced by protective interactive noncondensing (PINC) polymers such as poly(vinylpyrrolidone) [1]. Current nonviral strategies employing a wide variety of macromolecules such as lipids and polymers in conjunction with a multitude of routes of administration to enhance DNA delivery have not proven to be as efficient as many viral vectors thus far. This is a consequence of their inherent reliability upon the transmembrane carrier mechanisms of the cell for successful uptake. Conversely, viral systems may not be so selective, and although their delivery efficiency may be higher, concerns about genomic integration and potential inability of repeated administrations as a consequence of immune system stimulation can be problematic. Viral vector manufacture and quality control are also typically more problematic than for nonviral systems. Such reliance upon the cell can be bypassed by utilizing physical means to transiently alter cellular permeability. These methods are highly versatile and are generally not limited to specific routes. Transient alteration of cellular membrane integrity provides a doorway for intracellular flux of the macromolecules, such as DNA plasmid. One key requirement is that the balance between efficient delivery and destruction of the cell be maintained.

This chapter aims to illustrate how effectively selected technologies such as needle-free gene delivery, electroporation, and sonoporation perform, and it describes many of their applications along with their potential limitations. This chapter is not a comprehensive discussion of every device available for gene delivery, but has been compiled to provide the reader with an objective overview of some physical delivery strategies.

II. NEEDLE-FREE PARTICLE-MEDIATED GENE DELIVERY

The message is that "needle-free is here to stay"! Injection by traditional syringe is responsible for several side effects such as pain, inflammation, injury associated with administration, and infection, which afflict an estimated one million patients in the United States alone. For those therapies where transient expression, and localized gene delivery to the tissue or even more specifically, a localized cell population, is therapeutically applicable, device-mediated needle-free delivery of DNA plasmid could be ideal. Needle-free gene administration presents advantages such as limiting toxicity, potential cell receptor–independent uptake, minimal DNA size restriction, and the potential for multiple treatments via a relatively uncomplicated administration modality, thus improving patient compliance. Administration via needle-free jet injection is typically pain free. As with any therapy, certain parameters such as the disease type, the tissue accessibility, and the route of treatment as well as required levels, control, and duration of transgene expression must be defined. From a commercial viewpoint, ballistic technology may offer companies extended patent life and creates product diversity.

A. Needle-Free Devices

The needle-free revolution began in 1987 when Sanford and coworkers invented the "ballistic" gene gun approach, employing gun powder to deliver genes to plant cells [2]. The refinement of this in vitro process enabled the delivery of gold particles coated with DNA or RNA to select tissues in vivo via a low-pressure helium stream. Although a number of delivery devices are currently in circulation, they are all based upon the same principle of accelerating DNA-coated microcarriers. Gold nanobeads or microbeads are coated with a specific copy number of nucleic acids (in the order of thousands) by simple nucleic acid precipitation. In the early-generation devices, helium, released by plunger-mediated rupture of a membrane, created a pressure wave that accelerated a plastic macrocarrier to which the gold-coated particles were adherent into a steel-mesh screen. The particles bombarded the target tissue while the screen and the microcarrier were left behind. Further modifications to the device made by Johnston led to the creation of a hand-held wand that could be placed directly at the target interface, while eliminating the need for a vacuum chamber at the target site [3]. The success of this device in stimulating gene expression and immune responses in vivo has been described in the literature [4] (Fig. 1). A hand-held device that is commercially available with similar specifications is the Helios gene gun (Bio-Rad, Hercules, CA). The specifications of this instrument define that an area as

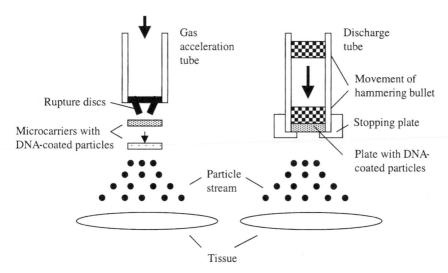

Figure 1 A diagrammatic representation of two different gene gun types: (left) via a microcarrier or by (right) a hammering bullet. The helium flow is represented by the arrows, demonstrating how the device functions, enabling release of a stream of particles that bombard a target tissue. (Adapted from Ref. 12.)

small as 2 cm^2 can be targeted. Gold-coated particles of micron size, with the capability of carrying 0.5–5.0 µg of DNA per particle, are adsorbed onto Gold-Coat™ tubing, cut into small cartridges, and inserted inside the device, up to 12 at a time. For instance, loading of multiple genes may enable multiepitope vaccination

Similarly, the Accell system from Agracetus/Powderject® initially employed an electrical discharge device mechanism to accelerate an aluminized membrane microcarrier. The delivery of gene constructs in vitro [5] and to liver, skin, and muscle in vivo [4,5] using this device have been described. The Accell device was a first-generation reusable gene gun device. The DNA-coated gold particles were adsorbed onto Tefzel tubing and loaded into a cartridge and inserted into the instrument. Release of the DNA takes less than 5 sec, in a penetration pattern 1 cm in diameter, covering a 2 cm^2 square target area as for the gene gun. Granulocyte–macrophage colony-stimulating factor cDNA was delivered and expressed in tumor cells in vitro via this method [6]. After vaccination with a transfected murine tumor cell line, protection to challenge with the wild-type tumor cell line was shown in mice. Other nucleic acid–based vaccines based on simian immunodeficiency virus (SIV) were successfully delivered to rhesus macaques, although the vaccines did not prevent infection [7]. Induction of an immune response from de novo antigen production after plasmid delivery in mouse epidermis was also shown [8–10]. Further redesign of the device replaced the electrical discharge and membrane with a helium discharge mechanism [11]. A comparison study of both devices showed that identical immune responses could be generated [10] in contrast to a study by Williams et al., who showed that the helium-driven device produced a fourfold increase in luciferase expression levels upon delivery to mouse skin [4] (Fig. 2).

Further developments were made to the device and were reported by Kuriyama et al. [12]. The high-pressure helium wave accelerates a hammering bullet, rather than a microcarrier, that makes contact with a vibration plate to which the DNA particles are affixed (see Fig. 1). The release of the particles into the target tissue results. Particle-mediated gene transfer with this device was reported in the liver using a 0.8-µg dose of lacZ plasmid. Expression levels were about 30-fold higher than background. Even the lobes of mice that were injected with 20 µg and electroporated with spoon-style electrodes and administered eight pulses of 100 V/cm intensity and 99 µsec duration did not express as highly as those treated via this gene gun protocol. It is possible that the particles accelerated into the tissue are not removed by afferent blood flow, which may be responsible for removal of other macromolecules, such as directly injected DNA.

Powderject technology focuses not only upon the supersonic delivery of small gold particles but also upon small molecules, peptides, or proteins that have been formulated into a powder. There are two generations of powder particle delivery devices: the Dermal Powderject ND, a single-use disposable device in-

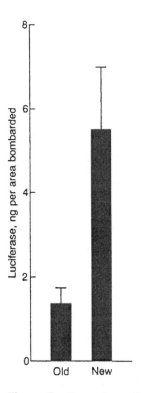

Figure 2 Comparison of luciferase expression after bombardment using a modified PDS-1000 (old)(electrical discharge mechanism) and the helium-driven apparatus (new). The new apparatus resulted in a fourfold increase in expression of the foreign gene in mouse skin. Values from mock bombardments (<0.3 pg per biopsy sample) were subtracted from each determination. Bars represent mean values ± SEM. (From Ref. 4.)

corporating a self-contained helium cylinder and the Dermal Powderject XR, a first-generation reusable device, which has been extensively used in initial human genetic vaccine trials. Unlike with small particle delivery, extracellular localization, but not cell-specific delivery, of the powder to the skin is achieved. In partnership with GSK Glaxo-Wellcome, a large genetic vaccine portfolio (Hepatitis B Prophylactic and Therapeutic, Influenza, HIV therapeutic) in both research and clinical development phases has been established.

The pioneers of needle-free liquid delivery were Weston Medical Group Plc., who introduced the Intraject® device in 1994. This device consists of an actuator mechanism enabling the release of sterile drug from the chamber. The drug capsule that contains the borosilicate glass chamber is biocompatible with many liquid drugs. Release of the drug from the chamber is enabled when the pressure of the device against the skin is optimal, thus priming release of the drug upon pressing the device, for example, against the leg. The expansion of compressed nitrogen enables device activation. As well as rapid administration of drug in a 1-mL volume in less than 60 msec, the device is cost effective ($\approx$ $1.50 per shot). Primarily designed for subcutaneous administration, the device is currently being modified for intramuscular use. The technology has been licensed to Medeva (flu vaccine); to Hoffmann La-Roche (Pegylated IFNa-2a (PEGASYS), currently in phase III trials for the treatment of hepatitis C and other indications); Pharmacia and Upjohn (Fragmin — a low molecular weight heparin), Sosei (osteoporosis product), GSK Glaxo-Wellcome (Imigran/Imitrex). Their target market includes "any drug that needs subcutaneous or intramuscular injection"; thus delivery of genetic vaccines using this approach may be possible. Other available jet injectors include the AdvantaJet® and GentleJet (Advantage Health Services, Inc.), the Biojector 2000 (Bioject, Inc.) and the Medijector™ (Medi-Ject).

B. Issues Associated with Needle-Free Delivery of DNA Plasmid

Delivery of "naked" DNA via these technologies is not possible, because the DNA is not dense enough for consistent delivery, and the forces required are sufficiently great to shear it. Gold particles are required to achieve the density necessary for intracellular delivery. The DNA dissociates from the gold particle once inside the cell and is expressed to produce the required protein. For extracellular delivery to the tissue, gold particles are not required. Excipients, even adjuvants, can be added to the formulation instead to enhance the physical properties of the DNA, enabling efficient delivery. Powder particles in the size range of 20–70 μm have sufficient density and size for optimal acceleration into the target tissue. Control over the level of penetration of the particles, the number of penetration events, and the amount of DNA delivered can be tightly regulated. For instance, the depth of tissue penetration is dependent on the particle size and the discharge voltage. Penetration depth is proportional to the momentum density (D) of an individual, particle:

$$D = Vp \cdot \frac{Mp}{Ap}$$

where Vp is particle velocity, Mp is mass, and A is particle surface area. Cell-specific uptake is not a viable option owing to direct intracellular transfer of the microbeads; however, problems associated with receptor-specific DNA uptake are bypassed.

It is also known that transgene expression is mainly localized in upper epidermal layers. Eisenbraun et al. demonstrated how penetration depth and expression are related by showing that particle delivery to lower dermal layers introduced disruption and lowered the overall response [8]. Increasing the discharge voltage from 15 to 25 kV, and thus increasing the depth of particle penetration, reduced expression of human growth hormone (hGH) from an hGH-expression system. The same study also highlighted the fact that minimal quantities of plasmid were required to initiate a substantial response. A dose-response between antibody titer and the amount of gold, and thus the number of plasmid copies administered was established, although the response did appear to plateau at doses of 0.16–0.8 mg gold per immunization at three different DNA:gold ratios. The delivery of about 300 copies of DNA plasmid per particle was calculated to be necessary to induce maximal transgene expression in an individual cell. This theory was proven by varying the ratio of plasmid:gold 500-fold, with no apparent change in expression levels. If <300 copies of plasmid were administered, expression was significantly reduced. In conjunction with this requirement, a defined number of cellular penetrations were shown to be required to elicit a reliable response. Maximal antibody titers were measured upon delivery of 10^7 particles to the epidermis for both priming and boost injections; however, by lowering the number of particles one logfold, a decrease in the immune response was observed, and no consistent dose response was apparent. No estimation of the efficacy of delivery could be given. In pig skin, an acceleration voltage of 25 kV enabled the diffuse and uniform delivery up to 60 µm deep into the wound, with ultimate expression of epidermal growth factor [13]. Another study that targeted the corneal epithelium of rabbits found that by maintaining pressure at a constant 300 psi, gold beads could be delivered to the squamous epithelium, but by increasing pressure, deposition in the lower wing or columnar basal cells occurred [14]. A higher pressure of 800 psi was required to target the particles to the lower epidermis in rat skin [15].

Peak expression in the epidermis has generally been followed by a variable decline that is dependent on the tissue and the expression system being transfected. Muscle demonstrated the slowest decline in expression in one study. This is consistent with the extended levels of transgene expression evident after direct intramuscular (IM) administration [16,17]. In contrast, expression following particle-mediated gene delivery is much more short term. A peak in luciferase expression was observed after particle-mediated gene delivery in the skin of the mouse ear at 24 hr, and expression persisted for up to 10 days. In the same study, expression of hGH peaked at 24 hr and was observed for up to 72 hr, with predominant transfection of keratinocytes [18]. Lin et al. demonstrated peak lacZ expression from a construct containing the involucrin promoter in mouse epidermis at 24 hr that decreased at 48 hr and was undetectable by 7 days, which is consistent with the fact that endogenous involucrin expression occurs only in this layer [19]. Locally acting growth factors were expressed for at least 5 days, with the wound fluid concentration decreasing with time [13]. Such transient expression is a useful inherent safety feature of biolistic particle delivery associated with the high turnover of epithelial tissue. Short term expression of lacZ plasmid also observed in mouse liver with the newly developed "hammering bullet" device peaked 2 days after administration and dropped one logfold by day 28. Immunohistochemical staining demonstrated that parenchymal hepatocytes were positive for transgene production. Owing to the lack of selectivity imposed upon particle delivery, it would be impossible to conclude that other cells were not also transfected. Unlike with direct injection where expression is located mainly along the needle tract, expression was dispersed randomly throughout the tissue [16]. Yang et al. showed that (β-galactosidase) expression was present in 20% of epidermal cells after ballistic delivery, which is in contrast with IM delivery, where only 1–3% cells stain positive [5]. Despite the somewhat low levels of expression obtained, a wide variety of mammalian cells and tissues such as the cornea have been targeted with the gene gun [14].

C. Applications for Needle-Free Injection Delivery

1. Genetic Vaccination

The most common route previously used for immunization is IM injection; however, owing to its accessibility, easy visualization, and immunocompetency, the skin provides a unique and efficient alternative for expressing foreign antigens. Residential professional antigen-processing cells (APCs), the bone-marrow–derived Langerhans cells comprise 3–4% of skin cells. In skin, the keratinocytes, the site of transfection, play a role in immune reactions. These cells secrete cytokines that induce the upregulation of various molecules present on the surface of Langerhans cells that promote antigen uptake and presentation; thus, their influence in immunization is key [13,18]. The presence and integrity of those target cells for 72 hr in the skin is critical for raising an immune response, unlike in the muscle, where ablation of target tissue 10 min after immunization failed to alter the induced antibody response. This indicates that the skin is a central antigen-presenting and processing factory; however, in muscle, rapid transport to distal sites provides the required response [18].

Delivery of the cDNA to the skin mimics vaccination via "live" vaccines producing in situ expression of the antigen, with higher antigen purity, minus the risks associated with pathogenicity of the vector. The continuous sloughing of the epidermal layer provides an additional safety mechanism for vaccinating immunocompromised patients and controlling transgene expression. The immune-rich nature of the epidermis dictates that smaller amounts of antigenic material are required, which is cost effective. Multivalent vaccines may be formulated. Needle-free devices have been accepted therapeutically: The World Health

Organization included a clause in their "Safety of injections in immunization programs" (revised in October 1998) indicating that " mono-dose prefilled cartridges" or "mono-dose cartridges for filling at the point of use" could be used in mass immunization programs. Many studies modeling various infectious disease states have shown the promise of genetic vaccination for prophylaxis.

Although both humoral and cell-mediated immunity are induced by injection with DNA vaccines, the specific nature of the immune response generated to antigen placed in the skin ultimately depends upon the method by which antigen is presented to the cells of the immune system. For example, a single direct administration of the antigen or antigen-expressing mediator to the skin or muscle produced mostly cell-mediated Th1 helper response, whereas particle administration provided a largely humoral Th2 response [20,21]. In one experiment, mice were injected intradermally (ID) or IM with 100 μg of plasmid encoding influenza A hemagglutinin or by gene gun (Accell electrical) with a 0.4- or 0.04- μg plasmid dose. The isotypic nature of the response did not change upon boosting the mice at 4 weeks with the same or the alternative immunization method or by challenging them with a viral vector. This work is of importance for developing vaccines and for the manipulation of T-helper cells in certain disease states, where a specific isotypic immune response may be required. Such a concept was highlighted in a recent study, concluding that for effective protection to Mycobacterium tuberculosis challenge, direct IM injection and not Helios gene gun targeting of Ag85A cDNA was necessary to elicit a Th1-protective immune response in mice [22]. Secretion of IL-2 and interferon-γ (INF-γ) was higher after IM immunization. A 50-fold lower dose of plasmid was required to elicit maximal antibody response with the gene gun than by IM injection in Balb/c mice, which demonstrated a bias toward developing a Th2 response. A much lower response to gene gun administration was observed in C57BL6 mice, confirming the existence of strain-specific effects.

The gene gun route of administration has also been proven to be effective in large animal models. Swine that received two doses of 250 ng HBsAg plasmid each elicited a response similar to that obtained upon injecting 20 μg of recombinant protein. Administering a higher number of doses of plasmid did not enhance the response or accelerate the kinetics further [23]. Rhesus macaques were immunized with DNA alone or with recombinant vaccinia vectors in a variety of schedules and were challenged with simian immunodeficiency virus via the gene gun [7]. Although viral loads (measured by polymerase chain reaction [PCR]) were reduced, the animals succumbed to symptoms of the disease. The study highlighted the potential for also developing an anti-AIDS vaccine. Like the skin, the oral mucosa can function as an immune center. Gene gun delivery of plasmid to both these sites in dogs indicated that cytokines, such as granulocyte-macrophage colony-stimulating factor (GM-CSF), murine interleukin-12 (mIL-12), and human IL-6 (hIL-6) were expressed in tissue extracts but not in serum, with peak expression levels on day 1 that decreased to background levels by days 4 and 7 [24]. Spontaneously forming oral and epidermal tumors could potentially be treated via this route.

The number of doses and dosing interval required to elicit a maximal response appears to vary between agents. For instance, a single administration of HBsAg promoted an equivalent response to giving two immunizations 4 weeks apart. Alternatively, two immunizations with the influenza NP (nucleoprotein) plasmid were required to elicit a peak antibody response. Priming of a memory response with the used dosing regimen was indicated by the induction of IgG isotype antibodies [10]. Anwer et al. showed a difference in immune response according to mode of injection: Only one dose of hGH plasmid was required to be administered via the Medi-jector device compared to two doses needed to elicit a similar response after IM immunization in dogs and pigs [25].

A number of studies employing each ballistic device type have been published and shown good induction of immunity. Johnston's group was the first to report an immune response after administration of DNA by a hand-held helium Biolistic gene gun device to the epidermis of mouse ears [26]. They observed a humoral immune response specific to hGH and human α- antitrypsin (hAAT) in mice upon administration of an initial, a second, and a boosting dose of 1–2 µg plasmid. Gene gun immunization offers an advantage over direct IM or ID injection by requiring a smaller amount of DNA to induce the same antibody titer. One of the initial studies compared the response induced in mice after administration via intramuscular injection in the quadriceps or by abdominal delivery via the Accell (discharge) device [10]. Only 16 ng of DNA (encoding either hGH or hAAT) was required to elicit the maximal titer response by gene gun administration. Increasing the dose 10- or 100-fold did not produce a stronger response even after boosting. In contrast, the responses measured after IM administration were much weaker even with injection of 5000-fold more plasmid. The gene gun effect on titer was enhanced a further 5- to 10-fold upon boosting.

More studies demonstrated proof of concept by eliciting protection to challenge by a pathogen. Fynan et al. compared the response induced by vaccinating with plasmid by direct IM, intravenous (IV), subcutaneous (SQ), intranasal, ID (foot pads) or intraperitoneal (IP) routes or by delivery to abdominal skin via the Accell device (electrical) at days 0 and 28 [9]. Lethal challenge with mouse-adapted influenza virus was made 10 days after the second plasmid administration. Protective immunity via each of the routes except IP, was achieved although the extent of protection, and thus the survival rate varied. In the survivors, varying levels of influenza symptoms were observed and overall those animals dosed via the IM, IV, or intranasal routes were well protected. A 95% protection rate was demonstrated upon delivery of only 0.4 µg DNA (a 250-fold lower dose of DNA than that used in the direct administrations) via the Accell device. A 10-fold reduction in DNA dose lowered the survival rate to 65%, whereas those mice dosed with a 100-fold less DNA did not survive the challenge. It is interesting to note that in previous studies, much lower doses of plasmid were required to produce a response. This discrepancy was explained by the differing components of the cytomegalovirus (CMV) promoter in each study [27,28].

An interesting and relatively new area is the potential for using DNA vaccines in neonates. In the year after birth, maternal IgG levels critical for initial protection to disease wane and the immune system of the infant is established and matures. Traditional immunization therapies employing recombinant proteins are often interfered with by the presence of maternal antibody; however, the direct intracellular transfection of DNA means that the maternal antibody would not interfere with the immunization protocol until de novo production of antigen occurs. Some groups have been able to demonstrate an immune response in the presence of maternal antibody (Ab) where others have not, and some have even demonstrated tolerance due to the infants' inability to recognize endogenously expressed antigen as nonself. One-day-old neonatal or young adult mice from naive or influenza-immune mothers were injected with either 50 µg of plasmid directly into the quadriceps or with 2 µg of plasmid via gene gun to test if a protective immune response could be raised upon challenge (29). One year after dosing, those mice from immunized mothers demonstrated a good IgG-specific response to nucleoprotein (NP) located intracellularly but not HA (hemagglutinin), which is located on the cell membrane, suggesting that the maternal antibody inhibited the IgG response to the HA but not the NP proteins. In young adults, IgGs to both HA and NP were produced when the level of maternal antibody was at background levels. The study also demonstrated the influence of the route and time of delivery and the presence of maternal antibody. In naïve neonatal mice, an immune response was observed 8 weeks after IM administration and 12 weeks after gene-gun dosing, respectively. Neonatal mice took 4 weeks longer to elicit a response than the adults did. Upon challenge with sublethal doses of influenza virus, the presence of the maternal antibody in neonates from flu-immune mothers abolished protection and 100% death was observed, whereas in adults, a 30% survival was observed in mice immunized with both HA and NP plasmid or just HA plasmid. In adult animals from naive mothers, an 80% survival rate was seen. In contrast with the abolition of an antibody response, the presence of a maternal antibody did not interfere with the cellular immune response; that is, antigen processing and presentation.

Cancer vaccines require priming of the body against a tumor cell. This can be achieved via a number of approaches, for example; the introduction of a gene encoding a class I or II major histocompatibility complex (MHC) antigen to a tumor cell line or a tumor-associated antigen, such as membrane glycoprotein gp100. When challenged with mouse melanoma cells IV, mice that were vaccinated with human gp100 (hgp100) reduced the formation of lung metastases by 55%. The number of tumor-free mice was also significantly enhanced upon ID challenge (30). The loss of pigmentation observed in 70% of the mice treated with hgp100 but not those treated with mgp100 or naive mice demonstrated that this tumor immune response was somewhat analogous to an autoimmune response. Antigen-specific protective immunity was observed in mice immunized with plasmid encoding the antigen ovalbumin (OVA). When challenged with a melanoma line cloned to express OVA, those animals that were immunized with OVA survived. In contrast, when challenged with the parent B16 melanoma cell line that did not express OVA, the animals were not protected and died [31]. In addition, the requirement for CD8$^+$ T cells in this model was shown in CD8$^+$-depleted mice, as they did not survive tumor challenge, although the role of CD4$^+$ cells was not investigated.

Gainer et al. demonstrated how different tumor lines have to be assessed for their ability to secrete transcribed genes, and showed that melanoma cells lines, for instance had a higher transfection efficiency than glioma cell lines [32]. The majority of immunogene therapy studies involve the ex vivo transfection of a gene, rendering the cells immunogenic. In this study, the levels of secreted IL-12 from each cell line did not vary but only if the cells were gamma radiated 2 hr posttransfection. Likewise, Mahvi et al. targeted mouse tumor cells and demonstrated gene gun–mediated uptake of plasmid encoding GM-CSF in pre-irradiated or post-irradiated cells [6]. The cytokine was detected in both serum and the tissue surrounding the site of administration, with peak expression at day 1 that decreased over a 10-day period, with higher levels being detected in the irradiated cells. Vaccination of mice with preirradiated gene gun–targeted cells protected 58% of mice upon challenge with nonirradiated, nontransfected tumor cells.

2. Gene Replacement

The route and mode of gene delivery frequently determines the level of transgene expression produced. For example, direct IM injection of plasmid encoding erythropoietin (Epo) produced long-term expression for up to 1 year in mice, with a corresponding significant and pathological rise in hematocrit (HCT) [33,34]. Such high Epo levels in the serum may actually be sufficient to initiate an immune response, which if expressed long-term may lead to tolerance. Removal of the muscle site may be necessary to combat such an effect. The use of gene gun administration of plasmid plus short-term gene expression resulting from natural sloughing of the epidermal layers may be exploited as a means for regulating the hematocrit. Klinman et al. reported short-term expression of Epo from a 3-µg plasmid dose that coincided with a transient increase in HCT to 60% 5 weeks later, which dropped to normal 20 weeks after treatment [35]. Interestingly, the HCT induction was maintained by administering the same dose of plasmid multiple times at 3-week intervals. At the same plasmid dose, gene gun administration was more efficient than IM administration in producing a rise in HCT. Plasmids leading to the secretion of systemically acting proteins are not the only therapeutics that can potentially be delivered via ballistic means. The localized delivery of growth factors to accelerate wound repair using the genegun has also been proven to be successful [13;36]. Platelet-derived growth factors in either homodimeric or heterodimeric forms, produced from the respective cDNAs, enhanced wound strength 7 and 14 days after transfection by particle-mediated delivery (helium Accell) in rats (36). Increasing skin thickness did not seem to present a problem, as a similar acceleration of wound repair lowering the healing time by 20% was demonstrated in porcine skin (electrical Accell) [13].

3. Immunomodulation

Systemic cytokine delivery of recombinant IL-12 (rIL-12) for example, is associated with dose-dependent side effects and toxicity. Cytokine therapy is needed for tumors that may not be resected in part owing to their association with other organs, their metastatic characteristics, and the avoidance of toxic side effects of currently available systemically administered therapies. One means for reducing these disadvantageous effects is by the localized or "paracrine"-like delivery of the cytokine agent to the tumor. The group of Yang and Sun have developed a gene therapy system requiring the gene gun technology, where syngeneic cell lines are injected subcutaneously in mice [37]. The mice are then treated by bombarding the skin over the tumor with the gene gun; for example, an Accell device to deliver a gene (10μg) encoding a cytokine. Epidermal cells are transfected as shown by histology. Typically, the animals are treated three to five times in a 7- to 10- day period. Rakhmilevich et al. demonstrated delivery of a plasmid encoding murine IL-12 via a helium-pulse Accell device 7 days after ID implantation of syngeneic cell lines in the abdominal area directly over and in three places adjacent to the tumor [11]. The levels of IL-12 were determined by cell proliferation bioassay, although serum levels were found to be below the limit of detection. Complete tumor regression was observed in four tumor models, showing that localized antitumor IL-12 therapy is feasible. A mainly CD8$^+$ T-cell response was initiated. Despite the epithelial localized delivery, an antitumor response can be achieved. Combination cytokine therapy is also possible and has been shown to be functional [37].

III. CELL MEMBRANE–PERMEABILIZING GENE DELIVERY TECHNIQUES

The primary challenge placed upon drug therapies has been to overcome the barriers to cellular delivery. The lipid bilayer cell membrane has a dual role maintaining cellular integrity by stabilizing ionic conductivity and gradients and electrolyte homeostasis, while also preventing cellular contents from spilling into the systemic-circulation. The generic ability of any cell to be transiently permeabilized by various physical techniques provides a means for cytoplasmic loading of transgenes. The next sections provide an overview of electroporation and sonoporation, parameter optimization, and issues pertinent to DNA delivery. Studies that encompass all these considerations in in vitro or in vivo settings will be reviewed.

A. Electroporation

The intracellular entry of many macromolecules typically impeded by the cell membrane has been made possible by electroporation. Owing to its purely physical nature, electroporation is applicable for transfecting many cell types in vitro with agents such as plasmid and anticancer agents such as bleomycin. Typically, prior to electroporation, the cells are exposed to or the tissue is perfused with the agent to be delivered, so that upon application of the electrical pulses, the macromolecules gain entry to the cell. Although the exact mechanism of membrane breakdown has not yet been defined, it has been overwhelmingly agreed that a change in transmembrane potential results in the formation of pores that subsequently reseal provided the applied field does not exceed the threshold maximum required for membrane integrity. Initially, the cellular and extracellularly localized ions orient themselves in the field creating a polarized environment. Once the field is high enough (0.5–2.0 V), the membrane structure becomes disturbed, conductivity increases, and the membrane breaks down. The hydrophobic pores that are created expand as a result of the osmotic movement of water or coalesce with other pores, and upon reaching a critical radius, transform via flip-flop transitions of membrane lipids into hydrophilic pores (Fig. 3).

The pore size range of 1–10 nm makes the explanation of plasmid uptake difficult, because the plasmid size is much larger than the size of the average pore. A number of theories have been proposed to explain the mechanism of DNA translocation. For instance, it is plausible that plasmids move by diffusion into the cell along with osmotic movement of water or they can "thread" their way through pores by means of their flexibility. Alternatively, uptake via membrane invaginations suggests "endocytic-like" uptake. The most accepted mechanism is the electrophoretic transport of the anion proposed in the light that the plasmid must be present prior to or at the time of pulsing [38]. The potential interaction of plasmid with lipids of a cationic nature in the membrane may also facilitate uptake by increasing DNA adsorption to the cell surface [39].

1. Regimen and Delivery Parameters

Some of the key parameters that have been identified in electroporation protocols are the pulse number, intensity and duration, and frequency of administration.

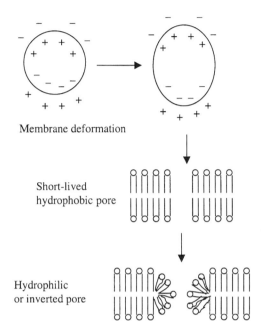

Figure 3 Representative model of pore formation. Under the influence of an electrical field, the ions adjacent to and within the cell orient themselves within the electrical field, which results in a series of membrane alterations that ultimately lead to the formation of an initially unstable hydrophobic and ultimately hydrophilic pore population.

Perhaps most importantly, the electroporation procedure should not affect the integrity or biological activity of the plasmid or molecule being delivered [40]. In turn, these parameters are dependent upon and are greatly influenced by the type of cell to be transfected, the medium in which the cells are resident, and their relationship and contact with adjacent cells [41]. For instance, a relatively homogeneous situation can be found in vitro enabling experimental predictability. The cell size influences the intensity of the pulse, and, in general, it has been determined that smaller cells require higher intensity pulses for optimal transfection [41]. This may not be true in vivo, where the tissue architecture deviates the behavior from the predictable and more homogeneous in vitro setting. For instance, myofibers are elongated rodlike cells. Upon exposure to an electrical pulse, the cell is exposed to a series of field lines. Those portions of the membrane orientated perpendicularly to the field lines are most susceptible to electrical discharge; however the portions of membrane where the radius of curvature has been changed are also prone to breakdown. Miklavcic et al. concluded that skeletal muscle is more electrically conductive with the fibers oriented longitudinally, so it was proposed that electrotransfection would be more efficient in this direction than if the electrodes were placed perpendicularly to the fibers [42,43]. Aihara and Miyazaki actually showed no difference in expression levels when the wire electrodes were placed in either orientation; however, the small size of the tissue and the fact that cells with small diameters were targeted in the transverse direction may have influenced this effect [44]. Tekle et al. demonstrated that pore formation is also dependent on the position of the cell relative to the electrode [45]. Those edges closest to the cathode or anode are more susceptible to depolarization and thus membrane breakdown, but the pores that form differ in size dependent upon which electrode lies closest. Most studies have employed transgene markers to determine how applicable the procedure is for gene delivery to a particular organ. By considering all these factors, transfection efficiency can be maximized, yet cell damage and ultimately cytolysis can be minimized.

The electric field generated by the discharge of a capacitor can be of exponential, square, or radiofrequency waveforms. Square-wave pulses are more efficient for electrotransfection than exponentially decaying waves provided that multiple pulses are administered [46]. In addition, the pulsing parameters are preset and theoretically the voltage increases steeply to the required setting, is maintained for the required duration, and then decreases rapidly. Presse et al. also suggested that membranes permeabilized with square waves recover more quickly, as the deformation structures that form are more labile [47].

The electrotransfection event follows a bell-shaped response curve. A threshold pulse intensity and duration are required to enhance macromolecular uptake, and uptake increases with increasing intensity up to a peak or optimum. If this is exceeded and the pulse is too intense or long, cytolysis and thus inhibition of expression is seen. In general, as the intensity and duration increase, loss of cells is inevitable, and the number of viable cells decreases. Voltage intensity and duration are related in such a way they can compensate for each other. For example, lowering the intensity and increasing the duration produces a similar response as using a high-intensity, short-duration pulse. Thus, both pore formation and resealing are critical events that must be balanced to ensure cellular patency. Despite the dependence of pore formation upon applied voltage, the time for resealing is not related [48]. Pore resealing is thus a critical event and must be balanced with pore formation. Pore sealing takes from seconds to minutes to occur, and is temperature dependent. Excellent reviews by Sukharev et al. [49] and Lurquin [50] describe the many physical issues that surround the "art" of electroporation and the formation of pores.

The time of electroporation is also critical as demonstrated by Rols et al., who showed that if electroporation was administered greater than 1 min after injection, transcription efficiency was decreased [51]. Klenchin actually concluded that DNA uptake was complete within 3 sec of pulsing. In general, as the pulse intensity and duration increase, loss of cells in inevitable, and the number of viable cells decreases [38]. In general, high-intensity, short-duration pulses are most effective for in vitro electrotransfer; however, the pulsing regimen required in vivo is tissue dependent and varies significantly. The number and size of the pores are related to the electrical field parameters, which are defined in terms of field intensity and pulse duration [52]. For example, short-duration, high-intensity pulses provide a large pore population of small size, whereas a regimen of longer duration, low-intensity pulses produces a smaller overall population with larger average diameters.

The highly efficient cytosolic delivery of an agent by electroporation means that smaller doses of plasmid are required to elicit the same response as that by a much higher dose in the absence of electroporation. As indicated already, the dose of plasmid required to elicit a maximal response without compromising cellular integrity must be evaluated in each cell or tissue [41]. Despite the small number of cells expressing, saturation in expression has been demonstrated [46,53]. In general, a linear dose-response–relating plasmid dose is apparent except when uptake saturates and transgene expression levels plateau. Manthorpe et al. showed that expression increased with increasing plasmid doses up to 25 µg, plateauing at higher doses without electroporation [53].

Potentially, it would be useful to find a formulation that enhanced transfection efficiency in combination with electroporation, while providing other beneficial properties. Most studies this far have focused on delivery of naked DNA in saline. For instance, damage to the cell membrane and joule heating are two side effects that can be alleviated. One means to do this is by chemical modification of the cell membrane by the addition of a triblock copolymer, a poloxamer, or addition of a surfactant [54]. Surfactants reduce surface tension, and their effects have been measured both in vitro and in vivo [55]. Thus, lower field strengths may be required. In addition, the concentration of surfactant must be below the critical micellar concentraction (CMC), or else it may extract lipids from the membrane. Klenchin et al. demonstrated that increased viscosity or reducing the charge on the anion lowered the transcription efficiency [38].

Finally, the resistance of the tissue is a key parameter that must be also considered carefully for successful electroporation. For example, if the osmotic content of the tissue is high, the current can reach levels that potentially cause damage. Many of the reports do not adequately define the parameters, and thus large variations in the pulsing regimens and transfection efficiency are often reported even though similar tissues are being targeted.

2. Electrode Type

Another critical parameter is the choice of electrode and the exposure of the cells and tissue to the electrical field. Multiple electrode device types such as parallel rectangular plate "caliper" electrodes, spoon electrodes, and needle arrays have been developed. The parallel plates are set up like calipers that ensure close contact of the plates can be made with the target tissue by adjusting the distance between them. The orientation of the parallel plates produces homogeneous field lines and thus an electrical field (Fig. 4). This is reflected by the high reproducibility and intensity of the response rates (70–85%). Despite their success, the plate electrodes offer some drawbacks. They are often difficult to position around a particular tissue such as a tumor or skin, and targeting regions of increasing depth and size may not be feasible, particularly in larger animals. Alternatively, the six-needle array (Genetronics Inc., San Diego, CA) comprises six stainless steel needles arranged equidistant to each other, thereby forming a circle of 0.5- or 1.0-cm diameter [56]. During any given pulse, only two opposite pairs of needles are charged, but with pulsing the charge sequentially rotates around the array (Fig. 4). The nature of the needle array electrodes means that the field strength will vary dependent on the position of an individual needle within the tissue. The point source nature of the electrode means that field intensity will be higher at the electrode surface, but will be less intense upon moving further into the tissue away from the needle [57]. Gehl et al. described a program to calculate how an electrical field is distributed in the vicinity of an electrode, with the assumptions that the medium is homogeneous and that electric potential and electrical charge are equivalent [58]. When a voltage of 0.53 kV/cm is applied with the parallel plates, approximately 85% of the tissue is exposed to a field with an intensity of 0.45 kV/cm or higher in contrast to the needle electrodes where only 12% of the area is exposed to a threshold field. One study showed that tumors re-formed in the areas where the field lay below the threshold required for electroporation [59]. In order to compensate for this, so that the tissue can be exposed to a minimal field, a pulse of greater intensity may be administered, and this may produce "arcing" and thus damage to the tissue. Other needle array arrangements have also been described, and a select few have been compared [60]. They showed that tumor volume was significantly reduced upon delivering bleomycin with a 2×2 needle array, and the mice lived longest compared to those animals electroporated with the parallel plates or a 3×3 array. Dev et al. showed that a six-needle array was most effective in antitumor electrochemotherapy with bleomycin [61]. A recent publication highlighted the release of ions from the needle electrodes, which potentially affects the tissue in terms of pH and pathology [62]. Efforts to minimize such degradation events, such as the electroplating of the electrodes with gold, are underway.

3. Cell and Tissue Electrotransfection

Owing to the physical nature of electroporation, the procedure can effectively be used to target any cell type with exogenous DNA. Electroporation has reproducibly delivered plasmid or other macromolecules in vivo to cells of multiple tissue types, including skeletal muscle [44,58,63–67], tumors [57,68], and cancers of

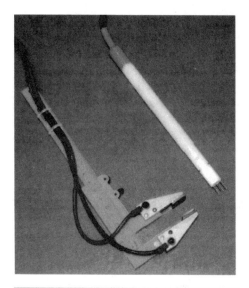

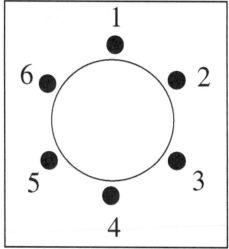

Figure 4 (Top) Photograph of the caliper electrodes and the 1-cm diameter six-needle array. (From Genetronics Inc.) (Bottom) The pulsing regimen used with the six-needle array is composed of six needles spaced at 60-degree intervals around a 1-cm diameter circle. The needles have been numbered from 1–6. The charge orientation of each needle changes with rotation of the field around the array so that only four needles (i.e., two pairs of oppositely charged needles) are charged during a single pulse. For instance, when needles 1 and 2 are positively charged, needles 4 and 5 are negative; the pulse progresses sequentially around the array with the next pulse so that needles 2 and 3 are positive, with needles 5 and 6 negative, and so on.

the head and neck [69,70], pancreas [71], prostate [72], liver [56,73–77], melanoma [51], cornea [78], skin [79–81], arteries [82,83], testes [84], and cardiac tissue [85]. Currently, there are three clinical studies using electroporation for macromolecular delivery: two for bleeding disorders (one using an ex vivo transfection protocol of bone marrow stromal cells (factor IX) and the other for ex vivo treatment of factor VIII deficiency using an AAV viral vector) and an anticancer trial for glioblastoma patients [61]. Another study by Transkaryotic Therapies Inc. (Cambridge, MA) is using electroporation to deliver plasmid encoding FVIII, although no results have yet been reported.

Electroporation of the Liver, Testes, and Cornea. Successful electrotransfection of a reporter gene in rat liver was achieved using different pulsing regimens. Heller et al. employed the six-needle array to deliver six pulses of 100 V/ cm intensity and 99 μsec duration after perfusion of a 25-μg dose of plasmid encoding β-gal [77]. Forty-eight hours later, 30–40% of the cells expressed β-gal, of which 5% still expressed 21 days after treatment. In contrast, without electroporation, overall expression levels were lower, and what little expression there was had disappeared by day 14. Suzuki et al. employed tweezer-trodes to administer eight pulses of 50 V/50 msec duration [76]. Green fluorescent (GFP) expression was measured 48 hr later. Low-voltage pulses (8, 40 V, 50 msec) proved to be the most efficient regimens for electrotransfecting rat corneas without inducing damage or inflammation. In mouse testes, Muramatsu et al. used a protocol delivering eight pulses of 25 V and 50 msec via the plate electrodes to enhance chloramphenicol acetyl transferase (CAT) expression most effectively, although a range of pulse intensities and durations were tested [84].

Skeletal Muscle. Low-intensity voltages are required for electrotransfer in skeletal muscle. This has been attributed to the fact that the gap junctions between myocytes amplify the changes in the transmembrane potential reducing the voltage required for membrane breakdown [86]. The ability of skeletal muscle to express an exogenous transgene for extended periods after direct injection has been shown in mice (19 months) and in nonhuman primates (4 months) [16,17,53,87]. Typically, however, expression is localized to the needle tract, with a very low and variable proportion of expressing cells. For instance, 30–40% of the cells in the vicinity of the needle tract were stained blue, whereas only 1.5% of the cells within the entire quadriceps muscle expressed β-gal [16]. The mechanism of uptake has not yet been clearly defined. With progression into higher animal models, expression levels decrease further [87]; therefore, the need to find a method that provides enhanced and sustained levels of transgene expression that lie within the therapeutically effective range is high. Formulating the plasmid with polymers such as poly(vinyl pyrrolidone) have successfully enhanced expression levels in mouse and rat muscle (5- to 10-fold) relative to DNA in saline [1]. Muscle-damaging agents such as bupivacaine have also been of benefit [88]. However, electroporation has been proven most effective in enhancing expression levels 100- 1000-fold [44,63–67,89]. Mathiesen showed that the number of cells expressing luciferase (50 μg) in the soleus or digitorum longus muscles increased from 1 to 10% by varying parameters such as pulsing frequency (10–1000 Hz), duration of each train (10–1000 msec), and duration of each pulse (100–4000 μsec) [66]. Silver wire electrodes were placed on either side of the muscle. High-intensity, long-duration trains and pulses were associated with necrosis, and no expression was observed in these regions. Electroporation, in general, enhances the overall number and density of myofibers expressing a transgene. Aihara and Miyazaki showed that unlike direct IM injection, expression of *lacZ* was broadly distributed and not limited to the needle tract [44].

Very thorough and comprehensive investigations on the effects of pulse intensity, duration, number, and frequency on luciferase expression in tibialis muscles electroporated with plate electrodes have been described [65,90]. MacLaughlin et al. demonstrated that low-intensity, long-duration pulses were optimal for enhancing transgene expression in skeletal muscle [90] (Fig. 5). An optimal regimen defined by Mir et al. used 8–16 pulses, of 200 V/cm intensity and 20 msec duration [65]. In a separate study, Mir et al. showed that decreasing the pulse duration from 20 to 1 msec lowered expression levels by at least one log, whereas increasing pulse intensity also produced a significant drop in luciferase levels [64]. In addition to delivering nonsecreted markers of transgene expression, studies employing biologically relevant markers have been completed. Angiogenic fibroblast growth factor (FGF-1) was successfully electrotransfected in both mouse and monkey muscle [65]. The muscle is also renowned for its ability to behave as a bioreactor secreting expressed proteins effectively into the systemic circulation. Serum murine IL-5 levels were enhanced from 0.2 to 20 ng/mL 5 days after delivery of 50 μg plasmid and electroporating the muscle with two wire electrodes, (three pulses of 100 V/cm and 50 msec) [44]. Expression peaked between days 5 and 7, but levels had decreased to about 10% of the peak by day 42. Long-term expression of Epo has been reported by Rizzuto et al. [91] and Kreiss et al. [92], whereas Fewell et al. [89] and Bettan et al. [93] showed sustained levels of Factor IX. Rizzuto et al. demonstrated that doses as low as 1 μg produced an elevated HCT of approximately 60% for up to 100 days, although peak Epo expression occurred at day 7 and decreased slowly thereafter [91]. Histological analysis revealed that 1 week after electroporation, more fibers with centrally localized nuclei were visible. The site of this region was not more than 10–20% of the total fiber number. Damage appeared to be localized to regions that experienced an overall higher current exposure. They also demonstrated a good dose-response in Epo expression along with HCT change. Bettan et al. reported a 150-fold increase in factor IX levels, with peak plasma concentrations of 220 ng/mL with administration of a 30-μg dose in SCID mice maintained for 2 months, and full functionality of the expressed protein was observed [93].

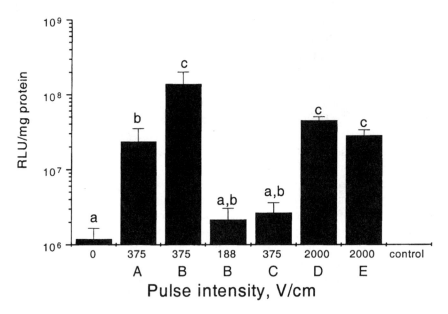

Figure 5 Effect of pulse duration of the electrotransfection on the level of transgene expresion. A = 4 pulses of 25 msec duration; B = 2/25 msec; C = 4/50 msec; D = 2/99 μsec, and E = 4/99 μsec. Plasmid dose was 10 μg plasmid formulated with 5% PVP in 50 μl. Luciferase levels were measured after 2 days. Each bar represents the mean relative light units and the standard error of the mean from four mice (eight muscles). Means denoted with different superscripts in the graph mark values that are statistically different from each other at (p ≤ .05). (From Ref. 90.)

In C57/B16 mice, expression decreased rapidly and antibodies to Factor IX were detected. A similar trend was observed upon delivery of secreted alkaline phosphatase (SEAP). Long-term expression could be maintained for 12 months in SCID mice, but a significant drop (60%) in expression levels in immunocompetent mice was seen between days 60 and 360.

Scale-up of the electroporation procedure has also been shown to be effective in larger animals. Mir et al. showed a two- to four- logfold enhancement in expression levels in rat and rabbit skeletal muscle after electroporating with either the plate or two-needle array electrodes [65]. MacLaughlin et al. have demonstrated enhanced GFP expression and also erythropoietin expression, with a transient change in HCT in swine [94]; and Draghia-Akli et al. targeted plasmid-encoding growth hormone releasing hormone also to swine [95].

The potential for enhancing expression via the IM route leads to the improved expression of antigens for DNA vaccination [96,97]. Widera et al. combined this concept and demonstrated an enhanced immunogenic response to the hepatitis B surface antigen in mouse muscle [96]. As already described, low-level expression of antigen may lead to requirements of multiple administrations to elicit even a modest immune response. The delivery method enhanced the potency of the antigen and its immunogenicity. The mechanism of this increased potency has not yet been clarified, although it has been postulated that APCs or dendritic cells could have been targeted. Alternatively, the procedure induced an inflammatory response. Some muscle damage is unavoidable, and may actually be beneficial, depending on the type of response, for instance, in immunization.

Tumors. The success of electroporation in combination with an anticancer agent, bleomycin, was realized with the publication by Okino and Mohri, who successfully killed tumors [98]. Since then, significant contributions to this field have culminated in a number of clinical trials [69]. Electroporation was induced via a six-needle array aimed primarily at treating tumors of the head and neck, and overall the procedure was well tolerated. In some instances, the muscle contracted and an eschar formed [61]. A number of electrochemotherapeutic studies for treating a human pancreatic [71], human larynx [70], or prostate tumor models [72] in mice and a liver tumor model in rabbits [75] have also been described. The six-needle array has been proven to be most effective for this application, because the needle array can be placed directly surrounding or in the affected tumor region. The heterogeneity of a tumor, for example, means that the current and power density vary depending on position within the tumor [59]. So far, the majority of studies have utilized bleomycin as the anticancer agent, although it is plausible that delivery of other agents could be enhanced. The pulsing regimens used are quite different. In general, high-intensity, low-duration pulsing regimens have been employed. An excellent review by Hofmann et al. describes how electrochemotherapy can be used for treatment of tumors [57]. Bleomycin is a potent anticancer agent that has severe side effects. In combination with electroporation, low doses of the agent are effective and severe toxicity is abrogated.

Rat livers that had been inoculated with tumor cells 10 days previously were dosed with bleomycin and electroporated with six pulses (1000 V/cm and 99 µsec). A reduction in size of 69% of tumors was observed [56]. After implantation of tumor cells 10 days prior to treatment, rat brains were electroporated with a two-needle array (0.5-cm diameter, eight pulses of 600 V/cm intensity and 99 µsec duration) placed on either side of the tumor, after which X-gal plasmid was administered intra-arterially (carotid). Three days later, high expression of X-gal in the border regions of the tumor was observed and in regions close to blood vessels, whereas normal surrounding brain cells did not express. To obtain high levels of expression without electroporation, 100 µg of plasmid must be administered; however, with electroporation, a dose of only 10 µg was required [83]. Heller et al. employed pulses of 1500 V/cm with 100-µsec duration and varying plasmid doses to treat hepatocellular carcinomas [74]. With higher voltages, visible pathological changes and damage were evident. A DNA dose response was observed. Expression was only observed in the electroporated region.

Aside from the studies using expression markers, Nishi et al. showed expression of human monocyte chemotactic protein (MCP-1) (a chemokine) in brain tumors 3 weeks after delivery [83,99]. Large numbers of macrophages and lymphocytes were evident in the tumor, although whether this had any conceivable effect on tumor growth is still unknown.

In contrast, Nishi et al. demonstrated perfusion of a solid subcutaneous tumor in a rat with plasmid-encoding green fluorescent protein by pulsing with a six-needle array (eight pulses, 150 V/cm and 10 msec) [100]. High levels of fluorescence in the region encompassed by the array were observed. Other reports of targeting a B16 murine region melanoma tumor model [10] or a mammary tumor line [101] are available.

Skin.   Like some of the tumor protocols, high-intensity pulses of shorter duration have been proven useful for permeabilizing the stratum corneum (15–25 μm). Long-duration pulses have not been reported in skin, as they are believed to induce irritation [61]. Delivering drugs via the transdermal route is desirable, as it is a noninvasive and painless delivery route that avoids first-pass circulation and gastrointestinal digestion. The skin is a protective layer that offers high resistance to uptake of large charged molecules; however, this protection mechanism can be overcome by the application of electrical fields [102]. Initial studies described the uptake of molecules such as analgesics, hormones, and anticoagulants. Two electrode types for this purpose have been described: caliper plate electrodes that grab and electroporate a skin fold and a meander-style electrode, a device that is simply laid on the skin surface [103]. Fluorescent microspheres of sizes 0.2, 4.0, and 45.0 μm were delivered to a 1-mm thick skin fold of a mouse, with electroporation via the caliper electrodes and three exponential decaying pulses (120 V and 1.2 msec). The presence of all three sizes of particles was observed in tissue cross sections, indicating that even inflexible solid particles of micron size are capable of crossing the stratum corneum. Intraepidermal delivery of liposomes placed on the skin surface followed by electroporation was also reported. The voltage necessary for breakdown of the stratum corneum is in the range 50–100 V, representing a field strength of about 20 kV/cm [61]. Delivery of plasmid to skin has also been reported. Dujardin and Preat reported GFP expression in skin after administering 10 square waves via the plate electrodes [104]. Three pulsing regimens were used (335 V/5 msec; 335 V/0.5 msec and 1000 V/100 μsec), although the diameter of the skinfold was not revealed. Two days after electroporation, fluorescence was observed in the epidermis.

A mixture of two plasmids was administered subcutaneously to the skin of newborn mice, and 10–40 minutes later a skin fold was electroporated with pulses in the range 400–600 V/cm and 100–300 μsec duration [81]. Fibroblasts were successfully transformed compared to skin of animals that were injected only or electroporated only. Skin electroporation in conjunction with other delivery devices may also be considered to be viable treatment routes in the near future.

IV. SONOPORATION

Sonoporation or ultrasound utilizes ultrasonic energy to permeabilize cell membranes enabling the uptake of plasmid. The energy is transferred via a hand-held device, the base of which is placed making contact with the targeted tissue or wells containing cells. There are three parameters that define sonoporation: the treatment time, the duty cycle, and the intensity. The duty cycle refers to the proportion of time relative to the overall treatment time that energy transfer is occurring. Typically, there are four available settings which are 10, 20, 30, or 100%, although this is dependent on the device model in use. This means that ultrasound is either pulsed or continuous. The intensity can be as high as 2 W/cm^2. Ultrasonic energy essentially is delivered as a sound wave. Like any wave, transmission through a heterogeneous medium results in the loss of intensity due to absorption and deflection or refraction within the medium or tissue. Characteristics such as the density, absorption coefficient, and intensity attenuation coefficient of the tissue dictate the wave path. Absorption of the wave itself varies with the protein, fat, and water content and the acoustic impedance of the tissue. Proteins absorb ultrasound well but fat and skin do not. The impedance of air is 0.0004, muscle is 1.64, and bone is 6×10^5 g/(cm^2)(sec). The interfaces formed with other media such as air or blood reflect the change of refraction of the wave. Provided the influences of each tissue upon a wave can be established, ultrasound can be focused on a target organ or region.

The mechanism of DNA translocation has been shown to be due to cavitation [105]. Cavitation is the process that arises when a pocket of air forms or an existing bubble is exposed to the ultrasonic energy. Inertial cavitation arises from microbubble formation as a consequence of the changing pressure within the ultrasonic field. These microbubbles rapidly increase and decrease in size in response to the field frequency, and they are ultimately destroyed. Alternatively, stable cavitation occurs when air bubbles already present in the tissue or medium oscillate in time with the field. It is the rapid oscillation of these bubbles that potentially disrupts the lipid bilayers of the cell membrane [106]. Membrane disruption is accompanied by a decrease in epidermal resistance. Upon application of low-frequency ultrasound, a 25-fold decrease in resistance was measured. Interestingly, the permeability of the molecule is now not related to the lipid coefficient, suggesting that aqueous "pores" or pathways were created. Diffusion of the molecule through water is much more rapid than through lipid.

Exposure of a tissue to an ultrasonic wave leads to a number of changes. Absorption of the energy generates heat within the tissue, although, in general, the effect is localized and does not induce damage as it diffuses away. As a result of radiation forces, acoustic streaming, or the movement of fluid through the ultrasonic field results. A direct by-product of this is altered sodium or calcium ion transfer that contributes to modified electrical activity of the tissue [107].

Efficient cytosolic loading of plasmid using ultrasound has been demonstrated both in vitro [105,108–111] and in vivo [112–114]. Ultrasound-mediated transfection has been demonstrated both in plated cells or cells in suspension. A low plasmid loading efficiency was shown in TK-deficient mouse L fibroblasts (LTK) producing only 10–35 transformed colonies/10^6 cells. Plasmid (50 μg) was formulated in a number of buffers of varying compositions [108]. This study appeared to be very preliminary and basic equipment was used. However, Kim et al. furthered this work and investigated the effects of the procedural parameters on transfection efficiency [105]. Primary rat fibroblasts or chondrocytes were exposed to ultrasound pulses of 1 MHz frequency and 20% or continuous duty cycle for a defined period of time. They demonstrated that plasmid integrity was maintained after 200 sec of sonication, although there is the remote possibility that more subtle changes at the molecular level occur. Wyber et al. also demonstrated the inocuous effect of ultrasound on DNA integrity [115]. Kim et al. indicated that exposure times of less than 1 min were necessary for maximal transfection, and longer treatment times were accompanied by cell death [105]. The transfection efficiency of chondrocytes was greater than fibroblasts. Wyber et al. described low-frequency (20 kHz) ultrasonic delivery of plasmid into yeast cells. The comprehensive study measured the effects of treatment time on transformant number and cell viability and related this directly to the number of cavitation events [115].

Formulated plasmid was successfully transferred into HeLa, C127I, or NIH3T3 cells in vitro. Liposomes were composed of dipalmitoylethylphosphocholine (DPEPC) and dioleoylphosphatidylethanolamine (DOPE) or 1,2-dimyristyloxypropyl-3-dimethylhydroxyethylammonium bromide (DMRIE-C) and cholesterol. Cell viability was retained if ultrasound was applied at a 10% duty cycle for 1 min even when the pulse intensity was raised. However, viability was significantly reduced when a 100% duty cycle was set and all other parameters remained the same. Transfection efficiency of CAT in HeLa cells was increased twofold by a 0.5 W/cm, 10% duty cycle pulse for 5 sec, although how these levels compared to those of plasmid/saline was not defined [111]. Unger described the newly developed FluoroGene™ vector (ImaRx Therapeutics, Tucson, AZ) that behaves as an acoustically active contrast agent and a delivery vehicle [116]. It is composed of a number of available cationic lipids with different halocarbons, such as perfluoropentane or perfluorohexane, and has been shown to enhance gene expression. The microparticles were administered IM or IV, and CAT expression was observed in the kidney and spleen, with levels as high as 37,500 pg/mg in muscle. The site of sonoporation was the targeted tissue.

A wide diversity of other cells have been permeabilized by sonoporation demonstrating how universal this technique can be. Red blood cells exposed to low-frequency ultrasound were permeabilized, the extent of which was measured by the release of hemoglobin. Greater exposure times led to greater degrees of permeabilization over a threshold of 100 msec, and this was accompanied by a reduction in cell viability. Liposome-formulated plasmid was delivered to vascular smooth muscle or endothelial cells [117]. A 60-sec treatment of 1 MHz frequency and 0.4 W/cm^2 intensity enhanced luciferase expression 7.5-fold 48 hr later.

One of the initial in vivo studies showed successful expression of β-gal and Factor IX in fetal and neonatal sheep liver via delivery through an umbilical vein with an adenoviral vector [112]. Huber and Pfisterer targeted a prostate cancer cell line with focused ultrasound in vitro. They demonstrated much greater-fold increases in transgene expression (200-fold) than in previous studies [114]. In vivo, only a 10-fold enhancement of β-gal expression with direct IT injection and ultrasound was observed relative to direct IT injection only in a prostate tumor model implanted in the thigh of a rat. No clear boundaries of gene expression were defined, but, in general, expression was observed in the area targeted by the field, with 5% of cells being stained blue. The lower success rate in vivo was attributed either to the low field strength applied or to the higher threshold for cavitation in vivo. Successful in vivo transfection was also demonstrated in murine colon carcinoma cells implanted subcutaneously in mice [110]. Plasmid encoding β-gal was infused through tumors implanted 14 days earlier, and the tumors were insonated with a wave of 20 W/cm^2 for 30 sec. A threefold increase in β-gal activity was measured. Anwer et al. administered plasmid encoding CAT formulated with N-[(1-(2–2-dioleyloxy)propyl)]-N-N-N-trimethyl-ammonium chloride (DOTMA) IV in tumor-bearing mice [113]. The tumor was the site of sonoporation. They also demonstrated enhanced IL-12 levels in tumor that produced a higher inhibition of tumor growth compared to administration of the lipid/plasmid complexes without sonoporation.

Overall, ultrasound is a diagnostic and therapeutic tool often used clinically in hospitals that has a high degree of efficacy and safety and is easy and simple to use. It is therefore readily available and applicable for enhancing gene delivery.

V. SUMMARY

The impressive facility of transmembrane delivery of plasmid via these described methods should be further exploited for ultimate use in the clinic. One of the rate-limiting steps for nonviral transfection of plasmids and large molecular weight macromolecules is uptake across the cell membrane, but by bombarding or permeabilizing the membrane, this barrier can be overcome. Rather than relying on cellular mechanisms, the parameters that define the procedure can be tailored to provide optimal delivery in a specific target tissue. There are certain criteria that a device must meet before being used in humans, and in some respects, this could lengthen the pathway to the clinic; however, the benefits that such devices can offer may outweigh the time delay for approval. By combining two highly effective and nontoxic means for plasmid uptake, nonviral plasmid systems in conjunction with physical delivery technologies will create many opportunities for the future of gene therapy.

REFERENCES

1. Mumper RJ, Duguid JG, Anwer K, et al. Polyvinyl derivatives as novel interactive polymers for controlled gene delivery to muscle. Pharm Res 13(5):701–709, 1996.
2. Klein TM, Wolf E, Wu R, Sanford JC. High-velocity microprojectiles for delivering nucleic acids into living cells. Nature 327:70–73; 1987.
3. Johnston SA. Biolistic transformation: microbes to mice. Nature 346:776–777, 1990.
4. Williams RS, Johnston SA, Riedy M, et al. Introduction of foreign genes into tissues of living mice by DNA-coated microprojectiles. PNAS 88:2726–2730, 1991.
5. Yang N-S, Burkholder J, Roberts B, et al. In vivo and in vitro gene transfer to mammalian somatic cells by particle bombardment. PNAS 87:9568–9572, 1990.
6. Mahvi DM, Burkholder JK, Turner J, et al. Particle-mediated gene transfer of granulocyte-macrophage colony-stimulating factor cDNA to tumor cells: implications for a clinically relevant tumor vaccine. Hum Gene Ther 7:1535–1543, 1996.
7. Fuller DH, Simpson L, Cole KS, et al. Gene gun-based nucleic acid immunization alone or in combination with recombinant vaccinia vectors suppresses virus burden in rhesus macaques challenged with a heterologous SIV. Immunol Cell Biol 75:389–396, 1997.
8. Eisenbraun MD, Fuller DH, Haynes JR. Examination of parameters affecting the elicitation of humoral immune responses by particle bombardment–mediated genetic immunization. DNA Cell Biol 12(9):791–797, 1993.
9. Fynan EF, Webster RG, Fuller DH, et al. DNA vaccines: protective immunizations by parenteral, mucosal and gene-gun inoculations. PNAS 90:11478–11482, 1993.
10. Pertmer TM, Eisenbraun MD, McCabe D, et al. Gene gun-based nucleic acid immunization: elicitation of humoral and cytotoxic T lymphocyte responses following epidermal delivery of nanogram quantities of DNA. Vaccine 13(15):1427–1430, 1995.
11. Rakhmilevich AL, Turner J, Ford MJ, et al. Gene gun-mediated skin transfection with interleukin 12 gene results in regression of established primary and metastatic murine tumors. PNAS 93:6291–6296, 1996.

12. Kuriyama S, Mitoro A, Tsujinoue H, et al. Particle-mediated gene transfer into murine livers using a newly developed gene gun. Gene Ther 7:1132–1136, 2000.

13. Andree C, Swains WF, Pages CP, et al. In vivo transfer and expression of a human epidermal growth factor gene accelerates wound repair. PNAS 91:12188–12192, 1994.

14. Tanelian DL, Barry MA, Johnston SA, et al. Controlled gene gun delivery and expression of DNA within the cornea. Biotechniques 23:484–488, 1997.

15. Ben SI, Whitsitt JS, Broadley KN, et al. Particle-mediated gene transfer with transforming growth factor-$\beta1$ cDNAs enhances wound repair in rat skin. J Clin Invest 98:2894–2902, 1996.

16. Wolff JA, Malone RW, Williams P, et al. Direct gene transfer into mouse muscle in vivo. Science 247:1465–1468, 1990.

17. Wolff JA, Ludtke JJ, Ascadi G, et al. Long-term persistence of plasmid DNA and foreign gene expression in mouse muscle. Hum Mol Genet 1(6):363–369, 1992.

18. Torres CAT, Iwasaki A, Barber BH, Robinson HL. Differential dependence on target site tissue for gene gun and intramuscular DNA immunizations. J Immunol 158:4529–4532, 1997.

19. Lin MTS, Pulkkinen L, Uitto J, Yoon K. The gene gun: current applications in cutaneous gene therapy. Int J Dermatol 39:161–170, 2000.

20. Lin MTS, Pulkkinen L, Uitto J, Yoon K. The gene gun: current applications in cutaneous gene therapy. Tech Note 2552, Bio-Rad, 1999.

21. Feltquate DM, Heaney S, Webster RG, Robinson HL. Different T Helper cell types and antibody isotypes generated by saline and gene gun DNA immunization. J Immunol 158:2278–2284, 1997.

22. Tanghe A, Denis O, Lambrecht B, et al. Tuberculosis DNA Vaccine encoding Ag85A is immunogenic and protective when administered by intramuscular needle injection but not by epidermal gene gun bombardment. Inject Immun 68(7):3854–3860, 2000.

23. Fuller JT, Macklin M, Drape B, et al. Gene gun-mediated DNA Immunization with HBsAg: efficacy in small and large animals. In: Vaccines. Cold Spring Harbor, NY: Cold Spring Harbor Laboratory Press, 1997, pp 157–161.

24. Keller ET, Burkholder JK, Shi F, et al. In vivo particle-mediated cytokine gene transfer into canine oral mucosa and epidermis. Cancer Gene Ther 3(3):186–191, 1996.

25. Anwer K, Earle K, Shi M, et al. Synergistic effect of formulated plasmid and needle-free injection for genetic vaccines. Pharm Res 16(6):889–95, 1999.

26. Tang D-C, De Vit M, Johnston SA. Genetic immunization in a simple method for eliciting an immune response. Natur 356:152–154, 1992.

27. Haynes JR, Fuller DH, McCabe D, et al. Induction and characterization of humoral and cellular immune responses elicited via gene gun-mediated nucleic acid immunization. Adv Drug Del Rev 21:3–18, 1996.

28. Haynes JR, McCabe DE, Swain WF, et al. Particle-mediated nucleic acid immunization. J Biotechnol 44:37–42, 1996.

29. Pertmer TM, Oran AE, Moser JM, et al. DNA Vaccines for influenza virus: differential effects of maternal antibody on immmune responses to hemagglutinin and nucleoprotein. J Virol 74(17):7787–7793, 2000.

30. Hawkins WG, Gold JS, Dyall R, et al. Immunization with DNA coding for gp100 results in CD4+ T-cell independent antitumor immunity. Surgery 128(2):273–280, 2000.

31. Condon A, Watkins SC, Celluzzi M, et al. DNA-based immunization by in vivo transfection of dendritic cells. Nat Med 2(10):1122–1128, 1996.

32. Gainer AL, Young ATL, Parney IF, et al. Gene gun transfection of human glioma and melanoma cell lines with genes encoding human IL-12 and GM-CSF. J Neuro-Oncol 47:23–30, 2000.

33. Tripathy SK, Goldwasser E, Lu M-M, et al. Stable delivery of physiologic levels of recombinant erythropoietin to the systemic circulation by intramuscular injection of replication-defective adenovirus PNAS 91:11557–11561, 1994.

34. Tripathy SK, Svensson EC, Black HB, et al. Long-term expression of erythropoietin in the systemic circulation of mice after intramuscular injection of a plasmid DNA vector. PNAS 93(20):10876–80, 1996.

35. Klinman DM, Conover J, Leiden JM, et al. Safe and effective regulation of hematocrit by gene gun administration of an erythropoietin-encoding DNA plasmid. Hum Gene Ther 10:659–665, 1999.

36. Eming SA, Whitsitt JS, He L, et al. Particle-mediated gene transfer of PDGF isoforms promotes wound repair. J Invest Dermatol 112:297–302, 1999.

37. Yang N-S, Sun WH. Gene gun and other non-viral approaches for cancer gene therapy. Nat Med 1(5):481–483, 1995.

38. Klenchin VA, Sukharev SI, Serov SM, et al. Electrically induced DNA uptake by cells is a fast process involving DNA electrophoresis. Biophys J 60:804–811, 1991.

39. Hristova NI, Tsoneva I, Neumann E. Sphingosine-mediated electroporative DNA transfer through lipid bilayers. FEBS Lett 415:81–86, 1997.

40. Bertling W, Hunger-Bertling K, Cline MJ. Intranuclear uptake and persistence of biologically active DNA after electroporation of mammalian cells. J Biochem Biophys Methods 14:223–232, 1987.

41. Knutson JC, Yee D. Electroporation: parameters affecting transfer of DNA into mammalian cells. Analyt Biochem 164:44–52, 1987.

42. Miklavcic D, Semrov D, Valencic V, et al. Tumor treatment by direct electric current: computation of electric current and power density distribution. Electro- and Magnetobiology 16(2):119–128, 1997.

43. Semrov D, Miklavcic D. Calculation of the electrical parameters in electrochemotherapy of solid tumours in mice. Comput Biol Med 28:439–448, 1998.

44. Aihara H, Miyazaki J-I. Gene transfer into muscle by electoporation in vivo. Nat Biotechnol 16:867–870, 1998.

45. Tekle E, Astumian RD, Chock PB. Selective and asymmetric molecular transport across electroporated cell membranes. PNAS 91:11512–11516, 1994.

46. Andreason GL, Evans GA. Optimization of electroporation for transfection of mammalian cell lines. Analyt Biochem 180:269–275, 1989.

47. Presse F, Quillet A, Mir L, et al. An Improved electrotransfection method using square shaped electric impulsions. Biochem Biophys Res Commun 151(3):982–990, 1988.

48. Bier M, Hammer SM, Canaday DJ, Lee RC. Kinetics of sealing for transient elec-

tropores in isolated mammalian skeletal muscle cells. Bioelectromagnetics 20:194–201, 1999.

49. Sukharev SI, Titomirov AV, Klenchin VA. Electrically-induced DNA transfer into cells. Electrotransfection in vivo. In: Wolff, ed. Gene Therapeutics: Methods and Applications of Direct Gene Transfer. Boston: Birkhauser, 1994, pp 210–232.
50. Lurquin PF. Gene transfer by electroporation. Mol Biotechnol 7:5–35, 1997.
51. Rols M-P, Delteil C, Golzio M, et al. In vivo electrically mediated protein and gene transfer in murine melanoma. Nat Biotechnol 16:168–171, 1998.
52. Weaver JC. Electroporation: a general phenomenon for manipulating cells and tissues. J Cell Biochem 51:426–435, 1993.
53. Manthorpe M, Cornefert-Jensen F, Hartikka J, et al. Gene therapy by intramuscular injection of plasmid DNA: studies in firefly luciferase gene expression in mice. Hum Gene Ther 4:419–431, 1993.
54. Troiano GC, Tung L, Sharma V, Stebe KJ. The reduction in electroporation voltages by the addition of a surfactant to planar lipid bilayers. Biophys J 75:880–888, 1998.
55. Lee RC, River LP, Pan F-S, et al. Surfactant-induced sealing of electropermeabilized skeletal muscle membranes in vivo. PNAS 89:4524–4528, 1992.
56. Jaroszeski MJ, Gilbert RA, Heller R. In vivo antitumor effects of electrochemotherapy in a hepatoma model. Biochim Biophys Acta 1334:15–18, 1997.
57. Hofmann GA, Dev SB, Nanda GS, Rabussay D. Electroporation therapy of solid tumors. Critical Revs Therapeut Drug Carrier Syst 16(6):523–529, 1999.
58. Gehl J, Sorensen TH, Nielsen K, et al. In vivo electroporation of skeletal muscle: threshold, efficacy and relation to electric field distribution. Biochim Biophys Acta 1428:233–240, 1999.
59. Miklavcic D, Beravs K, Semrov D, et al. The importance of electric field distribution for effective in vivo electroporation of tissues. Biophys J 74:2152–2158, 1998.
60. Gilbert RA, Jaroszeski MJ, Heller R. Novel electrode designs for electrochemotherapy. Biochim Biophys Acta 1334:8–14, 1997.
61. Dev SB, Rabussay DP, Widera G, Hofmann GA. Medical applications of electroporation. IEEE Trans Plasma Sci 28(1):206–223, 2000.
62. Tomov T, Tsoneva I. Are the stainless steel electrodes inert? Bioelectrochemistry 51:207–209, 2000.
63. Mir LM, Bureau MF, Rangara R, et al. Long-term high level in vivo gene expression after electric pulses-mediated gene transfer into skeletal muscle. Sixth European Working Group on Human Gene Therapy, Israel, November 21, 1998.
64. Mir LM, Bureau MF, Rangara R, et al. Long-term, high level in vivo gene expression after electric pulse-mediated gene transfer into skeletal muscle. C R Acad Sci 321:893–899, 1998.
65. Mir LM, Bureau MF, Gehl J, et al. High-efficiency gene transfer into skeletal muscle mediated by electric pulses. PNAS 96:4262–4267, 1999.
66. Mathiesen I. Electropermeabilization of skeletal muscle enhances gene transfer in vivo. Gene Ther 6:508–514, 1999.
67. Vicat JM, Boisseau S, Jourdes P, et al. Muscle transfection by electroporation with high voltage and short-pulse currents provides high-level and long-lasting gene expression. Hum Gene Ther 11:909–916, 2000.

68. Goto T, Nishi T, Tamura T, et al. Highly efficient electrogene therapy of solid tumor by using an expression plasmid for the herpes simplex virus thymidine kinase gene. PNAS 97(1):354–359, 2000.
69. Panje WR, Hier MP, Garman GR, et al. Electroporation therapy of head and neck cancer. Ann Otol Rhinol Laryngol 107:779–785, 1998.
70. Nanda GS, Sun FX, Hofmann GA, et al. Electroporation therapy of human larynx tumors Hep-2 implanted in nude mice. Anticancer Res 18:999–1004, 1998.
71. Nanda GS, Sun FX, Hofmann GA, et al. Electroporation enhances therapeutic efficacy of anticancer drugs: treatment of human pancreatic tumor in animal model. Anticancer Res 18:1361–1366, 1998.
72. Nanda GS, Merlock RA, Hofmann GA, Dev SB. A novel and effective therapy for prostate cancer. 89th Annual Symposium, American Association for Cancer Research, New Orleans, LA, 1998.
73. Jaroszeski MJ, Gilbert R, Nicolau C, Heller R. In vivo gene delivery by electroporation. Adv Drug Del Rev 35:131–137, 1999.
74. Heller L, Jaroszeski MJ, Coppola D, et al. Electrically mediated plasmid DNA delivery to hepatocellular carcinomas in vivo. Gene Ther 7:826–829, 2000.
75. Ramirez LH, Orlowski S, An D, et al. Electrochemotherapy on liver tumors in rabbits. Br J Cancer 77(12):2104–2111, 1998.
76. Suzuki T, Shin B-C, Fujikara K, et al. Direct gene transfer into rat liver cells by in vivo electroporation. FEBS Lett 425:436–440, 1998.
77. Heller R, Jaroszeski M, Atkin A, et al. In vivo electroinjection and expression in rat liver. FEBS Lett 389:225–228, 1996.
78. Oshima Y, Sakamoto T, Yamanaka I, et al. Targeted gene transfer to corneal endothelium in vivo by electric pulse. Gene Ther 5:1347–1354, 1998.
79. Gallo SA, Oseroff AR, Johnson PG, Hui SW. Characterization of electric-pulse-induced permeabilization of porcine skin using surface electrodes. Biophysical J 72:2805–2811, 1997.
80. Pliquett U, Gusbeth Ch. Skin electroporation: synergistic effect of electric field and heating (Abstr). Proc Int Symp Cont Rel Bioact Mater 26:5138, 1999.
81. Titomirov AV, Sukharev S, Kistanova E. In vivo electroporation and stable transformation of skin cells of newborn mice by plasmid DNA. Biochem Biophys Acta 1088:131–134, 1991.
82. Giordanoy FJ, Dev SB, Adams M, et al. In vivo gene delivery to the rabbit carotid by electroporation (Abstr). J Am Coll Cardiol 27(2):289a, 1996
83. Nishi T, Yoshizato K, Yamashiro S, et al. High-efficiency in vivo gene transfer using intraarterial plasmid DNA injection following in vivo electroporation. Cancer Res 56:1050–1055, 1996.
84. Muramatsu T, Shibata O, Ryoki S, et al. Foreign gene expression in the mouse testis by localized in vivo gene transfer. Biochem Biophys Res Commun 233:45–49, 1997.
85. Harrison RL, Byrne B, Tung L. Electroporation-mediated gene transfer in cardiac tissue. FEBS Lett 435:1–5, 1998.
86. Gaylor DC, Prakah-Asanta K, Lee RC. Significance of cell size and tissue structure in electrical trauma. J Theoret Biol 133:223–237, 1988.

87. Jiao S, Williams P, Berg RK, et al. Direct gene transfer into nonhuman primate myofibers in vivo. Hum Gene Ther 3:21–33, 1992.
88. Wells DJ. Improved gene transfer by direct plasmid injection associated with regeneration in mouse skeletal muscle. FEBS Lett 332(1–2):179–82, 1993.
89. Fewell JG, MacLaughlin FC, Mehta V, et al. Gene therapy for the treatment of hemophilia B using PINC™ formulated plasmid delivered to muscle with electroporation. Mol Ther Apr 3(4):574–583.
90. MacLaughlin FC, Li S, Li Y, et al. Plasmid gene delivery: advantages and limitations. In: Targeting of Drugs. Gregoriadis G, McCormack B, eds. Amsterdam: IOS Press, 2000, pp 81–91.
91. Rizzuto G, Cappelletti M, Maione D, et al. Efficient and regulated erythropoietin production by naked DNA injection and muscle electroporation. PNAS 96:6417–6422, 1999.
92. Kreiss P, Bettan M, Crouze J, Scherman D. Erythropoietin secretion and physiological effect in mouse after intramuscular plasmid DNA electrotransfer. J Gene Med 1:245–250, 1999.
93. Bettan M, Emmanuel F, Darteil R, et al. High-level protein secretion into blood circulation after electric pulse-mediated gene transfer into skeletal muscle. Mol Ther 2(3):204–210, 2000.
94. MacLaughlin FC, Draghia-Akli R, Gondo MM, et al. Scale-up of the electroporation procedure in larger animal species in conjunction with intramuscular delivery of plasmid encoding a variety of transgene products. In preparation.
95. Draghia-Akli R, Fiorotto M, Hill LA, et al. Myogenic expression of an injectable protease-resistant growth hormone-releasing hormone augments long-term growth in pigs. Nat Biotechnol 17:1179–1183, 1999.
96. Wider G, Austin M, Rabussay D, et al. Increased DNA vaccine delivery and immunogenicity by electroporation in vivo. J Immunol 164:4635–4640, 2000.
97. Mendiratta SK, Thai G, Eslahi NK, et al. Therapeutic tumor immunity induced by polyimmunization with melanoma antigens gp100 and TRP-2. Cancer Res 61:859–863, 2001.
98. Okino M, Mohri M. Effects of a high voltage electrical impulse and an anticancer drug on in vivo growing tumors. Jpn J Cancer Res 78(12}:1319–1321, 1987.
99. Nishi T, Dev SB, Yoshizato K, et al. Treatment of cancer using pulsed electric field in combination with chemotherapeutic agents or genes. Human Cell 10(1):81–85, 1997.
100. Nishi T, Goto T, Yoshizato K, et al. High efficiency gene transfer into solid tumors using in vivo electroporation. Proc Am Assoc for Cancer Res 39:404, 1998.
101. Wells JM, Li LH, Sen A, et al. Electroporation-enhanced gene delivery in mammary tumors. Gene Ther 7:541–547, 2000.
102. Vanbever R, Pliquett UF, Preat V, Weaver JC. Comparison of the effects of short, high-voltage and long, medium-voltage pulses on skin electrical and transport properties. J Control Rel 69:35–47, 1999.
103. Hofmann GA, Rustrum WV, Suder KS. Electroincorporation of microcarriers as a method for the transdermal delivery of large molecules. Bioelectrochem Bioenerg 38:209–222, 1995.

104. Dujardin N, Preat V. Localization and expression of GFP plasmid by in vivo skin electroporation in rat (abst). Proceed Int Symp Control Rel Bioact Mater 26:316, 1999.

105. Kim HJ, Greenleaf JF, Kinnick RR, et al. Ultrasound mediated transfection of mammalian cells. Hum Gene Ther 7:1339–1346, 1996.

106. Mitragotri S, Blankschtein D, Langer R. Transdermal drug delivery using low frequency sonophoresis. Pharm Res 13(3):411–420, 1996.

107. Miller DL. A review of the ultrasonic bioeffects of microsonation, gas-body activation, and related cavitation-like phenomena. Ultrasound Med Biol 13:443–470, 1987.

108. Fechheimer M, Boylan JF, Parker S, et al. Transfection of mammalian cells with plasmid DNA by scrape loading and sonication loading. PNAS 84:8463–8467, 1987.

109. Bao S, Thrall BD, Miller DL. Transfection of a reporter plasmid into cultured cells by sonoporation in vitro. Ultrasound Med Biol 23(6):953–959, 1997.

110. Manomer Y, Nakamura M, Ohno T, Furuhata H. Ultrasound facilitates transduction of naked plasmid DNA into colon carcinoma cells in vitro and in vivo. Hum Gene Ther 11:1521–1528, 2000.

111. Unger EC, McCreery TP, Sweitzer RH. Ultrasound enhances gene expression of liposomal transfection. Invest Radiol 32(12):723–727, 1997.

112. Themis M, Schneider H, Kiserud T, et al. Successful expression of β-galactosidase and factor IX transgenes in fetal and neonatal sheep after ultrasound-guided percutaneous adenovirus vector administration into the umbilical vein. Gene Ther 6:1239–1248, 1999.

113. Anwer K, Kao G, Proctor B, et al. Ultrasound enhancement of cationic lipid-mediated gene transfer to primary tumors following systemic administration. Gene Ther 7(21):1833–1839, 2000.

114. Huber PE, Pfisterer P. In vitro and in vivo transfection of plasmid DNA in the Dunning prostate tumor R3327-AT1 is enhanced by focused ultrasound. Gene Ther 7:1516–1525, 2000.

115. Wyber J, Andrews J, D'Emanuele A. The use of sonication for the efficient delivery of plasmid DNA into cells. Pharm Res 14(6):750–756, 1997.

116. Unger E. Ultrasound-mediated drug delivery with acoustically active carriers and Sonoporation™ with Fluorogene™ Gene. 5th Abstract at http://www.kuhp.kyoto-u.ac.jp/Official/intd/abstract/9701/Unger.html, 1997.

117. Lawrie A, Brisken AF, Francs SE, et al. Ultrasound enhances reporter gene expression after transfection of vascular cells in vitro. Circulation 99:2617–2620, 1999.

11

Nonviral Approaches for Cancer Gene Therapy

Vivian Wai-Yan Lui
Duke University Medical Center, Durham, North Carolina, U.S.A.

Leaf Huang
University of Pittsburgh School of Pharmacy, Pittsburgh, Pennsylvania, U.S.A.

I. INTRODUCTION

Cancer is by far the most common targeted disease for gene therapy. Conventional cancer treatments such as surgery, chemotherapy, radiotherapy, and hormone therapy all suffer from various shortcomings, although they could be effective for some cancers. These include limited specificity, nonresponsiveness to therapy, nonspecific side effects, and recurrence due to incomplete eradication of cancerous cells. All these limitations emphasize the need for alternative therapy with high effectiveness, specificity, and minimal side effects. Cancer gene therapy is believed to meet these requirements.

Tumor biology studies have introduced a tremendous list of potential therapeutic genes for cancer gene therapy. With the advance in gene discovery technology, more will be added to the list. For safe translation to clinical settings, in vivo evaluation of the therapeutic effect and potential toxicity associated with the candidate gene is needed. Therefore, the success of cancer gene therapy will greatly depend on the development of gene delivery vectors with high efficiency and minimal toxicity.

II. NONVIRAL VECTORS FOR CANCER GENE THERAPY

Preclinical and clinical studies suggested that the success of a cancer gene therapy is largely dependent on the development of an ideal delivery vector. An ideal vector for cancer gene therapy should have the following characteristics: specificity for the target cell, efficient gene expression, stability, unrestricted size limitation for DNA, ability to be repeatedly administered, minimal toxicity, and easy preparation with upscaling capacity and low cost. Because of the cytotoxic nature of most of the cancer gene therapy approaches, the issue of long-term gene expression is not as important as in the case of genetic correction of inherited diseases. Once the tumor cell is killed, the expression of a therapeutic gene is not needed. Although such an ideal vector is yet to be developed, these parameters should serve as a guideline for the improvement of current vectors.

In general, there are two types of gene delivery vectors: viral and nonviral vectors. The use of viruses for gene delivery has several merits over the currently available nonviral vectors, including a high level and prolonged gene expression. However, the death of a relatively healthy young patient associated with adenoviral vector gene therapy raised serious public concern about the safety of viral vectors [1]. This was the first patient in gene therapy trials who died of the therapy itself. The toxicities associated with viral vector include high inflammatory activity and immunogenicity, integration into the host genome, and mutation to and/ or contamination of the wild-type virus. In addition, the preparation and upscaling of vector production are difficult to achieve for some viral vectors, including retrovirus (the most widely used vector in gene therapy clinical trials) and adeno-associated virus (AAV). Therefore, a nonviral vector becomes an attractive alternative. Nonviral vectors are known to have low immunogenicity, and they are simple to use with easy large-scale production capability. Exciting nonviral delivery methods have been recently reported to achieve transfection efficiency approaching that of viral vectors. In fact, an increasing number of preclinical and clinical trials are employing nonviral vectors.

A. Clinical Trial Update for Nonviral Vectors in Cancer Gene Therapy

Cancer gene therapy accounts for 65.6% of all gene therapy clinical protocols worldwide (with a total of 425 as updated on May 25, 2000). Among the 279 cancer gene therapy protocols, nonviral vectors are being used in about 30% of cases, which is a total of 76 protocols. The other 70% employs various viral vectors including retrovirus, AAV, adenovirus, herpes simplex virus, poxvirus, and vaccinia virus.

Among the 76 protocols, lipidic vectors represent approximately 80% of all the nonviral vector protocols for cancer, followed by naked DNA (10.5%), gene gun (5.3%), and electroporation (2.6%) (Table 1). For lipidic vectors, 3β-

Table 1 Nonviral Vectors for Cancer Gene Therapy Clinical Trials[a]

Nonviral vector	No. of clinical trials	Phase	Route of administration	Gene	Cancer type
Naked DNA	8	I or II	i.t. or i.m.	p53, CD, CEA, and specific anti-idiotype	Liver cancer, breast cancer, leukemia, colorectal cancer, non–Hodgkin's lymphoma
Gene gun	4	I or II	ex vivo	IL-7, IL-12, GM-CSF	Melanoma, sarcoma
Electroporation	2	I	s.c.	IL-7, IL-2, and TGF-β antisense	Melanoma, glioma, renal cell cancer, lymphoma
RNA	1	I	ex vivo	CEA	Breast cancer, ovarian cancer
Lipofection	61	I, II or III	i.t., i.p. or ex vivo	HLA-B7, HLA-B7/b2m, CD80 (B7-1), IL-2, GM-CSF, E1A, IGF-1 antisense, NeoR Enterotoxin B	Melanoma, renal cell cancer, ovarian cancer, cervical cancer, prostate cancer, sarcoma, head and neck cancer, glioblastoma, small cell and non–small cell lung cancer, colorectal cancer, non–Hodgkin's lymphoma

[a] Information (updated on May 25, 2000) was obtained from www.wiley.co.uk/genmed.

[N-(N′,N′-dimethylaminoethane)carbamoyl]cholesterol(DC-chol)/dioleoylphos-phatidylethanolamine (DOPE) liposome, dimyristoyl oxypropyl dimethyl hydro-xyethyl ammonium bromide (DIMRIE)/DOPE liposome, and N-[1-(2,3-dioleoy-loxy)propyl]-N,N,N,-trimethylammonium chloride (DOTMA)/chol liposome have been approved for use in clinical trials. Most of the genes that are delivered by nonviral vectors are for immunomodulation, which include interleukins, IL-7 and IL-12, granulocyte-macrophage colony-stimulating factor (GM-CSF), interferon-α (IFN-α), human leukocyte antigen-B7 (HLA-B7), with IL-2 alone or in combination with HLA-B7 being the most widely used therapeutic genes. This is probably due to the limited efficiency of the vector, because the delivery of cytotoxic genes or antiproliferative genes requires much higher transfection efficiency. Other therapeutic genes delivered by nonviral vectors include p53, carcinoembryonic antigen (CEA), gp100, E1A, antisense for transforming growth factor β2 (TGF-β2), and insulin-like growth factor-1 (IGF1). Patients with vari-ous kinds of cancer are being evaluated. In most cases, local treatment by intratu-mor (i.t.) injection is being used, which is believed to be one of the safest routes of administration. The ex vivo approach is the second most common approach. For naked DNA gene transfer and electroporation, intramuscular (i.m.) or subcu-taneous (s.c.) injections are being used most frequently. So far, there is only one protocol approved for phase III clinical trial (for melanoma using the HLA-B7/b2m gene). All the other 75 protocols are at either phase I or phase II stage. Having the merits of being clinically safer than viral vectors, the use of nonviral vectors in gene therapy is likely to grow.

III. CANCER GENE THERAPY APPROACHES

There are currently three kinds of cancer gene therapy approaches with the com-mon goal of eradicating the cancer cells. These include targeting specific genetic defects that endow cancer cells with growth advantage, immunocancer gene ther-apy, and sensitization of cancer cells to drug treatment. Each individual approach has its own merits, and a combination of different approaches may have better clinical outcome [2,3]. Nonviral vectors are mainly employed for immunocancer gene therapy in humans owing to the limited transfection efficiency. However, with further improvements, nonviral vectors can be potentially used for all these anticancer approaches in humans.

A. Targeting Genetic Defects That Lead to Growth Advantage

When somatic cells are deposited with inherited or acquired genetic defects, which alter the balance between proliferation and cell death, uncontrolled cell growth or cancer results. Upregulation of growth signals and downregulation of negative growth regulators are the most common observations in cancer. Various therapeutic strategies have been designed to target these specific genetic defects (Table 2).

1. Downregulation of Oncogene or Proto-oncogene

A number of oncogenes and proto-oncogenes of either cellular or viral origin have been shown to promote uncontrolled cellular growth. These include many growth factors, growth factor receptors, and signaling molecules. Gene amplification, mutational activation, or overexpression is clinically correlated with the disease stage and prognosis in some cancers. Different strategies have been developed to abrogate the expression of these genes. Specific downregulation or inhibition could be achieved at DNA, RNA, or protein levels. Theoretically, correction of an oncogenic mutation could also be achieved by the use of, for example, antisense oligonucleotide (ODN), antisense RNA, ribozyme, triplex helix–forming oligonucleotide (TFO), and peptide nucleic acid (PNA), and by expression of either a transcriptional repressor or protein that inhibits oncoprotein function. Some, but not, all of these strategies have been successful in animal models and are being or awaited to be evaluated in human clinical trials.

Antisense Sequence. Downregulation by antisense ODN or antisense RNA expressed from a vector is by far the most common approach owing to its sequence specificity and simplicity in design. Antisense is a DNA or RNA sequence designed with an exact base complementary sequence to target mRNA. Upon entry into the cell, the antisense sequence would bind to the target mRNA with high specificity and affinity which results in the inhibition of gene expression by multiple mechanisms (Table 3). These include the blockade of ribosome attachment to mRNA, inhibition of mRNA splicing or RNase-H–mediated cleavage of the target mRNA with a DNA/RNA heteroduplex [4,5]. Besides synthetic antisense ODN, the antisense sequence could also be expressed as RNA from various mammalian expression systems in vivo. Many antisense sequences show beneficial therapeutic effects in vivo (i.e., reduction in tumor growth). These include antisense against the human papillomavirus (HPV) E7 protein (6), fibroblast growth factor-2 (FGF-2) (7), c-Jun N-terminal kinase-1/-2 (JNK-1/-2) (8), and insulin-like growth factor-I receptor (IGF-IR) (9). Currently, many antisense against oncogenes are evaluated in clinical trials (phase I or II), which include IGF-1, k-*ras*, c-*myc*, c-*fos*, and TGF-β, protein kinase Cα (PKC-α), c-*raf*, Ha-*ras*, and R1-α subunit of protein kinase A [10,11].

The application of antisense is not without problems. First, owing to the inability to predict accurately the RNA structure and accessible site in vivo, active and therapeutic antisense sequences are largely empirically selected. However, with the recent advances in microarray technology and screening methods, an active antisense sequence may be obtained more easily for therapeutic evaluation [12,13]. Second, phosphodiester ODN is very susceptible to nuclease degrada-

Table 2 Cancer Gene Therapy Strategies Targeting Genetic Defects Which Lead to Growth Advantage of Cancer Cells[a]

Strategy	Target genes or molecules
Downregulation of oncogene/proto-oncogene	IGF-1, *erbB-2*, EGFR, *k-ras*, *c-myc*, *c-fos*, *c-met*, FGF-2, IGF-IR, TGF-β, PKC-α *c-raf*, Ha-*ras*, k-*ras*, JNK-1, JNK-2, PKA R1-α subunit, HPV-E7, *bcl-2*, telomerase
Restoration of growth regulation	
Introduction of tumor-suppressor gene	P53, RB, BRCA1
Introduction of apoptosis-inducing gene	Cytochrome-p450; TP53; *bax*; *bcl-xs*; CARD; caspase-3,-8 and -9; Apaf-1; apoptin; p21, E1A; granzyme B/perforin; FasL/Fas; TNF/TNFR1; Apo3L/DR3; TRAIL/DR4 or -5
Downregulation of antiapoptotic gene	*bcl-2*, survivin, XIAP, IAP-1, IAP-2
Inhibition of angiogenesis	
Suppression of angiogenic factor	VEGF, PDGF, TGF-α/β, TNF-α, bFGF, IGF-1, EGF, HGF, pleiotropin, IL-6, IL-8
Introduction of angiogenesis inhibitor	Soluble factor 4, TSP-1, antagonist of uPA/uPAR, angiostatin, endostatin, TIMPs
Inhibition of metastasis	MMPs, TF, nm23, integrin $\alpha_v\beta_3$, AAC-11

[a] Therapeutic targets or potential therapeutic targets are shown for each strategy.

Table 3 Small Therapeutic Nucleic Acids That Can Be Potentially Used for Cancer Gene Therapy

Type	Target	Possible mechanism(s) of action
Antisense ODN	mRNA	1. Blocking translation 2. Inhibition of RNA splicing 3. RNase-mediated degradation of target
Antisense RNA	mRNA	1. Blocking translation 2. Inhibition of RNA splicing
Ribozyme	mRNA	1. Target RNA cleavage 2. Site-specific correction of target mRNA
PNA	DNA or RNA	1. Inhibiting gene expression by duplex or triplex formation 2. Inhibiting RNA function by binding
TFO	dsDNA	1. Inhibiting gene expression by triplex formation
RNA/DNA hybrid	mRNA	1. Site-direction correction at genomic level
DNA enzyme	mRNA	1. Target RNA cleavage
DNA or RNA aptamer	Protein, DNA or RNA	1. Alteration of protein, DNA, or RNA conformation by binding
Decoy	Protein	1. Inhibition of transcription factor 2. Inhibiting the activity of RNA- or DNA-binding protein

tion. Various chemical modifications have been applied for antisense to improve nuclease resistance with phosphorothioate oligonucleotide (PS-ODN) being the most commonly used in clinical trials. However, some PS-ODNs induce immuno-stimulatory effects in animals and humans [5].

Ribozyme. Ribozyme is a catalytically active RNA with sequence-specific cleavage activity. Therefore, ribozyme may be more efficient than the antisense sequence. Based on the structure, different types of ribozymes have been developed. These include hammerhead, hairpin, tRNA, and group I intron ribozymes, with the hammerhead ribozyme being the most commonly used. Similar to antisense, the design of therapeutic ribozyme is largely empirical. Moreover, to express a ribozyme sequence in vivo (and also short antisense RNA) is challenging, because most of the commonly used promoters are not efficient in expressing short and stable RNA. A number of endogenous promoters have been tested for this purpose. These include the transfer RNA (tRNA) and small nuclear RNA (snRNA), and heat shock protein 70 (HSP70) promoters [6,9]. Among these promoters, the U6 RNA promoter is one of the most efficient. We have recently developed a hammerhead ribozyme against the proto-oncogene c-*neu* and expressed it under the U6 promoter in ovarian cancer cells overexpressing c-*neu*. The expression level of the ribozyme is very high (5×10^6 copies/cell). Moreover, nonviral delivery of the ribozyme results in specific downregulation of c-*neu* as well as significant growth inhibition of ovarian cancer cells both in vitro [14].

Successful ribozyme constructs with in vivo therapeutic effects have been reported against many oncogenes besides c-*neu* [15,16]. These include ribozymes against k-*ras* [17], deletion mutant of epidermal growth factor receptor (EGFR) [18], PKC-α [19], and many others. Downregulation of these oncogenes resulted in growth inhibition, increase in chemosensitivity, or induction of apoptosis. However, no anti-oncogene ribozyme is being evaluated in clinical trials. Recently, the ability of ribozymes to carry out site-specific correction for genetic diseases has been successfully demonstrated [20]. Thus, the potential of ribozymes in correcting oncogenic mutations should be explored.

PNA.   PNA is a DNA-mimetic that can form very stable duplex or triplex structures with DNA, RNA, or PNA with higher specificity than DNA itself [21]. The A/T/C/G bases of DNA are retained, whereas the deoxyribose phosphodiester backbone of DNA is replaced by a pseudopeptide backbone with no sugar residue or phosphate group. The neutral backbone of PNA imposes difficulty in delivery to cells. PNA has an extremely high affinity to form sequence-specific duplexes or triplexes with DNA, thus inhibiting gene expression by the antisense effect (the PNA/DNA duplex cannot activate RNase-H). The underlying mechanism of its antisense effect is not fully understood. Antisense inhibition of β-galactosidase and telomerase has been demonstrated in *Escherichia coli* and in a cell-free system, respectively [22,23]. A recent study showed that a PNA targeting to the RNA component of telomerase in human cancer cells could result in shortened telomeric and arrested cell proliferation after a lag period of 5–30 cell generations [24]. With further improvement in PNA delivery methods, it is likely to be developed as a therapeutic drug in cancer gene therapy.

Triplex Helix-Forming Oligonucleotide.   A triplex helix forming oligonucleotide (TFO) is a short single-stranded ODN that can recognize and bind to a stretch of oligopurine–oligopyrimidine sequence in a double-stranded DNA by Hoogsteen hydrogen bonding. This results in a triplex formation. TFO can inhibit protein binding to DNA, gene expression or replication, and direct site-specific DNA damage, enhance recombination and induce site-directed mutagenesis [25, 26]. A TFO against c-*myc* has been shown to reduce the c-*myc* mRNA level in HeLa cells [27]. Owing to the limited sequence-recognition properties of TFO, no in vivo or therapeutic effect has been shown [28–31].

Intracellular Antibody.   Intracellular antibody is a single-chain antibody (scFv) expressed intracellularly to interfere with the intracellular processing, mobilization, or the function of the native oncoprotein. Theoretically, similar effects could be achieved by expressing dominant negative mutants of the oncogene [32,33]. The small size of scFv allows efficient antibody expression within the cell after gene transfer. Being expressed intracellularly will avoid the problem of eliciting a serious immune response as in the case of extracellular antibodies. Many intracellular antibodies against oncoproteins have been successfully developed. These include anti–*erbB*-2, anti-EGFR, anti–*bcl*-2 and anti-*Ras* scFv [33]. The expression of scFv resulted in specific downregulation of target protein, cytotoxicity, induction of apoptosis, or enhanced chemosensitivity. Among these intracellular antibodies, anti–*erbB*-2 scFv is being evaluated in clinical trials [33].

Transcription Repression. Oncogene expression could also be downregulated by suppressing its transcription. The number of transcriptional repressor genes used in cancer gene therapy is limited, but the clinical and preclinical results are encouraging. In vivo gene transfer of the adenovirus 5 E1A gene, semian virus 40 (SV40) large T antigen and polyomavirus enhancer activator 3 (PEA3) have resulted in downregulation of *erbB*-2 proto-oncogene expression, induction of apoptosis, and prolonged survival rate in animals [34]. An E1A phase I clinical trial for breast and ovarian cancer has recently been completed. Administration of E1A plasmid DNA complexed with DC-Chol liposomes resulted in erbB-2 protein downregulation in cancer cells with reduction in cancer cell numbers in some patients. Multiple mechanisms, may contribute to the tumor-suppressing and metastasis-suppressing effects of E1A. It can indirectly suppress the expression of *erbB*-2 by binding to CBP/p300, a transcriptional coactivator, that is required for efficient *erbB*-2 expression. Other mechanisms include the induction of apoptosis by p53-dependent and p53-independent pathways and the repression of various proteases involved in metastatic invasion [34,35].

Many other small nucleic acid therapeutics, such as RNA/DNA hybrid, DNA enzyme, aptamer, and decoy, have the potential of developing into anticancer drugs. However, their therapeutic effects in cancer have not been demonstrated (see Table 3).

2. Restoration of Growth Regulation

Normally, proliferation of cells is under vigorous check and balance by cellular regulators such as tumor suppressor, cell cycle checkpoint regulators, regulators of apoptosis, and regulations of DNA repair. Inactivation mutation, deletion, or the loss of these regulatory genes will result in uncontrolled proliferation and malignancy. Therefore, different strategies have been explored to restore the wild-type regulator function in cancer cells, which may reverse the transforming phenotype (see Table 2).

Gene Replacement. Tumor-suppressor genes and apoptosis-inducing genes have been employed for gene replacement in cancer gene therapy. Currently, there are more than 24 known tumor-suppressor genes. In preclinical studies, introduction of wild-type tumor-suppressor genes could reverse the malignant phenotype and, in some cases, induce apoptosis. Clinical trials with p53, RB, or BRCA1 have been initiated, with p53 being the most common one for gene replacement. The p53 gene encodes a transcription factor that regulates the cell cycle checkpoint, DNA repair, and cell death. It exerts a protective function in response to DNA damage by inhibiting cell cycle progression, thus allowing DNA repair to occur. On the other hand, p53 could initiate cell death if the damage is nonrepairable. Mutation of p53 is frequently observed in cancer and is associated with resistance to chemotherapy and radiation therapy [36,37]. Replacement of a mutant p53 gene resulted in growth suppression in prostate cancer, non–small cell lung carcinoma, head and neck cancer, and hepatocarcinoma [7,38–40]. In some cases, induction of apoptosis, enhanced sensitivity to chemotherapy, and radiation therapy was also observed [41]. Moreover, combination of p53 with another tumor-suppressor gene, p16, exhibits a better therapeutic effect than p53 alone [42].

Moreover, tumor cells could be eradicated by the introduction of apoptosis-inducing genes. These include many ligand/death receptors like granzyme B/perforin, FasL/Fas, TRAIL/DR4 (or DR5) and other apoptosis-inducers such as cytochrome c, E1A, bax, caspases, p21, and many others (see Table 2) [32,43,44]. Transfer of these genes has resulted in the induction of apoptosis. Given the complexity of apoptotic pathways and possible compensatory mechanisms in the cell, more effort is needed before effective apoptosis-inducing therapeutics could be developed.

Downregulation of Anti-apoptotic Molecule. Besides the loss of tumor-suppressor genes, overexpression of anti-apoptotic genes has been observed in some cancers. The excess anti-apoptotic function leads to inappropriate growth of cancer cells. One example is overexpression of *bcl*-2. *Bcl*-2 overexpression is believed to protect tumor cells from undergoing apoptosis induced by chemotherapeutic drugs [40]. Different strategies have been employed to abrogate *bcl*-2 expression in cancer. Antisense PS-ODN against *bcl*-2 has, shown a promising effect in non-Hodgkin's lymphoma [45]. In fact, an anti–*bcl*-2 PS-ODN has entered a phase II clinical trial. Anti–*bcl*-2 hammerhead ribozyme administration sensitized prostate cancer cells to apoptosis-inducing agent [46]. Recent studies in breast and ovarian cancer cells showed that anti–*bcl*-2 scFv expression could result in a chemosensitizing effect with cisplatin and taxol, respectively [47]. Recently, many inhibitors of apoptosis (IAPs) have been discovered such as survivin, XIAP, IAP-1, and IAP-2. The potential therapeutic effects of these molecules are being studied.

3. Inhibition of Angiogenesis

Another growth advantage of cancer is the ability to induce angiogenesis, which is essential for their growth, progression, and dissemination. Angiogenesis means the development of new blood vessels. Therapy targeting to tumor vasculature has four distinct advantages over direct targeting to cancer cells: (1) Tumor vasculature is known to be different from normal vessels regarding cellular component, tissue integrity, permeability and the regulation of proliferation and apoptosis [48]. Therefore, specific targeting and killing of tumor vasculature may be feasible (2). The genetic instability of cancer cells easily gives rise to the development of resistance to therapeutic intervention. However, endothelial cells with high genetic stability are not likely to develop resistance to treatment (3). Antiangiogenic intervention should be highly efficient, because the killing of one vessel would probably result in simultaneous killing of many tumor cells. However, direct killing of cancer cells probably requires gene delivery to almost every single cancer cell (4). Tumor endothelial cells are readily accessible from the circulation, whereas accessing tumor cells requires extravasation of the therapeutic gene.

Suppression of Angiogenic Factors. Inhibition of tumor angiogenesis could be achieved by the suppression of growth factors and signaling molecules required for angiogenesis. These include vascular endothelial growth factor (VEGF), TGF-α/-β, basic fibroblast growth factor (bFGF), epidermal growth factor (EGF), platelet-derived growth factor (PDGF), and many others (see Table 2). VEGF is one of the most potent stimulatory factors for angiogenesis. It can stimulate endothelial cell proliferation, increase vascular permeability, and induce procoagulation factors and matrix metalloproteinases required for metastasis [49]. Downregulation of VEGF expression by antisense RNA has resulted in reduced vessel formation in tumors [50]. Moreover, the therapeutic effect could also be achieved by VEGF receptor downregulation using ribozymes [51].

Inhibitors of Angiogenesis. Another approach is the introduction of angiogenesis inhibitors. Soluble platelet factor 4, thrombospondin-1 (TSP-1), antagonist of urokinase plasminogen activator/receptor (uPA/uPAR), angiostatin, endostatin, and tissue inhibitors of metalloproteinases (TIMPs) are known to inhibit endothelial cell proliferation and migration. The therapeutic potential of this approach has been successfully demonstrated in a number of in vivo studies. Human breast cancer cell transfected with the TSP-1 gene has resulted in the reduction of tumor growth and metastatic potential and angiogenesis [48]. Angiostatin gene transfer has been shown to suppress tumor growth and metastasis in fibrosarcoma and breast cancer [52,53]. Administration of the endostatin gene in a metastatic model of lung carcinoma greatly reduced metastasis [48].

Although with great therapeutic potential, this approach suffers from some intrinsic problems. First, complete eradication of tumor cells is unlikely to be achieved, because the dependence on angiogenesis decreases as tumor size is reduced by the treatment. Therefore, combination with some cytotoxic approach is needed to achieve complete tumor eradication. Another possibility is to maintain long-term expression of anti-angiogenic genes so that prolonged suppression of tumor growth or metastasis can be obtained. However, this is difficult to achieve with the current gene delivery vectors. Second, effective therapy will be dependent on specific targeting of these genes to tumor endothelial cells or to the metastatic tumor. Until now, attempts to direct gene transfer to tumors has only met with limited success.

4. Inhibition of Metastasis

The metastatic ability of cancer has made local treatment ineffective; thus making malignant diseases highly fatal. Metastasis is a complicated process involving multiple steps including degradation of the extracellular matrix, migration of tumor cells, tumor cell adhesion to a secondary site, and finally the stimulation of angiogenesis and tumor establishment. Some genes, although not many, have been implicated in tumor invasion and metastasis. These include matrix metalloproteinases (MMPs), tissue factor (TF), AAC-11, integrin $\alpha_v\beta_3$, and nm23 [49,54,55]. However, their roles in metastasis are not fully understood.

B. Immunocancer Gene Therapy

The establishment of a tumor implies the failure of the host immune system to recognize and eliminate the tumor. The mechanism underlying the tumor's ability to escape from the immune surveillance of the body is not fully understood. There are several proposed mechanisms including tumor-induced immunosuppression, stimulation of suppressor cells, lack of tumor antigen expression on tumor cells, defective antigen presentation due to defective binding to major histocompatibility complex (MHC) molecules, or lack of transporter-associated protein (TAP) for antigen presentation. However, the ability of the immune system to eradicate cancer cells should not be underestimated, especially for treatment of metastases. Appropriate stimulation of the host defense may allow immune effectors to search and eradicate disseminated tumor cells. There are currently four different strategies for immunocancer gene therapy. These are (1) enhancement of tumor immunogenicity, (2) stimulation of the cellular components of immune system, (3) reversal of tumor-induced immunosuppression, and (4) immunization with tumor-associated antigen. Specific genetic modification of the immune system should induce a more specific antitumor response with less side effects as compared with conventional immunotherapy.

1. Enhancement of Tumor Immunogenicity

This strategy is based on the hypothesis that tumor tolerance or anergy could be reversed if tumor immunogenicity is increased. Theoretically, this could be achieved by genetically modifying the tumor cells with a highly antigenic tumor antigen. Introduction of the B7 allele of the HLA molecule into tumor cells has resulted in some therapeutic effect in a clinical trial [56]. This was aimed at generating T-cell response against the unmodified tumor cells by enhancing antigen presentation. In a clinical trial for melanoma, liposomal complexes with HLA-B7 plasmid DNA were delivered by repeated intratumor injections. Gene transfer resulted in the stimulation of an immune response to HLA-B7 and autologous tumors. Moreover, tumor regression was observed at the injection site and at distant sites [56]. These promising results prompted more clinical trials with HLA-B7 gene delivery alone or in combination with IL-2. In addition, the antigenicity of tumor cells could be restored by the introduction of TAP-expressing genes. The TAP complex is essential for the translocation of processed antigenic peptide for presentation with MHC class I molecules. Loss of TAP in many cancers suggests that its downregulation may be involved in the escape of immune surveillance of tumor cells. Introduction of TAP gene into TAP-deficient small cell lung cancer cells improved specific cytotoxic T lymphocyte (CTL) recognition in vitro accompanied by prolonged survival of tumor bearing mice in vivo [57].

2. Modification of Immune Effectors

Genetic modification of the immune effector cells is aimed at augmenting their ability and capacity to recognize and eradicate tumor cells. This is currently being achieved by (1) ex vivo modification of immune effector cells isolated from patients and (2) introduction of cytokine gene to stimulate immune effector cells. However, with a better understanding of the host's antitumor responses, new strategies are emerging.

Tumor-Infiltrating Lymphocytes. Mononuclear cells derived from the inflammatory infiltrate in solid tumor are called tumor-infiltrating lymphocytes (TILs). These include CTLs and natural killer cells (NKs), and ex vivo modification of TIL with the class I gene Kb gene and TNF-α gene have been performed with the aim of enhancing their antitumor activity [58,59]. However, owing to the limited number of CTLs and NKs in the isolated TILs, ex vivo expansion with IL-2 is necessary. Furthermore, administration of IL-2 to sustain TIL activity in vivo may result in high toxicity.

Possible Modification of Other Immune Effectors. Recent advances in tumor immunology give valuable insights for the development of new immuno-cancer gene therapy strategies. One such exciting area is dendritic cell (DC)–mediated antigen presentation. The DC is believed to be one of the most potent antigen-presenting cells (APCs). Recently, the DC has been shown to uptake and present antigens derived from apoptotic tumor cells. It is possible to envision a combination therapeutic approach using an apoptosis-inducing gene or drug (to kill some tumor cells) in combination with DC stimulation or proliferation by gene therapy. In addition, one could deliver a gene-expressing tumor antigen to

DCs for efficient presentation to the immune system. However, the success of this strategy will largely depend on our ability to improve the efficiency of DC transfection.

3. Introduction of Cytokine Gene

Many immune effector cells are known to be stimulated by cytokines. A number of cytokines are found to have antitumor activity when deliver ex vivo, in vitro, or in vivo. These include IL-2, IL-4, IL-6, IL-7, IL-12, IL-18, interferon-γ, (IFNγ) GM-CSF, and TNF-α [60–62]. Many of these cytokines have multiple antitumor effects. For instance, IL-12 is believed to be one of the best antitumor cytokines. Its antitumor activity is manifold, including stimulation of NK and CTL activities and inhibition of angiogenesis. IL-12 gene delivery has shown encouraging therapeutic effects in many animal cancer models and in human clinical trials.

Cytokine gene transfer has resulted in enhanced immunity against the challenged tumors [61,62] and even regression of established local or metastatic tumors [63]. In some cases, long-term remission and antitumor memory were achieved. However, systemic effects due to the production of secretable cytokines from engineered cells or direct in vivo gene transfer may result in severe toxicity. In order to reduce the systemic side effects, local administration or expression of membrane-bound forms of the cytokines were performed [64].

Combination of different cytokine genes could result in synergistic antitumor activity with lower toxicity [65–67]. Indeed, cancer gene therapy clinical trials have been initiated with this combination approach using IL-7/IL-12, IL-7/IL-2, IL-7/IL-12/GM-CSF, IFN-γ/IL-2, IL-2/GM-CSF, IL-2/lymphotactin, IL-2/TGF-β, and IL-12/IFN-α.

4. Reversal of Tumor-Induced Immunosuppression

Tumor-induced immunosuppression is mediated by many cytokines, hormones, or other molecules with immunomodulatory properties [10]. These include IL-10, TGF-β, and prostaglandin E2(PGE2). They can counteract the immunostimulatory activity of antitumor cytokines, inhibit NK activity, inhibit lymphokine-activated killer cell (LAK)–mediated cytotoxicity, inhibit B-cell activity, or suppress antigen presentation.

Among the known tumor-induced immunosuppressors, TGF-β is the most potent. In addition, it is known to have other activities that could promote tumor progression and metastasis such as stimulation of angiogenesis through PDGF-β and regulation of collagen that is involved in tumor migration and adhesion [68]. Therefore, anti–TGF-β strategies should be very effective. Indeed, antibody neutralization of TGF-β can result in complete abrogation of tumor growth and reduced metastatic lesions in many different animal models [69,70]. These results suggest that TGF-β is a good target for cancer gene therapy. In rat gliosarcoma, downregulation of TGF-β by antisense resulted in prolonged survival [71].

5. Immunization Against Tumor Antigen

Genetic vaccination can induce humoral and cellular immunity against antigen encoded by the gene. Although it is not aimed at treating but prevention of the disease, much success has been reported for immunization against different kinds of cancer in animal models [72]. For vaccination, delivery of a small amount of DNA to the host would be enough to elicit protective immunity. Immunization could be achieved by ex vivo or in vivo transfection of DC or tumor cells.

Many different tumor antigens have been explored for genetic vaccination including CEA [73] a melanoma antigen, MART1 [37], and many others. Transfection or gene gun delivery of HPV E7 antigen has resulted in effective immunization against cervical carcinoma challenge [74]. With the advancement in tumor antigen discovery, more tumor antigens will certainly be tested for vaccination.

C. Sensitization to Chemotherapy

The nonresponsiveness of cancer cells to chemotherapy and the development of drug resistance are two big problems in conventional cancer therapy. Therefore, new strategies aiming at enhancing the killing effect of chemotherapy heve been developed, including (1) introduction of toxic gene to cancer cells or cells that grant support to the tumor, (2) introduction of genes for chemosensitization, and (3) introduction of a chemoprotective gene to normal cells to allow more effective killing of cancer cells at a higher drug dose. Currently, many clinical trial protocols have been approved for these strategies. This approach is particularly attractive, as the nonspecific side effects of conventional chemotherapy will be minimized by enhancing the drug sensitivity of cancer cells. Moreover, this should allow the killing of drug-resistant tumor cells. Indeed, many successful preclinical studies have demonstrated the promising therapeutic potential of this approach.

1. Introduction of a Toxic Gene

A suicide gene and a toxic gene are introduced to render cytotoxicity to tumor cells. A suicide gene encodes for an enzyme that converts a nontoxic prodrug to a toxic metabolite in the transfected cells, whereas a toxic gene exerts direct tumoricidal effect upon introduction to a cancer cell. Both are very efficient in killing the transduced cell no matter if it is a tumor cell or a normal cell. Therefore, tumor-specific killing should be achieved by targeting delivery in order to avoid nonspecific side effects.

Many potential suicide genes that could be employed for tumor ablation such as genes encoding for herpes simplex virus thymidine kinase (HSV-TK), cytosine deaminase (CD), cytochrome P450, deoxycytidine kinase (CK), and many others. The corresponding prodrug for each of the enzyme is listed on Table 4 [75–81].

Table 4 Summary of Genes Encoding Enzymes That Could Be Used for Cancer Gene Therapy

Tumor cell killing		Chemoprotection	
Toxic gene	Prodrug	Protection gene	Chemotherapeutic drug
HSV-TK	Ganciclovir, acyclovir	P-glycoprotein	Vinca alkaloids, taxol
Deoxycytidine kinase	Cytosine arabinoside	GST	Mechlorethamine
Cytosine deaminase	5-Fluorouracil	Cytosine deaminase	Ara-C
Cytochrome P450	Procarbizine, CPA, IFA	Dihydrofolate reductase	Methotrexate, trimetrexate
Carboxylesterase	CPT-11	Thymidine synthase	5-Fluorouracil, folate analog
UPRT	5-Fluorouracil	AGT	Alkylating agents, BG
Carboxypeptidase G2	CMDA	Ribonucleotide reductase	Hydroxyurea
β-Glucuronidase	DOX-G3A	Aldehyde dehydrogenase	Cyclophosphamide
Thd Pase	dFUr	Metallothionen	Cisplatin, chlorambucil
Linamarase cas 5	Linamarin		
Penicillin-V amidase	DPO		
E. coli PNP	MeP-dR, Fara AMP		
E. coli gpt	6-Thioxanthine		
E. coli nitroreductase	CB 1954		
Clostridium acetobutylicum ETS	Metronidazole		

HSV-TK, herpes simplex virus thymidine kinase; UPRT, uracil phosphoribosyltransferase; CMDA, 4-[(2-chloroethyl)(2-mesyloxyethyl)amino]benzoyl-L-glutamic acid; DOX-G3A, N-[4-doxorubicin-N-carbonyl(-oxymethyl)phenyl]O-β-glucuronyl carbamate; Thd Pase, thymidine phosphorlyase; DPO, doxorubicin-N-p-hydroxyphenoxyacetamide; PNP, purine nucleoside phosphorylase; gpt, guanine phosphoribosyl transferase; ETS, electron transport system; GST, glutathione-S-transferase; AGT, O⁶-alkylguanine-DNA alkyltransferase; CPA, cyclophosphamide; IFA, ifosfamide; dFUr, 5′-deoxy-5-fluorouridine; MeP-dR 6-methylpurine-2′-deoxyriboside; Fara AMP, arabinofuranosyl-2-fluoroadenine monophosphate; CB 1954, 5(-aziridine-l-yl)-2,4-dinitrobenzamide; BG, O⁶-benzylguanine.
Source: Data partially taken from Ref. 81.

Transfection of 100% of tumor cells with the suicide gene was thought to be required for effective therapy. However, upon the addition of prodrug, untransfected cells in the vicinity of the transduced cells were also killed. Many suicide gene systems are known to have this beneficial "bystander effect" for amplified killing. The mechanism of the bystander effect is not fully understood. Diffusion of toxic metabolites across the gap junction is believed to be one of the mechanisms involved [82,83].

Of all the known suicide genes, only HSV-TK and CD are being currently evaluated in clinical trials. HSV-TK is given with the prodrug ganciclovir (GCV) or acyclovir. HSV-TK can preferentially monophosphorylate GCV compared to normal mammalian TK. The monophosphorylated GCV is then further phosphorylated to GCV triphosphate. Since the final metabolite lacks a 3′-OH terminus, incorporation of it will result in chain termination in DNA synthesis, thus cell death [32,81]. Microbial CD is capable of converting the prodrug 5-fluorocytosine to the active antitumor drug, 5-fluorouracil (5-FU) by deamination. CD is not expressed in mammalian cells; therefore, nonspecific side effects should be minimal. Therapeutic effects of these suicide genes have been successfully demonstrated in many animal models. However, it is necessary to be alert to the possible toxicity of these systems. Intravenous adenoviral gene transfer of HSV-TK has resulted in liver toxicity contributed by the gene. Mutant enzymes with higher specificity to prodrugs have been selected. Expression of these mutants should allow more efficient killing of the transfected tumor cells with a lower dose of prodrug [81], thus resulting in less side effects.

Bacterial toxin, mainly diphtheria toxin A chain (DT-A), has been used to exert cytotoxic killing of cancer cells. Gene transfer by liposomal DNA complex or electroporation has resulted in tumor killing in hepatoma and glioma models [84,85]. In addition, tumor-specific expression of bacterial toxin has been shown by using the human α-fetoprotein promoter [84,86].

2. Introduction of a Chemosensitizing Gene

Drug resistance is mediated by many different mechanisms that are poorly understood. However, the expression of some genes could contribute drug resistance to the transfected cells. These include the multiple drug-resistant gene (MDR), dihydrofolate reductase (DHFR), and the anti-apoptotic genes bcl-2 and bcl-X$_L$. Some oncogenes are also known to contribute to drug resistance. For instance, overexpression of erbB-2 may contribute to Paclitaxel resistance in breast cancer [34] and constitutive expression of c-H-Ras resulted in the inhibition of doxorubicin-induced apoptosis [33]. Antisense, ribozyme, or intracellular antibody has been employed to abrogate the expression or function of these drug-resistance genes, and chemosensitization has been achieved many cases [33]. Besides, p53 gene transfer also resulted in enhancement of drug-induced apoptosis [41].

Alternatively, sensitization can be achieved with cytochrome P450 gene transfer. This is aimed at enhancing metabolic conversion of chemotherapeutic drug within a tumor. Human breast cancer cells were sensitized with cytochrome P450 gene transfer for enhanced killing by cyclophosphamide [87]. Recently, introduction of *bax* gene also resulted in a chemosensitizing effect in cancer cells [88].

3. Introduction of a Chemoprotective Gene to Normal Cell

The drug-induced toxicity in normal cells has limited the drug dose tolerable by cancer patients. Therefore, protection of normal cells should allow administration of higher dose of chemotherapeutic drug, thus resulting in an improved antitumor effect. This strategy has been explored for the protection of hematopoietic cells and the gastrointestinal tract [89]. Many drug-resistance genes and genes encoding drug-metabolizing enzymes have been investigated for chemoprotection. These include MDR, DHFR, thymidylate synthase, and many others (see Table 4) [81]. Mutants with improved drug-resistant properties have been developed for some, which should provide enhanced chemoprotection to the patient. However, contamination or transfection of tumor cells with the chemoprotective genes would be problematic.

IV. NONVIRAL DELIVERY VECTOR FOR CANCER GENE THERAPY

A. Naked DNA

Wolff et al. demonstrated that simple administration of naked DNA (free DNA) by intramuscular injection resulted in high level of expression. Thereafter, naked DNA gene transfer was shown in many other tissues such as liver, lung, heart, and kidney. The mechanism of in vivo naked DNA transfection is not fully understood. Possible mechanisms include large membrane disruption, small membrane pore formation, and receptor-mediated endocytosis [90]. At present, most data support the idea of receptor mediated endocytosis of naked DNA.

Naked DNA can be delivered via various routes and methods. These include i.m., intradermal (i.d.), intraveneous (i.v.), intra-arterial and intratumoral (i.t.) injections and electroporation (EP). Owing to the high-level and prolonged expression, i.m. administration of DNA vaccine, including tumor vaccine, has been popular. Intramuscular naked DNA vaccination can result in long-lasting humoral and cellular immunity for more than a year [91]. A similar result, if not better, has been reported for i.d. naked DNA vaccination [92]. Interestingly, simple topical application of naked DNA in aqueous solution can also result in immunization via hair follicles [93]. Currently, there are not many tumor vaccination studies performed with naked DNA. However, one such study showed the promise of naked DNA tumor vaccine. White et al. showed that upon i.m. or i.d. vaccination with naked DNA encoding the tumor antigen CEA, antitumor immunity was observed [73].

Recently, Liu et al. showed that i.v. injection of naked DNA in a large volume of saline (about 120% of blood volume) can result in a very high level of gene expression in the liver [94]. Other highly perfused organs such as the lung, heart, kidney, and spleen were also transfected with a lower efficiency. Although this gene delivery method is difficult to be translated for clinical application in humans in its present form, it can be a very useful technique for the study of in vivo function or antitumor effect of a potential therapeutic gene. Indeed, we have recently employed this method for the in vivo delivery of the IL-12 gene for the treatment of human HPV DNA-positive tumor in an animal model (unpublished data) [95]. An extremely high level (10–20 µg/mL of IL-12 in serum) of IL-12 gene expression is observed, and a single gene transfer is efficient enough to result in complete tumor regression. Interestingly, an initial injection of the IL-12 gene results in the inhibition of expression of a subsequently delivered gene. This phenomenon is not observed with a reporter gene such as cytomegalovirus (CMV)–luciferase, indicating it is specific to IL-12. We found that IFN-γ is the main mediator of this transgene inhibition effect of IL-12, as no inhibition was observed in IFN-γ knockout mice [96]. Our results demonstrate that this gene transfer method could be an excellent model for the study of the in vivo effects of any potential therapeutic genes.

Our laboratory has also recently improved this procedure so that delivery of a much smaller volume can result in reasonable gene transfer in the liver [97]. Wolff et al. showed that occlusion of blood flow by surgical procedures could further increase the gene expression in liver in the rodent and dog [98]. In the same study, naked DNA delivery through surgically accessible vessels such as the hepatic vein or bile duct is also feasible [98].

In an attempt to transfect muscle in bigger animals, Wolf et al. showed that rapid intra-arterial injection of naked DNA (through femoral arteries) with occlusion could result in a high level of gene expression [98]. The level of gene expression was higher than that obtained by i.m. injection [99]. Actually, muscle could be used as a secretory organ for therapeutic proteins such as antitumor cytokines.

Recently, intratumoral injection of naked DNA has been shown to result in a reasonable level of gene expression. Favrot et al. demonstrated i.t. injection of naked DNA carrying bax or p53 gene into established subcutaneous tumor could inhibit tumor growth in a human non–small cell carcinoma xenograft model [100]. This method of intratumor gene transfer could be widely used due to its simplicity.

Besides injection, EP can also be used for naked DNA delivery. Exogenous molecules can be introduced into cells by high-voltage electrical pulses. It is commonly used for the transfection of cultured cells. Besides, EP has also been used to enhance the uptake, thus the cytotoxicity, of chemotherapeutic drugs in animal models [92] and in clinical trials for the treatment of s.c. solid tumor [101]. Recently, EP has been used for the delivery of naked DNA encoding toxic genes (HSV-TK and DT-A) into s.c. solid tumors of colon cancer and gliomas in animal [85,102]. Intratumor injection of naked DNA was followed by EP. Substantial reduction in tumor size was observed in the electroporated group compared with the group without EP. This form of gene delivery is mostly limited to localized transfection in skin, muscle, liver, or s.c. tumor.

B. Gene Gun

Gene gun or particle-mediated gene delivery is a highly efficient physical method that forces DNA to penetrate a target tissue. DNA is coated on very fine gold beads or tungsten particles. Then a compressed shock wave of helium gas of variable pressure accelerates the beads for efficient penetration into the tissue. The gene gun was originally designed to deliver genes to plant cells. Recently, it has been shown to be effective in delivering genes to various tissues in vivo including skin, muscle, liver, spleen, and many other organs [72,103]. A transient burst of gene expression is usually observed. The amount of DNA needed to achieve a high level of transfection is very small (ng), and there is practically no limitation to the size of plasmid that can be administered. Since DNA is forced to penetrate the membrane and bypass endosomes, lysosomes, and even the nuclear barrier, there is only minimal degradation of DNA. However, the bombardment of solid particles to the tissue will inevitably cause some tissue damage.

The gene gun has been used in genetic vaccination against tumors. DNA-encoding cytokines, costimulatory molecules, or tumor-associated antigens have been delivered to tumor cells in ex vivo or in vivo approaches. These include IL-2, IL-6, IL-12, TNF-α, IFN-γ, B7–1, GM-CSF, and E7 of HPV-16 [74,104–107]. One outstanding tumor vaccination study is the gene gun–mediated vaccination against cervical carcinoma with naked DNA encoding HPV-16 E7 antigen. Significant reduction of liver and lung metastases was obtained in the treatment group [74]. Vaccination in larger animals such as the dog, pig, horse and monkey is also effective in eliciting humoral and cellular immunity. Besides the vaccination approach, gene gun delivery of a therapeutic gene is also effective in the eradication of tumors. Skin transfection of IL-12 by gene gun resulted in regression of established primary and metastatic tumors in mice [108–110]. Besides IL-12, gene gun transfection with other cytokine genes such as GM-CSF, IL-6, TNF-α, and IFN-γ are also effective in inhibiting tumor growth in vivo [111].

C. Lipidic Vectors

Lipidic vectors such as liposomes and micelles have been used as drug carriers for many years. They are known to be very safe and easy to produce in a large scale. Ligands could be easily conjugated chemically for targeted delivery. Prolonged circulation could be achieved by the addition of polyethylene glycol (PEG). All these factors make lipidic vectors very attractive for DNA delivery. Currently, several liposomal formulations have been used as DNA carriers in clinical trials for cystic fibrosis and cancer [112–114].

Lipoplex. Cationic lipid is the most commonly used lipidic vector for DNA delivery (Fig. 1). It can form DNA/lipid complexes (lipoplexes) with DNA efficiently by charge interaction. The formation of some liposomes requires the addition of a helper lipid, which can stabilize the structure of liposomes. Complex formation with lipids could protect DNA from attack by nucleases. The size of these self-assembled complexes is usually quite large, as DNA cannot be efficiently condensed by lipid alone, especially when monovalent cationic lipid is used. This large size of lipoplex imposes problems for systemic delivery of DNA. The problem can be resolved by adding polycationic polymer such as polylysine or protamine. This ternary complex comprises DNA/cationic lipid/cationic polymer is known as a lipopolyplex. A lipopolyplex has smaller size, improved homogeneity, and transfection efficiency. For instance, i.v. administration of polycation-condensed DNA (LPD) resulted in efficient gene transfer to most highly perfused organs, with transfection efficiency in the lung being the highest. Other lipidic vectors such as micelle and reconstituted chylomicron remnant (RCR) have been reported to have good in vivo transfection activity [115].

To achieve a high level of transfection efficiency, a lipidic vector has to overcome several cellular barriers. The complex has to (1) enter into the cell, (2) escape into cytoplasm, and (3) overcome the nuclear barrier to enter into the nucleus for transcription by cellular machinery. Although the lipidic vectors still need to be improved, they are by and large the most widely used nonviral vectors in gene therapy.

Entry into Cell. The lipid/DNA complex is believed to bind to the cell surface by charge interaction and is then internalized by endocytosis. The complex can be prepared with a different lipid to DNA ratio, thus resulting in complexes of different net charges. The fact that a neutral complex cannot efficiently transfect cells is believed to be due to poor charge interaction with the negatively charged cell membrane. The negative charge on cell membrane is mainly contributed by proteoglycans. Therefore, a positively charged complex can readily bind to the cell membrane by charge interaction. Indeed, cells with higher negative charge density on the cell surface can be transfected better in vitro [116,117].

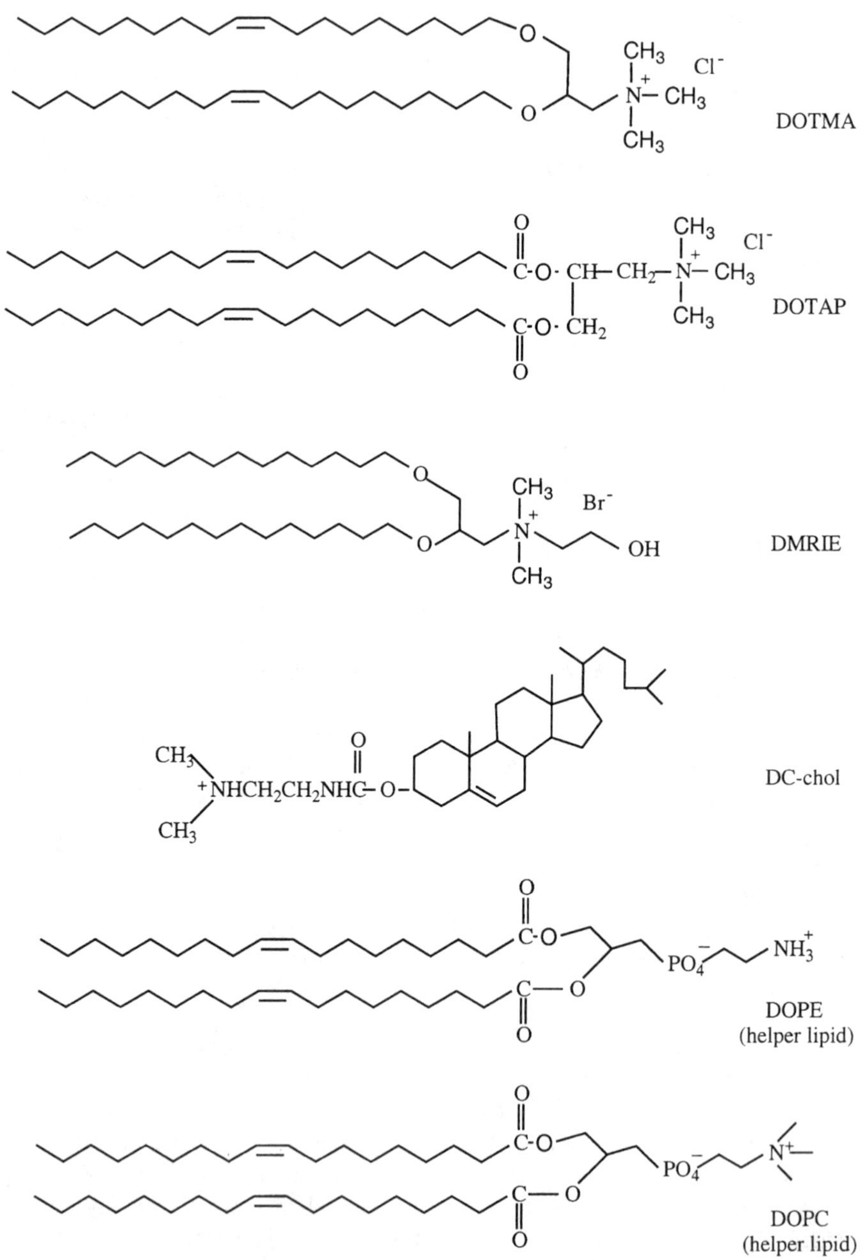

Figure 1 Structures of some common cationic lipids and helper lipids used for gene delivery.

However, surprising discrepancies do occur between in vitro an
tion conditions. Interaction with serum components greatly cha
bility, and transfection activity of the complex [118]. Therefore
lipid composition becomes very important. Li et al. found that the use of choles
terol instead of DOPE as the helper lipid in a DOTAP formulation would result
in higher stability in serum, thus leading to higher in vivo transfection efficiency.

Escape into Cytoplasm. Upon endocytosis, the complex is found to be
located in the endosomal compartment. Therefore, in order to enter into the nucleus, either the DNA or the complex has to escape from the endosome into the
cytoplasm first. Otherwise, the DNA in the endosome will be degraded by lysosomal nucleases. Szoka et al. has proposed a model to explain the release mechanism of the cationic liposome/DNA complex [119]. Anionic lipids are usually
located in the cytoplasmic monolayer of the cell membrane. When the positively
charged complex is endocytosed, it destabilizes the endosomal membrane and
results in the flip-flop of anionic phospholipids from the cytoplasmic monolayer.
The anionic lipids diffuse into the complex and form a charge neutral ion pair
with the cationic lipids in the complex. The anionic lipids thus compete with
DNA for interaction with cationic lipids and results in displacement and release
of DNA into the cytoplasm.

It has been shown that the addition of certain helper lipids into the complex
could facilitate endosomal escape. Besides helper lipids, fusogenic peptide is
known to confer high endosomolytic activity (i.e., lysis of endosome) to the complex, thereby enhancing the release of DNA into the cytoplasm. Fusogenic peptides form a random coil at physiological pH. At low pH, they change into an
α-helix structure and introduce pores in the endosomal membrane, thus resulting
in endosomal lysis. Indeed, we found that the addition of a small amount of
fusogenic peptide usually could enhance transfection efficiency to about 10-fold
or more.

Entry into Nucleus. Plasmid DNA entrance into the nucleus is a rate-
limiting step. The small size of the nuclear pore (~ 55 Å) allows free diffusion
of small molecules (≤ 40 kD) such as small ODNs. For large molecules, they
have to enter through the nuclear pore complex (NPC) by a facilitated process. It has been shown that DNA of size larger than 1 kb can easily enter into
the nucleus. Proteins carrying a nuclear localization signal (NLS) can bind to
and then be transported through the NPC. Direct conjugation of NLS peptide to
DNA, however, did not significantly enhance DNA uptake into the nucleus
[116].

Several studies showed that gene transfer by a lipoplex is highly dependent
on the cell cycle stage of the cell [121]. Lipofection performed at S or G2 phase
resulted in expression 30- or 500-fold higher compared to transfection at G1
phase. On the other hand, transfection efficiency of adenoviral vectors does not

depend on the cell cycle stages of the cells. As the nuclear barrier disappears at mitosis, synchronization or induction of mitosis may increase nuclear uptake of DNA. Another alternative is to promote and allow transcription in the cytoplasm instead of in the nucleus. A T7 cytoplasmic expression system was designed for this purpose [122,123].

D. Polymer

Similar to cationic lipidic vectors, polycationic polymers interact with DNA by charge interaction. The complex formed by polycationic polymer with DNA is called a polyplex. Various polycationic polymers have been explored for DNA delivery. Polycationic polymer could enhance cellular uptake of DNA and compact DNA into very small particles (< 100 nm) owing to its high density of positive charge. Moreover, targeting ligands could be covalently conjugated with the polymer for targeted delivery of DNA. The small particle size and the protection from attack by nucleases greatly facilitate the systemic delivery of DNA and extravasation to the target tissue. Similar to a lipidic vector, a polyplex has to overcome many barriers, and the strategies used to improve polyplex-mediated transfection are rather similar to that of lipofection and will not be discussed in details.

The most common polymers for DNA delivery include poly-L-lysine (PLL), protamine, polyethylenimine (PEI), dendrimer, and poly(L-lysine) graft–hyaluronic acid copolymer (PLL-graft-HA) [37,124–126] (Fig. 2). The transfection efficiency of these polymers is quite high, with PEI being one of the best. Recently, various biocompatible controlled-release polymers have been introduced for DNA delivery, including poly(D,L-lactide-co-glycolide) (PLGA) [127], gelatin, and chitosan [128]. However, the transfection efficiency both in vitro and in vivo is not very high.

PEI is a water-soluble polymer with ethylamine as the repeating units. The high cationic charge density is contributed by amino nitrogen, which appears at every third atom in the polymer. It is available in both linear and branched forms with various molecular weights. The in vitro transfection efficiencies of linear and branched PEI are comparable. The molecular weight has been shown to affect transfection efficiency. It seems that low molecular weight PEI with a low degree of branching has higher transfection activity [129]. The unusually high transfection efficiency is believed to be due to the "proton sponge effect" [126]. The high buffering capacity of PEI results in massive proton accumulation in the endosome accompanied by osmotic swelling and subsequent endosome disruption. Therefore, the endosomal escape of the PEI/DNA complex is very efficient. Moreover, the PEI/DNA complex has been shown to have an efficient nuclear entry with an unknown mechanism [126].

Tumor targeting of polyplexes (mostly with PLL) have been demonstrated in a number of studies [130]. Conjugation of transferrin, bFGF, folate, anti-

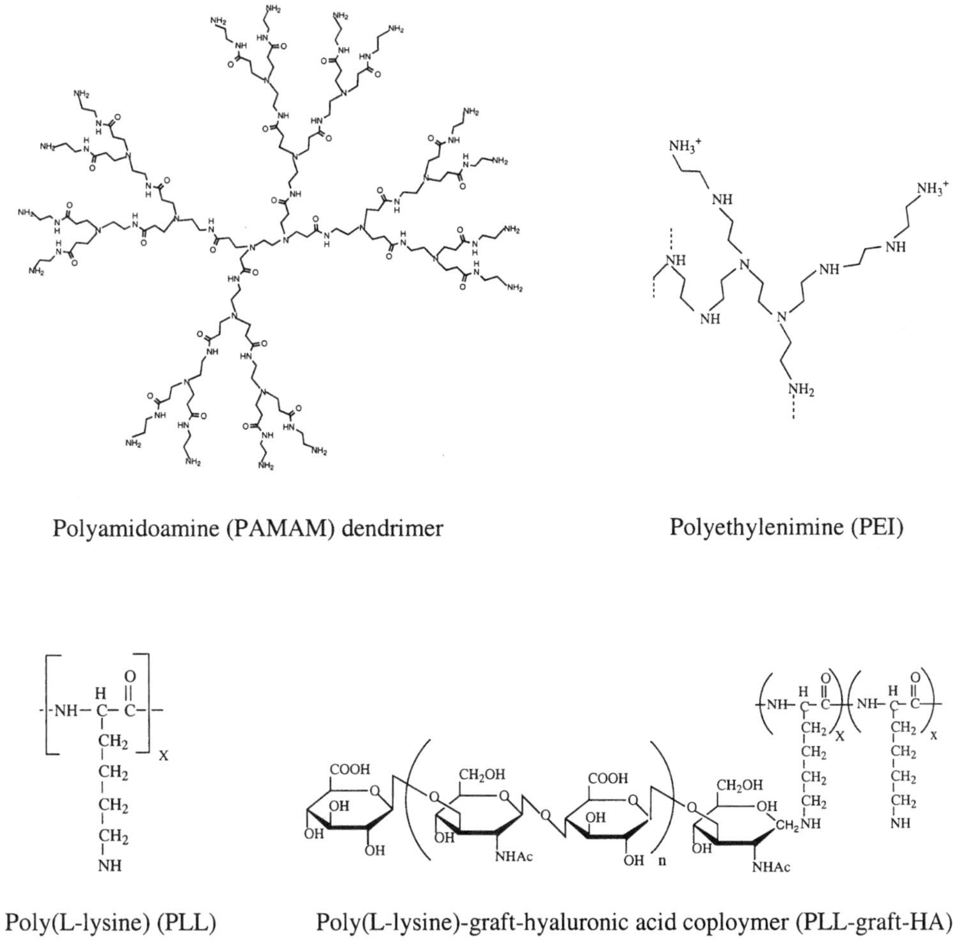

Polyamidoamine (PAMAM) dendrimer   Polyethylenimine (PEI)

Poly(L-lysine) (PLL)   Poly(L-lysine)-graft-hyaluronic acid coploymer (PLL-graft-HA)

Figure 2 Structures of some common polymers for gene delivery.

EGFR, and anti–*erbB*-2 antibodies enhance DNA uptake into tumor cells in vitro. Local injection of DNA/PEI–transferrin into s.c. tumors showed that the conjugated complex was 10- to 100-fold more efficient than naked DNA [130,131]. However, most of the time, ligand conjugation enhances the specificity of DNA delivery with compromised transfection efficiency in vivo. This is probably due to the alteration of polyplex characteristics by the ligand. Therefore, more effort is needed to improve the efficiency of targeting for higher gene expression in the tumor.

V. POTENTIAL PROBLEMS OF NONVIRAL VECTORS

In order to be more effective for clinical evaluation of therapeutic genes, nonviral vectors would still require some improvements. These include (1) further enhancing transfection efficiency, (2) prolonging gene expression, (3) specific targeting, and (4) further minimizing toxicity. Some potential strategies for nonviral vector improvement regarding these concerns are discussed.

A. Improving Transfection Efficiency

Most nonviral vectors are less efficient than viruses for gene transfer. In order to elicit cytotoxicity for complete eradication of cancer cells, high tranfection efficiency is definitely required. The only exception is the delivery of tumor antigen for vaccination. Improvements can be achieved at the level of both the delivery method and the DNA.

Repeated administration is one of the most common practices to improve transfection efficiency of nonviral vectors. Unlike some viral vectors such as adenovirus and retrovirus, the low toxicity profile of most nonviral vectors allows this to be done.

As mentioned before, the delivery of naked DNA by hydrodynamics-based gene transfer is one of the major breakthroughs. The fact that the achievable level of gene expression could be as high as that of viral vector is exciting. Transient, but reversible, damage associated with this method has been reported. Although difficulties exist with its translation to the clinical settings in its present form, several recent reports demonstrated improvement of the technique, such that clinical application may be feasible (97,98). A better understanding of the mechanism behind this gene transfer method is likely to allow further improvements.

Another strategy to improve transfection efficiency is to prevent the degradation of DNA in biological fluids. Especially for systemic delivery, various methods have been introduced to protect DNA from nuclease attack to allow a higher bioavailability to the target. Most lipidic and polymeric vectors can provide some sort of protection for DNA. Moreover, incorporation of PEG on the surface of the lipidic vector could not only protect the encapsulated DNA from degradation but also prolong the circulation time of the lipoplex. Therefore, more cells could be exposed to the vector.

Manipulation of the DNA molecule could also be performed. The finding that herpes simplex type 1 (HSV-1) VP22 protein can spread from cell to cell has given new insights for nonviral vector development [132,133]. HSV–VP22 is a tegument protein of HSV particles with high affinity for chromatin [134]. It has the ability to jump from cell to cell and accumulate in the nucleus of neighboring cells. Phelan et al. demonstrated that the expression of a p53–VP22 chimeric protein in a cell could result in the accumulation in neighboring cell and exert cytotoxic effects in p53-negative human osteosarcoma cells [133]. The same group also reported the intercellular spreading of a functionally active chimeric protein, VP22–TK, upon gene delivery [135]. These studies validate the use of HSV–VP22 for enhancing transfection efficiency in cancer gene therapy.

As discussed before, increasing the amount of DNA entering into nucleus should improve the transfection efficiency. However, direct conjugation of a nuclear localization signal peptide to DNA did not greatly enhance nuclear entry. Another approach is to induce mitosis of the cell to allow more DNA to enter into the nucleus.

B. Prolonging Gene Expression

Prolonging the expression of a therapeutic gene in tumor cell or tumor vasculature (in the case of antiangiogenic approach) would allow an enhanced therapeutic effect. Unlike gene therapy for inherited diseases, long-lasting gene expression in cancer is usually not required for cancer gene therapy; however, a longer gene expression may allow the therapeutic gene to exert an improved therapeutic effect.

The expression of a gene could be prolonged by the insertion of an integration sequence. Insertion of a long terminal repeat (LTR) sequence of AAV into DNA has resulted in longer gene expression. Another new strategy is the incorporation of a transposon into the vector to achieve stable chromosomal integration of the gene. Transposons are naturally occurring genetic elements capable of moving from one chromosomal location to another. The use of naked plasmid DNA with a transposon element has resulted in prolonged gene expression (up to 6 months) in vivo. The fact that the resultant transposition efficiency is highly comparable with many integrating viral vectors is very exciting. However, one concern is the possible genomic alteration in the transfected cell. Since tumor cells are the primary targets in cancer gene therapy, as long as the cytotoxic effect results in killing of the tumor cell, genomic integration should not be a main issue.

Alternatively, controlled release of DNA from polymer could prolong gene expression by releasing DNA to the host in an extended period of time.

C. Specific Targeting

Specific targeting to tumor or tumor vasculature could be achieved by the incorporation of a tumor-specific receptor or ligand into the vector. Transferrin is one of the most common examples. Other tumor-specific ligands include folate, bFGF, anti–erbB-2 antibody, and anti-EGFR antibody [130].

Another alternative is the use of tumor-specific promoters for the restriction of gene expression in cancer cells, thus minimizing the side effects on normal cells [137,138]. Many promoters are known to be constitutively active in cancer cells, including the PSA, erbB-2, CEA, α-fetoprotein, muc-1, α-lactalbumin, hexokinase II, and others (Table 5) [80,138–146]. In addition, there are also known promoters specific to tumor endothelium. Therefore, targeting of anti-angiogenic genes can be achieved with higher selectivity. There are also treatment-responsive promoters such as mdr-1 and erg-1 promoters, and their activities could be induced by chemotherapeutic drug or radiation treatment. Therefore, they are potentially useful for the treatment of drug-resistant or radiation-resistant cancer cells. In addition, cell cycle–regulated promoters such as E2F-1 and cyclin A promoters could also be used. Because tumor cells are highly proliferative, their cell cycle–regulated promoters are likely to be more active than those of normal cells. This molecular approach is very attractive; however, most of these promoters have a basal level of leakiness, which means they could also have promoter activity in normal cells. Therefore, in order to increase the targeting specificity, further improvement is needed either to reduce the background activity or to direct better the gene carrier to the tumor cells. Another problem is the reduced strength of the tumor-specific promoters. Many such promoters are about one-tenth or lower in activity as compared to the CMV promoter and often are insufficient for therapeutic purposes.

D. Minimizing Toxicity

High immunogenicity is the main contributor of toxicity associated with viral vectors. Although partial or complete removal of viral protein could result in lower antigenicity, those viral vectors usually suffer from compromised activity or low production titer. On the other hand, nonviral vectors are known to be much less immunogenic and thus safer to use.

Although being safer, nonviral vectors are not without toxicity. Recently, several groups reported that in vivo administration of nonviral vectors could enhance the nonspecific inflammatory response induced by DNA [147–149]. Plasmid DNA obtained from bacteria, as well as viral DNA, contains an unmethylated "CpG motif," which is immunostimulatory [150]. Certain CpG sequences are more potent than others. The most potent immunostimulatory CpG (CpG-S) has the conserved sequence of RRCGYY (CpG is flanked by two purines in the 5′ side and two pyrimidines in the 3′ side). When B cells, monocytes, macrophages, and DCs take up DNA containing CpG-S, they are stimulated to produce proinflammatory cytokines such as IL-6, IL-12, IFN-γ, and TNF-α. Moreover, cell surface costimulatory molecules such as class II MHC, B7–1, B7–2, and intercellular adhesion molecule-1 (ICAM-1) are upregulated, which may enhance antigen presentation [150,151]. The result is the activation of both innate and acquired immune responses.

Table 5 Summary of Target-Specific Promoters That Could Be Used for Therapeutic Gene Expression in Target Tissues

Promoter	Cancer type	Promoter	Target specificity
Tumor-specific		**Inducible or treatment-responsive**	
erbB-2	Breast and ovarian cancer	mdr-1	Chemotherapy-induced
Osteocalcin	Osteosarcoma and others	Metallothionein	Cisplatin-resistant tumor
Ornithine decarboxylase	Tumor overexpresses n-myc	Thyroid hormone T3	Thyroid hormone T3
Nestin	Glioma and glioblastoma	VCAM-1	Cytokine (TNF-α)
HPV	HPV-positive cancer	erg-1	Radiation-induced
Telomerase	Cancer	HSP70	Heat-induced
PSA	Prostate cancer	HRE	Hypoxic solid tumor
CEA	Many adenocarcinomas		
AFP	Liver cancer	**Tumor endothelium specific**	
muc-1	Breast cancer	KDR	Tumor endothelium
α-Lactalbumin	Breast cancer	E-selectin	Tumor endothelium
β-Lactoglobulin	Breast cancer	Endoglin	Tumor endothelium
SLP1	Ovarian and cervical cancer		
grp78	Tumor with glucose starvation	**Cell cycle specific**	
L-plastin	Cancer	E2F-1	Proliferating cell
Hexokinase II	Cancer	Cyclin A	Proliferating cell

HPV, human papillomavirus; PSA, prostate-specific antigen; CEA, carcinoembryonic antigen; AFP, alpha-fetoprotein; muc-1, mucin-1; SLP1, secretory leukoprotease inhibitor; HRE, hypoxia response element; grp78, glucose-starvation inducible promoter; VCAM-1, vascular cell adhesion molecule-1; KDR, kinase insert domain containing receptor; mdr-1, multiple drug resistance; erg-1, early growth response-1 gene; hsp 70, heat shock protein 70.
Source: Data partially taken from Ref. 138.

Nevertheless, the immunostimulatory effects of DNA may actually be beneficial in cancer gene therapy, provided that it is tolerated by the patient. Indeed, it has been demonstrated that when an empty plasmid is complexed with a cationic lipid and administered intravenously, significant antitumor activity is observed in a lung metastasis model [147,149]. However, if the dose is too high, lethal toxicity could occur. Therefore, there is still a need to minimize toxicity associated with CpG-S.

Recently, Cheng et al. eliminated 270 of 526 CpG dinucleotides from a reporter plasmid. The inflammatory response of the lipoplex made from this plasmid was greatly reduced, whereas the level of gene expression was not affected [152]. However, if CpG is in the coding sequence, mutation may alter the activity of the encoded protein. Alternatively, using polymerase chain reaction (PCR) fragment carrying the coding sequence can avoid the immunostimulation resulting from unnecessary sequencing of the plasmid. Another strategy is the introduction of immune neutralizing motifs (CpG-N). It is known that the mutation of a CG to GC sequence would eliminate or oppose the immunostimulatory activity of CpG motif. Therefore, codelivery of CpG-N with the DNA may minimize the nonspecific immune responses. Besides, the nonspecific immune response could also be suppressed by the administration of an immunosuppressive drug [152,153].

Another undesirable effect contributed by the nonspecific stimulation of the immune response is the shutting down of gene expression, which may prohibit effective repeated administrations. We observed a nonresponsive refractory period of about 10 days or longer after the administration of LPD. The reason for this gene-silencing effect is not clear but proinflammatory cytokines seemed to be involved [154]. In fact, pretreatment with an immunosuppressive drug could result in prolonged gene expression and shorter refractory period for LPD [153]. This gene-silencing effect is more serious with many viral vectors. A better understanding of the immunostimulatory effect of CpG and the gene-silencing effect on the transgene would lead to further improvement of both viral and nonviral vectors.

VI. SUMMARY

The use of nonviral vectors for cancer gene therapy is promising. The improvement of current nonviral vectors would allow better evaluation of the therapeutic effect of a candidate gene. Combination of different therapeutic strategies has shown some success in improving the clinical outcome. However, interaction of the delivered genes or the vector with the host cannot be overlooked. A better understanding of the host–vector interaction, the gene-silencing effect, and the gene transfer mechanism of the existing vectors would certainly contribute to the development of impoved and, hopefully, versatile vectors for cancer gene therapy.

REFERENCES

1. Marshell E. Gene therapy death prompts review of adenovirus vector. Science 1999; 26:2244–2245.
2. Lai SL, Perng RP, Hwang J. p53 Gene status modulates the chemosensitivity of non-small cell lung cancer cells. J Biomed Sci 2000; 7:64–70.
3. Martinet O, Ermekova V, Qiao JQ, Sauter B, Mandeli J, Chen L, Chen SH. Immunomodulatory gene Therapy with interleukin 12 and 4–1BB ligand: long-term remission of liver metastates in a mouse model. J Natl Cancer Inst 2000; 92:931–936.
4. Myers KJ, Dean NM. Sensible use of antisense: how to use oligonucleotides as research tools. Trends Pharmacol Sci 2000; 21:19–23.
5. Koller E, Gaarde WA, Monia BP. Elucidating cell signaling mechanisms using antisense technology. Trends Pharmacol Sci 2000; 21:142–148.
6. He YK, Lui VW, Baar J, Wang L, Shurin M, Almonte C, Watkins SC, Huang L. Potentiation of E7 antisense RNA-induced antitumor immunity by co-delivery of IL-12 gene in HPV16 DNA-positive mouse tumor. Gene Ther 1998; 5:1462–1471.
7. Qian C, Drozdzik M, Caselmann WH, Prieto J. The potential of gene therapy in the treatment of hepatocellular carcinoma. J Hepatol 2000; 32:344–351.
8. Bost F, McKay R, Dean N, Mercola D. The JUN kinase/stress–activated protein kinase pathway is required for epidermal growth factor stimulation of growth of human A549 lung carcinoma cells. J Biol Chem 1997; 272:33422–33429.
9. Nakamura K, Hongo A, Kodama J, Miyagi Y, Yoshinouchi M, Kudo T. Down-regulation of the insulin-like growth factor I receptor by antisense RNA can reverse the transformed phenotype of human cervical cancer cell lines. Cancer Res 2000; 60:760–765.
10. Wojtowicz-Praga S. Reversal of tumor-induced immunosuppression: a new approach to cancer therapy. J Immunother 1997; 20:165–177.
11. Flanagan WM. Antisense comes of age. Cancer Metastasis Rev 1998; 17:169–176.
12. Milner N, Mir KU, Southern EM. Selecting effective antisense reagents on combinatorial oligonucleotide arrays. Nat Biotechnol 1997; 15:537–541.
13. Ho SP, Bao Y, Lesher T, Malhotra R, Ma LY, Fluharty SJ, Sakai RR. Mapping of RNA accessible sites for antisense experiments with oligonucleotide libraries. Nat Biotechnol 1998; 16:59–63.
14. Lui VW, He Y, Huang L. Specific down-regulation of HER-2/neu mediated by a chimeric U6 hammerhead ribozyme results in growth inhibition of human ovarian carcinoma. Mol Ther 2001; 3(2):169–177.
15. Hsieh SS, Malerczyk C, Aigner A, Czubayko F. ERbB-2 expression is, rate-limiting for epidermal growth factor-mediated stimulation of ovarian cancer cell proliferation. Int J Cancer 2000; 86:644–651.
16. Tang CK, Concepcion XZ, Milan M, Gong X, Montgomery E, Lippman ME. Ribo-

zyme-mediated down-regulation of ErbB-4 in estrogen receptor- positive breast cancer cells inhibits proliferation both in vitro and in vivo. Cancer Res 1999; 59: 5315–5322.

17. Funato T, Ishii T, Kambe M, Scanlon KJ, Sasaki T. Anti-K-ras ribozyme induces growth inhibition and increased chemosensitivity in human colon cancer cells. Cancer Gene Ther 2000; 7:495–500.

18. Halatsch ME, Schmidt U, Botefur IC, Holland JF, Ohnuma T. Marked inhibition of glioblastoma target cell tumorigenicity in vitro by retrovirus-mediated transfer of a hairpin ribozyme against deletion-mutant epidermal growth factor receptor messenger RNA. J Neurosurg 2000; 92:297–305.

19. Leirdal M, Sioud M. Ribozyme inhibition of the protein kinase C alpha triggers apoptosis in glioma cells. Br J Cancer 1999; 80:1558–1564.

20. Rossi JJ. Ribozymes to the rescue: repairing genetically defective mRNAs. Trends Genet 1998; 14:295–295.

21. Nielsen PE, Egholm M. An introduciton to PNA. In: Nielsen PE, Egholm M, eds. Peptide Nucleic Acids: Protocols and Applications. Wymondhan, UK: Horizon Scientific Press, 1999, p 1–19.

22. Raney KD, Hamilton SE, Corey DR. Interactions of peptide nucleic acids with DNA processing enzymes. In: Nielsen PE, Egholm M, eds. Peptide Nucleic Acids: Protocols and Applications. Wyomondham, UK: Horizon Scientific Press, 1999; pp 241–251.

23. Good L, Nielsen PE. Antisense effects in Escherichia coli. In: Nielsen PE, Egholm M,eds. Peptide Nucleic Acids: Protocols and Applications. Wymondham, UK: Horizon Scientific Press, 1999, pp 213–220.

24. Shammas MA, Simmons CG, Corey DR, Reis RJ. Telomerase inhibition by peptide nucleic acids reverses 'immortality' of transformed human cells. Oncogene 1999; 18:6191–6200.

25. Wang G, Seidman MM, Glazer PM. Mutagenesis in mammalian cells induced by triple helix formation and transcription-coupled repair. Science 1996; 271:802–805.

26. Vasquez KM, Wang G, Havre PA, Glazer PM. Chromosomal mutations induced by triplex-forming oligonucleotides in mammalian cells. Nucleic Acids Res 1999; 27:1176–1181.

27. Postel EH, Flint SJ, Kessler DJ, Hogan ME. Evidence that a triplex-forming oligodeoxyribonucleotide binds to the c- myc promoter in HeLa cells, thereby reducing c-myc mRNA levels. Proc Natl Acad Sci USA 1991; 88:8227–8231.

28. Barre FX, Ait-Si-Ali S, Giovannangeli C, Luis R, Robin P, Pritchard LL, Helene C, Harel-Bellan A. Unambiguous demonstration of triple-helix-directed gene modification. Proc Natl Acad Sci USA 2000; 97:3084–3088.

29. Giovannangeli C, Perrouault L, Escude C, Nguyen T, Helene C. Specific inhibition of in vitro transcription elongation by triplex- forming oligonucleotide-intercalator conjugates targeted to HIV proviral DNA. Biochemistry 1996; 35:10539–10548.

30. Klysik J, Kinsey BM, Hua P, Glass GA, Orson FM. A 15-base acridine conjugated oligodeoxynucleotide forms triplex DNA with its IL-2R alpha promoter target with greatly improved avidity. Bioconjug Chem 1997; 8:318–326.

31. Gee JE, Yen RL, Hung M-C, Hogan ME. Triplex formation at the rat neu oncogene promoter. Gene 1994; 149:109–114.

32. Gomez-Navarro J, Curiel DT, Douglas JT. Gene therapy for cancer. Eur J Cancer 1999; 35:2039–2057.

33. Piche A, Rancourt C. A role for intracellular immunization in chemosensitization of tumor cells? Gene Ther 1999; 6:1202–1209.

34. Hung MC, Wang SC, Hortobagyi G. Targeting HER-2/neu-overexpressing cancer cells with transcriptional repressor genes delivered by cationic liposome. In: Huang L, Hung MC, Wagner E, eds. Nonviral Vectors for Gene Therapy. San Diego Academic Press, 1999, pp 357–377.

35. Yu D, Hung M-C. The erbB2 gene as a cancer therapeutic target and the tumor- and metastasis-supressing function of E1A. Cancer Metastasis Rev 1998; 17:195–202.

36. Lowe SW, Bodis S, McClatchey A, Remington L, Ruley HE, Fisher DE, Housman DE, Jacks T. p53 status and the efficacy of cancer therapy in vivo. Science 1994; 266:807–810.

37. Kouraklis G. Progress in cancer gene therapy. Acta Oncol 1999; 38:675–683.

38. Clayman GL, el-Naggar AK, Lippman SM, Henderson YC, Frederick M, Merritt JA, Zumstein LA, Timmons TM, Liu TJ, Ginsberg L, Roth JA, Hong WK, Bruso P, Goepfert H. Adenovirus-mediated p53 gene transfer in patients with advanced recurrent head and neck squamous cell carcinoma. J Clin Oncol 1998; 16:2221–2232.

39. Roth JA, Swisher SG, Merritt JA, Lawrence DD, Kemp BL, Carrasco CH, El-Naggar AK, Fossella FV, Glisson BS, Hong WK, Khurl FR, Kurie JM, Nesbitt JC, Pisters K, Putnam JB, Schrump DS, Shin DM, Walsh GL. Gene therapy for non-small cell lung cancer: a preliminary report of a phase I trial of adenoviral p53 gene replacement. Semin Oncol 1998; 25:33–37.

40. Gjertsen BT, Logothetis CL, McDonnell TJ. Molecular regulation of cell death and therapeutic strategies for cell death induction in prostate carcinoma. Cancer Metastasis Rev 1999; 17:345–351.

41. Chang FL, Ling YF, Lai MD. Exogenous mutant p53 DNA enhanced cisplatin-induced apoptosis in TSGH- 8301 human bladder cancer cells. Anticancer Res 2000; 20:329–336.

42. Sandig V, Brand K, Herwig S, Lukas J, Bartek J, Strauss M. Adenovirally transferred p16INK4/CDKN2 and p53 genes cooperate to induce apoptotic tumor cell death. Nat Med 1997; 3:313–319.

43. Yamabe K, Shimizu S, Ito T, Yoshioka Y, Nomura M, Narita M, Saito I, Kanegae Y, Matsuda H. Cancer gene therapy using a pro-apoptosic gene, caspase-3. Gene Ther 1999; 6:1952–1959.

44. Tsao YP, Huang SJ, Chang JL, Hsieh JT, Pong RC, Chen SL. Adenovirus-mediated p21((WAF1/SDII/CIP1)) gene transfer induces apoptosis of human cervical cancer cell lines. J Virol 1999; 73:4983–4890.

45. Webb A, Cunningham D, Cotter F, Clarke PA, di Stefano F, Ross P, Corbo M, Dziewanowska Z. BCL-2 antisense therapy in patients with non-Hodgkin lymphoma. Lancet 1997; 349:1137–1141.

46. Dorai T, Olsson CA, Katz AE, Buttyan R. Development of a hammerhead ribozyme against bcl-2. I. Preliminary evaluation of a potential gene therapeutic agent for hormone refractory human prostate cancer. Prostate 1997; 32:246–258.

47. Piche A, Grim J, Rancourt C, Gomez-Navarro J, Reed JC, Curiel DT. Modulation of Bcl-2 protein levels by an intracellular anti-Bcl-2 single-chain antibody increases drug-induced cytotoxicity in the breast cancer cell line MCF-7. Cancer Res 1998; 58:2134–2140.

48. Folkman J, Cotran R. Relation of vascular proliferation to tumor growth. Int Rev Exp Pathol 1976; 16:207–248.

49. Taniguchi T, Rigg A, Lemoine NR. Targeting angiogenesis: genetic intervention which strikes at the weak link of tumorigenesis [editorial]. Gene Ther 1998; 5: 1011–1013.

50. Oku T, Tjuvajev JG, Miyagawa T, Sasajima T, Joshi A, Joshi R, Finn R, Claffey KP, Blasberg RG. Tumor growth modulation by sense and antisense vascular endothelial growth factor gene expression: effects on angiogenesis, vascular permeability, blood volume, blood flow, fluorodeoxyglucose uptake, proliferation of human melanoma intracerebral xenografts. Cancer Res 1998; 58:4185–4192.

51. Pavco PA, Bouhana KS, Gallegos AM, Agrawal A, Blanchard KS, Grimm SL, Jensen KL, Andrews LE, Wincott FE, Pitot PA, Tressler RJ, Cushman C, Reynolds MA, Parry TJ. Antitumor and antimetastatic activity of ribozymes targeting the messenger RNA of vascular endothelial growth factor receptors. Clin Cancer Res 2000; 6:2094–2103.

52. Sacco MG, Caniatti M, Cato EM, Frattini A, Chiesa G, Ceruti R, Adorni F, Zecca L, Scanziani E, Vezzoni P. Liposome-delivered angiostatin strongly inhibits tumor growth and metastatization in a transgenic model of spontaneous breast cancer. Cancer Res 2000; 60:2660–2665.

53. Cao Y, O'Reilly MS, Marshall B, Flynn E, Ji RW, Folkman J. Expression of angiostatin cDNA in a murine fibrosarcoma suppresses primary tumor growth and produces long-term dormancy of metastases [published erratum appears J Clin Invest 1998 Dec 1; 102(11):2031]. J Clin Invest 1998; 101:1055–1063.

54. Kanitakis J, Euvrard S, Bourchany D, Faure M, Claudy A. Expression of the nm23 metastasis-suppressor gene product in skin tumors. J Cutan Pathol 1997; 24:151–156.

55. Kim JW, Cho HS, Kim JH, Hur SY, Kim TE, Lee JM, Kim IK, Namkoong SE. AAC-11 overexpression induces invasion and protects cervical cancer cells from apoptosis. Lab Invest 2000; 80:587–594.

56. Nabel GJ, Nabel EG, Yang ZY, Fox BA, Plautz GE, Gao X, Huang L, Shu S, Gordon D, Chang AE. Direct gene transfer with DNA-liposome complexes in melanoma: expression, biologic activity, and lack of toxicity in humans. Proc Natl Acad Sci USA 1993; 90:11307–11311.

57. Alimonti J, Zhang QJ, Gabathuler R, Reid G, Chen SS, Jefferies WA. TAP expression provides a general method for improving the recognition of malignant cells in vivo. Nat Biotechnol 2000; 18:515–520.

58. Hwu P, Yannelli J, Kriegler M, Anderson WF, Perez C, Chiang Y, Schwarz S, Cowherd R, Delgado C, Mule J, et al. Functional and molecular characterization of tumor-infiltrating lymphocytes transduced with tumor necrosis factor-alpha cDNA for the gene therapy of cancer in humans. J Immunol 1993; 150:4104–4115.

59. Fushimi T, Kojima A, Moore MA, Crystal RG. Macrophage inflammatory protein 3alpha transgene attracts dendritic cells to established murine tumors and suppresses tumor growth. J Clin Invest 2000; 105:1383–1393.

60. Giezeman-Smits KM, Okada H, Brissette-Storkus CS, Villa LA, Attanucci J, Lotze MT, Pollack IF, Bozik ME, Chambers WH. Cytokine gene therapy gliomas: induction of reactive CD4+ T cells by interleukin-4– transferred 9L gliosarcoma is essential for protective immunity. Cancer Res 2000; 60:2449–2457.

61. Fearon ER, Pardoll DM, Itaya T, Golumbek P, Levitsky HI, Simons JW, Karasuyama H, Vogelstein B, Frost P. Interleukin-2 production by tumor cells bypasses T helper function in the generation of an antitumor response. Cell 1990; 60:397–403.

62. Gansbacher B, Bannerji R, Daniels B, Zier K, Cronin K, Gilboa E. Retroviral vector-mediated gamma-interferon gene transfer into tumor cells generates potent and long lasting antitumor immunity. Cancer Res 1990; 50:7820–7825.

63. Porgador A, Bannerji R, Watanabe Y, Feldman M, Gilboa E, Eisenbach L. Antimetastatic vaccination of tumor-bearing mice with two types of IFN-gamma gene-inserted tumor cells. J Immunol 1993; 150:1458–1470.

64. Kim YS, Sonn CH, Bothwell ALM. Tumor cells expressing membrane-bound form of IL-4 induce antitumor immunity. Gene Therapy 2000; 7:837–843.

65. Hirschowitz EA, Naama HA, Evoy D, Lieberman MD, Daly J, Crystal RG. Regional treatment of hepatic micrometastasis by adenovirus vector-mediated delivery of interleukin-2 and interleukin-12 cDNAs to the hepatic parenchyma. Cancer Gene Ther 1999; 6:491–498.

66. Cao X, Wang Q, Ju DW, Tao Q, Wang J. Efficient inducation of local and systemic antitumor immune response by liposome-mediated intratumoral co-transfer of interleukin-2 gene and interleukin-6 gene. J Exp Clin Cancer Res 1999; 18:191–200.

67. Nagai H, Hara I, Horikawa T, Fujii M, Kurimoto M, Kamidono S, Ichihashi M. Antitumor effects on mouse melanoma elicited by local secretion of interleukin-12 and their enhancement by treatment with interleukin-18. Cancer Invest 2000; 18:206–213.

68. Khew-Goodall Y, Gamble JR, Vadas MA. Regulation of adhesion and adhesion molecules in endothelium by transforming growth factor-beta. Currt Top Microbiol Immunol 1993; 184:187–199.

69. Arteaga CL, Carty-Dugger T, Moses HL, Hurd SD, Pietenpol JA. Transforming growth factor beta 1 can induce estrogen-independent tumorigenicity of human breast cancer cells in athymic mice. Cell Growth Differ 1993; 4:193–201.

70. Arteaga CL, Hurd SD, Winnier AR, Johnson MD, Fendly BM, Forbes JT, Anti-transforming growth factor (TGF)-beta antibodies inhibit breast cancer cell tumorigenicity and increase mouse spleen natural killer cell activity. Implications for a possible role of tumor cell/host TGF- beta interactions in human breast cancer progression. J Clin Invest 1993; 92:2569–2576.

71. Fakhrai H, Dorigo O, Shawler DL, Lin H, Mercola D, Black KL, Royston I, Sobol RE. Eradication of established intracranial rat gliomas by transforming growth factor beta antisense gene therapy. Proc Natl Acad Sci USA 1996; 93:2909–2914.

72. Yang NS, Hogge GS, MacEwen EG. Particle-mediated gene delivery applications to canine and other larger animial systems. In: Huang L, Hung MC, Wagner E, eds. Nonviral Vectors for Gene Therapy. San Diego: Academic Press, 1999, pp 171–190.

73. White SA, LoBuglio AF, Arani RB, Pike MJ, Moore SE, Barlow DL, Conry RM. Induction of anti-tumor immunity by intrasplenic administration of a carcinoembryonic antigen DNA vaccine. J Gene Med 2000; 2:135–140.

74. Chen CH, Ji H, Suh KW, Choti MA, Pardoll DM, Wu TC. Gene gun-mediated DNA vaccination induces antitumor immunity against human papillomavirus type 16 E7-expressing murine tumor metastases in the liver and lungs. Gene Ther 1999; 6:1972–1981.

75. Fairbairn LJ, Rafferty JA, Lashford LS. Engineering drug resistance in human cell. Bone Marrow Transplant 2000; 25(Suppl2):S110–S113.

76. Beausejour CM, Le NL, Letourneau S, Cournoyer D, Momparler RL. Coexpression of cytidine deaminase and mutant dihydrofolate reductase by a bicistronic retroviral vector confers resistance to cytosine arabinoside and methotrexate. Hum Gene Ther 1998; 9:2537–2544.

77. Lu JY, Lowe DA, Kennedy MD, Low PS. Folate-targeted enzyme prodrug cancer therapy utilizing penicillin-V amidase and a doxorubicin prodrug. J Drug Target 1999; 7:43–53.

78. Stribbling SM, Friedlos F, Martin J, Davies L, Spooner RA, Marais R, Springer CJ. Regressions of established breast carcinoma xenografts by carboxypeptidase G2 suicide gene therapy and the prodrug CMDA are due to a bystander effect. Hum Gene Ther 2000; 11:285–292.

79. Kawamura K, Tasaki K, Hamada H, Takenaga K, Sakiyama S, Tagawa M. Expression of Escherichia coli uracil phosphoribosyltransferase gene in murine colon carcinoma cells augments the antitumoral effect of 5- fluorouracil and induces protective immunity. Cancer Gene Ther 2000; 7:637–643.

80. Pawlik CA, Iyengar RV, Krull EJ, Mason SE, Khanna R, Harris LC, Potter PM, Danks MK, Guichard SM. Use of the ornithine decarboxylase promoter to ahieve N-MYC–mediated overexpression of a rabbit carboxylesterase to sensitize neuroblastoma cells to CPT-11. Mol Ther 2000; 1:457–463.

81. Encell LP, Landis DM, Loeb LA. Improving enzymes for cancer gene therapy. Nat Biotechnol 1999; 17:143–147.

82. Vrionis FD, Wu JK, Qi P, Waltzman M, Cherington V, Spray DC. The bystander effect exerted by tumor cells expressing the herpes simplex virus thymidine kinase (HSVtk) gene is dependent on connexin expression and cell communication via gap junctions. Gene Ther 1997; 4:577–585.

83. Touraine RL, Ishii-Morita H, Ramsey WJ, Blaese RM. The bystander effect in the HSVtk/ganciclovir system and its relationship to gap junctional communication. Gene Ther 1998; 5:1705–1711.

84. Kunitomi M, Takayama E, Suzuki S, Yasuda T, Tsutsui K, Nagaike K, Hiroi S, Tadakuma T. Selective inhibition of hepatoma cells using diphtheria toxin A under the control of the promoter/enhancer region of the human alpha- fetoprotein gene. Jpn J Cancer Res 2000; 91:343–350.

85. Yoshizato K, Nishi T, Goto T, Dev SB, Takeshima H, Kino T, Tada K, Kimura T, Shiraishi S, Kochi M, Kuratsu JI, Hofmann GA, Ushio Y. Gene delivery with optimized electroporation parameters shows potential for treatment of gliomas. Int J Oncol 2000; 16:899–905.

86. Murayama Y, Tadakuma T, Kunitomi M, Kumai K, Tsutsui K, Yasuda T, Kitajima M. Cell-specific expression of the diphtheria toxin A-chain coding sequence under the control of the upstream region of the human alpha- fetoprotein gene. J Surg Oncol 1999; 70:145–149.

87. Waxman DJ, Chen L, Hecht JE, Jounaidi Y. Cytochrome P450-based cancer gene therapy: recent advances and future prospects. Drug Metab Rev 1999; 31:503–522.

88. Xiang J, Gomez-Navarro J, Arafat W, Liu B, Barker SD, Alvarez RD, Siegal GP, Curiel DT. Pro-apoptotic treatment with an adenovirus encoding Bax enhances the effect of chemotherapy in ovarian cancer. J Gene Med 2000; 2:97–106.

89. Mclvor RS. Gene therapy of genetic diseases and cancer. Pediatric Transplantation 1999; 3:116–121.

90. Budker V, Budker T, Zhang G, Subbotin V, Loomis A, Wolff JA. Hypothesis: naked plasmid DNA is taken up by cells in vivo by a receptor-mediated process. J Gene Med 2000; 2:76–88.

91. Rhodes GH. Immune pathways used in nucleic acid vaccination. In: Huang L, Hung MC, Wagner E, eds. Nonviral Vectors for Gene Therapy. San Diego: Academic Press, 1999, pp 379–408.

92. Raz E, Carson DA, Parker SE, Parr TB, Abai AM, Aichinger G, Gromkowski SH, Singh M, Lew D, Yankauckas MA, et al. Intradermal gene immunization: the possible role of DNA uptake in the induction of cellular immunity to viruses. Proc Natl Acad Sci USA 1994; 91:9519–9523.

93. Fan H, Lin Q, Morrissey GR, Khavari PA. Immunization via hair follicles by topical application of naked DNA to normal skin. Nat Biotechnol 1999; 17:870–872.

94. Liu F, Song Y, Liu D. Hydrodynamics-based transfection in animals by systemic administration of plasmid DNA. Gene Ther 1999; 6:1258–1266.

95. Lui VW, He Y, Falo L, Huang L. Systemic administration of naked DNA encoding interleukin 12 for the treatment of human papillomavirus DNA-positive tumor. Hum Gene Ther 2002; 13(2):177–185.

96. Lui VW, Falo LD Jr, Huang L. Systemic production of IL-12 by naked DNA mediated gene transfer-toxicity and attenuation of transgene expression in vivo. J Gene Med 2001; 3(4):384–393.

97. Liu F, Huang L. Improving plasmid DNA-mediated liver gene transfer by prolonging its retention in the hepatic vasculature. J Gene Med 2001; 3(6):569–576.

98. Zhang G, Vargo D, Budker V, Armstrong N, Knechtle S, Wolff JA. Expression of naked plasmid DNA injected into the afferent and efferent vessels of rodent and dog livers. Hum Gene Ther 1997; 8:1763–1772.

99. Budker V, Zhang G, Danko I, Williams P, Wolff J. The efficient expression of intravascularly delivered DNA in rat muscle. Gene Ther 1998; 5:272–276.

100. Coll JL, Negoescu A, Louis N, Sachs L, Tenaud C, Girardot V, Demeinex B, Brambilla E, Brambilla C, Favrot M. Antitumor activity of bax and p53 naked gene transfer in lung cancer: in vitro and in vivo analysis. Hum Gene Ther 1998; 9: 2063–2074.

101. Belehradek M, Domenge C, Luboinski B, Orlowski S, Belehradek J Jr, Mir LM. Electrochemotheraphy, a new antitumor treatment. First clinical phase I-II trial. Cancer 1993; 72:3694–3700.

102. Goto T, Nishi T, Tamura T, Dev SB, Takeshima H, Kochi M, Yoshizato K, Kuratsu J, Sakata T, Hofmann GA, Ushio Y. Highly efficient electro-gene therapy of solid tumor by using an expression plasmid for the herpes simplex virus thymidine kinase gene. Proc Natl Acad Sci USA 2000; 97:354–359.

103. Yang N-S, Burkholder J, Roberts B, Martinell B, McCabe D. In vivo and in vitro

gene transfer to mammalian somatic cells by particel bombardment. PNAS 1990; 87:9568–9572.

104. Tuting T, Gambotto A, Baar J, Davis ID Storkus WJ, Zavodny PJ, Narula S, Tahara H, Robbins PD, Lotze MT. Interferon-alpha gene therapy for cancer: retroviral transduction of fibroblasts and particle-mediated transfection of tumor cells are both effective strategies for gene delivery in murine tumor models. Gene Ther 1997; 4: 1053–1060.

105. Albertini MR, Emler CA, Schell K, Tans KJ, King DM, Sheehy MJ. Dual expression of human leukocyte antigen molecules and the B7–1 costimulatory molecule (CD80) on human melanoma cells after particle mediated gene transfer. Cancer Gene Ther 1996; 3:192–201.

106. Hogge GS, Burkholder JK, Culp J, Albertini MR, Dubielzig RR, Yang NS, MacEwen EG. Preclinical development of human granulocyte-macrophage colony-stimulating factor-transfected melanoma cell vaccine using established canine cell lines and normal dogs. Cancer Gene Ther 1999; 6:26–36.

107. Hogge GS, Burkholder JK, Culp J, Albertini MR, Dubielzig RR, Keller ET, Yang NS, MacEwen EG. Development of human granulocyte-macrophage colony-stimulating factor-transfected tumor cell vaccines for the treatment of spontaneous canine cancer. Hum Gene Ther 1998; 9:1851–1861.

108. Rakhmilevich AL, Turner J, Ford MJ, McCabe D, Sun WH, Sondel PM, Grota K, Yang NS. Gene gun-mediated skin transfection with interleukin 12 gene results in regression of established primary and metastatic murine tumors. Proc Natl Acad Sci USA 1996; 93:6291–6296.

109. Wang C, Quevedo ME, Lannutti BJ, Gordon KB, Guo D, Sun W, Paller AS. In vivo gene therapy with interleukin-12 inhibits primary vascular tumor growth and induces apoptosis in a mouse model. J Invest Dermatol 1999; 112:775–781.

110. Oshikawa K, Ishii Y, Hamamoto T, Sugiyama Y, Kitamura S, Kagawa Y. Particle-mediated gene transfer of murine interleukin-12 cDNA suppresses the growth of Lewis lung carcinoma. In Vivo 1999; 13:397–402.

111. Sun WH, Burkholder JK, Sun J, Culp J, Turner J, Lu XG, Pugh TD, Ershler WB, Yang NS. In vivo cytokine gene transfer by gene gun reduces tumor growth in mice. Proc Natl Acad Sci USA 1995; 92:2889–2893.

112. Caplen NJ, Alton EW, Middleton PG, Dorin JR, Stevenson BJ, Gao X, Durham SR, Jeffery PK, Hodson ME, Coutelle C, et al. Liposome-mediated CFTR gene transfer to the nasal epithelium of patients with cystic fibrosis [see comments] published erratum appears in Nat Med 1995 Mar; 1(3):272]. Nat Med 1995; 1:39–46.

113. Nabel GJ, Nabel E, Yang ZY, Fox BA, Plautz G, Gao X, Huang L, Shu S, Gordon D, Chang AE. Direct gene transfer with DNA-liposome complexes in melanoma: Expression, biologic activity, and lack of toxicity in humans. PNAS 1993; 90: 11307–11311.

114. Hui KM, Ang PT, Huang L, Tay SK. Phase I study of immunotheraphy of cutaneous metastases of human carcinoma using allogeneic and xenogeneic MHC-DNA-liposome complexes. Gene Ther 1997; 4:783–790.

115. Hara T, Tan Y, Huang L. In vivo delivery to the liver using reconstituted chylomicron remnants as a novel non-viral vector. PNAS 1997; 94:14547–14552.

116. Mounkes LC, Zhong W, Cipres-Palacin G, Heath TD, Debs RJ. Proteoglycans me-

diated cationic liposome-DNA complex-based gene delivery in vitro and in vivo. J Biol Chem 1998; 273:26164–26170.

117. Mislick KA, Baldeschwieler JD. Evidence for the role of proteoglycans in cation-mediated gene transfer. PNAS 1996; 93:12349–12354.

118. Li S, Rizzo MA, Bhattacharya S, Huang L. Characterization of cationic lipid protamine-DNA (LPD) complexes for intravenous gene delivery. Gene Ther 1998; 5: 930–937.

119. Xu Y, Szoka FC. Mechanism of DNA release from cationic liposome/DNA complexes used in cell transfection. Biochemistry 1996; 35:5616–5623.

120. Sebestyen MG, Ludtke JJ, Bassik MC, Zhang G, Budker V, Lukhtanov EA, Hagstrom JE, Wolff JA. DNA vector chemistry: the covalent attachment of signal peptides to plasmid DNA. Nat Biotechnol 1998; 16:80–85.

121. Brunner S, Sauer T, Carotta S, Cotten M, Saltik M, Wagner E. Cell cycle dependence of gene transfer by lipoplex, polyplex and recombinant adenovirus. Gene Ther 2000; 7:401–407.

122. Brisson M, He Y, Li S, Yang JP, Huang L. A novel T7 RNA polymerase autogene for efficient cytoplasmic expression of target genes. Gene Therapy 1999; 6:263–270.

123. Gao X, Huang L. Cytoplasmic expression of a reporter gene by co-delivery of T7 RNA polymerase and T7 promoter sequence with cationic liposomesc. Nucleic Acids Res 1993; 21:2867–2872.

124. Asayama S, Nogawa M, Takei Y, Akaike T, Maruyama A. Synthesis of novel polyampholyte comb-type copolymers consisting of a poly(L-lysine) backbone and hyaluronic acid side chains for a DNA carrier. Bioconjug Chem 1998; 9:476–481.

125. Li S, Huang L. Nonviral gene therapy: promises and challenges. Gene Ther 2000; 7:31–34.

126. Kichler A, Behr J-P, Erbacher P. Polyethylenimines: a family of potent polymers for nucleic acid delivery. In: Huang L, Hung MC, Wagner E, eds. Nonviral Vectors for Gene Therapy. San Diego: Academic Press, 1999; pp 191–206.

127. Luo D, Saltzman WM. Synthetic DNA delivery systems. Nat Biotechnol 2000; 18: 33–37.

128. Leong KW. Biopolymer-DNA nanospheres. In: Huang L, Hung MC, Wagner E, eds. Nonviral Vectors for Gene Therapy. San Diego: Academic Press, 1999, pp 267–287.

129. Fischer D, Bieber T, Li Y, Elsasser HP, Kissel T. A novel non-viral vector for DNA delivery based on low molecular weight, branched polyethylenimine: effect of molecular weight on transfection efficiency and cytotoxicity. Pharm Res 1999; 16:1273–1279.

130. Wagner E. Ligand-polycation conjugates for receptor-targeted gene transfer. In: Huang L, Hung MC, Wagner E, eds. Nonviral Vectors for Gene Therapy. San Diego: Academic Press, 1999, pp 207–227.

131. Kircheis R, Schuller S, Brunner S, Ogris M, Heider KH, Zauner W, Wagner E. Polycation-based DNA complexes for tumor-targeted gene delivery in vivo. J Gene Med 1999; 1:111–120.

132. Aints A, Dilber MS, Smith CI. Intercellular spread of GFP-VP22. J Gene Med 1999; 1:275–279.

133. Phelan A, Elliott G, O'Hare P. Intercellular delivery of functional p53 by the herpesvirus protein VP22. Nat Biotechnol 1998; 16:440–443.

134. Elliott G, O'Hare P. Cytoplasm-to-nucleus translocation of a herpesvirus tegument protein during cell division. J Virol 2000; 74:2131–2141.

135. Dilber MS, Phelan A, Aints A, Mohamed AJ, Elliott G, Smith CI, O'Hare P. Intercellular delivery of thymidine kinase prodrug activating enzyme by the herpes simplex virus protein, VP22. Gene Ther 1999; 6:12–21.

136. Yant SR, Meuse L, Chiu W, Ivics Z, Izsvak Z, Kay MA. Somatic integration and long-term transgene expression in normal and haemophilic mice using a DNA transposon system. Nat Genet 2000; 25:35–41.

137. Modlich U, Pugh CW, Bicknell R. Increasing endothelial cell specific expression by the use of heterologous hypoxic and cytokine-inducible enhancers. Gene Ther 2000; 7:896–902.

138. Nettelbeck DM, Jerome V, Muller R. Gene therapy: designer promoters for tumour targeting. Trends Genet 2000; 16:174–181.

139. Anderson LM, Krotz S, Weitzman SA, Thimmapaya B. Breast cancer-specific expression of the Candida albicans cytosine deaminase gene using a transcriptional targeting approach. Cancer Gene Ther 2000; 7:845–852.

140. Koeneman KS, Kao C, Ko SC, Yang L, Wada Y, Kallmes DF, Gillenwater JY, Zhau HE, Chung LW, Gardner TA. Osteocalcin-directed gene therapy for prostate-cancer bone metastasis. World J Urol 2000; 18:102–110.

141. Vandier D, Calvez V, Massade L, Gouyette A, Mickley L, Fojo T, Rixe O. Transactivation of the metallothionein promoter in cisplatin-resistant cancer cells: a specific gene therapy strategy. J Natl Cancer Inst 2000; 92:642–647.

142. Modlich U, Pugh CW, Bicknell R. Increasing endothelial cell specific expression by the use of heterologous hypoxic and cytokine-inducible enhancers. Gene Ther 2000; 7:896–902.

143. Shillitoe EJ, Noonan S. Strength and specificity of different gene promoters in oral cancer cells. Oral Oncol 2000; 36:214–220.

144. Koga S, Hirohata S, Kondo Y, Komata T, Takakura M, Inoue M, Kyo S, Kondo S. A novel telomerase-specific gene therapy: gene transfer of caspase-8 utilizing the human telomerase catalytic subunit gene promoter. Hum Gene Ther 2000; 11: 1397–1406.

145. Martin V, Cortes ML, de Felipe P, Farsetti A, Calcaterra NB, Izquierdo M. Cancer gene therapy by thyroid hormone-mediated expression of toxin genes. Cancer Res 2000; 60:3218–3224.

146. Chen X, Zhang D, Dennert G, Hung G, Lee AS. Eradication of murine mammary adenocarcinoma through HSVtk expression directed by the glucose-starvation inducible grp78 promoter. Breast Cancer Res Treat 2000; 59:81–90.

147. Dow SW, Fradkin LG, Liggitt DH, Willson AP, Heath TD, Potter TA. Lipid-DNA complexes induce potent activation of innate immune responses and antitumor activity when administered intravenously. J Immunol 1999; 163:1552–1561.

148. Yew NS, Wang KX, Przybylska M, Bagley RG, Stedman M, Marshall J, Scheule RK, Cheng SH. Contribution of plasmid DNA to inflammation in the lung after administration of cationic lipid:pDNA complexes. Hum Gene Ther 1999; 10:223–234.

149. Whitmore M, Li S, Huang L. LPD lipopolyplex initiates a potent cytokine response and inhibits tumor growth. Gene Ther 1999; 6:1867–1875.
150. Krieg AM. Direct immunologic activities of CpG DNA and implications for gene therapy. J Gene Med 1999; 1:56–63.
151. Cho HJ, Takabayashi K, Cheng PM, Nguyen MD, Corr M, Tuck S, Raz E. Immuno-stimulatory DNA-based vaccines induce cytotoxic lymphocyte activity by a T-helper cell-independent mechanism. Nat Biotechnol 2000; 18:509–514.
152. Yew NS, Zhao H, Wu I-H, Song A, Tousignant JD, Przybylska M, Cheng SH. Reduced inflammatory response to plasmid DNA vectors by elimination and inhibi-tion of immunostimulatory CpG motifs. Molecular Therapy 2000; 1:1–8.
153. Tan Y, Li S, Pitt BR, Huang L. The inhibitory role of CpG immunostimulatory motifs in cationic lipid vector-mediated transgene expression in vivo [see com-ments]. Hum Gene Ther 1999; 10:2153–2161.
154. Li S, Wu SP, Whitmore M, Loeffert EJ, Wang L, Watkins SC, Pitt BR, Huang L. Effect of immune response on gene transfer to the lung via systemic administration of cationic lipidic vectors. Am J Physiol 1999; 276:L796–804

12
Replicating Adenoviral Vectors for Cancer Therapy

Murali Ramachandra, John A. Howe, and G. William Demers
Canji Inc., San Diego, California, U.S.A.

I. INTRODUCTION

Adenoviral vectors are one of the most efficient means of delivering foreign genes into mammalian cells both in cell culture and in vivo [1]. The ease with which adenoviral vectors can be constructed, produced, and purified in large scale have contributed to the wide use of adenoviral vectors in gene therapy studies. Because transient high-level expression of proteins aimed at killing cancer cells and augmentation of the immune response can be achieved with adenoviral vectors, they may be particularly well suited for cancer therapy. Vectors encoding tumor-suppressor genes, immunomodulatory cytokines, or prodrug-activating suicide enzymes have been generated and are being evaluated for use in cancer therapy [2]. However, early clinical studies with replication-deficient adenoviral vectors carrying various therapeutic transgenes have shown that delivery of adequate viral particles within the tumor tissue to treat effectively the majority of cells is challenging [3].

One potential way to overcome the delivery issues is to take advantage of the natural ability of adenoviruses to replicate, lyse infected cells, and reinfect neighboring cells thereby spreading the virus throughout the tumor. In addition, virus released from cells could travel via the circulation or lymphatic drainage to infect tumor cells at distant tumor site(s). Several clinical trials were performed a few decades ago in which replication-competent wild-type viruses, including adenoviruses, were administered to patients with various malignancies [4,5].

Such therapies were not pursued aggressively because of the limited and unpredictable response and the development of more active chemotherapeutic agents. However, there has been a renewed interest in the use of replicating vectors for cancer therapy based on recent clinical results [6–8] and the availability of improved technologies to produce, purify [9], and characterize [10] these vectors [1].

To prevent deleterious effects during therapy with the replicating virus-based approaches, it is highly desirable that viral replication is attenuated in normal cells. Investigators have manipulated the genome with the objective to create selective viruses that are attenuated in normal cells while maintaining viral replication in tumor cells. Interpretation of selectivity is performed using in vitro assays by measuring the response of cells to viral infection, usually measuring cell killing or viral yield [11]. Unfortunately, the lack of a permissive host, other than humans, has limited the evaluation of the effects of replicating adenoviruses on normal cells in vivo. Strategies evaluated to confer selectivity to replicating adenoviruses include deletion of E1B-55K, deletion of the complete E1B region, partial or complete deletion of E1A, and expression of the E1A gene controlled by tumor-specific promoters. Therefore, a brief review of E1 proteins is presented before reviewing specific strategies.

II. FUNCTIONS ENCODED IN ADENOVIRAL E1 REGION

The E1 region encodes two transcription units, E1A and E1B (Fig. 1), that are essential for productive adenoviral infection, and can also act together to bring about the oncogenic transformation of primary cells in vitro [for reviews, see Refs. 12 and 13]. The E1A gene is the first viral gene expressed in the viral life cycle and encodes polypeptides that activate expression of the delayed early viral transcription units. E1A proteins also stimulate infected cells, which are normally differentiated and quiescent, to reenter the cell cycle. The mitogenic ability of E1A helps facilitate an optimal environment for viral DNA replication during infection, and is also required to immortalize primary cells during E1-mediated transformation. Genetic mapping studies have shown that the two related proteins produced by E1A, 243R and 289R, alter or inhibit the function of a number of cellular regulatory proteins via specific binding domains [12]. The 289R protein contains a powerful transcriptional activation domain that stimulates the adenoviral early promoters by facilitating the interaction of transcription factors with basal transcription machinery to activate RNA polymerase II transcription [13]. The amino-terminal region, common to both 243R and 289R, contains functional domains that can induce cell cycle progression through interaction with the retinoblastoma protein (Rb) and p300/CBP family members. Interestingly, E1A binding to either Rb or p300/CBP is sufficient to induce cellular DNA synthesis, but

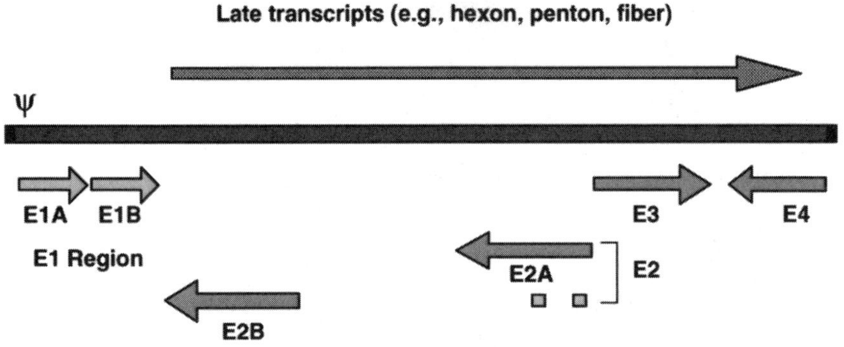

Late transcripts (e.g., hexon, penton, fiber)

Figure 1 Transcription map of adenovirus type 5. Arrows indicate direction of transcription. Early transcripts are shown below the genome and late transcripts are indicated above the genome. Many of the arrows denote families of alternatively spliced and/or polyadenylated mRNAs. The small arrows at the termini represent the inverted terminal repeats, and ψ indicates the packaging signal.

binding to both is required to induce further cell cycle progression and mitosis [14]. Mechanistically, E1A binding to Rb results in the dissociation of Rb/E2F complexes and the stimulation of a number of cellular genes by E2F that are required for DNA synthesis and cell cycle regulation [for a review, see Ref. 15]. The ability of E1A to stimulate S-phase by binding p300/CBP likely involves the ability of p300/CBP to influence p53 and E2F by acetylation. p300/CBP enhances the transcriptional activity of p53, but the interaction of E1A with p300/CBP prevents acetylation of p53 and represses p53-induced cell cycle arrest [16]. Recently, it has been shown that p300/CBP can also activate E2F by acetylation [17]. It is possible that E1A can also induce the acetytransferase activity of p300/CBP and stimulate E2F activity [18].

The proteins encoded by E1B transcription unit are required during the early phase of the replication cycle, and during oncogenic transformation, to inhibit E1A-induced apoptosis and to regulate the cellular tumor-suppressor p53. E1B-19K is a functional homolog of the cellular *Bcl*-2 gene and protects against programmed cell death that is induced as a response to disruption of growth control pathways by E1A [for review, see Refs. 19 and 20]. In transformed cells, the E1B-55K protein efficiently binds p53 and represses the transcriptional activity of p53 [21]. It is not known if repression of p53 transcriptional activity plays a role in the lytic cycle, but E1B-55K does form a complex with the adenovirus E4orf6 that can target p53 for degradation [22]. In addition, the E1B-55K/E4orf6 complex also plays an important role in the late phase of the infection cycle to induce the preferential accumulation and translation of viral mRNAs [23].

III. E1B-55K–DELETED VECTORS FOR CANCER THERAPY

The role of the E1B-55K to regulate p53 during adenoviral infection and transformation led to the hypothesis that an E1B-55K mutated virus would replicate efficiently in p53-defective tumor cells but not in normal cells in which wild-type p53 might block viral replication [24]. In agreement with this hypothesis, an E1B-55K–deleted virus *dl*1520, also called ONYX-015/CI-1042, was reported to replicate as efficiently as wild-type virus in p53 pathway–deficient tumor cells, but was attenuated in normal cells or tumor cells that expressed functional p53 [24,25]. However, in a series of subsequent reports in which a wider range of cell lines and assay conditions were used, replication of ONYX-015 in vitro was shown to be attenuated in tumor cells independent of p53 status and in most normal cells [reviewed in Ref. 26]. The replication of viral DNA is not affected by the E1B-55K deletion in ONYX-015 [27], but viral late protein synthesis and blockage of host protein synthesis are likely attenuated in most cell types regardless of p53 status [23,28]. These findings suggest that attenuation of ONYX-015 results primarily from the function of E1B in the late phase of the infection cycle. However, antitumor efficacy with ONYX-015 has been demonstrated in a number of preclinical tumor models with intratumoral or intravenous delivery [25,29,30]. Despite the limited tumor-selective growth of ONYX-015 in vitro, this virus has been used in a variety of phase I and II clinical trials and is reported to show an acceptable safety profile (see below).

IV. COMPLEMENTATION OF E1A FUNCTIONS BY DEREGULATED E2F IN TUMOR CELLS

A number of viruses containing E1A mutations that are defective for the cell cycle effects of E1A but retain the ability to stimulate the transcription of the viral early genes have been studied as potential selectively replicating vectors for cancer treatment [31–34]. The rationale for this approach is that most tumor cell lines contain abnormalities in the control of E2F activity as a result of mutations in *RB1* or *CDKN2A*, or overexpression of the CCND/CDK4 complex [for review, see Ref. 35]. Therefore, the role of E1A to deregulate E2F and stimulate the G1/S-phase transition, which is necessary for successful infection of quiescent cells, may not be required in tumor cells.

Two groups have reported studies with viruses in which the Rb-binding domain of E1A is abolished as a result of different deletions in the CR2 region of E1A [33,34]. In one study, Fueyo and coworkers found that adenovirus Δ24 (deletion of amino acids 121–128 of E1A) did not induce viral cytopathic effects in quiescent lung fibroblasts but replicated efficiently in a number of glioma tumor lines known to contain mutations in *RB1* or *CDKN2A*. A virus with a slightly different E1A deletion, *dl*922-947 (deletion of amino acids 122–129 of E1A), was reported to be defective for viral growth in nonproliferating cells of epithelial and endothelial origin but propagated at levels comparable or higher than wild-type virus in tumor cells. Comparison between *dl*922-947 and *dl*1520/ ONYX.015 showed that *dl*922-947 was more efficacious for tumor cell killing in vitro and in a number of tumor models in mice [34]. Although both Δ24 and *dl*922-947 were defective for viral growth in nonproliferating cells, *dl*922-947 was shown to replicate more efficiently than wild-type virus in proliferating normal cells [34]. This finding raises important safety issues with regard to the use of mutants such as Δ24 and *dl*922-947 in humans.

Another virus with E1A mutations, *dl*01/07, has been tested as a potential selectively replicating vector [31,32]. This mutant adenovirus, like Δ24 and *dl*922-947, produces an E1A protein defective for binding Rb, but also contains an additional mutation that inhibits interaction with p300/CBP (Fig. 2). The rationale for abolishing E1A binding to both p300 and Rb is that wild-type E1A can induce S-phase in quiescent normal cells by associating with either p300/CBP or Rb but not when binding to both is abolished. *dl*01/07 has been shown to have attenuated cytopathic effects in both quiescent and growing fibroblast lines and in an extensive range of proliferating normal cells including fibroblasts and epithelial and endothelial cells. However, E1A*dl*01/07 is almost as effective as wild-type virus in killing a wide range of tumor cells. It is therefore likely that the inability of E1A protein encoded in *dl*01/07 to bind p300 and Rb is complemented in tumor cells where E2F activity cannot be properly regulated (Fig. 3).

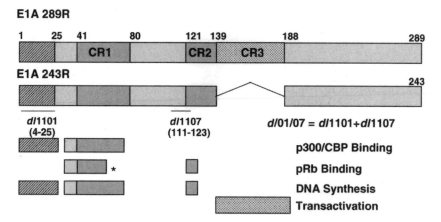

Figure 2 Schematic diagram of the E1A proteins. The regions of the E1A 289R and 243R proteins involved in binding to p300/CBP and Rb, transactivation, and DNA synthesis are shown. Specific deletions in *dl*01/07 that abolish interaction of E1A with p300/ CBP and Rb are also shown.

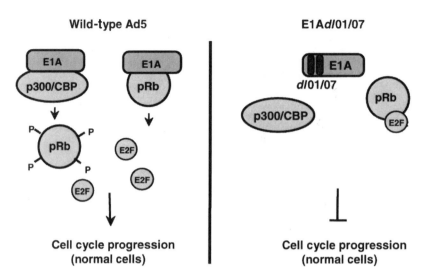

Figure 3 Basis of selectivity in *dl*01/07. *dl*01/07 produces an E1A protein defective for binding to p300/CBP and Rb. Since binding of E1A to either p300/CBP or Rb can induce S-phase and efficient viral replication, *dl*01/07 is attenuated for replication in normal cells. In contrast, *dl*01/07 can replicate well in tumors because of the presence of deregulated E2F activity.

Importantly, it has been demonstrated that *dl*01/07 has an antitumor effect that is comparable to wild-type virus in mouse models of human cancer [32]. In addition, *dl*01/07 has enhanced tumor-specific cell killing and antitumor effects when compared to an E1B-55K mutant [32] and to *dl*1520/ONYX-015 (G.W.D., J.H., and M.R., unpublished observations). Interestingly, the tumor cell killing effect of E1A*dl*01/07 has been enhanced by engineering the E3 region to overexpress E3-11.6K, a viral protein that appears to augment viral spread in the late phase of the viral infection cycle (see below) [31]. *dl*01/07 displays the desired characteristics of a selectively replicating virus; attenuation for viral growth in normal cells and potent antitumor effects in mouse models, but it is yet to be seen if this potential demonstrated in the laboratory will translate into utility for cancer therapy in human disease.

V. TUMOR-SPECIFIC EXPRESSION OF E1A AND OTHER EARLY VIRAL PROTEINS

Conditional expression of E1A or other early viral proteins using tissue or tumor-specific promoters is another strategy examined to confer selectivity to replicating vectors. Efforts at Calydon, Inc. (Sunnyvale, CA) were focused on prostatic tis-sue-specific promoters. Insertion of the prostate-specific antigen (PSA) promoter into adenovirus type 5 to drive expression of E1A has resulted in a virus desig-nated CN706 that is competent for replication in PSA-positive prostatic cancer cells [36]. Variations of CN706 include CV763 in which expression of E1A is controlled by the human kallikrein promoter, which is active in prostatic cancer cells. The kallikrein promoter was used because unlike PSA, which is not ex-pressed in all prostatic cancers, kallikrein is expressed in every prostatic cancer with an incremental increase in protein expression from benign epithelium to high-grade prostatic intraepithelial neoplasia to adenocarcinoma. To further atten-uate the growth of such prostate cancer–targeted viruses in nonprostatic cancer cells, both transcriptional regulatory regions were used in combination in the vector CV764 [37]. The PSA promoter was inserted upstream of E1A and human kallikrein 2 promoter was used to drive E1B in CV764 [37]. Although these vectors efficiently killed PSA-positive tumor cell in vitro, they could not elimi-nate distant xenograft tumors. Further engineering with incorporation of rat pro-basin prostate-specific promoter driving E1A and the PSA promoter driving the E1B gene with fully restored E3 region resulted in CV787 [38]. As a result of restored E3 region, CV787 was highly effective in the destruction of PSA-posi-tive cells and was capable of eliminating subcutaneous prostatic tumor xenografts following intravenous dosing of the virus [38].

For potential use in therapy of hepatocellular carcinoma (HCC), a replica-tion-restricted adenoviral vector was constructed by introducing the α-fetoprotein (AFP) gene promoter to drive expression of E1A [39]. AFP promoter was chosen because the AFP gene is highly expressed in 70–80% of patients with HCC but not in normal adults. The resulting vector (AvE1a04i) was reported to replicate well in human AFP-producing HCC cell lines, but attenuated in non–AFP-pro-ducing human cell lines, as well as primary cultures of normal human lung epithe-lial and endothelial cells. In addition. AvE1a04i was shown to prevent tumor growth of an ex vivo transduced AFP-expressing HCC cell line but not a non–AFP-expressing cell line.

Based on the observations that the DF3/MUC1 antigen is aberrantly over-expressed in nearly 80% of primary human breast and other carcinomas, an adenoviral vector (Ad.DF3-E1) in which the DF3/MUC1 promoter drives expres-sion of E1A was constructed [8]. Ad.DF3-E1 was reported to replicate selectively in MUC1-positive breast cancer cells [8]. To enhance the antitumor activity, the tumor necrosis factor (TNF) transgene was inserted in the E3 region under the control of cytomegalovirus (CMV) promoter. Compared to a replication-defec-tive vector encoding TNF, infection of cells with Ad.DF3-E1/CMV-TNF resulted in ~100-fold higher TNF in MUC1-positive cells. Furthermore, TNF production by Ad.DF3-E1/CMV-TNF was also ~100-fold higher in MUC1-positive cells compared to MUC1-negative cells. Ad.DF3-E1 was effective in inhibiting the growth of human breast tumor xenografts in nude mice and expression of TNF transgene in the vector potentiated the in vivo antitumor activity.

Similar transcriptional targeting approaches have also been undertaken using tissue-specific promoters. Using a fragment of the pS2 promoter containing two estrogen response elements to control expression of E1A and E4 proteins, a virus designated Ad5ERE2 was constructed [40]. This was reported to replicate preferentially in breast cancer cell lines expressing estrogen receptors. In addition, replication of the virus was severely affected in cell culture in the presence of an antiestrogen such as tamoxifen.

With the availability of human genomic sequence and improved technologies to identify new tumor-specific promoters, it is likely that replicating adenoviruses regulated by tumor-specific promoters will continue to be the focus of research efforts in the field. Further modifications of the transcriptional regulatory sequences with incorporation of insulator regions [41] to prevent leakiness will likely enhance the specificity.

VI. Complementation of E1A Functions by Cellular Factors

A number of cellular proteins are known to exhibit E1A-like transactivating functions. These proteins include factors present in embryonic stem cells [42], heat shock proteins [43], and nuclear factor–interleukin-6 (NF–IL-6) [44]. The presence of such E1A-like factors in tumor cells should therefore allow replication of E1A-deleted vectors. NF–IL-6, an IL-6–regulated human nuclear factor of the C/EBP family [44], was shown to complement E1A deletion in a number of cancer cell lines with a varying degree of efficiency to render them replication competent on induction with recombinant human IL-6 [45]. Based on these findings, it was suggested that Ad5dl312, an E1A-deleted adenovirus, could be used as a cytotoxic agent selectively to kill tumor cells responsive to or possessing an IL-6 autocrine loop [45].

VII. CONDITIONAL EXPRESSION OF DOMINANT-NEGATIVE INHIBITOR TO CONFER SELECTIVITY

As opposed to using tumor cell functions to complement viral gene inactivation or to drive viral gene expression, regulatory pathways that are fully functional in normal cells can be exploited to attenuate the viral growth in normal cells. Because of the significant heterogeneity and deregulated pathways in tumors compared to normal cells, strategies that take advantage of normal cell functions to inhibit viral replication might be useful for treatment of a broad range of tumors.

Forced expression of a dominant-negative inhibitor of viral replication using a p53-responsive promoter was recently examined as a strategy to confer selectivity [46]. A fusion protein comprising the DNA-binding domain of E2F and the transrepression domain of the retinoblastoma protein (Rb) was used as the dominant negative inhibitor. Such E2F-Rb fusion proteins are potent repressors of E2F-dependent promoters [47]. In addition to viral E2 and E1A promoters, many of the cellular S-phase genes are regulated by E2F. Therefore, repression of E2F-dependent promoters can lead to inhibition of critical events required for viral replication such as viral gene expression and S-phase induction. Most tumor cells are believed to have defects in p53 function because of alterations in p53 itself, mdm2, or p14^{ARF}, or interaction with viral proteins such as human papillomavirus E6 protein [48]. Thus, when E2F-Rb is under the control of the p53-responsive promoter, expression of E2F-Rb and inhibition of viral replication occur in normal cells (Fig. 4).

p53-Dependent expression of E2F-Rb and repression of E2F-dependent promoters are expected selectively to attenuate the virus in normal cells, but not

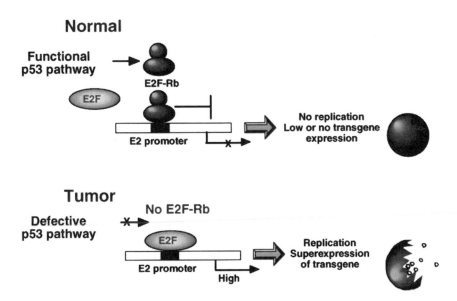

Figure 4   p53-Dependent expression of E2F-Rb to selectively attenuate viral replication in normal cells. Expression of E2F-Rb, a potent repressor of E2F-dependent promoters, occurs in normal cells with functional p53 pathway but not in tumor cells with defective p53 pathway. Repression of E2F-dependent promoters leads to inhibition of critical events required for viral replication such as early viral gene expression and S-phase induction, resulting in attenuation of virus growth (see text for details).

in tumor cells, even those with wild-type p53. E2F-responsive promoters are more active in tumor cells than in normal cells [49]. Therefore, even in tumor cells with functional p53, transcription from E2F-dependent viral E1A and E2 promoters could occur prior to the expression of E2F-Rb resulting in replication of the virus. E2F-Rb may be less effective in inhibiting adenoviral early promoters when E2F activity is upregulated in tumor cells, because inhibition by E2F-Rb is dependent upon competition with endogenous E2F activity. In contrast, p53-dependent E2F-Rb expression would be effective in attenuating the virus in normal cells because of both functional p53 and tightly regulated E2F pathways.

To further enhance tumor cell killing and viral spread, an extra copy of the viral major late promoter was introduced in the E3 region to drive overexpression of viral E3-11.6K. A resulting vector encoding E3-11.6K and capable of expressing E2F-Rb in a p53-dependent manner (01/PEME) was found to be attenuated in normal cells but efficacious in killing tumor cells in both in vitro cell cultures and in vivo tumor models [46].

Unlike other strategies that rely on the introduction of mutation of viral genes or expression of viral genes under the control of tumor-specific promoters, the dominant negative inhibitor strategy makes use of an exogenous cassette to confer selective attenuation to adenoviral vectors. Although E2F-Rb and a p53-responsive promoter were used as the components to confer selectivity, the strategy could be extended to different inhibitors and promoters regulated by other cellular regulatory pathways.

VIII. MODIFICATION OF THE VIRAL GENOME TO ENHANCE ONCOLYTIC ACTIVITY

Efficient killing of tumor cells and rapid spread of the vector within a tumor is essential for maximum efficacy with replicating vectors. In adenovirus-infected cells, E1B-19K gene product prevents premature apoptosis of infected cells [50], whereas E3-11.6K [51,52] and E4orf4 [53–55] promote apoptosis late in the infection cycle. Consistent with the role of E1B-19K, viruses deleted in 19K induce apoptosis very efficiently, resulting in the faster spread of the virus than the wild-type. Duque and coworkers constructed and evaluated a vector (AdH5 dL118) that lacks expression of both E1B-19K and E1B-55K proteins but with wild-type E1A. AdH5 dL118 showed striking cytopathic effect on malignant cell lines in vitro and a partial reduction of tumorigenicity in vivo [56]. The virus also enhanced chemosensitivity to DNA-damaging agents in vitro and in vivo. Similar findings, including early release of the virus, increased cell-to-cell spread, enhanced tumor cell killing, and augmented chemosensitivity were also reported using another E1B-19K–deleted vector, Ad337 [57].

Overexpression of E3-11.6K late in the infection as a means to enhance the cytopathicity yet allow for adequate replication has also been investigated

[31,46]. Strategies using E3-11.6K overexpression may offer an advantage over approaches involving E1B-19K mutations, because E1B-19K mutations also affect viral replication in some tumor cells [23]. E3-11.6K, also known as adenovirus death protein, is a nuclear membrane glycoprotein required at late stages of viral infection for efficient cell lysis and release of virions from cells [51]. E3-11.6K is expressed at low levels at early times, but is synthesized in large amounts from mRNAs derived from the major late promoter [58]. To achieve overexpression of E3-11.6K, Doronin and coworkers removed all other open reading frames in the E3 region except that of E3-11.6K in an adenoviral vector with dl01/07 deletion in the E1A gene [31]. The resulting vectors, KD1 and KD3, showed overexpression of E3-11.6K, lysed cells, and spread from cell to cell more efficiently than the parental dl01/07. Increased efficacy of KD1 and KD3 was also apparent in human tumor xenograft experiments.

As described earlier in this chapter, as an alternate strategy to obtain E3-11.6K overexpression, an extra copy of the viral major late promoter was introduced in the E3 region along the E3-11.6K coding sequence after deleting all other E3 coding sequences [46]. Since the major late promoter becomes active after the onset of DNA replication, the use of major late promoter couples gene expression to viral DNA replication and thus provides another level of control in replicating vectors. Vectors thus constructed exhibited enhanced tumor cell killing ability in both in vitro and in vivo experiments. Increased efficacy with E3-11.6K–overexpressing vectors is consistent with the ability of E3-restored vectors to inhibit tumor growth following intravenous administration [38]. Potential in vivo advantages of E3-11.6K–overexpressing viruses in the absence of other E3 proteins and in the context of intravenous delivery have not yet been demonstrated.

IX. THERAPEUTIC GENES IN REPLICATING ADENOVIRAL VECTORS

Replicating adenoviruses cannot only serve as therapeutic agents themselves but also can deliver therapeutic transgenes to cancer cells to enhance the oncolytic effect. It is believed that the controlled replication of the virus provides an added specificity to the therapy. Replicating adenoviruses encoding prodrug-activating enzymes, cytokines, or proapoptotic genes have been constructed and evaluated.

Because of significant heterogeneity within the tumor, replicating vectors may not be effective if they are engineered to target specific pathway defects. To enhance the potency and to expand the spectrum of tumor cell type, genes encoding prodrug-activating suicide enzymes in replicating vectors have been studied. Bystander effects from prodrug conversion could add to the oncolytic activities of the adenoviral vectors. Since toxic metabolites resulting from widely used prodrug-activating enzymes such as cytosine deaminase (CD) and herpes simplex virus type-1 thymidine kinase (HSV-TK) interfere in the replication of adenoviruses, combination of prodrug-activating systems with replicating vectors would also allow to control replication and spread of the vector. However, to increase the potency of replicating vectors encoding these enzymes, the timing and frequency of prodrug administration need to be optimized.

To improve both the safety and efficacy of a replicating adenoviral vector, an E1B-55K–deleted vector encoding a CD/HSV-TK fusion gene, designated FGR, was constructed [59]. The CD/HSV-TK fusion gene therapy has also been shown to sensitize cells to radiation treatment. In vitro studies showed that the FGR virus exhibited the same tumor cell specificity and replication kinetics as the E1B-55K–deleted virus. Incorporation of the CD/HSV-TK fusion gene enhanced the cytopathic effect of the virus and sensitized infected cells to radiation following prodrug addition. Additionally, depending on the frequency and levels of prodrug addition, it was possible to block viral replication, which could potentially be used to control viral replication. Based on these results, Freytag and coworkers suggested that inclusion of the CD/HSV-TK fusion gene in a lytic, replication-competent adenovirus might not only improve efficacy but also improve the safety with replicating viruses [59].

Morris and coworkers have evaluated HSV-TK in adenoviral vectors without or with E1B-55K deletion. An E1B-55K deleted AdTK^{RC} was constructed by introducing in the E1 region an expression cassette comprising the HSV-TK gene transcriptionally linked to E1A and E1B-19K genes under the control the CMV-IE promoter. In melanoma, cervical, and colon tumor xenograft models, AdTK^{RC} showed increased efficacy in combination with ganciclovir administration starting 3 days after viral injection over vector without ganciclovir or a replication-defective TK encoding vectors [60,61]. In contrast, the efficacy of EIB-55K–positive vectors encoding HSV-TK was not further improved with ganciclovir treatment [62]. Unlike E1B-55K–deleted vectors, E1B-55K–containing vectors replicate more efficiently, and therefore, the inhibition of viral replication and oncolysis by ganciclovir may have compensated the increased cytotoxicity of the prodrug-activating system [63].

A replicating adenovirus encoding the human consensus interferon gene inserted in the E3 region after introducing a 1.9-kb E3 deletion was shown to be more effective than the wild-type virus in the human myelogenous leukemic cell line K562 and hamster melanoma RPMI 1846 tumor models [64]. However, this virus was not engineered to prevent cytopathic effects on normal cells.

X. CAPSID MODIFICATION TO ALTER THE TROPISM OF REPLICATING ADENOVIRAL VECTORS

Most of the published studies have focused on altering the viral genome to change the viral regulatory pathways after the virus enters the cell in order to make them tumor targeted. However, if the virus cannot infect target cells within the tumor efficiently, all such efforts would be futile during therapy. Tumor targeting of conventional E1-deleted adenoviruses has been achieved using antibodies or other ligands such as epidermal growth factor and basic fibroblast growth factor linked to viral capsid [65–67]. As these targeting strategies are refined, it is anticipated that such modifications will be combined with alterations in the replication pathways in developing highly tumor-specific vectors.

It has been shown that incorporation of small peptides, such as stretches of lysines, to bind heparin sulfate and polyanionic cellular receptors as well as peptides containing the RGD motif to bind α_v integrins results in altered tropism and enhanced infectivity [68]. Therefore, Shinoura et al. constructed an adenoviral vector (Adv-E1AdB-F/K20) with E1B-55K deletion and with a fiber modification, F/K20, with a stretch of 20 lysine residues added at the carboxy-terminus of the fiber [68]. The addition of 20 lysines to the fiber resulted in an increase in the transduction efficiency of gliomas. The oncolytic activity of Adv-E1AdB-F/K20 was markedly higher compared to a similar vector with the native fiber on glioblastoma cell lines in vitro and in vivo.

Suzuki and coworkers recently modified the fiber of a replicating adenoviral vector to make the entry of coxsackievirus-adenovirus receptor (CAR) independent as a strategy to enhance the oncolytic potency against a broad range of tumors [69]. They incorporated a RGD peptide by genetic modification into the H1 loop of the fiber knob in an adenovirus with a partial deletion in the E1A gene, which abolishes the ability of E1A protein to bind Rb protein. As discussed earlier in the chapter, such deletion in E1A was reported to result in an adenovirus capable of replicating selectively in tumor cells [33]. The vector with fiber modification showed enhanced infectivity and oncolytic potency in vitro and in vivo [69].

XI. INDUCED REPLICATION OF E1-DELETED VECTORS BY COMPLEMENTATION OF E1 PROTEINS IN *TRANS* WITH REPLICATING ADENOVIRUSES

Replicating adenoviruses have also been evaluated to complement E1 deficiency of conventional E1-deleted adenoviral vectors encoding therapeutic transgenes. Such codelivery strategies can be very useful in evaluating therapeutic transgenes to be used in combination with replicating vectors. However, the use of two vector systems may not work optimally at low concentrations and pose additional hurdles in developing these agents for use in the clinic. To provide increased capacity to carry therapeutic transgenes, Alemany and coworkers [70] split the viral genome into two defective viruses that complement each other. One of those vectors contained only the minimal viral sequences required in *cis* for replication and packaging and the E1 genes controlled by the human AFP promoter. Since this vector contained minimal viral sequences, foreign DNA of up to 30 kb in length could be inserted in this vector. Replication of this vector occurred in the presence of the second vector that carried all adenoviral genes except E1. In vitro studies showed that coinfection of AFP-positive hepatocarcinoma cells resulted in viral replication and cell killing. Intratumoral injection of human hepatocarcinoma xenografts in immunodeficient mice resulted in antitumor activity.

While using replicating adenoviruses as helper viruses, codelivery of first-generation vectors encoding cytotoxic genes can cause premature death of the infected cells resulting in attenuation of viral replication. Therefore, Motai and coworkers [71] proposed that genes with indirect antitumor effects such as cytokine genes and the antiangiogenesis gene would be better candidates for codelivery with replicating adenoviruses. Codelivery of an E1B-55K–deleted replicating vector (AxE1AdB) with a conventional replication-defective adenoviral vector encoding IL-2 and IL-12 were evaluated. Both IL-2 and IL-12 have strong antitumor properties, and IL-2 is clinically used to treat melanoma, renal cell carcinoma, and colon carcinoma [72,73]. Coinfection of AxE1AdB with replication-defective IL-2 or IL-12 vectors resulted in a more significant increase in cytokine production than those infected with replication-defective IL-2 or IL-12 vector alone. Enhanced efficacy with such codelivery strategies using replication-defective IL-2 or IL-12 vector was also observed in human tumor xenograft models.

When a replicating vector with both E1 and E4 genes controlled by an estrogen-responsive promoter was used in combination with an E1-deleted vector carrying proapoptotic gene *Bcl-xs*, a dramatic increase in the ability of both viruses to induce cell death was observed [40]. Since E1 and E4 protein expression was controlled by an estrogen-responsive promoter, the cytopathic effects of the combination of viruses were shown to be modulated by the antiestrogen tamoxifen.

XII. CLINICAL TRIALS WITH AN E1B-55K–DELETED VECTOR BY ONYX PHARMACEUTICALS

Clinical trials with the E1B–deleted virus, ONYX-015, have shown the virus to be safe by number of routes of administration. In 22 patients with recurrent head and neck cancer given intratumoral dosing, no dose-limiting toxicities were observed [74]. Tumor sites were injected with 1×10^7 to 1×10^{11} pfu of ONYX-015 in the dose-escalation phase I study. The observed symptoms were flulike, including grades 1 and 2 fever, chills, and nausea. Eight patients received more than one cycle of intratumoral injections. There was no evidence of virus in the plasma of patients on days 3, 8, 15, 22, and 29 posttreatment polymerase chain reaction (PCR). Although no objective responses by conventional criteria were seen, 5 of the 22 patients were reported to have a response at the injected tumor site. There was no correlation among responders and p53 tumor status or preexisting adenoviral antibodies.

Based on the tolerability of single intratumoral injections of ONYX-015, phase II studies treating patients with head and neck cancer by intratumoral injection of 5×10^{10} particles on 5 consecutive days [6,75] was performed. An additional arm of the studies included patients given injections twice daily for 5 consecutive days and repeated the following week. The toxicities of flulike symptoms including fever, chills, nausea, and injection site pain were similar to the profile from single injections. Biopsies taken at varying time points posttreatment were assessed for viral replication by in situ hybridization, E1A immunohistochemical staining, and electron microscopy. The presence of ONYX-015 in the tumor could be detected between days 1 and 10 in 7 of 11 samples. Virus was not detected in surrounding normal tissue. ONYX-015 was detected in the blood by PCR in 41% of the patients 24 hr after the last injection and in 9% of the patients 10 days after the treatment, suggesting viral replication was responsible for the viremia. Complete or partial tumor responses were reported in 14% of the patients. A correlation between tumor response and the p53 status of the tumor was reported; 7 of 12 p53-altered tumors responded, whereas 0 of 7 wild-type p53 tumors responded. There was no correlation between tumor responses and preexisting antibodies against adenovirus.

A phase II trial of intratumoral administration of ONYX-015 in combination with chemotherapy (cisplatin and 5-fluorouracil) reported 36% of recurrent head and neck carcinoma patients with a partial response and 27% patients achieved a complete response [8]. No correlation between tumor response and preexisting adenoviral antibodies, p53 status, tumor size, or prior treatment was seen. As in previous trials with ONYX-015, adverse events were flulike symptoms and injection site pain. Biopsies analyzed for adenoviral DNA by in situ hybridization revealed evidence of replication in four of seven biopsies taken between days 5 and 15. Adenoviral DNA was not detected in normal tissue. In 11 patients with two tumors, the largest tumor was injected with ONYX-015 leaving the second tumor as a control noninjected tumor. In 6 of 11 patients, only the injected tumor responded. In 3 of 11 patients, both tumors responded. There were no responses in either tumor in 2 of 11 patients. The assessment of the combination of chemotherapy with ONYX-015 to provide local control in head and neck cancer awaits the phase III clinical trial.

Other clinical studies with ONYX-015 including intratumoral injection for pancreatic carcinoma [76] and liver metastases [77] have been performed. The ONYX-015 virus has also been administered by intraperitoneal delivery in ovarian cancer patients [78] and by intrahepatic artery administration to treat patients with liver metastases [79]. ONYX-015 has been used in a mouthwash to treat oral dysplasia [80]. Cancer patients have also been dosed by intravenous administration of ONYX-015 [81].

XIII. CLINICAL TRIALS WITH PROSTATIC CANCER–TARGETED ADENOVIRUSES BY CALYDON, INC.

As discussed earlier in the chapter, prostate cancer is the target of a series of adenoviral vectors from Calydon, Inc. [36–38]. These constructs in which prostate-specific promoters control E1 expression were reported to limit viral replication to prostatic cancer. One of these viruses, CN706, has been used in a phase I trial to treat patients with locally recurrent prostatic cancer [82,83]. Delivery of CN706 to the tumor site was achieved safely by ultrasound-guided intratumoral injection. Late gene expression and nuclear inclusion bodies were observed in biopsy samples, suggesting CN706 replication occurred in vivo. At the two highest dose levels, 5 of 11 patients had a partial response of over 50% reduction in PSA concentration for at least 4 weeks, and in 3 of those patients, the duration of PSA reduction lasted 9 months [84].

XIV. CLINICAL STUDIES WITH REPLICATING ADENOVIRUSES WITH THERAPEUTIC TRANSGENES

Therapeutic transgenes incorporated in replicating viruses to enhance oncolytic activity include interferon [64], HSV-TK [60–63], and a CD/HSV-TK fusion gene [59]. A replicating vector with the CD/HSV-TK fusion gene (Ad5-CD/TK$_{rep}$) has been used in the clinic to treat prostatic cancer [85]. Two days after intraprostatic injections of Ad5-CD/TK$_{rep}$, patients were treated with the prodrugs 5-fluorocytosine and ganciclovir for 7 days.

XV. CLINICAL STUDIES WITH OTHER ONCOLYTIC VIRUSES

In addition to the clinical trials with replication-competent adenoviruses, replication-competent herpes virus has been used to treat glioblastoma patients. An attenuated herpes virus G207 was dosed by intracranial administration to 21 patients with no adverse events attributable to the therapy [86]. Another study performed with herpes virus 1716 treated nine patients with no reactivation of latent HSV and no encephalitis [87]. The demonstration of the safety in a number of clinical trials using the strategy of replication-competent viruses as the therapeutic entity has elicited the proposal of a wide range of other virus candidates to be anticancer therapeutics including reovirus [88], Newcastle disease virus [89], vaccinia virus [90], and vesicular stomatitis virus [91].

XVI. SUMMARY

Various adenoviral vectors have been generated by manipulating the viral genome and shown to be selective in in vitro cell culture assays. However, evaluation of their selectivity in vivo has not been possible in animal models, since human adenoviruses do not replicate or replicate very inefficiently in nonhuman cells. Clinical studies with a few of the conditionally replicating viruses such as ONYX-015 and Calydon viruses have demonstrated acceptable safety profiles of these vectors. Although most of the vectors published so far have proven to be highly efficacious in immunodeficient animal models, clinical studies have shown only limited anecdotal responses following therapy. Various factors such as persistence in the blood, physical barriers that affect delivery and spread of the vectors within the tumor, tumor-specific replication, and interaction with the host immune system contribute to the activity of replicating viruses in humans. Development of suitable models and evaluation of strategies to overcome these hurdles are essential to improve replicating adenoviral vectors for cancer therapy.

ACKNOWLEDGMENTS

We thank Dr. Robert Ralston and Paul Shabram for helpful discussions.

REFERENCES

1. Hitt MM, Addison CL, Graham FL. Human adenovirus vectors for gene transfer into mammalian cells. Adv Pharmacol 1997; 40:137–206.
2. Seth P, Katayose U, Rakkar ANS. Adenoviral vectors for cancer therapy. In: Seth P, ed. Adenoviruses: Basic Biology to Gene Therapy. Austin, TX: Landes, 1999: 103–120.
3. Vile RG, Russell SJ, Lemoine NR. Cancer gene therapy: hard lessons and new courses. Gene Ther 2000; 7:2–8.
4. Smith RR, Huebner RJ, Rowe WP, Schatten WE, Thomas LB. Studies on the use of viruses in the treatment of carcinoma of the cervix. Cancer 1956; 9:1211–1218.
5. Southam CM. Present status of oncolytic virus studies. Trans NY Acad Sci 1960; 22:657–673.
6. Nemunaitis J, Ganly I, Khuri F, Arseneau J, Kuhn J, McCarty T, Landers S, Maples P, Romel L, Randlev B, Reid T, Kaye S, Kim D. Selective replication and oncolysis

in p53 mutant tumors with ONYX-015, an E1B-55kD gene-deleted adenovirus, in patients with advanced head and neck cancer: a phase II trial. Cancer Res 2000; 60: 6359–6366.

7. Nemunaitis J, Swisher SG, Timmons T, Connors D, Mack M, Doerksen L, Weill D, Wait J, Lawrence DD, Kemp BL, Fossella F, Glisson BS, Hong WK, Khuri FR, Kurie JM, Lee JJ, Lee JS, Nguyen DM, Nesbitt JC, Perez-Soler R, Pisters KM, Putnam JB, Richli WR, Shin DM, Walsh GL, et al. Adenovirus-mediated p53 gene transfer in sequence with cisplatin to tumors of patients with non-small-cell lung cancer. J Clin Oncol 2000; 18:609–622.

8. Khuri FR, Nemunaitis J, Ganly I, Arseneau J, Tannock IF, Romel L, Gore M, Ironside J, MacDougall RH, Heise C, Randlev B, Gillenwater AM, Bruso P, Kaye SB, Hong WK, Kim DH. A controlled trial of intratumoral ONYX-015, a selectively-replicating adenovirus, in combination with cisplatin and 5-fluorouracil in patients with recurrent head and neck cancer. Nat Med 2000; 6:879–885.

9. Huyghe BG, Liu X, Sutjipto S, Sugarman BJ, Horn MT, Shepard HM, Scandella CJ, Shabram P. Purification of a type 5 recombinant adenovirus encoding human p53 by column chromatography. Hum Gene Ther 1995; 6:1403–1416.

10. Shabram PW, Giroux DD, Goudreau AM, Gregory RJ, Horn MT, Huyghe BG, Liu X, Nunnally MH, Sugarman BJ, Sutjipto S. Analytical anion-exchange HPLC of recombinant type-5 adenoviral particles. Hum Gene Ther 1997; 8:453–465.

11. Barker DD, Berk AJ. Adenovirus proteins from both E1B reading frames are required for transformation of rodent cells by viral infection and DNA transfection. Virology 1987; 156:107–121.

12. Bayley ST, Mamryk JS. Adenovirus E1A proteins and transformation (review). Int J Oncol 1994; 5:425–444.

13. Shenk T. Adenoviridae: the viruses and their replication. In: Fields BN, Knipe DM, Howley PM, eds. Fields Virology, 3rd ed., Vol. 2. Philadelphia: Lippincott-Raven, 1996:2111–2148.

14. Howe JA, Mymryk JS, Egan C, Branton PE, Bayley ST. Retinoblastoma growth suppressor and a 300-kDa protein appear to regulate cellular DNA synthesis. Proc Natl Acad Sci USA 1990; 87:5883–5887.

15. Dyson N. The regulation of E2F by pRB-family proteins. Genes Dev 1998; 12: 2245–2262.

16. Chakravarti D, Ogryzko V, Kao HY, Nash A, Chen H, Nakatani Y, Evans RM. A viral mechanism for inhibition of p300 and PCAF acetyltransferase activity. Cell 1999; 96:393–403.

17. Martinez-Balbas MA, Bauer VM, Neilsen SJ, Brehon A, Kouzarides T. Regulation of E2F1 activity by acetylation. EMBO J 2000; 19:662–671.

18. Ait-Si-Ali S, Ramirez S, Barre FX, Dkhissi F, Magnaghi-Jaulin L, Girault JA, Robin P, Knibiehler M, Pritchard LL, Ducommun B, Trouche D, Harel-Bellan A. Histone acetyltransferase activity of CBP is controlled by cycle-dependent kinases and onco-protein E1A. Nature 1998; 396:184–186.

19. Branton PE. Early gene expression. In: Seth P, ed. Adenoviruses: Basic Biology to Gene Therapy. Vol. 1. Austin, TX: Landes, 1999:39–58.

20. White E. Regulation of apoptosis by adenovirus E1A and E1B oncogenes. Semin Virol 1998; 8:505–513.

21. Yew PR, Berk AJ. Inhibition of p53 transactivation required for transformation by adenovirus early 1B protein. Nature 1992; 357:82–85.

22. Querido E, Marcellus RC, Lai A, Charbonneau R, Teodoro JG, Ketner G, Branton PE. Regulation of p53 levels by the E1B 55-kilodalton protein and E4orf6 in adeno-virus-infected cells. J Virol 1997; 71:3788–3798.

23. Pilder S, Moore M, Logan J, Shenk T. The adenovirus E1B-55K transforming poly-peptide modulates transport or cytoplasmic stabilization of viral and host cell mRNAs. Mol Cell Biol 1986; 6:470–476.

24. Bischoff JR, Kirn DH, Williams A, Heise C, Horn S, Muna M, Ng L, Nye JA, Sampson-Johannes A, Fattaey A, McCormick F. An adenovirus mutant that repli-cates selectively in p53-deficient human tumor cells. Science 1996; 274:373–376.

25. Heise C, Sampson-Johannes A, Williams A, McCormick F, Von Hoff DD, Kirn DH. ONYX-015, an E1B gene-attenuated adenovirus, causes tumor-specific cytolysis and antitumoral efficacy that can be augmented by standard chemotherapeutic agents. Nat Med 1997; 3:639–645.

26. Alemany R, Balague C, Curiel DT. Replicative adenoviruses for cancer therapy. Nat Biotechnol 2000; 18:723–727.

27. Goodrum FD, Ornelles DA. p53 status does not determine outcome of EIB 55–kilodalton mutant adenovirus lytic infection. J Virol 1998; 72:9479–9490.

28. Harada JN, Berk AJ. p53-Independent and -dependent requirements for E1B-55K in adenovirus type 5 replication. J Virol 1999; 73:5333–5344.

29. Heise CC, Williams AM, Xue S, Propst M, Kirn DH. Intravenous administration of ONYX-015, a selectively replicating adenovirus, induces antitumoral efficacy. Cancer Res 1999; 59:2623–2628.

30. Heise CC, Williams A, Olesch J, Kirn DH. Efficacy of a replication-competent adenovirus (ONYX-015) following intratumoral injection: intratumoral spread and distribution effects. Cancer Gene Ther 1999; 6:499–504.

31. Doronin K, Toth K, Kuppuswamy M, Ward P, Tollefson AE, Wold WS. Tumor-specific, replication-competent adenovirus vectors overexpressing the adenovirus death protein. J Virol 2000; 74:6147–6155.

32. Howe J, Demers GW, Johnson DE, Neugebauer SEA, Perry ST, Vaillancourt MT, Faha B. Evaluation of E1-mutant adenoviruses as conditionally replicating agents for cancer therapy. Mol Ther 2000; 2:485–495.

33. Fueyo J, Gomez-Manzano C, Alemany R, Lee PS, McDonnell TJ, Mitlianga P, Shi YX, Levin VA, Yung WK, Kyritsis AP. A mutant oncolytic adenovirus targeting the Rb pathway produces anti-glioma effect in vivo. Oncogene 2000; 19:2–12.

34. Heise C, Hermiston T, Johnson L, Brooks G, Sampson-Johannes A, Williams A, Hawkins L, Kirn D. An adenovirus E1A mutant that demonstrates potent and selec-tive systemic anti-tumoral efficacy. Nat Med 2000; 6:1134–1139.

35. Sherr CJ. Cancer cell cycles. Science 1996; 274:1672–1677.

36. Rodriguez R, Schuur ER, Lim HY, Henderson GA, Simons JW, Henderson DR. Prostate attenuated replication competent adenovirus (ARCA) CN706: a selective cytotoxic for prostate-specific antigen-positive prostate cancer cells. Cancer Res 1997; 57:2559–2563.

37. Yu DC, Sakamoto GT, Henderson DR. Identification of the transcriptional regula-tory sequences of human kallikrein 2 and their use in the construction of calydon

virus 764, an attenuated replication competent adenovirus for prostate cancer therapy. Cancer Res 1999; 59:1498–1504.

38. Yu DC, Chen Y, Seng M, Dilley J, Henderson DR. The addition of adenovirus type 5 region E3 enables calydon virus 787 to eliminate distant prostate tumor xenografts. Cancer Res 1999; 59:4200–4203.

39. Hallenbeck PL, Chang YN, Hay C, Golightly D, Stewart D, Lin J, Phipps S, Chiang YL. A novel tumor-specific replication-restricted adenoviral vector for gene therapy of hepatocellular carcinoma. Hum Gene Ther 1999; 10:1721–1733.

40. Hernandez-Alcoceba R, Pihalja M, Wicha MS, Clarke MF. A novel, conditionally replicative adenovirus for the treatment of breast cancer that allows controlled replication of E1a-deleted adenoviral vectors. Hum Gene Ther 2000; 11:2009–2024.

41. Bell AC, Felsenfeld G. Stopped at the border: boundaries and insulators. Curr Opin Genet Dev 1999; 9:191–198.

42. La Thangue NB, Rigby PW. An adenovirus E1A-like transcription factor is regulated during the differentiation of murine embryonal carcinoma stem cells. Cell 1987; 49: 507–513.

43. Imperiale MJ, Kao HT, Feldman LT, Nevins JR, Strickland S. Common control of the heat shock gene and early adenovirus genes: evidence for a cellular E1A-like activity. Mol Cell Biol 1984; 4:867–874.

44. Spergel JM, Hsu W, Akira S, Thimmappaya B, Kishimoto T, Chen-Kiang S. NF-IL6, a member of the C/EBP family, regulates E1A-responsive promoters in the absence of E1A. J Virol 1992; 66:1021–1030.

45. Rancourt C, Piche A, Gomez-Navarro J, Wang M, Alvarez RD, Siegal GP, Fuller GM, Jones SA, Curiel DT. Interleukin-6 modulated conditionally replicative adenovirus as an antitumor/cytotoxic agent for cancer therapy. Clin Cancer Res 1999; 5:43–50.

46. Ramachandra M, Rahman A, Zou A, Vaillancourt M, Howe J, Sugarman B, Demers WG, Engler H, Johnson D, Shabram P. Reengineering adenovirus regulatory pathways to enhance oncolytic specificity and efficacy. Nat Biotechnol 2001; 19:1035–1041.

47. Antelman D, Gregory RJ, Wills KN. Retinoblastoma fusion polypeptides. US Patent and Trademark Office. USA: Canji, Inc. (San Diego, CA), 2000.

48. Prives C, Hall PA. The p53 pathway. J Pathol 1999; 187:112–126.

49. Parr MJ, Manome Y, Tanaka T, Wen P, Kufe DW, Kaelin WG Jr, Fine HA. Tumor-selective transgene expression in vivo mediated by an E2F-responsive adenoviral vector. Nat Med 1997; 3:1145–1149.

50. White E, Faha B, Stillman B. Regulation of adenovirus gene expression in human W138 cells by an E1B-encoded tumor antigen. Mol Cell Biol 1986; 6:3763–3773.

51. Tollefson AE, Ryerse JS, Scaria A, Hermiston TW, Wold WS. The E3-11.6-kDa adenovirus death protein (ADP) is required for efficient cell death: characterization of cells infected with adp mutants. Virology 1996; 220:152–162.

52. Tollefson AE, Scaria A, Hermiston TW, Ryerse JS, Wold LJ, Wold WS. The adenovirus death protein (E3-11.6K) is required at very late stages of infection for efficient cell lysis and release of adenovirus from infected cells. J Virol 1996; 70:2296–2306.

53. Lavoie JN, Nguyen M, Marcellus RC, Branton PE, Shore GC. E4orf4, a novel adenovirus death factor that induces p53-independent apoptosis by a pathway that is not inhibited by zVAD-fmk. J Cell Biol 1998; 140:637–645.

54. Marcellus RC, Chan H, Paquette D, Thirlwell S, Boivin D, Branton PE. Induction of p53-independent apoptosis by the adenovirus E4orf4 protein requires binding to the Balpha subunit of protein phosphatase 2A. J Virol 2000; 74:7869–7877.

55. Marcellus RC, Lavoie JN, Boivin D, Shore GC, Ketner G, Branton PE. The early region 4 orf4 protein of human adenovirus type 5 induces p53-independent cell death by apoptosis. J Virol 1998; 72:7144–7153.

56. Duque PM, Alonoso C, Sanchez-Prieto R, Lleonart M, Martinez C, de Buitrago GG, Cano A, Quintanilla M, Ramon y Cajal S. Adenovirus lacking the 19-kDa and 55-kDa E1B genes exerts a marked cytotoxic effect in human malignant cells. Cancer Gene Ther 1999; 6:554–563.

57. Sauthoff H, Heitner S, Rom WN, Hay JG. Deletion of the adenoviral E1b-19kD gene enhances tumor cell killing of a replicating adenoviral vector. Hum Gene Ther 2000; 11:379–388.

58. Tollefson AE, Scaria A, Saha SK, Wold WS. The 11,600-MW protein encoded by region E3 of adenovirus is expressed early but is greatly amplified at late stages of infection. J Virol 1992; 66:3633–3642.

59. Freytag SO, Rogulski KR, Paielli DL, Gilbert JD, Kim TH. A novel three-pronged approach to kill cancer cells selectively: concomitant viral, double suicide gene, and radiotherapy. Hum Gene Ther 1998; 9:1323–1333.

60. Wildner O, Blaese RM, Morris JC. Therapy of colon cancer with oncolytic adenovirus is enhanced by the addition of herpes simplex virus–thymidine kinase. Cancer Res 1999; 59:410–413.

61. Wildner O, Morris JC, Vahanian NN, Ford H Jr, Ramsey WJ, Blaese RM. Adenoviral vectors capable of replication improve the efficacy of HSVtk/GCV suicide gene therapy of cancer. Gene Ther 1999; 6:57–62.

62. Wildner O, Morris JC. Therapy of peritoneal carcinomatosis from colon cancer with oncolytic adenoviruses. J Gene Med 2000; 2:353–360.

63. Wildner O, Morris JC. The role of the E1B 55 kDa gene product in oncolytic adenoviral vectors expressing herpes simplex virus-tk: assessment of antitumor efficacy and toxicity. Cancer Res 2000; 60:4167–4174.

64. Zhang JF, Hu C, Geng Y, Selm J, Klein SB, Orazi A, Taylor MW. Treatment of a human breast cancer xenograft with an adenovirus vector containing an interferon gene results in rapid regression due to viral oncolysis and gene therapy. Proc Natl Acad Sci USA 1996; 93:4513–4518.

65. Curiel DT. Strategies to adapt adenoviral vectors for targeted delivery. Ann NY Acad Sci 1999; 886:158–171.

66. Krasnykh VN, Douglas JT, van Beusechem VW. Genetic targeting of adenoviral vectors. Mol Ther 2000; 1:391–405.

67. Wickham TJ. Targeting adenovirus. Gene Ther 2000; 7:110–114.

68. Shinoura N, Yoshida Y, Tsunoda R, Ohashi M, Zhang W, Asai A, Kirino T, Hamada H. Highly augmented cytopathic effect of a fiber-mutant E1B-defective adenovirus for gene therapy of gliomas. Cancer Res 1999; 59:3411–3416.

69. Suzuki K, Fueyo J, Krasnykh V, Reynolds PN, Curiel DT, Alemany R. A conditionally replicative adenovirus with enhanced infectivity shows improved oncolytic potency. Clin Cancer Res 2001; 7:120–126.

70. Alemany R, Lai S, Lou YC, Jan HY, Fang X, Zhang WW. Complementary adeno-viral vectors for oncolysis. Cancer Gene Ther 1999; 6:21–25.

71. Motoi F, Sunamura M, Ding L, Duda DG, Yoshida Y, Zhang W, Matsuno S, Hamada H. Effective gene therapy for pancreatic cancer by cytokines mediated by restricted replication-competent adenovirus. Hum Gene Ther 2000; 11:223–235.

72. Lotze MT, Matory YL, Rayner AA, Ettinghausen SE, Seipp CA, Rosenberg SA. Clinical effects and toxicity of interleukin-2 in patients with cancer. Cancer 1986; 58:2764–2772.

73. Haluska FG, Multani PS. Melanoma. Cancer Chemother Biol Response Modif 1999; 18:470–488.

74. Ganly I, Kirn D, Eckhardt SG, Rodriguez GI, Soutar DS, Otto R, Robertson AG, Park O, Gulley ML, Heise C, Von Hoff DD, Kaye SB. A phase I study of Onyx-015, an E1B attenuated adenovirus, administered intratumorally to patients with re-current head and neck cancer. Clin Cancer Res 2000; 6:798–806.

75. Nemunaitis J, Khuri F, Ganly I, Arseneau J, Posner M, Vokes E, Kuhn J, McCarty T, Landers S, Blackburn A, Romel L, Randlev B, Kaye S, Kirn D. Phase II trial of intratumoral administration of Onyx-015, a replication-selective adenovirus, in patients with refractory head and neck cancer. J Clin Oncol 2001; 19:289–298.

76. Hecht JR, Bedford R, Abbruzzese JL, Lahoti S, Lee M, Kirn DH. A phase I/II trial of intratumoral endoscopic ultrasound (EUS) injection of Onyx-015 with intravenous gemcitabine in unresectable pancreatic carcinoma. Proc Am Soc Clin Oncol 2000; A1039.

77. Bergsland E, Mani S, Kirn D, Fell S, Heise C, Maack C, Venook A, Warren R. Intratumoral injection of Onyx-015 for gastrointestinal tumors metastatic to the liver: a phase I trial. Proc Am Soc Clin Oncol 1998; A814.

78. Vasey PA, Seiden M, O'Neill V, Campo S, Johnston S, Davis J, Kirn D, Kaye SB, Shulman LN. Phase I trial of intraperitoneal Onyx-015 adenovirus in patients with recurrent ovarian cancer. Proc Am Soc Clin Oncol 2000; A1512.

79. Reid T, Galanis E, Abbrezzese J, Sze D, Romel L, Hatfield M, Rubin J, Kirn D. Hepatic artery infusion of Onyx-015, a selectively-replicating adenovirus in combi-nation with 5-FU/Leucovorin for gastrointestinal carcinoma metastatic to the liver: a phase I/II trial. Proc. Am Soc Clin Oncol 2000; A953.

80. Rudin CM, Recant W, Bennet P, Sulzen L, Kirn DH, Vokes EE. Preliminary report: adenovirus Onyx-015 administered by mouthwash as a chemopreventative agent for the treatment of oral dysplastic lesions. Proc Am Soc Clin Oncol 1999; A1715.

81. Nemunaitis J, Cunningham C, Edelman G, Berman B, Blackburn A, Buchanan A, Hatfield M, Kirn D, Crowley M. Phase I dose escalation trial of intravenous (IV) infusion of Onyx-015 in patients with refractory cancer. Proc Am Soc Clin Oncol 2000; A724.

82. Simmons JW, Mikhak B, vanderPoel HG, DeMarzo AM, Rodriguez R, Goemann MM, Nelson WG, Li S, Detorie N, Hamper UM, Ramakrisha N, DeWeese T. Molec-ular and clinical activity of CN706, a PSA selective oncolytic Ad5 vector in a phase I trial in locally recurrent prostate cancer following radiation therapy. Proc Am Soc Clin Oncol 2000; A1804.

83. DeWeese TL, Mickhak B, Ramakrishna N, DeMarzo A, Rodriguez R, Goemann MA, Drew R, Li S, Hamper UM, DeJong MR, Detorie N, Simmons JW. Bioactivity

of CN706, a PSA specific oncolytic adenoviral vector: a phase I trial of in vivo gene therapy for locally-recurrent prostate cancer following radiation therapy. Proc Am Assoc Cancer Res 2000; A783.

84. Henderson DR. Results of a Phase I/II study of CN706, a replication-competent cytolytic adenovirus, for the treatment of prostate cancer (PCa) that is locally recurrent following radiation therapy. Gene Ther Cancer IX 2000; A29rev.

85. Freytag SO, Kim JH, Khil M, Nafziger D, Menon M, Peabody J, Stricker H, Deperalta-Venturina M, Pegg J, Aguilar-Cordova E. Phase I study of replication-competent adenovirus-mediated double suicide gene therapy for local recurrence of prostate cancer after definitive radiation therapy. Cancer Gene Therapy IX 2000; A59.

86. Markert JM, Medlock MD, Rabkin SD, Gillespie GY, Todo T, Hunter WD, Palmer CA, Feigenbaum F, Tornatore C, Tufaro F, Martuza RL. Conditionally replicating herpes simplex virus mutant, G207 for the treatment of malignant glioma: results of a phase I trial. Gene Ther 2000; 7:867–874.

87. Rampling R, Cruickshank G, Papanastassiou V, Nicoll J, Hadley D, Brennan D, Petty R, MacLean A, Harland J, McKie E, Mabbs R, Brown M. Toxicity evaluation of replication-competent herpes simplex virus (ICP 34.5 null mutant 1716) in patients with recurrent malignant glioma. Gene Ther 2000; 7:859–866.

88. Norman KL, Lee PW. Reovirus as a novel oncolytic agent. J Clin Invest 2000; 105: 1035–1038.

89. Sinkovics JG, Horvath JC. Newcastle disease virus (NDV): brief history of its oncolytic strains. J Clin Virol 2000; 16:1–15.

90. Timiryasova TM, Li J, Chen B, Chong D, Langridge WH, Gridley DS, Fodor I. Antitumor effect of vaccinia virus in glioma model. Oncol Res 1999; 11:133–144.

91. Stojdl DF, Lichty B, Knowles S, Marius R, Atkins H, Sonenberg N, Bell JC. Exploiting tumor-specific defects in the interferon pathway with a previously unknown oncolytic virus. Nat Med 2000; 6:821–825.

13
Cardiovascular Gene Therapy

Mikko P. Turunen, Mikko O. Hiltunen, and Seppo Ylä-Herttuala
University of Kuopio, Kuopio, Finland

I. INTRODUCTION

Cardiovascular diseases remain one of the leading causes of death and disability in the Western world despite considerable preventive and therapeutic efforts. Therefore, gene therapy has become an altenative method of treatment for cardiovascular diseases during the past decade. Research has focused mainly on the treatment of restenosis and therapeutic angiogenesis, but as the knowledge of mechanisms associated with multifactorial disease, such as atherosclerosis, proceeds, the possibilities for the treatment of other cardiovascular diseases become more attractive. However, there are two main areas of research which need to progress before gene therapy can fulfill its expectations in the cardiovascular field: (1) Underlying pathological processes need to be better understood and (2) low efficiency and poor targeting of current gene transfer methods must be improved.

II. VECTOR REQUIREMENTS FOR VASCULAR GENE TRANSFER

Gene transfer vectors have been extensively discussed elsewhere in this book. Therefore, we present only some viewpoints for the requirements of vectors for cardiovascular gene transfer. Blood vessels are among the easiest targets for gene therapy because of ease of access. The possible target cells in arteries are endothelial cells, smooth muscle cells, macrophages, T cells, and fibroblasts [1]. Other cell types as potential targets for cardiovascular gene therapy are liver cells (e.g., low = density lipoprotein [LDL] receptor deficiency) [2], myoblasts (myocardial gene transfer) [3], and skeletel muscle cells (therapeutic angiogenesis) [4]. Even though cell proliferation is present in neointimal formation, (pseudo)retrovirus-mediated gene transfer has been disappointing [5]. For example, direct in vivo gene transfer to rabbit carotid artery with retroviruses leads only to a 0.05 % transfection efficiency. By using adenoviruses, a much higher transduction efficiency (5–10%) is achieved but the transduction is transient, lasting only for a few days or weeks [6]. Other viral systems, such as adeno-associated viruses or lentiviruses, could provide longer gene expression in vivo. On the other hand, direct in vivo gene transfer, even with the local delivery methods, is likely to cause some biodistribution of the vectors to various tissues [7]. Furthermore, in many cardiovascular gene therapy applications, only a temporary transgene expression is required [8]. Thus, it will be very important to develop further nonviral gene delivery methods for cardiovascular gene transfer.

Unfortunately, arterial gene transfer efficiency of liposomes has been low [5]. The transfection efficiency has been improved by using cationic polymers, such as fractured dendrimers and polyethylenimine (PEI) [9]. New electroporation systems are also interesting alternatives for cardiovascular gene delivery of plasmids and oligonucleotides [10–13]. Various physical methods, like ultrasound [14] and pressure-mediated gene transfer [15], have also been succesfully used for cardiovascular gene delivery.

III. GENE DELIVERY METHODS

A. Local Intravascular Gene Delivery

Intravascular gene transfer is easily performed during angioplasty, stenting, and other intravascular manipulations [1]. Limitations of intravascular gene transfer are the presence of anatomical barriers, such as the internal elastic lamina and atherosclerotic lesions [8,16,17] and the presence of the blood complement system which efficiently inactivates many gene transfer vectors [18]. Several catheters are commercially available which can be used for gene delivery [19,20]. A double-balloon catheter is made of two latex balloons which, when inflated in a target arterial segment, isolate a transfection chamber of varying length into which gene transfer solution can be infused. This catheter was used for catheter-based arterial gene transfer a decade ago [21]. The major limitations of this catheter type are seized blood flow and likely leakage through arterial side branches. The Dispatch Delivery catheter (Boston Scientific, Maple Grove, MN) is an infusion–perfusion balloon catheter which forms a separate compartment adjacent to the target vessel wall when the catheter is inflated. Prolonged gene transfer vector infusion can be performed because blood flows through a central core of the catheter. This system has been succesfully used to achieve substantial gene delivery into the endothelium and superficial medial layers of both normal and atherosclerotic rabbit and human arteries [6,22]. Porous and microporous catheters have also been used for arterial delivery of marker genes. The channeled balloon catheter has 24 longitudinal channels, each containing 100-μm pores, and also allows continuous blood flow into peripheral tissues [23]. Iontophoretic catheters are catheters that use an electroporation technique in combination with a balloon catheter, New catheters are continuously being developed for intravascular injections. These include the infusasleeve, the transport catheter, the stented porous balloon catheter and the nipple infusion catheter.

B. Perivascular Gene Delivery

When a gene transfer vector is administered on the adventitial surface of the vessel wall, it can stay in close contact with arterial cells for a long time. Adventitial gene transfer can be used for the delivery of therapeutic genes into the arterial wall during bypass operations, prosthetic and anastomosis surgery, and endarterectomies [1]. Adventitial gene delivery can be performed with Silastic [9,24] or a biodegradable collar [25], biodegradable gel [26], or direct injection into adventitia [27]. The limitation of this technique is that the gene transfer vector (or secreted/diffusible gene product) has to reach target cells, which in the case of intimal cells, is difficult to accomplish.

C. Systemic Gene Delivery

So far, only adenoviral vectors have been shown consistently to produce high levels of transgene expression following systemic administration to animals. Systemically administered gene transfer vectors usually transfect liver cells [28]. Therefore, diseases in which the liver is the target organ could be treated by systemic gene transfer. The inhibitory effect of blood components for many gene transfer vectors currently limit the use of systemic gene transfer [18]. If a vector is administered systemically, it should, owing to safety aspects, either have specificity to target cells or bear a specific promoter which is expressed only in certain cell types or diseased tissue. Current peptide libraries can provide new targeting possibilities for future gene transfer vectors [29].

IV. GENE THERAPY FOR CARDIOVASCULAR DISEASE

A. Therapeutic Angiogenesis

Several different angiogenic growth factors have been used to stimulate the development of collateral arteries [1]. The main focus of therapeutic angiogenesis has been the treatment of critical limb and myocardial ischemia, and the positive results of preclinical studies have led to several clinical trials. Below we will describe the most commonly used growth factors in therapeutic angiogenesis.

Vascular endothelial growth factors (VEGFs) are angiogenic growth factors that have significant roles in development, angiogenesis, and cancer [30]. Some VEGFs stimulate endothelial cell proliferation, increase endothelial permeability, regulate vascular tone, act as an endothelial "survival factor" in retinal vessels, and inhibit apoptosis in endothelial cells by inducing expression of antiapoptotic genes [31,32]. In addition to direct angiogenic effects, some VEGFs also induce release of vasoactive molecules, such as nitric oxide and prostacyclin from vascular endothelium [24].

VEGF-A gene is a distant relative of platelet-derived growth factor (PDGF) [30]. Five VEGF-A isoforms ($VEGF_{121}$, $VEGF_{145}$, $VEGF_{165}$, $VEGF_{189}$, and $VEGF_{206}$) are generated by alternative splicing from a single VEGF gene that is mapped to chromosome 6p21, and they are distinguished by their heparan sulfate–binding properties [1,30]. All splice variants are glycosylated and expressed as dimeric proteins. They also possess a secretion signal sequence, but only $VEGF_{121}$, $VEGF_{145}$, and $VEGF_{165}$ are secreted. In addition to mRNA splicing, expression of VEGF-A is regulated by proteolytic processing and hypoxia [30,33,34]. In response to hypoxia, the levels of transcription factor hypoxia-inducible factor-1 elevate and stimulate VEGF mRNA synthesis. In addition, the stability of VEGF mRNA is increased by hypoxia. Most of these splice variants bind to two tyrosine kinase receptors VEGFR-1 (flt-1) and VEGFR-2 (KDR/flk-1), which share 44% homology with each other and are expressed almost exclusively on endothelial cells. The mitogenic effects of VEGF are mediated through VEGFR-2 but not VEGFR-1. Neuropilin-1 and Neuropilin-2 are nonprotein tyrosine kinase receptors that also bind $VEGF_{165}$ [35]. VEGFR-2 mediates the mitogenic effect of VEGF by activating signaling pathways, like mitogen-activated protein kinase (MAPK) in endothelial cells. VEGFR-1 lacks this mitogenic capacity, and its function in endothelial cells is mainly unknown, although the transduction pathway suggests that activation of VEGFR-1 induces endothelial cell migration. Recently, VEGF has also been shown to regulate apoptosis of endothelial cells [36–38].

Other members of the VEGF family have also been characterized. VEGFR-1 binds VEGF-B, which share similar mitogenic characteristics as VEGF-A and is able to form heterodimers with VEGF-A [39,40]. VEGFR-2 and VEGFR-3 (flt-4) bind VEGF-C and VEGF-D [41]. VEGFR-3 differs from other VEGF receptors by its characteristic expression and angiogenic effect in lymphatic vessels [42]. The newest member of the growing VEGF family is secreted dimer VEGF-E, which has similar functional characteristics as $VEGF_{165}$, but binds only to VEGFR-2 [43],[44]. VEGF-E is encoded by the Orf virus.

A single dose of recombinant VEGF protein in the blood stream or locally in the vessel wall has the capacity to accelerate reendothelization in balloon-injured rat carotid artery [45]. Also, recombinant VEGF-C has the capacity to induce angiogenesis in vivo [46,47]. Injection of VEGF plasmid in ischemic rabbit hind limbs and the adventitial surface of rabbit carotid arteries has been shown to improve the status of the treated vessels [24,48]. Preliminary results from the beneficial effects of $VEGF_{165}$ gene transfer in human peripheral arteries have also been reported [49].

It has been shown that VEGF is expressed in tumors promoting tumor angiogenesis and that tumor growth can be inhibited with anti-VEGF antibodies [50,51,52]. Thus, it is possible that VEGF, if released to the systemic circulation during gene transfer, may promote tumor growth elsewhere in the body.

Fibroblast growth factor (FGF) acts directly on vascular cells and induces endothelial cell growth and angiogenesis [53–57] In a model of stress-induced myocardial ischemia, intracoronary injection of recombinant adenovirus expressing human FGF-5 resulted in improved stress-induced function and increased blood flow 2 weeks after the gene transfer [53]. These improvements were associated with evidence of angiogenesis. Adenovirus-mediated expression of the secreted form of FGF-2 has been shown to induce cellular proliferation and angiogenesis in mice, suggesting its use for therapeutic angiogenesis in vivo [54]. Also, direct injection of recombinant FGF protein into myocardium has led to angiogenesis in the human heart [57].

Angiopoietin-1 (Ang1) is an angiogenic factor that signals through endothelial cell–specific Tie2 receptor tyrosine kinase [58]. Like VEGF, Ang1 is essential for normal vascular development, and its overexpression leads to increased angiogenesis [59]. Ang2 has been shown to be a naturally occurring antagonist for Ang1 and Tie2, and overexpression of Ang2 disrupts blood vessel formation in the mouse embryo [60].

Platelet-derived growth factor (PDGF) participates in angiogenic process by paracrine recruitment of mural cell precursors to the vessel wall and by the autocrine stimulation of endothelial cells. So-called scatter factor/hepatocyte growth factor (SF/HGF) is known to be a pleiotropic growth factor that stimulates proliferation and migration of endothelial cells via the c-Met receptor present on endothelial cells and smooth muscle cells [61]. Recent studies have shown that application of recombinant SF/HGF in vivo to an ischemic rabbit hind limb model is associated with significant improvements in collateral formation and regional blood flow [61].

There are numerous factors contributing to angiogenesis and arteriogenesis [62], and in the future, factors like monocyte chemotactic protein-1 (MCP-1) [63,64] or developmentally regulated endothelial cell locus-1 (DEL-1) (Valentis, Inc., Burlingame, CA) could provide new possibilities for therapeutic collateral formation.

B. Restenosis

Neointimal hyperplasia and arterial remodeling cause restenosis in 20–30% of patients following percutaneous transluminal coronary angioplasty (PTCA). There has been so far very little success in reducing the frequency of restenosis. Smooth muscle cells (SMCs) in the blood vessel wall are normally present in a relatively quiescent state. The change in the SMC phenotype to the proliferative state is an important mechanism in the development of restenosis. A number of growth factors are involved in the induction of restenosis such as PDGF, acidic and basic FGFs and transforming growth factor-β (TGF-β). PDGFs were initially discovered as growth factors released from platelets at the sites of vascular injury. Later on it was found that PDGFs are also produced by macrophages and SMCs in atherosclerotic lesions. Currently, PDGFs are known to be among the most potent chemoattractants and mitogens for SMCs. PDGF also promotes malignant transformation and is a survival factor that protects cells from apoptosis. In addition, PDGF stimulates production of extracellular matrix components. PDGF β-receptor antisense oligonucleotides have been used to block the PDGF signaling in the rat carotid artery denudation model [65,66].

Intimal hyperplasia after vascular manipulations can be limited with gene transfer. Gene delivery of thymidine kinase combined with ganciclovir medication, apoptosis regulator bcl-x, and arterial transfection with tumor-suppressor genes p53 or retinoblastoma limit intimal hyperplasia after denudation in normal and atherosclerotic arteries [67–70]. Decoy and antisense oligonucleotides against cellular targets, such as E2F and kinases, inhibit neointimal proliferation [71,72]. It has been shown that a growth-arrest homeobox gene (gax) is rapidly downregulated in vascular SMCs after mitogen stimulation in vitro and after vascular injury in vivo, and that gax-mediated gene transfer efficiently blocks SMC proliferation in a restenosis model in rats [73,74]. Increased expression of tissue inhibitor of metalloproteinases (TIMP) and endothelial nitric oxide synthase (eNOS) reduces intimal thickening [75,76], and administration of antibodies against agents that cause cell proliferation, such as PDGF and FGF, has been shown to limit the neointimal response to balloon injury in rats [77]. Proliferation of endothelial cells stimulated by VEGF leads to improved reendothelization after injury, which may reduce neointimal proliferation [24].

Promising results in the prevention of SMC proliferation have been accomplished in animal models by using the VEGF-A gene [78]. Inhibition of SMC proliferation is caused at least partly by NO production from the endothelium [24]. It might also be possible to improve the effect of VEGF by combining it with other angiogenic factors, such as angiopoietins. In the future, combination treatments may become useful for gene therapy of restenosis and angiogenesis.

Various antisense strategies have been used to block cell cycle genes for the treatment of restenosis [79–81] However, oligonucleotides may have unspecific biological effects [11,82]. In order to develop novel and effective therapies for restenosis, the function and regulatory mechanism of different growth factors and vascular mediators must be better understood.

C. Atherosclerosis and Hyperlipidemia

Several cardiovascular risk factors contribute to the development of atherosclerosis. Risk factors for atherosclerosis include hypercholesterolemia, hypertension, smoking, diabetes, male gender, positive family history, and possibly infections. Atherosclerosis is the principal underlying cause responsible for the development of myocardial and cerebral infarction, angina pectoris, and peripheral artery disease. During atherogenesis, accumulation of lipids and lipoproteins together with SMC proliferation in the arterial wall cause obstructive lesions and clinically manifesting disease. The prevention and treatment includes dietary restrictions together with lipid-lowering medication and percutaneous and surgical operations. Even though gene therapy may not be useful for the primary prevention of atherosclerosis, some forms of hyperlipidemias might be amenable to gene therapy.

Liver gene therapy has become a promising new alternative for the treatment of LDL receptor deficiency and other inherited liver diseases. The Watanabe heritable hyperlipidemic (WHHL) rabbit can be used as an animal model for familial hypercholesterolemia (FH) gene therapy, since these rabbits have a 12-nucleotide deletion in their LDL receptor gene [83] leading to a similar dyslipidemia as in human FH. Both ex vivo [84] and in vivo [85,86] approaches have been used in liver-directed gene therapy. Other gene therapy options for various types of hyperlipidemias include very low-density lipoprotein (VLDL) receptor, lipoprotein lipase, hepatic lipase, apolipoprotein A-I, and apolipoprotein E [2].

D. Stents

Intravascular stents are devices that are deployed to open the vessel and prevent closure. Neointimal formation and thrombosis after stent implantation are major disadvantages of this technology [87,88]. Gene therapy approaches have been utilized to prevent these pathological processes [89]. Polymer-coated stents have been used for delivery of plasmid DNA [90]. Ex vivo gene transfer has been performed with stents covered with transduced endothelial cells [91]. Unfortunately, balloon expansion is traumatic to seeded cells, and cells are easily lost in a flow enviroment. Tissue plasminogen activator (tPA), a thrombolytic proteinase, can be overexpressed by seeded endothelial cells in stents [91]. However, protease activity of upregulated tPA on the matrix might decrease the implantation efficiency of seeded cells. Also, stents are impractical in treating very little arteries and arteries with complicated anatomy. Further developments in stenting are needed, and gene therapy could lead to a better outcome of the procedure in the future.

V. ANIMAL MODELS AND NEW METHODS FOR STUDYING CARDIOVASCULAR DISEASE

Genetically modified animal models are useful for cardiovascular studies, and they provide many useful ways to study human diseases. For example, atherosclerosis is a multifactorial disease which develops slowly and is difficult to study using standard laboratory animals. Gene knockout technologies have provided researchers with a powerful tool for studying the effects of single genes in cardiovascular disease. The mechanisms of apolipoprotein functions have been studied using knockout mice, and these animals can also be used for gene therapy experiments.[1]

Many restenosis studies are carried out in a rat carotid artery model where common carotid arteries are subjected to balloon injury creating intimal accumulation of SMCs. Also, similar models have been used in rabbits and pigs. A cholesterol diet can be used further to enhance the restenotic process. These models typically produce thick neointima which responds readily to treatment genes [68,76,92–99].

DNA array is a tool to detect events at the transcriptional level during disease progression and to identify gene polymorphisms predisposing to or protecting from disease [100,101]. Currently, the expression of 10–15% of human genes can be analyzed simultaneously in a single experiment using cDNA or an oligonucleotide-based format of DNA array. Alternatively, smaller DNA arrays with a limited number of selected genes, such as cytokines, growth factors, or transcription factors, can be used. Better understanding of molecular mechanisms, such as gene expression profiling of vascular wall thickening, may lead to more specific treatment strategies for the prevention of these diseases.

The role of endothelial dysfunction in the pathogenesis of atherosclerosis is well recognized. Several studies have suggested that oxidatively modified LDL (Ox-LDL) is a key component in the cause of endothelial injury [102]. DNA array technology and serial analysis of gene expression (SAGE) [103] have been successfully used to identify differentially regulated genes in endothelial cells which, in response to systemic risk factors, contribute to the dysfunctional phenotype of the endothelium [104].

VI. HUMAN GENE THERAPY TRIALS FOR VASCULAR DISEASES

A. Peripheral Vascular Disease

The goal for the treatment of peripheral vascular disease is to induce angiogenesis in ischemic tissues and prevent restenosis after angioplasty. The first human vascular gene therapy protocols have been started with patients suffering from chronic critical leg ischemia. Naked VEGF cDNA has been introduced in patients with leg ischemia via hydrogel balloon catheter to create neovessels in the ischemic peripheral tissues [105]. Recently, high-dose intramuscular naked VEGF plasmid injections have been shown to cause beneficial clinical effects, such as reliefing pain and healing of ischemic ulcerations, in patients with critical limb ischemia. [4] These findings suggest that VEGF is a potent factor for gene therapy in the treatment of peripheral vascular disease. Also, FGF-1,-2, and-4 have shown positive results in preclinical experiments and are now being tested in several multicenter trials for treatment of peripheral vascular disease [1].

B. Coronary Heart Disease

Intramyocardial injections of naked VEGF plasmid have been performed via mini thoracotomy in patients with end-stage coronary heart disease [106]. Intramyocardial VEGF$_{121}$ adenoviral injections have been performed during coronary artery bypass surgery [107]. The first clinical studies with FGF recombinant protein suggest that direct FGF-1 injections into the myocardium during coronary bypass surgery lead to the formation of neovessels in the hearts of bypass patients [57]. Clinical trials have also been started using intracoronary FGF-4 adenoviral injections and E2F decoy oligonucleotides [81]. Catheter-mediated VEGF plasmid/liposome gene transfer in human coronary arteries after PTCA has been well tolerated and is potentially applicable for the prevention of restenosis and myocardial ischemia [108].

VII. SAFETY ASPECTS

It has been recognized that ectopic transgene expression is a possible drawback of gene therapy. After catheter-mediated intra-arterial delivery in rabbits [7] and after tail vein injection of adenovirus in mice [28], it has been demonstrated by sensitive polymerase chain reaction (PCR) methods that viral DNA can be found in various tissues, such as the liver, lung, kidney, and testis. The infectivity and expression of adenoviral genes is dependent on the presence of specific receptors, promoter elements, and local immune defense [109,110]. Safety aspects and biodistribution of the vectors need to be better understood before safe tools for clinical treatments can be produced.

The use of viral vectors in gene therapy has raised questions concerning their immunogenity, oncogenic properties, and unknown long-term effects. First-generation adenoviruses produce inflammatory and immunological reactions, but hopefully improved designs (second-generation, gutless vectors) will solve these problems [111–113].

Nonviral delivery of plasmid DNA has been considered to be relatively safe. Even though some nonviral vehicles have cellular toxicity, the major problem has been the low efficacy of these systems [9]. There is great interest in developing more efficient and targeted nonviral gene delivery vehicles. Gene transfer with plasmid DNA results only in transient expression, which should be sufficient for therapeutic effects in many cardiovascular applications [1,114].

VIII. SUMMARY

It is important to characterize further the mechanisms of angiogenesis, restenosis, and other atherosclerosis-related diseases so that new therapeutic strategies can be identified and evaluated in animal models. Local gene transfer into the vascular wall offers a promising alternative to treat atherosclerosis-related diseases. Blood vessels are among the easiest targets for gene therapy because of the availability of percutaneous, catheter-based treatment methods. On the other hand, gene transfer to the artery wall can also be accomplished from adventitia, and in some situations, intramuscular gene delivery is a possibility. Furthermore, it is likely that delivering only one gene may not be enough to treat advanced atherosclerotic diseases and ischemic tissues [1]. In the future, several factors might be combined, and these "gene cocktails" are expected to produce enhanced therapeutic effects in vascular gene therapy.

   Development of new vectors for gene transfer will provide more efficient and safer methods for gene delivery. For example, plasmids and viruses coding for more than one protein, and bearing regulatory elements, would be useful for future gene therapy applications. In conditions in which stable expression of therapeutic proteins are needed, it is necessary to develop better ex vivo and in vivo gene transfer protocols. Also, production of viruses that can efficiently transfect nondividing cells will be important for the future applications of vascular gene therapy.

   Current clinical trials have shown that gene therapy has provided a promising new approach for the treatment of tissue ischemia, restenosis, and other vascular diseases [1]. Better understanding of molecular mechanisms, such as gene expression profiling, of vascular wall thickening may lead to more specific treatment strategies for the prevention of these diseases [115].

ACKNOWLEDGMENT

This study was supported by Kuopio University Hospital (Evo grant 5130).

REFERENCES

1. Ylä-Herttuala S, Martin JF. Cardiovascular gene therapy. Lancet 2000; 355:213–222.

2. Pakkanen T, Ylä-Herttuala S. Gene therapy for dyslipidaemias. Eur Heart J Suppl 2000; D62–D64.
3. Kornowski R, Fuchs S, Leon MB, Epstein SE. Delivery strategies to achieve therapeutic myocardial angiogenesis. Circulation 2000; 101:454–458.
4. Baumgartner I, Pieczek A, Manor O, Blair R, Kearney M, Walsh K, et al. Constitutive expression of phVEGF$_{165}$ after intramuscular gene transfer promotes collateral vessel development in patients with critical limb ischemia. Circulation 1998; 97: 1114–1123.
5. Laitinen M, Pakkanen T, Donetti E, Baetta R, Luoma J, Lehtolainen P, et al. Gene transfer into the carotid artery using an adventitial collar: comparison of the effectiveness of the plasmid- liposome complexes, retroviruses, pseudotyped retroviruses, and adenoviruses. Hum Gene Ther 1997; 8:1645–1650.
6. Laitinen M, Mäkinen K, Manninen H, Matsi P, Kossila M, Agrawal RS, et al. Adenovirus-mediated gene transfer to lower limb artery of patients with chronic critical leg ischemia. Hum Gene Ther 1998; 9:1481–1486.
7. Hiltunen MO, Turunen MP, Turunen AM, Rissanen TT, Laitinen M, Kosma V-M, et al. Biodistribution of adenoviral vector to non-target tissues after in vivo gene transfer to arterial wall using intravascular and periadventitial gene delivery methods. FASEB J. 2000; 14:2230–2236.
8. Ylä-Herttuala S. Vascular gene transfer. Curr Opin Lipidol 1997; 8:72–76.
9. Turunen MP, Hiltunen MO, Ruponen M, Virkamäki L, Szoka FCJ, Urtti A, et al. Efficient adventitial gene delivery to rabbit carotid artery with cationic polymer-plasmid complexes. Gene Ther 1999; 6:6–11.
10. Aihara H, Miyazaki J. Gene transfer into muscle by electroporation in vivo. Nat Biotechnol 1998; 16:867–870.
11. Flanagan WM, Wagner RW. Potent and selective gene inhibition using antisense oligodeoxynucleotides. Mol Cell Biochem 1997; 172:213–225.
12. Mir LM, Bureau MF, Gehl J, Rangara R, Rouy D, Caillaud JM, et al. High-efficiency gene transfer into skeletal muscle mediated by electric pulses. Proc Natl Acad Sci USA 1999; 96:4262–4267.
13. Vicat JM, Boisseau S, Jourdes P, Laine M, Wion D, Bouali-Benazzouz R, et al. Muscle transfection by electroporation with high-voltage and short-pulse currents provides high-level and long-lasting gene expression. Hum Gene Ther 2000; 11: 909–916.
14. Lawrie A, Brisken AF, Francis SE, Tayler DI, Chamberlain J, Crossman DC, et al. Ultrasound enhances reporter gene expression after transfection of vascular cells in vitro. Circulation 1999; 99:2617–2620.
15. Budker V, Zhang G, Danko I, Williams P, Wolff J. The efficient expression of intravascularly delivered DNA in rat muscle. Gene Ther 1998; 5:272–276.
16. Rome JJ, Shayani V, Flugelman MY, Newman KD, Farb A, Virmani R, et al. Anatomic barriers influence the distribution of in vivo gene transfer into the arterial wall. Modeling with microscopic tracer particles and verification with a recombinant adenoviral vector. Arterioscler Thromb 1994; 14:148–161.
17. Feldman LJ, Isner JM. Gene therapy for the vulnerable plaque. J Am Coll Cardiol 1995; 26:826–835.
18. Plank C, Mechtler K, Szoka-FC J, Wagner E. Activation of the complement system

by synthetic DNA complexes: a potential barrier for intravenous gene delivery. Hum. Gene Ther 1996; 7:1437–1446.

19. Willard JE, Landau C, Glamann DB, Burns D, Jessen ME, Pirwitz MJ, et al. Genetic modification of the vessel wall. Comparison of surgical and catheter-based techniques for delivery of recombinant adenovirus. Circulation 1994; 89:2190–2197.

20. Plautz GE, Nabel EG, Fox B, Yang ZY, Jaffe M, Gordon D, et al. Direct gene transfer for the understanding and treatment of human disease. Ann NY Acad Sci 1994; 716:144–153.

21. Nabel EG, Plautz G, Nabel GJ. Site-specific gene expression in vivo by direct gene transfer into the arterial wall. Science 1990; 249:1285–1288.

22. Tahlil O, Brami M, Feldman LJ, Branellec D, Steg PG. The dispatch(tm) catheter as a delivery tool for arterial gene transfer. Cardiovasc Res 1997; 33:181–187.

23. Hong MK, Wong SC, Farb A, Mehlman MD, Virmani R, Barry JJ, et al. Feasibility and drug delivery efficiency of a new balloon angioplasty catheter capable of performing simultaneous local drug delivery. Coron Artery Dis 1993; 4:1023–1027.

24. Laitinen M, Zachary I, Breier G, Pakkanen T, Häkkinen T, Luoma J, et al. Vegf gene transfer reduces intimal thickening via increased production of nitric oxide in carotid arteries. Hum Gene Ther 1997; 8:1737–1744.

25. Pakkanen TM, Laitinen M, Hippeläinen M, Hiltunen MO, Alhava E, Ylä-Herttuala S. Periadventitial lacZ gene transfer to pig carotid arteries using a biodegradable collagen collar or a wrap of collagen sheet with adenoviruses and plasmid-liposome complexes. J Gene Med 2000; 2:52–60.

26. Stephan DJ, Yang ZY, San H, Simari RD, Wheeler CJ, Felgner PL, et al. A new cationic liposome dna complex enhances the efficiency of arterial gene transfer in vivo. Hum Gene Ther 1996; 7:1803–1812.

27. Rios CD, Ooboshi H, Piegors D, Davidson BL, Heistad DD. Adenovirus-mediated gene transfer to normal and atherosclerotic arteries. A novel approach. Arterioscler Thromb Vasc Biol 1995; 15:2241–2245.

28. Ye XH, Gao GP, Pabin C, Raper SE, Wilson JM. Evaluating the potential of germ line transmission after intravenous administration of recombinant adenovirus in the C3H mouse. Hum Gene Ther 1998; 9:2135–2142.

29. Koivunen E, Arap W, Valtanen H, Rainisalo A, Medina OP, Heikkilä P, et al. Tumor targeting with a selective gelatinase inhibitor. Nat Biotechnol 1999; 17: 768–774.

30. Ferrara N, Bunting S. Vascular endothelial growth factor, a specific regulator of angiogenesis. Curr Opin Nephrol Hypertens 1996; 5:35–44.

31. Murohara T, Horowitz JR, Silver M, Tsurumi Y, Chen D, Sullivan A, et al. Vascular endothelial growth factor/vascular permeability factor enhances vascular permeability via nitric oxide and prostacyclin. Circulation 1998; 97:99–107.

32. Gerber HP, Dixit V, Ferrara N. Vascular endothelial growth factor induces expression of the antiapoptotic proteins Bcl–2 and A1 in vascular endothelial cells. J Biol Chem 1998; 273:13313–13316.

33. Risau W. Mechanisms of angiogenesis. Nature 1997; 386:671–674.

34. Plouet J, Moro F, Bertagnolli S, Coldeboeuf N, Mazarguil H, Clamens S, et al. Extracellular cleavage of the vascular endothelial growth factor 189- amino acid

form by urokinase is required for its mitogenic effect. J Biol Chem 1997; 272: 13390–13396.

35. Neufeld G, Cohen T, Gengrinovitch S, Poltorak Z. Vascular endothelial growth factor (VEGF) and its receptors. FASEB J 1999; 13:9–22.

36. Gerber HP, McMurtrey A, Kowalski J, Yan M, Keyt BA, Dixit V, et al. Vascular endothelial growth factor regulates endothelial cell survival through the phosphatidylinositol 3-kinase/Akt signal transduction pathway. Requirement for Flk-1/KDR activation. J Biol Chem 1998; 273:30336–30343.

37. Gupta S, Gorla GR, Irani AN. Hepatocyte transplantation: emerging insights into mechanisms of liver repopulation and their relevance to potential therapies. J Hepatol 1999; 30:162–170.

38. Nor JE, Christensen J, Mooney DJ, Polverini PJ. Vascular endothelial growth factor (VEGF)-mediated angiogenesis is associated with enhanced endothelial cell survival and induction of Bcl-2 expression. Am J Pathol 1999; 154:375–384.

39. Olofsson B, Korpelainen E, Pepper MS, Mandriota SJ, Aase K, Kumar V, et al. Vascular endothelial growth factor B (VEGF-B) binds to VEGF receptor-1 and regulates plasminogen activator activity in endothelial cells. Proc Natl Acad Sci USA 1998; 95:11709–11714.

40. Olofsson B, Pajusola K, Kaipainen A, von Euler G, Joukov V, Saksela O, et al. Vascular endothelial growth factor B, a novel growth factor for endothelial cells. Proc Natl Acad Sci USA 1996; 93:2576–2581.

41. Achen MG, Jeltsch M, Kukk E, Makinen T, Vitali A, Wilks AF, et al. Vascular endothelial growth factor D (VEGF-D) is a ligand for the tyrosine kinases VEGF receptor 2 (Flk1) and VEGF receptor 3 (Flt4). Proc Natl Acad Sci USA 1998; 95: 548–553.

42. Paavonen K, Puolakkainen P, Jussila L, Jahkola T, Alitalo K. Vascular endothelial growth factor receptor-3 in lymphangiogenesis in wound healing. Am J Pathol 2000; 156:1499–1504.

43. Ogawa S, Oku A, Sawano A, Yamaguchi S, Yazaki Y, Shibuya M. A novel type of vascular endothelial growth factor, VEGF-E (NZ-7 VEGF), preferentially utilizes KDR/Flk-1 receptor and carries a potent mitotic activity without heparin-binding domain. J Biol Chem 1998; 273:31273–31282.

44. Meyer M, Clauss M, Lepple-Wienhues A, Waltenberger J, Augustin HG, Ziche M, et al. A novel vascular endothelial growth factor encoded by orf virus, VEGF-E, mediates angiogenesis via signalling through VEGFR-2 (KDR) but not VEGFR-1 (Flt-1) receptor tyrosine kinases. EMBO J 1999; 18:363–374.

45. Asahara T, Bauters C, Pastore C, Kearney M, Rossow S, Bunting S, et al. Local delivery of vascular endothelial growth factor accelerates reendothelialization and attenuates intimal hyperplasia in balloon-injured rat carotid artery. Circulation 1995; 91:2793–2801.

46. Cao Y, Linden P, Farnebo J, Cao R, Eriksson A, Kumar V, et al. Vascular endothelial growth factor C induces angiogenesis in vivo. Proc Natl Acad Sci USA 1998; 95:14389–14394.

47. Hiltunen MO, Turunen MP, Turunen AM, Rissanen TT, Laitinen M, Kosma V-M, et al. Intravascular adenovirus-mediated VEGF-C gene transfer reduces neointima formation in balloon-denuded rabbit aorta. Circulation 2000; 102:2262–2268.

48. Tsurumi Y, Kearney M, Chen DF, Silver M, Takeshita S, Yang JH, et al. Treatment of acute limb ischemia by intramuscular injection of vascular endothelial growth factor gene. Circulation 1997; 96:382–388.

49. Baumgartner I, Isner JM. Stimulation of peripheral angiogenesis by vascular endothelial growth factor (VEGF). Vasa 1998; 27:201–206.

50. Nicosia RF. What is the role of vascular endothelial growth factor-related molecules in tumor angiogenesis? Am J Pathol 1998; 153:11–16.

51. Ferrara N. Molecular and biological properties of vascular endothelial growth factor. J Mol Med 1999; 77:527–543.

52. Ferrara N, Alitalo K. Clinical applications of angiogenic growth factors and their inhibitors. Nat Med 1999; 5:1359–1364.

53. Giordano FJ, Ping P, McKirnan MD, Nozaki S, DeMaria AN, Dillmann WH, et al. Intracoronary gene transfer of fibroblast growth factor-5 increases blood flow and contractile function in an ischemic region of the heart. Nat Med 1996; 2:534–539.

54. Ueno H, Li JJ, Masuda S, Qi Z, Yamamoto H, Takeshita A. Adenovirus-mediated expression of the secreted form of basic fibroblast growth factor (FGF-2) induces cellular proliferation and angiogenesis in vivo. Arterioscler Thromb Vasc Biol 1997; 17:2453–2460.

55. Tabata H, Silver M, Isner JM. Arterial gene transfer of acidic fibroblast growth factor for therapeutic angiogenesis in vivo: critical role of secretion signal in use of naked DNA. Cardiovasc Res 1997; 35:470–479.

56. Nabel EG, Yang Z, Liptay S, San H, Gordon D, Haudenschild CC, et al. Recombinant platelet-derived growth factor B gene expression in porcine arteries induce intimal hyperplasia in vivo. J Clin Invest 1993; 91:1822–1829.

57. Schumacher B, Pecher P, von Specht BU, Stegmann T. Induction of neoangiogenesis in ischemic myocardium by human growth factors: first clinical results of a new treatment of coronary heart disease. Circulation 1998; 97:645–650.

58. Davis S, Aldrich TH, Jones PF, Acheson A, Compton DL, Jain V, et al. Isolation of angiopoietin-1, a ligand for the TIE2 receptor, by secretion-trap expression cloning. Cell 1996; 87:1161–1169.

59. Suri C, Jones PF, Patan S, Bartunkova S, Maisonpierre PC, Davis S, et al. Requisite role of angiopoietin-1, a ligand for the TIE2 receptor, during embryonic angiogenesis. Cell 1996; 87:1171–1180.

60. Maisonpierre PC, Suri C, Jones PF, Bartunkova S, Wiegand SJ, Radziejewski C, et al. Angiopoietin-2, a natural antagonist for Tie2 that disrupts in vivo angiogenesis. Science 1997; 277:55–60.

61. Van Belle E, Witzenbichler B, Chen D, Silver M, Chang L, Schwall R, et al. Potentiated angiogenic effect of scatter factor/hepatocyte growth factor via induction of vascular endothelial growth factor: the case for paracrine amplification of angiogenesis. Circulation 1998; 97:381–390.

62. Carmeliet P. Mechanisms of angiogenesis and arteriogenesis. Nat Med 2000; 6:389–395.

63. Ito WD, Arras M, Winkler B, Scholz D, Schaper J, Schaper W. Monocyte chemotactic protein-1 increases collateral and peripheral conductance after femoral artery occlusion. Circ Res 1997; 80:829–837.

64. Salcedo R, Ponce ML, Young HA, Wasserman K, Ward JM, Kleinman HK, et al. Human endothelial cells express CCR2 and respond to MCP-1: direct role of MCP-1 in angiogenesis and tumor progression. Blood 2000; 96:34–40.

65. Sirois MG, Simons M, Edelman ER. Antisense oligonucleotide inhibition of PDGFR-beta receptor subunit expression directs suppression of intimal thickening [see comments]. Circulation 1997; 95:669–676.

66. Hart CE, Kraiss LW, Vergel S, Gilbertson D, Kenagy R, Kirkman T, et al. PDGFbeta receptor blockade inhibits intimal hyperplasia in the baboon. Circulation 1999; 99:564–569.

67. Ohno T, Gordon D, San H, Pompili VJ, Imperiale MJ, Nabel GJ, et al. Gene therapy for vascular smooth muscle cell proliferation after arterial injury. Science 1994; 265:781–784.

68. Guzman RJ, Hirschowitz EA, Brody SL, Crystal RG, Epstein SE, Finkel T. In vivo suppression of injury-induced vascular smooth muscle cell accumulation using adenovirus-mediated transfer of the herpes simplex virus thymidine kinase gene. Proc Natl Acad Sci USA 1994; 91:10732–10736.

69. Pollman MJ, Hall JL, Mann MJ, Zhang L, Gibbons GH. Inhibition of neointimal cell bcl-x expression induces apoptosis and regression of vascular disease. Nat Med 1998; 4:222–227.

70. Steg PG, Tahlil O, Aubailly N, Caillaud JM, Dedieu JF, Berthelot K, et al. Reduction of restenosis after angioplasty in an atheromatous rabbit model by suicide gene therapy. Circulation 1997; 96:408–411.

71. Morishita R, Higaki J, Tomita N, Ogihara T. Application of transcription factor "decoy" strategy as means of gene therapy and study of gene expression in cardiovascular disease. Circ Res 1998; 82:1023–1028.

72. Suzuki J, Isobe M, Morishita R, Aoki M, Horie S, Okubo Y, et al. Prevention of graft coronary arteriosclerosis by antisense cdk2 kinase oligonucleotide. Nat Med 1997; 3:900–903.

73. Skopicki HA, Lyons GE, Schatteman G, Smith RC, Andres V, Schirm S, et al. Embryonic expression of the Gax homeodomain protein in cardiac, smooth, and skeletal muscle. Circ Res 1997; 80:452–462.

74. Smith RC, Branellec D, Gorski DH, Guo K, Perlman H, Dedieu JF, et al. p21CIP1-mediated inhibition of cell proliferation by overexpression of the gax homeodomain gene. Genes Dev 1997; 11:1674–1689.

75. Baker AH, Zaltsman AB, George SJ, Newby AC. Divergent effects of tissue inhibitor of metalloproteinase-1, -2, or -3 overexpression on rat vascular smooth muscle cell invasion, proliferation, and death in vitro. TIMP-3 promotes apoptosis. J Clin Invest 1998; 101:1478–1487.

76. von der Leyen HE, Gibbons GH, Morishita R, Lewis NP, Zhang L, Nakajima M, et al. Gene therapy inhibiting neointimal vascular lesion: in vivo transfer of endothelial cell nitric oxide synthase gene. Proc Natl Acad Sci USA 1995; 92:1137–1141.

77. Rutherford C, Martin W, Salame M, Carrier M, Anggard E, Ferns G. Substantial inhibition of neo-intimal response to balloon injury in the rat carotid artery using a combination of antibodies to platelet-derived growth factor-BB and basic fibroblast growth factor. Atherosclerosis 1997; 130:45–51.

78. Isner JM, Walsh K, Rosenfield K, Schainfeld R, Asahara T, Hogan K, et al. Arterial gene therapy for restenosis. Hum Gene Ther 1996; 7:989–1011.

79. Mann MJ, Gibbons GH, Tsao PS, von der Leyen HE, Cooke JP, Buitrago R, et al. Cell cycle inhibition preserves endothelial function in genetically engineered rabbit vein grafts. J Clin Invest 1997; 99:1295–1301.

80. Mann MJ, Whittemore AD, Donaldson MC, Belkin M, Conte MS, Polak JF, et al. Ex-vivo gene therapy of human vascular bypass grafts with E2F decoy: the PRE-VENT single-centre, randomised, controlled trial. Lancet 1999; 354:1493–1498.

81. Mann MJ, Whittemore AD, Donaldson MC, Belkin MA, Orav EJ, Polak J, et al. Preliminary experience with genetic engineering of human vein grafts: evidence for target gene inhibition (abst). Circulation 1997; 96:I–4, 20.

82. Burgess TL, Fisher EF, Ross SL, Bready JV, Qian YX, Bayewitch LA, et al. The antiproliferative activity of c-myb and c-myc antisense oligonucleotides in smooth muscle cells is caused by a nonantisense mechanism. Proc Natl Acad Sci USA 1995; 92:4051–4055.

83. Yamamoto T, Bishop RW, Brown MS, Goldstein JL, Russell DW. Deletion in cysteine-rich region of LDL receptor impedes transport to cell surface in WHHL rabbit. Science 1986; 232:1230–1237.

84. Grossman M, Raper SE, Kozarsky K, Stein EA, Engelhardt JF, Muller D, et al. Successful ex vivo gene therapy directed to liver in a patient with familial hyper-cholesterolaemia. Nat Genet 1994; 6:335–341.

85. Branchereau S, Calise D, Ferry N. Factors influencing retroviral-mediated gene transfer into hepatocytes in vivo. Hum Gene Ther 1994; 5:803–808.

86. Pakkanen TM, Laitinen M, Hippeläinen M, Kallionpää H, Lehtolainen P, Leppänen P, et al. Enhanced plasma cholesterol lowering effect of retrovirus- mediated LDL receptor gene transfer to WHHL rabbit liver after improved surgical technique and stimulation of hepatocyte proliferation by combined partial liver resection and thymidine kinase ganciclovir treatment. Gene Ther 1999; 6:34–41.

87. Bai H, Masuda J, Sawa Y, Nakano S, Shirakura R, Shimazaki Y, et al. Neointima formation after vascular stent implantation. Spatial and chronological distribution of smooth muscle cell proliferation and phenotypic modulation. Arterioscler Thromb 1994; 14:1846–1853.

88. Mintz GS, Hoffmann R, Mehran R, Pichard AD, Kent KM, Satler LF, et al. In-stent restenosis: the Washington Hospital Center experience. Am J Cardiol 1998; 81:7E–13E.

89. Rade JJ, Schulick AH, Virmani R, Dichek DA. Local adenoviral-mediated expression of recombinant hirudin reduces neointima formation after arterial injury. Nat Med 1996; 2:293–298.

90. Riessen R, Rahimizadeh H, Blessing E, Takeshita S, Barry JJ, Isner JM. Arterial gene transfer using pure DNA applied directly to a hydrogel-coated angioplasty balloon. Hum Gene Ther 1993; 4:749–758.

91. Dichek DA, Neville RF, Zwiebel JA, Freeman SM, Leon MB, Anderson WF Seeding of intravascular stents with genetically engineered endothelial cells. Circulation 1989; 80:1347–1353.

92. Feldman LJ, Tahlil O, Steg G. Perspectives of arterial gene therapy for the prevention of restenosis. Cardiovasc Res 1996; 32:194–207.

93. Groves PH, Banning AP, Penny WJ, Lewis MJ, Cheadle HA, Newby AC. Kinetics of smooth muscle cell proliferation and intimal thickening in a pig carotid model of balloon injury. Atherosclerosis 1995; 117:83–96.

94. Guzman LA, Mick MJ, Arnold AM, Forudi F, Whitlow PL. Role of intimal hyperplasia and arterial remodeling after balloon angioplasty—an experimental study in the atherosclerotic rabbit model. Arterioscler Thromb Vasc Biol 1996; 16:479–487.

95. Hanke H, Oberhoff M, Hanke S, Hassenstein S, Kamenz J, Schmid KM, et al. Inhibition of cellular proliferation after experimental balloon angioplasty by low-molecular-weight heparin. Circulation 1992; 85:1548–1556.

96. Harrell RL, Rajanayagam S, Doanes AM, Guzman RJ, Hirschowitz EA, Crystal RG, et al. Inhibition of vascular smooth muscle cell proliferation and neointimal accumulation by adenovirus-mediated gene transfer of cytosine deaminase. Circulation 1997; 96:621–627.

97. Stadius ML, Rowan R, Fleischhauer JF, Kernoff R, Billingham M, Gown AM. Time course and cellular characteristics of the iliac artery response to acute balloon injury. An angiographic, morphometric, and immunocytochemical analysis in the cholesterol-fed New Zealand white rabbit. Arterioscler Thromb 1992; 12:1267–1273.

98. Steg PG, Feldman LJ, Scoazec JY, Tahlil O, Barry JJ, Boulechfar S, et al. Arterial gene transfer to rabbit endothelial and smooth muscle cells using percutaneous delivery of an adenoviral vector. Circulation 1994; 90:1648–1656.

99. Yan ZQ, Yokota T, Zhang W, Hansson GK. Expression of inducible nitric oxide synthase inhibits platelet adhesion and restores blood flow in the injured artery. Circ Res 1996; 79:38–44.

100. Duggan DJ, Bittner M, Chen Y, Meltzer P, Trent JM. Expression profiling using cDNA microarrays. Nat Genet 1999; 21:10–14.

101. Hacia JG. Resequencing and mutational analysis using oligonucleotide microarrays. Nat Genet 1999; 21:42–47.

102. Rosenfeld ME, Palinski W, Ylä-Herttuala S, Carew TE. Macrophages, endothelial cells, and lipoprotein oxidation in the pathogenesis of atherosclerosis. Toxicol Pathol 1990; 18:560–571.

103. de Waard V, van den Berg BM, Veken J, Schultz-Heienbrok R, Pannekoek H, van Zonneveld AJ. Serial analysis of gene expression to assess the endothelial cell response to an atherogenic stimulus. Gene 1999; 226:1–8.

104. Ross R. Mechanisms of disease—atherosclerosis—an inflammatory disease. N Engl J Med 1999; 340:115–126.

105. Isner JM, Pieczek A, Schainfeld R, Blair R, Haley L, Asahara T, et al. Clinical evidence of angiogenesis after arterial gene transfer of phVEGF$_{165}$ in patient with ischaemic limb. Lancet 1996; 348:370–374.

106. Losordo DW, Vale PR, Isner JM. Gene therapy for myocardial angiogenesis. Am Heart J 1999; 138:132–141.

107. Rosengart TK, Lee LY, Patel SR, Sanborn TA, Parikh M, Bergman GW, et al. Angiogenesis gene therapy : phase I assessment of direct intramyocardial administration of an adenovirus vector expressing VEGF121 cDNA to individuals with clinically significant severe coronary artery disease. Circulation 1999; 100:468–474.

108. Laitinen M, Hartikainen J, Hiltunen MO, Eränen J, Kiviniemi M, Närvänen O, et al. Catheter-mediated vascular endothelial growth factor gene transfer to human coronary arteries after angioplasty. Hum Gene Ther 2000; 11:263–270.

109. Svensson U, Persson R. Entry of adenovirus 2 into HeLa cells. J Virol 1984; 51: 687–694.

110. Hay RT. Origin of adenovirus DNA replication. Role of the nuclear factor I binding site in vivo. J Mol Biol 1985; 186:129–136.

111. Newman KD, Dunn PF, Owens JW, Schulick AH, Virmani R, Sukhova G, et al. Adenovirus-mediated gene transfer into normal rabbit arteries results in prolonged vascular cell activation, inflammation, and neointimal hyperplasia. J Clin Invest 1995; 96:2955–2965.

112. Tripathy SK, Black HB, Goldwasser E, Leiden JM. Immune responses to transgene-encoded proteins limit the stability of gene expression after injection of replication-defective adenovirus vectors. Nat Med 1996; 2:545–550.

113. Wilson JM. Adenoviruses as gene-delivery vehicles. N Engl J Med 1996; 334: 1185–1187.

114. Finkel T, Epstein SE. Gene therapy for vascular disease. FASEB J 1995; 9:843–851.

115. Hiltunen MO, Niemi M, Yla-Herttuala S. Functional genomics and DNA array techniques in atherosclerosis research. Curr Opin Lipidol 1999; 10:515–519.

14

Pulmonary Gene Therapy

Jane C. Davies, Duncan M. Geddes, and Eric W. F. W. Alton
*Imperial College at the National Heart and Lung Institute
and Royal Brompton Hospital
London, England*

I. INTRODUCTION

In some respects, the lung is an ideal candidate organ for gene therapy. Two diseases, cystic fibrosis (CF) and α_1-antitrypsin (α_1-AT) deficiency, are relatively common single-gene disorders for which the genetic basis is known and for which current treatment strategies are not curative. CF in particular was an obvious initial target for gene therapy, but it is becoming clear that acquired diseases, including cancer, may also be amenable to this approach. The lung can be accessed by several routes including topically to the airways, intravascular administration, and direct percutaneous injection. Early studies have, however, demonstrated a range of barriers to gene expression in this organ, which are to some extent limiting current success. The background to these barriers, methods being employed to overcome them, and the clinical progress in CF, α_1-AT deficiency, lung cancer, and other disease states will be discussed in this chapter.

II. ROUTE OF ADMINISTRATION

Delivery of any drug to the lung can be achieved via several routes. In the case of gene therapy, the choice of route will be influenced by a variety of factors including the cells being targeted and the distribution and function of the transgene product. Additionally, as discussed later in the section on CF (Sec. IV. A), clinical trials suggest that the route of administration may influence the host immune response, a factor which may be of relevance, particularly if repeated application is required.

A. A Topical Administration to the Airways

Delivery to the airways is a common route of administration for many drugs, and advances in both nebulizer and inhaler technology enable particles to be delivered out to distal airways. However, the smallest airways and alveolar regions may be poorly accessible [1], and disease processes, for example, plugging of the CF airway with mucus, may hinder this approach. This route of administration is likely to be most useful when the gene administered is targeted to the superficial airway epithelium. This route has been used successfully in both CF and α_1-AT deficiency, although there are certain limitations, which are discussed in the following section. Recently, several studies have examined the feasibility of topical administration to the fetal airway via the amniotic fluid. One potential advantage of this approach could be evasion of the host immune responses, which are immature at this stage [2]. Successful in utero *CFTR* gene transfer has been reported with both adenovirus [3] and adeno-associated virus (AAV) [4].

B. Intravascular Administration

Vascular delivery can be viewed in two contexts, with the first of these being a route to transfect cells of the lung via their circulation and the second for distant expression of a secreted product. This route could potentially overcome some of the barriers to topical administration via the airway, but requires the gene transfer agent to survive within the blood stream, to cross the endothelium, and then to reach the desired site within the lung. Studies have shown that intravascular administration leads to distal expression at the level of the alveolus rather than within the epithelium of the conducting airways [5]. This would be a major drawback in diseases such as CF in which expression within the airway is required, but it could be useful for alveolar or endothelium-expressed proteins. A second, alternative context is the intravascular injection of a gene encoding a secreted protein. This could result in transfection of a distant organ such as the liver, which would then act as a "factory" producing the desired protein which would reach the lung via the circulation. This approach would be feasible for α_1-AT deficiency or for inflammatory diseases.

C. Intramuscular Injection

Intramuscular (IM) gene transfer has met with success in nonpulmonary diseases such as hemophilia. Advantages of this approach include access, ease of transduction, the large muscle mass for repeated application, and the relatively noninvasive nature of the injection. For diseases of the lung, this route is obviously only suitable in the case of secreted proteins, for example, α_1-AT deficiency and inflammatory diseases.

D. Direct Injection

Direct injection may be particularly attractive when a small or discrete area of the lung, for example, a tumor or the pleural space, is the site of desired gene expression. Advantages to this approach include the ability to target accurately the site of required expression (particularly if using radiological guidance), with a potential therefore to limit side effects. Disadvantages are the invasive nature of administration and possibly the small area able to be accessed. Direct injection has been used with success in a variety of human lung cancer trials.

E. Ex Vivo Gene Transfer

Unlike with some of the gene transfer approaches to bone marrow cells [6], the cells of the respiratory tract do not lend themselves easily to removal, ex vivo correction, and replacement. However, there is the potential for alternative approaches. For example, the lung contains large numbers of bone marrow or blood-derived cells; genetically modified leukocytes which could home to sites of inflammation and produce anti-inflammatory agents would be an attractive prospect. Alternatively, there has been much interest in the recently identified plasticity of both bone marrow and solid organ stem cells, with increasing numbers of reports of bone marrow stem cells [BMSCs] differentiating into cells of other lineages, including skeletal muscle [7], brain [8], liver [9], cardiac muscle and vascular endothelium [10], and most recently lung [11]. The study reporting success in the lung infused a highly selected population of BMSCs into radiation-myeloablated mice, and demonstrated donor-derived epithelial cells in both the alveolar region and, to a lesser extent, the conducting airway. An alternative to allogeneic [donor] BMSCs would of course be ex vivo correction and reinfusion of the patient's own BMSCs. This would remove the need for immunosuppression, and has been performed with success with bone marrow stem cells in inherited immune deficiencies [6]. This is an exciting and rapidly developing field, which has the potential to provide therapeutic options for diseases in a wide variety of organs.

III. BARRIERS TO GENE EXPRESSION IN THE AIRWAY

The lung poses multiple barriers to gene expression, some of which are applicable to topically delivered gene transfer agents only and others which apply to any route of administration.

A. Mucociliary Clearance System

Both components of the mucociliary clearance system, namely, airway secretions and the cell surface, pose significant barriers to exogenous gene transfer. The normal, healthy respiratory epithelium produces a thin layer of mucus which is required for normal ciliary function, but which inhibits gene transfer via cationic liposomes [12]. This inhibition may be exaggerated by particular disease states. For example, CF patients produce excess mucus of increased viscosity related to the underlying ion transport defects. Airway infection and inflammation ensue at an early age, and the surface epithelium becomes covered in a layer of thick, tenacious sputum with a high DNA content [13]. In vitro work has shown that this sputum is a highly efficient barrier to both viral and nonviral gene transfer vectors [12]. Mucolytic drugs such as rhDNase are used in clinical practice to improve expectoration [14], which could be a useful method to overcome this barrier. In contrast to worries over its potential nucleolytic action on the exogenously applied plasmid cDNA, in vitro work has demonstrated that this is not a major problem [15]. There is also evidence from animal models that preexisting nonspecific inflammation, such as would be likely in the CF airway, may be a barrier to gene transfer. Reporter gene expression was reduced significantly in mice with pseudomonas-induced airway inflammation [16], and CF bronchoalveolar lavage fluid was shown to inhibit AAV-mediated gene transfer in vitro, with the levels of inhibition being directly related to elastase content [17].

B. Cell Surface and Receptors

Once through the barriers imposed by such secretions, or following nontopical routes of administration, the gene transfer agent has to enter the cell. Both the structure of the glycocalyx and the rate of endocytosis have been identified as limiting factors in gene transfer efficiency. Evidence for the cell glycocalyx as a significant barrier to gene transfer comes from an in vitro study by Pickles et al. [18]. Cells in which the adenoviral receptor, CAR, was retargeted to the apical surface demonstrated significant adenovirus-mediated reporter gene expression only after the cell surface glycocalyx was modified by the removal of sialic acid with neuraminidase. In an attempt to increase the rate of endocytosis, Drapkin et al. identified a cell surface receptor for urokinase plasminogen activator uPAR [19]. By coupling a ligand for this receptor to either adenoviral or adeno-associated viral vectors, gene expression was significantly enhanced. Similar approaches are being utilized by other groups. Several vectors require a specific cell surface receptor in order to enter the cell. For example, although adenovirus naturally infects respiratory epithelia, the CAR receptor to which it binds is located basolaterally [20]. In vitro studies have demonstrated increased gene transfer after breakdown of intercellular tight junctions [21], permitting access to these receptors. Concerns over such an approach, which could lead to the loss of epithe-lial integrity, predisposition to invasive infection, or the development of airway edema [22], need to be addressed in order that the clinical applicability can be determined. With regard to receptor-mediated endocytosis, viral serotype differ-ences may require further research. For example, recent data have confirmed that AAV-2, a vector used in several airway gene transfer studies with limited success, does not possess a receptor on the apical cell membrane, whereas the AAV-5 serotype can enter via binding to specific sialic acid residues [23].

C. A Intracellular Processes

Several intracellular processes pose barriers to successful gene expression, in-cluding the endosomal system, cytoplasmic trafficking, and nuclear entry [24]. Although mechanisms involved in movement of the vector from the cell surface to the nucleus remain incompletely understood, preliminary data with AAV sug-gest that the use of proteasome inhibitors [25] and endosomolytic peptides [26] may improve gene expression. Once in the cytoplasm, both size [27] and instabil-ity [28] of DNA and the cell cytoskeleton [12] may further limit the efficiency of nuclear delivery. In the majority of cases, the final step required for transgene expression is nuclear entry. In the absence of cell division, nuclear entry of physi-ological proteins such as transcription factors is thought to be mediated via nu-clear localization signals [NLSs] and to occur only via the nuclear pore complex [NPC]. The uptake of DNA into the nucleus is not part of normal physiology and is currently incompletely understood. Recent data have confirmed that entry of plasmid DNA involves the coupling to NLS-containing DNA-binding proteins in the cytoplasm and entry via the NPC [29]. Further understanding of these processes may increase nuclear entry and improve levels of gene expression [30]. An alternative to this would be the use of a vector capable of cytoplasmic expres-sion such as the paramyxoviruses. An example of this family, the murine Sendai virus, has recently shown high levels of reporter gene expression in the respira-tory tract [31].

D. Immune System

This complex area has been covered in detail in a previous chapter of this book. In brief, the host immune response has been implicated in problems with both single and repeated application of gene transfer agents to the lung. With respect to the safety of single application, in early studies of single-dose adenovirus-mediated *CFTR* gene transfer to the lungs, an inflammatory response was induced by high titers of the vector, which was, in some cases, responsible for adverse clinical sequelae [32]. More recent trials have confirmed that this problem is minimized by the use of lower doses of vector, modification of the viral genome, and possibly by the route of nebulization rather than direct instillation. With regard to nonviral gene transfer agents, the presence of unmethylated CpG motifs on plasmid DNA has been suggested as a cause for an observed inflammatory response [33], and efforts are being aimed at either removing or selectively methylating such structures. For many applications of pulmonary gene therapy, including CF and cancer, repeated application is likely to be required for long-term benefit. Recognition of viral coat proteins and the production of neutralizing antibodies has limited efficacy in some trials of repeated application of virus-mediated gene transfer [34], although this appears to be less of a problem with synthetic vectors [35]. In addition to the host response to the vector, the expressed protein, especially if previously unencountered, may be immunogenic, although to date this does not seem to be a concern. Paradoxically, in certain contexts, such as described in the section on lung cancer, an immune response to the transgene product is in fact the aim of the approach.

IV. PROGRESS IN SPECIFIC LUNG DISEASES

A. Cystic Fibrosis

CF is the disease that has received the most attention as a potential target for pulmonary gene therapy, and hence it is discussed here in detail. The disease affects approximately $1:2500$ white newborns, making it the commonest recessively inherited lethal disease in this population [36]. CF results from a variety of mutations in the gene encoding the cystic fibrosis transmembrane conductance regulator (CFTR) protein, a cAMP-regulated chloride channel in epithelial cells [37]. Dysfunction of CFTR is either due to lack of production, failure of the protein to reach its site of action on the apical membrane of the cell, or to defects in function [38]. The end result is impaired chloride transport at the apical surface of epithelial cells. Although CFTR protein is expressed in many organs, the clinical picture is dominated by disease in the respiratory, gastrointestinal, and reproductive tracts, relating in each of these sites to impaired clearance and obstruction by viscous secretions. Of these sites, pulmonary disease accounts for most of the morbidity associated with CF, and is the cause of death in over 90% of patients [39].

The relationship between the underlying molecular defect and the clinical picture of CF is complex and incompletely understood. CFTR has been shown to have functions in addition to chloride ion transport, which may play a role in disease pathogenesis, including regulation of a variety of other channels. Inhibition of the epithelial sodium ion channel, ENaC, leads to sodium hyperabsorption (and thus water down its osmotic gradient) in the CF airway [40]. CFTR has also been shown to regulate the activity of the outwardly rectifying chloride channel [41], possibly through ATP transport, and there is evidence that the protein plays roles in the regulation of basolateral K^+ channels [42] and aquaporins [43]. CFTR function has also been shown to be important in the sulfation of mucins [44] and sialylation of cell surface glycoconjugates [45]. The asialylated form of cell surface glycolipids, which are found more frequently on the surface of CF cells, act as receptors for both *Pseudomonas aeruginosa* and *Staphylococcus aureus* [46], which are pathogens important in the progression of CF lung disease.

The hallmarks of CF lung disease are early, severe, and sustained neutrophil-mediated inflammation and persistent infection with a relatively narrow range of bacteria [47]. The exact role of CFTR dysfunction in the sequence of events leading to chronic airway infection and irreversible damage is controversial and incompletely understood. The lungs in CF are thought to be normal at birth [48,49]. Recent murine studies have, however, postulated a role for CFTR in the fetal period, the investigators reporting amelioration of the lethal effects of CF following in utero *CFTR* gene transfer [3], and in a subsequent report, significant lung pathology in wild-type littermates in whom exogenous *CFTR* administration led to abnormally high levels of protein [50]. In human disease, the airways begin to show microscopic inflammatory changes in early infancy [51]. Some studies have detected such inflammation in the absence of culturable pathogens [52], which has led to much interest in the possibility that the CFTR defect in itself is somehow proinflammatory. In support of this hypothesis, studies on human fetal CF airways in a xenograft model have shown excess inflammatory cytokine expression and mucosal leukocyte migration both before and after bacterial infection [53]. Against the hypothesis though are human infant bronchoalveolar lavage (BAL) studies clearly demonstrating normal cytokine levels before infection, but an exaggerated inflammatory response once infection has been detected [54]. Whether or not the underlying defect is per se proinflammatory is of major importance, not only to assist further understanding of disease pathogenesis but also as much interest has focused recently on an anti-inflammatory approach to treatment. Further studies are needed to resolve this debate.

The second hallmark of CF lung disease, bacterial infection, occurs early and is difficult to eradicate. Several explanations for the predisposition of the CF airway to infection have been postulated, with the simplest being those related to airway surface liquid. However, two opposing theories exist regarding the nature and effects of airway surface liquid (ASL) abnormalities. In one of these, the low-volume hypothesis, the thin layer of ASL is depleted as a request of sodium and water hyperabsorption, leading to defective mucociliary clearance [55]. In the other, the high-salt theory, the levels of both chloride and sodium in the ASL are high with respect to plasma [56], leading to dysfunction of salt-sensitive antibacterial defense proteins including B-defensins, lysozyme, and lactoferrin. Available in vivo data in CF subjects favors the low-volume hypothesis, with ASL electrolyte levels not differing from non-CF values [57]. Other possible explanations (which may not be mutually exclusive) include defective sialylation processes within CF cells leading to increased numbers of bacterial receptors as described above [46] and defective internalization and elimination of organisms by epithelial cells [58]. However, the relative importance of each of these postulated mechanisms is unclear.

V. PROBLEMS WITH *CFTR* GENE TRANSFER

In addition to the barriers discussed earlier in the chapter, several issues are specific to CF. These relate both to the fact that every cell within the respiratory epithelium is abnormal (unlike the situation with, for example, cancer) and that unlike secreted protein such as α1-AT, CFTR is membrane bound. This increases the challenges in both achieving normal protein expression and measuring success.

A. Which Cells to Target

Intriguingly, the site of earliest pathology in CF, the distal small airways, is not the area of the respiratory tract which in normal health maximally expresses CFTR. Although the protein can be detected in epithelial cells at all levels of the distal airways [59], the highest levels are found in the submucosal glands of the proximal cartilaginous airways [60]. This raises important questions over which are the most relevant cells to target for gene transfer. Topical application via nebulization is likely to target successfully the surface epithelium, but is less likely to reach submucosal glands. Whether or not gene transfer to these cells will be necessary for clinical effect remains to be determined. Furthermore, surface epithelial cells are probably terminally differentiated [61]; even if an integrating vector were safely designed, gene expression would be lost with the death of the cell, which has an estimated life span of approximately 120 days. Certain putative stem cells of the respiratory tract, such as the basal cells of the larger airways, are not exposed on the airway lumen and therefore may possibly be difficult to target. More distally, the Clara cell, which is exposed on the surface of smaller airways, has recently been identified as a progenitor cell [62]. Targeting of such cells could have the potential to effect long-term CFTR expression in progeny.

B. Required Degree of Gene Transfer

Levels of expressed CFTR in the lungs overall are encouragingly low compared to other organs such as the kidney [63], and the normal phenotype of CF heterozygotes implies that CFTR levels of 50% are adequate. In fact, evidence from both in vitro and in vivo studies suggests that significantly lower levels than this may be sufficient for a normal phenotype, with less than 10% of wild-type cells correcting the chloride transport abnormalities of CF monolayers [64]. Interbreeding experiments with CF mouse models have confirmed the nonlinear relationship between gene activity and phenotype, demonstrating that 5% of normal *CFTR* gene expression in each cell results in approximately 50% of normal chloride ion transport [65]. Importantly, this level of expression conferred protection from intestinal disease with 100% survival.

Different levels may be required to restore the various functions of CFTR. It is notable that none of the clinical trials to date has reported correction of the sodium ion hyperabsorption despite encouraging evidence of increased chloride ion transport. Indeed, it has been demonstrated in vitro that almost 100% of cells are required to express wild-type CFTR before the sodium ion hyperabsorption is normalized [66]. The levels of transfection required for other functions of CFTR to be restored, for example, sulfation/sialylation defects or transport of other molecules, remains unknown, and appears to depend on the cell type transduced. Zhang et al. demonstrated that cationic liposome-mediated *CFTR* transfer corrected the mucus sulfation defect but not the chloride ion transport defect in vitro, whereas the converse was seen with adenovirus, which despite higher levels of expression, did not transduce goblet cells [67]. Encouragingly, certain mutations such as A455E, associated with approximately 5% CFTR function, appear to confer a mild pulmonary phenotype [68]. Taken together, these data support the concept that incomplete gene transfer will result in correction of the cellular defect, but the level required for normal function remains to be determined. However, when considering this issue, it is important to bear in mind that in both the interbred mice and the humans with the A455E genotype, *each* cell would express low levels of CFTR (5–10%). The same degree of phenotypic correction may not be seen if 5–10% of cells expressed higher levels of protein, as might be the case with application of exogenous *CFTR*.

C. Assessment of Efficacy

As for any study of gene transfer, measured levels of both transgene mRNA and protein will demonstrate the degree of successful transfection. However, *CFTR* is expressed in low levels in the healthy lung, suggesting that levels of transgene mRNA and protein will be similarly low and thus difficult to detect. Importantly also, detection of either mRNA or CFTR protein provides no evidence of functional correction. In fact, in several clinical trials, such molecular endpoints have failed to correlate with more relevant functional outcome measures.

Restoration of ion transport is assessed most readily in vivo by measurement of transepithelial potential difference (PD). The baseline PD is abnormally negative in CF patients, owing largely to the hyperabsorption of positively charged sodium ions by epithelial cells. CF patients also demonstrate an exaggerated response to the sodium channel blocker amiloride and a reduced response to attempts to stimulate chloride ion secretion with low chloride solutions or isoprenaline, a cAMP-agonist [69,70]. These measurements have been well defined for the nasal epithelium, are used diagnostically in problem cases, and have been employed as endpoints in nasal trials of both *CFTR* gene transfer and novel pharmocological agents [71]. The practicalities of performing PD measurements has, until recently, limited their use to the nasal epithelium. With the onset of clinical trials of *CFTR* gene administration to the lower airway, the requirement for functional endpoints led to the development of methods to assess tracheal and bronchial PD [72]. These measurements are performed bronchoscopically and under general anesthesia to avoid the interference of topical anesthetic agents. The feasibility of such an approach has been confirmed in adults and, more recently, in children [73], and the technique was employed as a functional endpoint in our recent trial of liposome-mediated *CFTR* gene therapy [72].

Using a similar rationale, the ion transport defects of CF cells can be detected ex vivo. Early studies described both single-cell patch-clamping experiments [74] and the use of tissue biopsies in Ussing chambers [75], although both techniques are time consuming and technically challenging, limiting their usefulness in the context of clinical trials. One technique which has been proven to be useful in several studies uses epifluorescence microscopy [76]. Small clumps of cells, such as those obtained by brushing or biopsy, are loaded with a halide-sensitive fluorescent marker such as 6-methoxy-N-(3-sulfopropyl)quinolinium (SPQ). Fluorescence is quenched by the application of iodide, and the quantifiable response of the cells to cAMP agonists such as forskolin indicates the degree of halide permeability: Non-CF cells and those successfully transfected with *CFTR* show fluorescence on release of iodide. Although time consuming, this method has the benefit of providing quantifiable data from large numbers of cells. These assays provide evidence of functional correction, although the link between ion transport and disease pathogenesis remains unclear. More clinically relevant endpoints are therefore being explored.

Asialoganglioside receptors for bacteria, such as *P. aeruginosa*, are known to be increased on the surface of CF cells [46], and we have shown that in vitro liposome-mediated *CFTR* gene transfer reduces binding of this organism to the cell surface [77], making such an assay potentially useful as a clinically relevant endpoint. Other groups are examining markers of inflammation, for example, in exhaled gases, as a noninvasive endpoint. These and other clinical measurements, such as mucociliary clearance, mucus rheology, pulmonary function, and frequency of respiratory exacerbations, will be worth considering in later phase studies looking for clinical benefit. The increased life expectancy of CF patients will limit the use of survival as an endpoint for such studies.

VI. PROGRESS TO CLINICAL STUDIES

A. Preclinical Studies

Only 1 year after the discovery of the *CFTR* gene, the first description of in vitro *CFTR* cDNA transfer was reported by two independent groups. Functional studies confirmed retrovirus-mediated gene transfer to a CF pancreatic cell line [78] and vaccinia-mediated delivery of normal *CFTR* cDNA to a CF respiratory epithelial cell line [79]. These promising results paved the way for the first animal experiments, being greatly facilitated by the development of transgenic CF mice [80–82], which allowed functional assessment of *CFTR* gene transfer in vivo.

Hyde et al. used null mice, shown to express no detectable CFTR mRNA, and to demonstrate greatly reduced cAMP-dependent chloride conductance in the airways [83]. Intratracheally administered *CFTR*-liposome complexes restored chloride ion permeability to a level comparable with that seen in wild-type mice. A study by our group assessed the effects of non-invasive nebulization of 3 beta-[N-(N′,N′-dimethylaminoethane) carbamoyl] cholesterol/dioeoyl-L-alpha-phosphatidyl-ethanolamine (DC-chol/DOPE)–cDNA complexes to a CF insertional mutant mouse [84]. Human *CFTR* mRNA was detectable by reverse transcriptase polymerase chain reaction (RT-PCR) in all animals, with nasal and tracheal PD demonstrating a mean 50% correction of the forskolin-induced chloride efflux. Early attempts to perform similar studies with adenoviral vectors were less successful owing to the relative inefficiency of this virus to infect respiratory epithelial cells [85]. Encouragingly, however, basal cells, which were successfully transduced following mechanical injury to the superficial epithelial layers, subsequently differentiated into columnar epithelial cells with persistent transgene expression [86]. Another group has recently demonstrated long-term expression and functional correction of CF mice with adenovirus-mediated CFTR transfer [87]. With respect to AAV-mediated *CFTR* gene transfer, animal models have demonstrated persistence of expression for up to 6 months [88]. However, unlike the integration seen in naturally occurring infection, gene expression appears to be episomal. Further attempts to manipulate the AAV vector and thus achieve stable chromosomal integration might lead to longer term expression.

B. Human Trials

1. Adenoviral Vectors

In the first human study, adenovirus (Ad)/*CFTR* was administered onto the nasal epithelium of three CF patients [89]. No viral replication was seen, and only a minor degree of local inflammation was reported. Baseline PD became normal in all three patients, and the β-adrenergic agonist terbutaline produced small responses similar to those seen in non-CF patients. However, the study was not controlled, inflammation is known to reduce baseline PD, and the protocol used to assess the ion transport changes may not reliably discriminate CF from non-CF patients, making these efficacy data difficult to interpret. Crystal et al. administered Ad/*CFTR* to both the nasal epithelium and lower airway epithelium via the bronchoscope [32]. One of the highest dose patients became unwell with fever, hypoxia, and pulmonary infiltrates. The problem was attributed to vector-induced inflammation, as no infective virus was detected. This study has led to a reduction in the doses of viral vectors administered, which, possibly together with nebulization rather than direct instillation, has to a large extent overcome the problems of acute inflammation. In terms of efficacy, one patient had detectable *CFTR* mRNA and protein from nasal administration, and another patient had detectable protein in the lung. Expression was transient (< 10 days) with no dose response. Another study [90] assessing dose response in the nasal epithelium showed no evidence of functional correction and, in those receiving the highest doses, significant local inflammatory reactions and the detection of neutralizing antibodies.

Ad/*CFTR* has been directly administered via the bronchoscope to a segmental bronchus of CF subjects in a dose escalation study [91]. Flulike symptoms were seen at a high vector dose, and an immune response, which was largely cell mediated, was observed. Efficacy was assessed by in situ hybridization to vector-specific DNA, which was positive at low levels 4 days after gene administration. Another trial of single escalating doses of Ad/*CFTR* to first the nasal and then the lung epithelium has been reported [92]. No acute toxic effects were seen at doses up to 5.4×10^8 PFU, and there were no changes in inflammatory parameters in BAL or serum. Interestingly, there was no rise in serum anti-viral antibodies, or appearance of antibodies in the BAL fluid. Molecular analysis confirmed successful gene transfer, although no functional endpoints were employed.

Much attention has focused on the potential problems with repeated administration in clinical trials addressing this issue in the nose and in the lower airway. In one nasal study, despite there being no detectable adverse effects, the ability of the Ad/*CFTR* to correct the abnormal chloride transport was reduced on subsequent application [34]. In the lower airway study, Ad/*CFTR* was administered via an endobronchial spray in three doses over a 9-month period to patients with CF with bronchoscopic assessment 3 and 30 days after administration [93]. Three days after the first administration, vector-derived mRNA was detected in a dose-related pattern, with those in the high dose group expressing levels in the 5% of wild-type range, which was demonstrated in vitro to be sufficient for phenotypic correction. Examination of distribution was unfortunately not possible, as the vector could not be identified with fluorescent in situ hybridization. mRNA levels were undetectable in all patients by day 30. The second administration resulted in some expression which was not dose related, and the third administration produced no expression in any sample. Similarly to the study reported above, these investigators detected no increase in serum neutralizing antibodies, although gene expression was not seen in patients with high preexisting levels of antibody. This study therefore confirms that although repeated administration of adenoviral vectors appears to be safe, efficacy is severely compromised, although the exact mechanism for this is uncertain. This group has also studied the initiation of an antiadenoviral antibody response following other routes of administration including intradermal, intramyocardial, and into liver metastases [94]. They conclude that the extent of the neutralizing antibody response depends on the levels of preexisting antibodies and the route of administration (low with airway, variable with myocardium and skin, and high with the hepatic route) rather than the dose of vector.

Thus, although expression of CFTR has been detected in both the upper and lower airway following adenovirus-mediated gene transfer, both the level and, to a lesser extent, duration of expression are suboptimal. Efforts are being focused on the development of strategies to increase efficacy and permit readministration.

2. Adeno-associated Viral Vectors

AAV-mediated *CFTR* transfer has been assessed in the maxillary sinus of CF patients using preexisting antrostomies for ease of administration and assessment [95]. Ten patients received escalating doses in an unblinded fashion with no significant inflammatory response. Molecular endpoints demonstrated gene transfer, with DNA being detected up to 41 days after administration, but assessment of expression was difficult. Functional assessment with PD responses to isoprenaline and amiloride demonstrated some changes, although numbers were small. Other groups are progressing through to clinical trials with AAV vectors. As mentioned above, there appear to be significant differences in the cell entry abilities of the different AAV serotypes, with an apical receptor (sialic acid) being identified for serotype 5, but no receptor-mediated cell entry via the apical surface was seen with serotype 2 [23].

3. Synthetic Vectors

The first placebo-controlled clinical trial of liposome-mediated *CFTR* gene transfer was reported by our group in 1995 [96]. This involved the direct application of DC-Chol/DOPE/*CFTR* to the nasal epithelium of CF patients. No adverse clinical effects were seen, and examination of nasal biopsies showed no histological changes. Plasmid DNA and vector-specific mRNA were detected in five of the eight patients receiving the *CFTR* gene. Functional assessment of nasal PD revealed approximately 20% correction of the low-chloride response, peaking at 3 days but disappearing by 7 days. Several other nasal trials of liposome-mediated *CFTR* gene transfer have since been reported. Gill et al. administered DC-Chol/ DOPE/*CFTR* or placebo to the nasal epithelium of 12 CF patients [97]. Six of the eight patients receiving the gene showed some evidence of changes in CFTR chloride channel function. In two patients, low-chloride PD responses were corrected into the non-CF range for approximately 7 days, and cells from five patients demonstrated significant restoration of chloride transport on SPQ fluorescence microscopy. Porteous et al. reported that following nasal administration of another liposome, 1,2-dioleoyl-3-trimethylammoniumpropane (DOTAP)/*CFTR*, transgene DNA was detectable in seven of eight patients, in two patients until 28 days, with mRNA being detected in two patients at 3 and 7 days [98]. Nasal PD responses from two patients showed changes consistent with a partial correction of the chloride defect, although none of the patients demonstrated functional correction on SPQ fluorescence microscopy. A study by Noone et al. failed to demonstrate either molecular or functional evidence of CFTR expression following nasal administration with another cationic lipid, 1-2-dimyristoyl-sn-glycero-3-ethylphosphatidylcholine (EDMPC) cholesterol, to 11 patients, although the complex appeared to be safe [99]. Unlike the reported problems with readministration of adenovirus-mediated gene transfer, a recent study using DC-Chol/ DOPE reported that repeated nasal administration was both well tolerated and as efficient as a single application [35]. Patients received a monthly dose of lipid/ *CFTR* or placebo for 3 months in a double-blind design. Four days after each dose, nasal cells were sampled for both toxicity and efficacy assays. There was no evidence of inflammation or of an anti-CFTR immune response, and functional endpoints including ex vivo halide efflux were positive in an average of six patients after each dose.

Our group recently reported a placebo-controlled trial of liposome-mediated *CFTR* to the lungs of CF patients [72]. For comparison with our previous nasal trial and to assess the degree to which the nasal epithelium mimics that of the lower airway, patients also received a nasal dose of the complex. Those in the placebo arm of the study received lipid 67 alone. Administration was well tolerated, but respiratory symptoms including mild chest tightness and cough were seen in both groups. This had not been observed following administration of lipid 67 alone to non-CF subjects [100], and may relate to the pulmonary inflammation present in the lungs of the CF patients. In addition, those in the treatment group reported mild influenzalike symptoms within the first 24 hr. The reason for this was unclear, but may relate to the presence of unmethylated CpG groups on the bacterially derived DNA [33]. Importantly, these symptoms were not reported after nasal administration, suggesting that for safety, at least, the nasal epithelium may not be a good surrogate site for such trials. The major efficacy endpoint was lower airway PD. In neither group was there any change in the parameters of sodium absorption (baseline or amiloride response). The treatment group, however, demonstrated a significant response to perfusion with low chloride and isoprenaline of approximately 25% of non-CF values. Supportive evidence of functional correction was obtained from both SPQ epifluorescence microscopy and ex vivo pseudomonas adherence. Broadly, similar degrees of correction of the chloride defect were seen in the nose, and there appeared to be a longer duration than seen in the first nasal trial using DC-Chol/DOPE [96].

In conclusion, clinical trials have demonstrated that *CFTR* gene transfer to the airway can be achieved safely. However, current levels of expression are unlikely to lead to clinical efficacy. Further work is focusing on both vector and host in an attempt to achieve this and allow for repeated application. Given the speed of progress since the discovery of the *CFTR* gene just over a decade ago, there is good reason for optimism in this field.

C. α_1-Antitrypsin Deficiency

Deficiency of α1-AT, the principal endogenous antiprotease, leads to both pulmonary emphysema and, in some cases, liver disease [101]. Similarly to CF, the disease is inherited in an autosomal recessive fashion, with several mutations having been identified resulting in either absent or severely low levels of circulating protein. The principal action of α_1-AT in the lung is to counterbalance the damaging effects of proteases such as neutrophilic (and, to a lesser extent, bacterial) elastases in the distal conducting airways and alveoli. Current therapy involves avoidance of damaging environmental triggers such as cigarette smoke and symptomatic treatments. Plasma-derived α_1-AT can be administered intravenously, but supply is limited, it is expensive, and it has a short half-life necessitating frequent administration [102]. Its purification from human serum also raises the possibility of viral transmission. Gene therapy has therefore been considered as an alternative approach. Unlike the situation with CF, the secreted nature of the deficient protein makes expression at a distant site a possibility for exogenous gene transfer and simplifies endpoint assays.

1. Preclinical Studies

Preclinical studies have explored a variety of routes including tail vein and portal vein injections, biliary infusion, injection into liver [103] and muscle [l04], and implantation of genetically modified myoblasts [105]. Topical administration to the respiratory epithelium has been performed in the rabbit [106]. These studies have demonstrated gene transfer but therapeutically inadequate levels of protein production.

2. Human Trials

To date, only one human trial of α_1-AT gene transfer has been reported [107]. Patients with α_1-AT deficiency received a single dose of cationic liposome–α_1-AT complex into one nostril, with the other nostril acting as a control. Protein was detected in nasal lavage fluid, with levels peaking at day 5 at approximately one-third of normal. This rise was not seen in fluid from the control nostril. In addition, levels of the proinflammatory cytokine, interleukin-8 (IL-8), were decreased in the treated nostril. Most interestingly, this anti-inflammatory effect was not observed when intravenously administered purified α_1-AT protein achieved normal nasal levels (assessed by lavage), leading the investigators to speculate that different routes of administration may lead to a variation in the sites of expression. This hypothesis has been supported by animal data from the same group showing different sites of localization of the protein within the lung comparing topical gene expression with intravenous (IV) protein administration [108]. Future studies will address this issue and assess administration to the lower airway.

D. Lung Cancer

Cancer may intuitively seem a less obvious choice than single-gene disorders for gene therapy. However, the field is growing rapidly, with several approaches demonstrating some success in preclinical and clinical studies.

1. Tumor-Suppressor Gene Therapy

Mutation of the tumor-supressor gene, p53, is one of the commonest findings in certain types of lung cancer (non–small cell), the presence of this mutation correlating with lack of response to conventional treatment with chemotherapy and radiotherapy [109]. A lack of normal p53 gene function leads to failure of apoptosis. In early studies, transfer of normal p53 into lung cancer cell lines increased apoptosis, an effect that was enhanced in the presence of the chemotherapeutic agent cisplatin. In animal models, regression of tumor bulk has suggested a bystander effect in which a subset of cells expressing the gene inhibits the growth of neighboring nontransfected cells.

    Several uncontrolled clinical trials of virus-mediated p53 have been reported within the last few years in patients with demonstrated p53 mutations who had failed to respond to conventional treatment. The gene transfer agent has been administered locally into the tumor either via the bronchus or percutaneously under computed tomographic (CT) guidance. In the first of these, which employed a retroviral vector, side effects were well tolerated, with evidence of gene expression at high doses [110]. Some of the patients showed stabilization or regression of growth in the injected tumors. A dose-escalation study [111] of adenovirus-mediated p53 (up to 10^{10} PFU) to 15 patients demonstrated gene transfer again only at the higher doses, with four of six successfully transfected tumors showing transient disease control. Neither of these studies showed any evidence of an effect in untreated sites. Swisher et al. administered monthly doses of adenovirus-mediated p53 for up to 6 months [112]. There was no increase in adverse events with repeated administrations, although fever was common. With regard to efficacy, 46% of patients had detectable p53 mRNA in at least one biopsy specimen, with some patients showing expression after several doses. A significant increase in apoptotic index was seen in 11 of 24 evaluable patients. A partial radiological response was seen in 2 patients and stabilization in a further 16 patients despite the generation of high titers of antiadenoviral antibodies.

Another study by the same group has assessed the effect of Ad/p53 gene transfer in combination with cisplatin, reporting similar levels of gene transfer and a partial response (disease stability) in 17 of 24 patients [113]. A recent study has assessed the immune response to adenovirus-mediated intratumor p53 injection [114]. All patients had antiadenoviral neutralizing antibodies before treatment, which increased in titer, in a non–dose-related fashion, but which did not prevent expression of a subsequent dose. Treatment led to an increase in lymphocyte proliferative responses. One patient who had pretreatment anti-p53 antibodies demonstrated a rise in titer after treatment, although no other patient developed a response. Intratumoral injection of p53 for lung cancer appears, on the basis of these studies, to be safe and repeatable. The studies which have reported a clinical response have shown this to be localized, and this may therefore prove to be most useful for patients with nonresectable disease.

2. Suicide Gene Therapy

The basis of suicide gene therapy is the transfer of a gene encoding an enzyme capable of converting a nontoxic agent to a toxic chemotherapeutic. This has the huge benefit of limiting toxicity to the site of gene expression and thus minimizing side effects. The most commonly used system has been the herpes simplex thymidine kinase (HSV*tk*) gene which initiates conversion of ganciclovir into a triphosphorylated derivative which is incorporated into DNA in place of guanosine triphosphate, leading to inhibition of cell replication [115]. Major advantages of this approach are the stimulation of a local immune response and a significant bystander effect. The approach is being utilized in trials of malignancies in many sites, including in patients with pleural mesothelioma. Intrapleural administration of Ad/HSV*tk* to 21 patients led to expression of transgene protein in just over 50% of patients [116]. Strong immune responses against adenovirus were initiated, although this was well tolerated. Partial regression of tumors has been reported in some patients.

3. Immunotherapy

Immunogenetic therapy is based on the transfer of genes encoding molecules involved in the host immune response in an attempt to enhance immune recognition and destruction of tumor cells. Such genes include various cytokines, interferon-γ, granulocyte-macrophage colony-stimulating factor (GM-CSF), and heat shock proteins. An alternative strategy has been to introduce a foreign gene, such as the bacterially derived β-galactosidase, in the hope that this will be recognized as foreign. A group which examined the feasibility of this did in fact demonstrate local tumor regression and the induction of strong antibody responses both to the adenoviral vector and the transgene [117,118]. Other studies have administered vaccinia virus–mediated IL-2 intrapleurally to patients with mesothelioma [119]. No clinical benefit has been shown despite increases in T-cell infiltrates. Preclinical studies are exploring the possibilities of using cytokine cocktails.

4. Oncolytic Adenovirus

ONYX-015 is a replication-specific adenovirus with a deletion in the E1B-55-kD gene region, which is required for inhibition of wild-type p53 function. Therefore, viral replication is prevented in cells with normal p53 function but permitted in tumor cells with mutant p53, where it leads to cell death and a significant bystander effect [120]. This effect has been confirmed in preclinical studies [121,122], with the latter study showing synergy with conventional chemotherapeutic agents. A recent clinical study confirmed both the safety and feasibility of intravenously administered ONYX-015 in patients with end-stage metastatic lung cancers [123]. Infusions were administered weekly with or without conventional chemotherapeutic agents. No dose-limiting toxicity was detected at doses up to 2×10^{13} particles, with mild adverse events being fevers, rigors, and transient elevation of liver enzymes. All patients showed an increase in neutralizing antibody levels and raised levels of circulating cytokines. Viral infection of tumor deposits was seen in one patient and at doses above 2×10^{12} circulating viral genome indicated replication. Replication-specific oncolytic viruses may therefore become a useful approach. Future studies may address the effects of transient immunosuppression, reduction in hepatic clearance of virus, and combination of this approach with cytotoxic drugs or therapeutic transgenes.

E. Miscellaneous Lung Diseases

1. Acute Lung Injury

Acute lung injury (ALI) (also called adult/acute respiratory distress syndrome [ARDS]) is the term given to the end result of a variety of insults, including severe sepsis, aspiration, trauma, near drowning, and pancreatitis [124]. The hallmarks of ALI include impaired oxygenation, often poorly responsive to invasive ventilation, and patchy infiltrates on chest x-ray in the absence of a raised pulmonary arterial wedge pressure or left atrial hypertension. Despite a reduced mortality over the past two decades with improvements in intensive care management, ALI carries a poor prognosis, particularly in the pediatric age group. No specific therapy exists currently, and management is largely supportive. The pathophysiology of ALI is an evolving process of immediate injury, exudative alveolar inflammation with edema, and finally fibroproliferative repair [125]. The recognition that the early stages of this process are characterized by generalized intravascular activation, endothelial damage, and high levels of proinflammatory cytokines has led to the development of novel gene and small molecule–based approaches to treatment. These have some theoretical advantages over the use of proteins, including the potential for cell specificity, the possibility of delivering intracellular proteins, duration of action, and possibly cost. Broadly, the approaches target inflammation or oxidative stress either directly via cytokines or by acting on alternative pathways involving antioxidants and other protective proteins. In addition, other studies have explored the potential for increasing fluid reabsorption and reducing pulmonary edema in ALI.

Anti-inflammatory Approaches. A decrease in inflammation can be achieved by either antagonizing proinflammatory cytokines or increasing the effects of anti-inflammatory cytokines. In the context of ALI, studies have been undertaken both with conventional gene transfer techniques and with antisense oligonucleotides, which reduce translation to protein by specific binding to mRNA. Ad-mediated tumor necrosis factor-α (TNF-α) receptor gene transfer was shown to reduce septic shock in mice injected with lipopolysaccharide (LPS) and to reduce pulmonary inflammation [126]. IM delivery of Ad/IL-10, the major anti-inflammatory cytokine, showed similar benefits, with a marked suppression of LPS-induced TNF-α and IL-6 production [127]. Both prostaglandin synthase and nitric oxide synthase genes have demonstrated a potential therapeutic benefit for ALI, with the former reducing LPS-induced injury in rabbit lung [128] and the latter inhibiting neointimal vascular lesions in rat artery [129]. Antisense technology has also been used in this context; oligonuleotides against intercellular adhesion molecule-1 (ICAM-1), a major intercellular adhesion molecule, reduced endotoxin-induced neutrophilic influx to the lung [130]. Hyperoxia has been implicated in lung damage seen in both ALI and infant respiratory distress syndrome (IRDS). The recent demonstration that this effect occurs via mitogen-activated protein kinase, p38MAPK (which also mediates LPS- and TNF-α–induced damage) [131], has led to another novel approach. This pathway is attenuated by CO, a by-product of heme degradation catalyzed by hemoxygenase. Ad expressing inducible hemoxygenase, HO-1, instilled intratracheally, increased the survival of rats with hyperoxic lung damage [132].

Targeting Pulmonary Edema. Alveolar edema is a recognized feature of ALI. In the healthy state, alveolar liquid is cleared by the basolaterally-situated Na^+/K^+ ATPase, which is upregulated during resolution of pulmonary edema [133]. Both adenovirus and cationic liposomes have been used to overexpress this ATPase in animal models of lung injury with pulmonary edema. Factor et al. used a hyperoxic rat model and showed Ad-mediated transfer of the $\beta1$ subunit increased alveolar fluid clearance by >300% and improved survival [134]. Stern et al. administered both the α and β subunits with cationic lipid to thiourea-treated mice, and again reported a significant reduction in pulmonary edema [135]. This approach may therefore have a role in the management of severe edema associated with ALI.

2. Asthma

Asthma is a disease of extremely high prevalence characterized by airway inflammation and hyperreactivity leading to wheezing, cough, breathlessness. Current treatment strategies are largely successful, consisting of bronchodilators and anti-inflammatory agents, most commonly corticosteroids [136]. However, there is a subgroup of patients for whom these treatments are inadequate, leading some groups to investigate novel approaches including gene therapy. T lymphocytes of the Th2 phenotype have been found to play a major role in this disease [137]. Redressing the imbalance between Th2 and Th1 (usually involved in the host response to infection) by administration of Th1-type cytokines such as interferon-γ (IFN-γ) [138] or IL-12 [139] has been shown to be effective in reducing airway eosinophils and hyperreactivity after allergen challenge. This led Dow at al. to treat ovalbumin-sensitized mice with the *IFN-γ* gene complexed with cationic lipid either intratracheally or IV [140]. Both routes of administration led to inhibition of airway hyperresponsiveness and eosinophilia, and the IV administration also led to lower IgE levels. Another group showed that vaccinia-mediated IL-12 administered intranasally prevented the development of both local and systemic eosinophilia and airway hyperreactivity in response to challenge [141]. Mathieu et al. have recently identified the glucocorticoid receptor gene as a potential candidate, showing that overexpression in vitro led to a reduction in transcription factors involved in the expression of proinflammatory cytokines [142]. An alternative approach has been postulated by Metzger and Nyce [143]. They hypothesize (a view echoed by others) that the multifactorial, polygenic basis of asthma makes conventional gene therapy unlikely to succeed. They highlight the importance of adenosine, which is present in high levels in the asthmatic lung, and which, when inhaled, causes bronchoconstriction. They developed an antisense against the adenosine receptor, which improved allergen-induced bronchoconstriction and reduced histamine responsiveness in a rabbit model.

Thus, in line with the multifactorial nature of the disease, many approaches are being explored for new gene therapies for asthma. Given the success of conventional treatment, however, it is likely that such therapies may be useful for a small minority of patients who are currently failing standard regimens [144].

3. Fibrotic Lung Disease

Lung fibrosis can occur either as an idiopathic isolated disease, as part of a multisystem disorder, or as a side effect of radiotherapy or certain drugs such as bleomycin [145]. In many cases, the prognosis is poor and therapy limited. Transforming growth factor-β (TGF-β) is thought to play an important role in the progressive nature of the disease [146], leading to this molecule as a target for novel therapies. Smad7, a TGF-β antagonist, was administered with an Ad vector intratracheally to bleomycin-treated mice [147]. The mice demonstrated reduced hydroxyproline and less morphological fibrosis than controls receiving Smad6. Another group has showed similar results with adenovirus-mediated transfer of decorin, an endogenous proteoglycan with anti–TGF-β activities [148]. The observation that transgenic mice deficient in plasminogen activatory inhibitor-1 (PAI-1) developed less fibrosis in response to bleomycin led to further investigation in this area. Bleomycin-pretreated mice given Ad-uPA (urokinase-type plasminogen activator) intratracheally had a significant lower lung hydroxy-proline content than controls [149]. Another group has demonstrated prevention of radiation-induced lung fibrosis and improved survival with liposome-medi-ated–manganese superoxide dismutase (MnSOD) [150]. Finally, a clinical study showing success with interferon-γ 1β protein therapy [151] and recent reports suggesting a role for Fas-mediated alveolar cell apoptosis [152] may lead to fur-ther new gene-based strategies for this disease.

4. Lung Transplantation

The success of lung transplantation is limited both by acute ischemia–reperfusion injury [153] and by the host response leading to organ rejection [154]. Strategies to attenuate these processes may therefore be of benefit. Within the context of solid organ transplantation, therapeutic genes could theoretically be administered (1) to the host prior to removal of the lungs, (2) to the lungs ex vivo prior to transplantation, or (3) to the recipient after surgery. Cassivi et al. demonstrated the feasibility of Ad–β-gal transfection either before procurement (via tracheos-tomy) or ex vivo after surgical removal [155]. At the critical time of reperfusion, they demonstrated significantly greater transgene levels in lungs transfected via the tracheostomy than in those transfected ex vivo. Levels in other organs were virtually absent, confirming limitation to the lungs. However, another study using liposomes demonstrated that ex vivo transfection was superior to intravenous injection of donors prior to organ harvesting [156]. The optimal route and timing of gene administration may therefore depend on the vector and the desired func-tion and site of transgene expression. Various mediators have been implicated in ischemia–reperfusion injury, including stimulated leukocytes and platelets, complement, proinflammatory cytokines, and oxidants [153]. Attempts to combat this process with exogenous recombinant proteins have been limited, thought largely to be related to the inability to achieve and maintain high local levels. Gene transfer may therefore be a more useful approach. Itano et al. administered Ad/*hIL-10* intravenously to rats 24 hr before organ harvest [157]. When assessed 24 hr after isotransplantation, dose-dependent IL-10 expression was observed along with significant improvements in gas exchange and neutrophilic sequestra-tion when compared with controls (Ad–*LacZ*). In the higher dose group, myelo-peroxidase and NO synthase were also decreased, suggesting that IL-10 may be of benefit in reducing injury at the time of reperfusion. Several groups have reported a reduction in rejection of donor lungs with a variety of methods. Schmid et al. administered lipid-mediated Fas ligand retrogradely through the pulmonary venous system prior to lung removal [158] along with a single dose of cyclosporine. Compared with controls, the rats receiving Fas-transfected lungs had better day 5 gas exchange and demonstrated significantly less histological evidence of acute rejection. Other groups have reported success with TGF-β both lipid-mediated ex vivo [159] and Ad IM into donor posttransplant [160].

VII. SUMMARY

The principle of gene therapy for lung disease has now been proven with clinical trials showing successful gene transfer in a variety of genetic and acquired disorders. The current major problems are poor efficiency, both in terms of low levels of transfer and short duration of expression, and the limitation by the host immune response of repeated application. With regard to efficiency, much research is going into both overcoming barriers to gene transfer and novel vector design, such as viral pseudotyping and the development of newer generation cationic liposomes. With regard to duration of expression, approaches being investigated include manipulation of either the vector or the host to permit repeated administration, the use of integrating vectors, and attempts to identify and target respiratory epithelial progenitor cells. Focused efforts on these areas are likely to lead to the development of clinically successful gene therapy approaches in the near future.

REFERENCES

1. Ganderton D. Targeted delivery of inhaled drugs: current challenges and future goals. J Aerosol Med 1999; 12(Suppl)1:S3–8.
2. Tran ND, Porada CD, Almeida-Porada G, Glimp HA, Anderson WF, Zanjani ED. Induction of stable prenatal tolerance to beta-galactosidase by in utero gene transfer into preimmune sheep fetuses. Blood 2001; 97:3417–3423.
3. Larson JE, Morrow SL, Happel L, Sharp JF, Cohen JC. Reversal of cystic fibrosis phenotype in mice by gene therapy in utero. Lancet 1997; 349:619–620.
4. Boyle MP, Enke RA, Adams RJ, Guggino WB, Zeitlin PL. In utero aav-mediated gene transfer to rabbit pulmonary epithelium. Mol Ther 2001; 4:115–121.
5. Muller DW, Gordon D, San H, Yang Z, Pompili VJ, Nabel GJ, Nabel EG. Catheter-mediated pulmonary vascular gene transfer and expression. Circ Res 1994; 75:1039–1049.
6. Cavazzana-Calvo M, Hacein-Bey S, Yates F, de Villartay JP, Le Deist F, Fischer A. Gene therapy of severe combined immunodeficiencies. J Gene Med 2001; 3:201–206.

7. Gussoni E, Soneoka Y, Strickland CD, Buzney EA, Khan MK, Flint AF, Kunkel LM, Mulligan RC. Dystrophin expression in the mdx mouse restored by stem cell transplantation. Nature 1999; 401:390–394.

8. Ono K, Takii T, Onozaki K, Ikawa M, Okabe M, Sawada M. Migration of exogenous immature hematopoietic cells into adult mouse brain parenchyma under GFP-expressing bone marrow chimera. Biochem Biophys Res Comm 1999; 262:610–614.

9. Alison MR, Poulsom R, Jeffery R, Dhillon AP, Quaglia A, Jacob J, Novelli M, Prentice G, Williamson J, Wright NA. Hepatocytes from non-hepatic adult stem cells. Nature 2000; 406:257.

10. Jackson KA, Majka SM, Wang H, Pocius J, Hartley CJ, Majesky MW, Entman ML, Michael LH, Hirschi KK, Goodell MA. Regeneration of ischemic cardiac muscle and vascular endothelium by adult stem cells. J Clin Invest 2001; 107:1395–1402.

11. Krause DS, Thiese ND, Collector MI, Henegariu O, Hwang S, Gardner R, Neutzel S, Sharkis SJ. Multi-organ, multi-lineage engraftment by a single bone marrow-derived stem cell. Cell 2001; 105:369–377.

12. Kitson C Angel B, Judd D, Rothery S, Severs NJ, Dewar A, Huang L, Wadsworth SC, Cheng SH, Geddes DM, Alton EW. The extra- and intracellular barriers to lipid and adenovirus-mediated pulmonary gene transfer in native sheep airway epithelium. Gene Ther 1999; 6:534–546.

13. Lethem MI, James SL, Marriott C, Burke JF. The origin of DNA associated with mucus glycoproteins in cystic fibrosis sputum. Eur Respir J 1990; 3:19–23.

14. Ranasinha C, Assoufi B, Shak S, Christiansen D, Fuchs H, Empey D, Geddes D, Hodson M. Efficacy and safety of short-term administration of aerosolised recombinant human DNase I in adults with stable stage cystic fibrosis. Lancet 1993; 342:199–202.

15. Stern M, Caplen NJ, Browning JE, Griesenbach U, Sorgi F, Huang L, Gruenert DC, Marriot C, Crystal RG, Geddes DM, Alton EW. The effects of mucolytic agents on gene transfer across a CF sputum barrier in vitro. Gene Ther 1998; 5:91–98.

16. van Heekeren A, Ferkol T, Tosi M. Effects of bronchopulmonary inflammation induced by pseudomonas aeruginosa on gene transfer to airway epithelial cells in mice. Gene Ther 1998; 5:345–351.

17. Virella-Lowell I, Poirier A, Chesnut KA, Brantly M, Flotte TR. Inhibition of recombinant adeno-associated virus (rAAV) transduction by bronchial secretions from cystic fibrosis patients. Gene Ther 2000; 7:1783–1789.

18. Pickles RJ, Fahrner JA, Petrella JM, Boucher RC, Bergelson JM. Retargeting the coxsackievirus and adenovirus receptor to the apical surface of polarized epithelial cells reveals the glycocalyx as a barrier to adenovirus-mediated gene transfer. J Virol 2000; 74:6050–6057.

19. Drapkin PT, O'Riordan CR, Yi SM, Chiorini JA, Cardella J, Zabner J, Welsh MJ. Targeting the urokinase plasminogen activator receptor enhances gene transfer to human airway epithelia. J Clin Invest 2000: 105:589–596.

20. Walters RW, Grunst T, Bergelson JM, Finberg RW, Welsh MJ, Zabner J. Basolateral localization of fiber receptors limits adenovirus infection from the apical surface of airway epithelia. J Biol Chem 1999; 274:10219–10226.

21. Parsons DW, Grubb BR, Johnson LG, Boucher RC. Enhanced in vivo airway gene transfer via transient modification of host barrier properties with a surface-active agent. Hum Gene Ther 1998; 9:2661–2672.
22. Schneeberger EE, Lynch RD. Structure, function, and regulation of cellular tight junctions. Am J Physiol 1992; 262:L647–661.
23. Walters RW, Yi SM, Keshavjee S, Brown KE, Welsh MJ, Chiorini JA, Zabner J. Binding of adeno-associated virus type 5 to 2,3-linked sialic acid is required for gene transfer. J Biol Chem 2001; 276:20610–20616.
24. Zabner J, Fasbender AJ, Moninger T, Poellinger KA, Welsh MJ. Cellular and molecular barriers to gene transfer by a cationic lipid. J Biol Chem 1995; 270:18997–19007.
25. Duan D, Yue Y, Yan Z, Yang J, Engelhardt JF. Endosomal processing limits gene transfer to polarized airway epithelia by adeno-associated virus. J Clin Invest 2000; 105:1573–1587.
26. Lim DW, Yeom YI, Park TG. Poly(DMAEMA-NVP)-b-PEG-galactose as gene delivery vector for hepatocytes. Bioconjug Chem 2000; 11:688–695.
27. Lukacs GL, Haggie P, Seksek O, Lechardeur D, Freedman N, Verkman AS. Size-dependent DNA mobility in cytoplasm and nucleus. J Biol Chem 2000; 275:1625–1629.
28. Lechardeur D, Sohn KJ, Haardt M, Joshi PB, Monck M, Graham RW, Beatty B, Squire J, O'Brodovich H, Lukacs GL. Metabolic instability of plasmid DNA in the cytosol: a potential barrier to gene transfer. Gene Ther 1999; 6:482–497.
29. Wilson GL, Dean BS, Wang G, Dean DA. Nuclear import of plasmid DNA in digitonin-permeabilized cells requires both cytoplasmic factors and specific DNA sequences. J Biol Chem 1999; 274:22025–22032.
30. Bremner KH, Seymour LW, Pouton CW. Harnessing nuclear localization pathways for transgene delivery. Curr Opin Mol Ther 2001; 3:170–177.
31. Yonemitsu Y, Kitson C, Ferrari S, Farley R, Griesenbach U, Judd D, Steel R, Scheid P, Zhu J, Jeffery PK, Kato A, Hasan MK, Nagai Y, Masaki I, Fukumura M, Hasegawa M, Geddes DM, Alton EW. Efficient gene transfer to airway epithelium using recombinant Sendai virus. Nat Biotech 2000; 18:970–973.
32. Crystal RG, McElvaney NG, Rosenfeld MA, Chu CS, Mastrangeli A, Hay JG, Brody SL, Jaffe HA, Eissa NT, Danel C. Administration of an adenovirus containing the human CFTR cDNA to the respiratory tract of individuals with cystic fibrosis. Nat Genet 1994; 8:42–51.
33. Schwarzt DA, Quinn TJ, Thorne PS, Sayeed S, Yi AK, Krieg AM. CpG motifs in bacterial DNA cause inflammation in the lower respiratory tract. J Clin Invest 1997; 100:68–73.
34. Zabner J, Ramsay BW, Meeker DP, Aitken ML, Balfour RP, Gibson RL, Launspach J, Moscicki RA, Richards SM, Standaert TA, et al. Repeat administration of an adenovirus vector encoding cystic fibrosis transmembrane conductance regulator to the nasal epithelium of patients with cystic fibrosis. J Clin Invest 1996; 97:1504–1511.
35. Hyde SC, Southern KW, Gileadi U, Fitzjohn EM, Mofford KA, Waddell BE, Gooi HC, Goddard CA, Hannavy K, Smyth SE, Egan JJ, Sorgi FL, Huang L, Cuthbert

AW, Evans MJ, Colledge WH, Higgins CF, Webb AK, Gill DR. Repeat administration of DNA/liposomes to the nasal epithelium of patients with cystic fibrosis. Gene Ther 2000; 7:1156–1165.

36. CF Foundation 1995. Patient Registry 1994 Annual Data Report, CF Foundation, Bethesda, MD.

37. Rommens JM, Iannuzzi MC, Kerem B-S, Drumm ML, Melmer G, Dean M, Rozmahel R, Cole JL, Kennedy D, Hidaka N, et al. Identification of the cystic fibrosis gene: chromosome walking and jumping. Science 1989; 245:1059–1065.

38. Welsh MJ, Smith AE. Molecular mechanisms of CFTR chloride channel dysfunction in cystic fibrosis. Cell 1993; 73:1251–1254.

39. Koch C, Hoiby N. Pathogenesis of cystic fibrosis. Lancet 1993; 341:1065–1069.

40. Ismailov II, Awayda MS, Jovov B, Berdiev BK, Fuller CM, Dedman JR, Kaetzel M, Benos DJ. Regulation of epithelial sodium channels by the cystic fibrosis transmembrane conductance regulator. J Biol Chem 1996; 271:4725–4732.

41. Gabriel SE, Clarke LL, Boucher RC, Stutts MJ. CFTR and outward rectifying chloride channels are distinct proteins with a regulatory relationship. Nature 1993; 363: 263–268.

42. Mall M, Kunzelmann K, Hipper A, Busch AE, Greger R. cAMP stimulation of CFTR-expressing Xenopus oocytes activates a chromanol-inhibitable K+ conductance. Pflugers Arch 1996; 432:516–522.

43. Schreiber R, Greger R, Nitschke R, Kunzelmann K. Cystic fibrosis transmembrane conductance regulator activates water conductance in Xenopus oocytes. Pflugers Arch 1997; 434:841–847.

44. Cheng P-W, Boat TF, Cranfill K, Yankaskas JR, Boucher RC. Increased sulfation of glycoconjugates by cultured nasal epithelial cells from patients with cystic fibrosis. J Clin Invest 1989; 84:68–72.

45. Dosanjh A, Lencer W, Brown D, Ausiello DA, Stow JL. Heterologous expression of ΔF508 CFTR results in decreased sialylation of membrane glycoconjugates. Am J Physiol 1994; 266:C360–C366.

46. Saiman L, Prince A. Pseudomonas aeruginosa pili bind to asialoGM1 which is increased on the surface of cystic fibrosis epithelial cells. J Clin Invest 1993; 92: 1875–1880.

47. Govan JR, Nelson JW. Microbiology of cystic fibrosis lung infections: themes and issues. J R Soc Med 1993; 86 S20:11–18.

48. Oppenheimer EH, Esterly JR. Pathology of cystic fibrosis; review of the literature and comparison with 146 autopsied cases. Perspect Pediatr Pathol 1975; 2:241–278.

49. Chow CW, Landau LI, Taussig LM. Bronchial mucous glands in the newborn with cystic fibrosis. Eur J Pediatr 1982; 139:240–243.

50. Larson JE, Delcarpio JB, Farberman MM, Morrow SL, Cohen JC. CFTR modulates lung secretory cell proliferation and differentiation. Am J Physiol Lung Cell Mol Physiol 2000; 279:L333–341.

51. Khan TZ, Wagener JS, Bost T, Martinez J, Accurso FJ, Riches DW. Early pulmonary inflammation in infants with cystic fibrosis. Am J Respir Crit Care Med 1995; 151:1075–1082.

52. Abman SH, Ogle JW, Harbeck RJ, Butler-Simon N, Hammond KB, Accurso FJ. Early bacteriologic, immunologic, and clinical courses of young infants with cystic fibrosis identified by neonatal screening. J Pediatr 1991; 119:211–217.
53. Tirouvanziam R, de Bentzmann S, Hubeau C, Hinnrasky J, Jacquot J, Peault B, Puchelle E. Inflammation and infection in naive human cystic fibrosis airway grafts. Am J Respir Cell Mol Biol 2000; 23:121–127.
54. Armstrong DS, Grimwood K, Carzino R, Carlin JB, Olinsky A, Phelan PD. Lower respiratory infection and inflammation in infants with newly diagnosed cystic fibrosis. BMJ 1995; 310:1571–1572.
55. Matsui H, Grubb BR, Tarran R, Randell SH, Gatzy JT, Davis CW, Boucher RC. Evidence for periciliary liquid layer depletion, not abnormal ion composition, in the pathogenesis of cystic fibrosis airways disease. Cell 1998; 95:1005–1015.
56. Smith JJ, Travis SM, Greenberg EP, Welsh MJ. Cystic fibrosis airway epithelia fail to kill bacteria because of abnormal airway surface fluid. Cell 1996; 85:229–236.
57. Knowles MR, Robinson JM, Wood RE, Pue CA, Mentz WM, Wager GC, Gatzy JT, Boucher RC. Ion composition of airway surface liquid of patients with cystic fibrosis as compared with normal and disease-control subjects. J Clin Invest 1997; 100:2588–2595.
58. Pier GB, Grout M, Zaidi TS, Olsen JC, Johnson LG, Yankaskas JR, Goldberg JB. Role of mutant CFTR in hypersusceptibility of cystic fibrosis patients to lung infections. Science 1996; 271:64–67.
59. Engelhardt JF, Zepeda M, Cohn JA, Yankaskas JR, Wilson JM. Expression of the cystic fibrosis gene in adult human lung. J Clin Invest 1994; 93:737–749.
60. Engelhardt JF, Yankaskas JR, Ernst SA, Yang Y, Marino CR, Boucher RC, Cohn JA, Wilson JM. Submucosal glands are the predominant site of CFTR expression in the human bronchus. Nat Genet 1992; 2:240–248.
61. Hong KU, Reynolds SD, Giangreco A, Hurley CM, Stripp BR. Clara cell secretory protein-expressing cells of the airway neuroepithelial body microenvironment include a label-retaining subset and are critical for epithelial renewal after progenitor cell depletion. Am J Respir Cell Mol Biol 2001; 24:671–681.
62. Warburton D, Wuenschell C, Flores-Delgado G, Anderson K. Commitment and differentiation of lung cell lineages. Biochem Cell Biol 1998; 76:971–995.
63. Wilson PD. Cystic fibrosis transmembrane conductance regulator in the kidney: clues to its role? Exp Nephrol 1999; 7:284–289.
64. Johnson LG, Olsen JC, Sarkadi B, Moore KL, Swanstrom R, Boucher RC. Efficiency of gene transfer for restoration of normal airway epithelial function in cystic fibrosis. Nat Genet 1992; 2:21–25.
65. Dorin JR, Farley R, Webb S, Smith SN, Farini E, Delaney SJ, Wainwright BJ, Alton EW, Porteous DJ. A demonstration using mouse models that successful gene therapy for cystic fibrosis requires only partial gene correction. Gene Ther 1996; 3:797–801.
66. Johnson LG, Boyles SE, Wilson J, Boucher RC. Normalization of raised sodium absorption and raised calcium-mediated chloride secretion by adenovirus-mediated expression of cystic fibrosis transmembrane conductance regulator in primary human cystic fibrosis airway epithelial cells. J Clin Invest 1995; 95:1377–1382.

67. Zhang Y, Jiang Q, Dudus L, Yankaskas JR, Engelhardt JF. Vector-specific profiles of two independent primary defects in cystic fibrosis airways. Hum Gene Ther 1998; 20:635–648.

68. Gan KH, Veeze HJ, van den Ouweland AM, Halley DJ, Scheffer H, van der Hout A, Overbeek SE, de Jongste JC, Bakker W, Heijerman HG. A cystic fibrosis mutation associated with mild lung disease. N Engl J Med 1995; 333:95–99.

69. Knowles MR, Paradiso AM, Boucher RC. In vivo nasal potential difference: techniques and protocols for assessing efficacy of gene transfer in cystic fibrosis. Hum Gene Ther 1995; 6:445–455.

70. Middleton PG, Geddes DM, Alton EWFW. Protocols for in vivo measurement of the ion transport defects in cystic fibrosis nasal epithelium. Eur Respir J 1994; 7: 2050–2056.

71. Wilschanski M, Famini C, Blau H, Rivlin J, Augarten A, Avital A, Kerem B, Kerem E. A pilot trial of the effect of gentamicin on nasal potential difference measurements in cystic fibrosis patients carrying stop mutations. Am J Respir Crit Care Med 2000; 161:860–865.

72. Alton EWFW, Stern M, Farley R, Jaffe A, Chadwick SL, Phillips J, Davies J, Smith SN, Browning J, Davies MG, Hodson ME, Durham SR, Li D, Jeffery PK, Scallan M, Balfour R, Eastman SJ, Cheng SH, Smith AE, Meeker D, Geddes DM. Cationic lipid-mediated CFTR gene transfer to the lungs and nose of patients with cystic fibrosis: a double-blind placebo-controlled trial. Lancet 1999; 353:947–954.

73. Davies JC, Davies MG, Jaffe A, Bush A, Scallon M, Geddes DM, Alton EWFW. Confirmation of abnormal chloride ion secretion in the lower airway of children with cystic fibrosis. Ped Pulmonol 2000; S20:293.

74. Welsh MJ An apical-membrane chloride channel in human tracheal epithelium. Science 1986; 232:1648–1650.

75. Alton EW, Rogers DF, Logan-Sinclair R, Yacoub M, Barnes PJ, Geddes DM. Bioelectric properties of cystic fibrosis airways obtained at heart-lung transplantation. Thorax 1992; 47:1010–1014.

76. Stern M, Munkonge FM, Caplen NJ, Sorgi F, Huang L, Geddes DM, Alton EW. Quantitative fluorescence measurements of chloride secretion in native airway epithelium from CF and non-CF subjects. Gene Ther 1995; 2:766–774.

77. Davies JC, Stern M, Dewar A, Caplen NJ, Munkonge FM, Pitt T, Sorgi F, Huang L, Bush A, Geddes DM, Alton EW. CFTR gene transfer reduces the binding of Pseudomonas aeruginosa to cystic fibrosis respiratory epithelium. Am J Respir Cell Mol Biol 1997; 16:657–663.

78. Drumm ML, Pope HA, Cliff WH, Rommens JM, Marvin SA, Tsui LC, Collins FS, Frizzell RA, Wilson JM. Correction of the cystic fibrosis defect in vitro by retrovirus-mediated gene transfer. Cell 1990; 62:1227–1233.

79. Rich DP, Anderson MP, Gregory RJ, Cheng SH, Paul S, Jefferson DM, McCann JD, Klinger KW, Smith AE, Welsh MJ. Expression of cystic fibrosis transmembrane conductance regulator corrects defective chloride channel regulation in cystic fibrosis airway epithelial cells. Nature 1990; 347:358–363.

80. Dorin JR, Dickinson P, Alton EW, Smith SN, Geddes DM, Stevenson BJ, Kimber WL, Fleming S, Clarke AR, Hooper ML, et al. Cystic fibrosis in the mouse by targeted insertional mutagenesis. Nature 1992; 359:211–216.

81. Snouwaert JN, Brigman KK, Latour AM, Iraj E, Schwab U, Gilmour MI, Koller
     BH. A murine model of cystic fibrosis. Am J Respir Crit Care Med 1995; 151:
     S59–64.
82. Delaney SJ, Alton EWFW, Smith S, Lunn DP, Farley R, Lovelock PK, Thomson
     SA, Hume DA, Lamb D, Porteous DJ, Dorin JR, Wainwright BJ. Cystic fibrosis
     carrying the missense mutation G551D replicated human genotype-phenotype cor-
     relations. EMBO J 1996; 15:955–963.
83. Hyde SC, Gill DR, Higgins CF, Trezise AE, MacVinish LJ, Cuthbert AW, Ratcliff
     R, Evans MJ, Colledge WH. Correction of the ion transport defect in cystic fibrosis
     transgenic mice by gene therapy. Nature 1993; 362:250–255.
84. Alton EWFW, Middleton PG, Caplen NJ, Smith SN, Steel DM, Munkonge FM,
     Jeffery PK, Geddes DM, Hart SL, Williamson R, et al. Non-invasive liposome-
     mediated gene delivery can correct the ion transport defect in cystic fibrosis mutant
     mice. Nat Genet 1993; 5:135–142.
85. Grubb BR, Pickles RJ, Ye H, Yankaskas JR, Vick RN, Engelhardt JF, Wilson JM,
     Johnson LG, Boucher RC. Inefficient gene transfer by adenovirus vector to cystic
     fibrosis airway epithelia of mice and humans. Nature 1994; 371:802–806.
86. Pickles RJ, Barker PM, Ye H, Boucher RC. Efficient adenovirus-mediated gene
     transfer to basal but not columnar cells of cartilaginous airway epithelia. Hum Gene
     Ther 1996; 7:921–931.
87. Scaria A St George JA, Jiang C, Kaplan JM, Wadsworth SC, Gregory RJ. Adenovi-
     rus-mediated persistent cystic fibrosis transmembrane conductance regulator ex-
     pression in mouse airway epithelium. J Virol 1998; 72:7302–7309.
88. Flotte TR, Afione SA, Conrad C, McGrath SA, Solow R, Oka H, Zeitlin PL, Gug-
     gino WB, Carter BJ. Stable in vivo expression of the cystic fibrosis transmembrane
     conductance regulator with an adeno-associated virus vector. Proc Natl Acad Sci
     USA 1993; 90:10613–10617.
89. Zabner J, Couture LA, Gregory RJ, Graham SM, Smith AE, Welsh MJ. Adenovi-
     rus-mediated gene transfer transiently corrects the chloride transport defect in nasal
     epithelia of patients with cystic fibrosis. Cell 1993; 75:207–216.
90. Boucher RC, Knowles MR, Johnson LG, Olsen JC, Pickles R, Wilson JM, Engel-
     hardt J, Yang Y, Grossman M. Gene therapy for cystic fibrosis using E1-deleted
     adenovirus: a phase I trial in the nasal cavity. The University of North Carolina at
     Chapel Hill. Hum Gene Ther 1994; 5:615–639.
91. Zuckerman JB, Robinson CB, McCoy KS, Shell R, Sferra TJ, Chirmule N, Magosin
     SA, Propert KJ, Brown-Parr EC, Hughes JV, Tazelaar J, Baker C, Goldman MJ,
     Wilson JM. A phase I study of adenovirus-mediated transfer of the human cystic
     fibrosis transmembrane regulator gene to a lung segment of individuals with cystic
     fibrosis. Hum Gene Ther 1999; 10:2973–2985.
92. Bellon G, Michel-Calemard L, Thouvenot D, Jagneaux V, Poitevin F, Malcus C,
     Accart N, Layani MP, Aymard M, Bernon H, Bienvenu J, Courtney M, Doring G,
     Gilly B, Gilly R, Lamy D, Levrey H, Morel Y, Paulin C, Perraud F, Rodillon L,
     Sene C, So S, Touraine-Moulin F, Pavirani A, et al. Aerosol administration of a
     recombinant adenovirus expressing CFTR to cystic fibrosis patients: a phase I clini-
     cal trial. Hum Gene Ther 1997; 8:15–25.
93. Harvey B-G, Leopold PL, Hackett NR, Grasso TM, Williams PM, Tucker AL,

Kaner RJ, Ferris B, Gonda I, Sweeney TD, Ramalingam R, Kovesdi I, Shak S, Crystal RG. Airway epithelial CFTR mRNA expression in cystic fibrosis patients after repetitive administration of a recombinant adenovirus. J Clin Invest 1999; 104:1245–1255.

94. Harvey BG, Hackett NR, El-Sawy T, Rosengart TK, Hirschowitz EA, Lieberman MD, Lesser ML, Crystal RG: Variability of human systemic humoral immune responses to adenovirus gene transfer vectors administered to different organs. J Virol 1999; 73:6729–6742.

95. Wagner JA, Reynolds T, Moran ML, Moss RB, Wine JJ, Flotte TR, Gardner P. Efficient and persistent gene transfer of AAV-CFTR in maxillary sinus. Lancet 1998; 351:1702–1703.

96. Caplen NJ, Alton EWFW, Middleton PG, Dorin JR, Stevenson BJ, Gao X, Durham SR, Jeffery PK, Hodson ME, Coutelle C, et al. Liposome-mediated CFTR gene transfer to the nasal epithelium of patients with cystic fibrosis. Nat Med 1995; 1: 39–46.

97. Gill DR, Southern KW, Mofford KA, Seddon T, Huang L, Sorgi F, Thomson A, MacVinish LJ, Ratcliff R, Bilton D, Lane DJ, Littlewood JM, Webb AK, Middleton PG, Colledge WH, Cuthbert AW, Evans MJ, Higgins CF, Hyde SC. A placebo controlled study of liposome-mediated gene transfer to the nasal epithelium of patients with cystic fibrosis. Gene Ther 1997; 4:199–209.

98. Porteous DJ, Dorin JR, McLachlan G, Davidson-Smith H, Davidson H, Stevenson BJ, Carothers AD, Wallace WA, Moralee S, Hoenes C, Kallmeyer G, Michaelis U, Naujoks K, Ho LP, Samways JM, Imrie M, Greening AP, Innes JA. Evidence for safety and efficacy of DOTAP cationic liposome mediated CFTR gene transfer to nasal epithelium of patients with cystic fibrosis. Gene Ther 1997; 4:210–218.

99. Noone PG, Hohneker KW, Zhou Z, Johnson LG, Foy C, Gipson C, Jones K, Noah TL, Leigh MW, Schwartzbach C, Efthimiou J, Pearlman R, Boucher RC, Knowles MR. Safety and biological efficacy of a lipid-CFTR complex for gene transfer in the nasal epithelium of adult patients with cystic fibrosis. Mol Ther 2000; 1:105–114.

100. Chadwick SL, Kingston HD, Stern M, Cook RM, O'Connor BJ, Lukasson M, Balfour RP, Rosenberg M, Cheng SH, Smith AE, Meeker DP, Geddes DM, Alton EW. Safety of a single aerosol administration of escalating doses of the cationic lipid GL-67/DOPE/DMPE-PEG$_{5000}$ formulation to the lungs of normal volunteers. Gene Ther 1997; 4:937–942.

101. Coakley RJ, Taggart C, O'Neill S, McElvaney NG. Alpha1-antitrypsin deficiency: biological answers to clinical questions. Am J Med Sci 2001; 321:33–41.

102. Pierce JA. Alpha1-antitrypsin augmentation therapy. Chest 1997; 112:872–874.

103. Kay MA, Graham F, Leland F, Woo SL. Therapeutic serum concentrations of human alpha-1-antitrypsin after adenoviral-mediated gene transfer into mouse hepatocytes. Hepatology 1995; 21:815–819.

104. Song S, Morgan M, Ellis T, Poirier A, Chesnut K, Wang J, Brantly M, Muzyczka N, Byrne BJ, Atkinson M, Flotte TR. Sustained secretion of human alpha-1-antitrypsin from murine muscle transduced with adeno-associated virus vectors. Proc Natl Acad Sci USA 1998; 95:14384–14388.

105. Bou-Gharios G, Wells DJ, Lu QL, Morgan JE, Partridge T. Differential expression

and secretion of alpha1 anti-trypsin between direct DNA injection and implantation of transfected myoblast. Gene Ther 1999; 6:1021–1029.

106. Canonico AE, Conary JT, Meyrick BO, Brigham KL. Aerosol and intravenous transfection of human alpha 1-antitrypsin gene to lungs of rabbits. Am J Respir Cell Mol Biol 1994; 10:24–29.

107. Brigham KL, Lane KB, Meyrick B, Stecenko AA, Strack S, Cannon DR, Caudill M, Canonico AE. Transfection of nasal mucosa with a normal α1-antitrypsin gene in α1-antitrypsin-deficient subjects: comparison with protein therapy. Hum Gene Ther 2000; 11:1023–1032.

108. Canonico AE, Parker RE, Meyrick BO, Gao X, Lane K, Cannon D, Wilson T, Brigham KL. Alpha-1 antitrypsin (AAT) gene therapy is superior to protein therapy in a piglet in situ lung model. Am J Respir Lung Mol Biol 1997; 14:385–390.

109. Rom WN, Hay JG, Lee TC, Jiang Y, Tchou-Wong K-M. Molecular and genetic aspects of lung cancer. Am J Respir Crit Care Med 2000; 161:1355–1367.

110. Roth JA, Nguyen D, Lawrence DD, Kemp BL, Carrasco CH, Ferson DZ, Hong WK, Komaki R, Lee JJ, Nesbitt JC, Pisters KM, Putnam JB, Schea R, Shin DM, Walsh GL, Dolormente MM, Han CI, Martin FD, Yen N, Xu K, Stephens LC, McDonnell TJ, Mukhopadhyay T, Cai D. Retrovirus-mediated wild-type p53 gene transfer to tumors of patients with lung cancer. Nat Med 1996; 2:985–991.

111. Schuler M, Rochlitz C, Horowitz JA, Schlegel J, Perruchoud AP, Kommoss F, Bolliger CT, Kauczor HU, Dalquen P, Fritz MA, Swanson S, Herrmann R, Huber C. A phase I study of adenovirus-mediated wild-type p53 gene transfer in patients with advanced non-small cell lung cancer. Hum Gene Ther 1998; 9:2075–2082.

112. Swisher SG, Roth JA, Nemunaitis J, Lawrence DD, Kemp BL, Carrasco CH, Connors DG, El-Naggar AK, Fossella F, Glisson BS, Hong WK, Khuri FR, Kurie JM, Lee JJ, Lee JS, Mack M, Merritt JA, Nguyen DM, Nesbitt JC, Perez-Soler R, Pisters KM, Putnam JB Jr, Richli WR, Savin M, Waugh MK, et al. Adenovirus-mediated p53 gene transfer in advanced non-small cell lung cancer. J Natl Cancer Inst 1999; 91:763–771.

113. Nemunaitis J, Swisher SG, Timmons T, Connors D, Mack M, Doerksen L, Weill D, Wait J, Lawrence DD, Kemp BL, Fossella F, Glisson BS, Hong WK, Khuri FR, Kurie JM, Lee JJ, Lee JS, Nguyen DM, Nesbitt JC, Perez-Soler R, Pisters KM, Putnam JB, Richli WR, Shin DM, Walsh GL, et al. Adenovirus-mediated p53 gene transfer in sequence with cisplatin to tumors of patients with non-small cell lung cancer. J Clin Oncol 2000; 18:609–622.

114. Yen N, Ioannides CG, Xu K, Swisher SG, Lawrence DD, Kemp BL, El-Naggar AK, Cristiano RJ, Fang B, Glisson BS, Hong WK, Khuri FR, Kurie JM, Lee JJ, Lee JS, Merritt JA, Mukhopadhyay T, Nesbitt JC, Nguyen D, Perez-Soler R, Pisters KM, Putnam JB Jr, Schrump DS, Shin DM, Walsh GL, Roth JA. Cellular and humoral immune responses to adenovirus and p53 protein antigens in patients following intratumoral injection of an adenovirus vector expressing wild-type P53 (Ad-p53). Cancer Gene Ther 2000; 7:530–536.

115. Smythe WR. Prodrug/drug sensitivity gene therapy: current status. Curr Oncol Rep 2000; 2:17–22.

116. Sterman DH, Treat J, Litzky LA, Amin KM, Coonrod L, Molnar-Kimber K, Recio A, Knox L, Wilson JM, Albelda SM, Kaiser LR. Adenovirus-mediated herpes sim-

plex virus thymidine kinase/ ganciclovir gene therapy in patients with localised malignancy: results of a phase I clinical trial in malignant mesothelioma. Hum Gene Ther 1998; 9:1083–1092.

117. Tursz T, Cesne AL, Baldeyrou P, Gautier E, Opolon P, Schatz C, Pavirani A, Courtney M, Lamy D, Ragot T, Saulnier P, Andremont A, Monier R, Perricaudet M, Le Chevalier T. Phase I study of a recombinant adenovirus-mediated gene transfer in lung cancer patients. J Natl Cancer Inst 1996; 88:1857–1863.

118. Gahery-Segard H, Molinier-Frenkel V, Le Boulaire C, Saulnier P, Opolon P, Lengagne R, Gautier E, Le Cesne A, Zitvogel L, Venet A, Schatz C, Courtney M, Le Chevalier T, Tursz T, Guillet JG, Farace F. Phase I trial of recombinant adenovirus gene transfer in lung cancer. Longitudinal study of the immune responses to transgene and viral products. J Clin Invest 1997; 100:2218–2226.

119. Mukherjee S, Haenel T, Himbeck R, Scott B, Ramshaw I, Lake RA, Harnett G, Phillips P, Morey S, Smith D, Davidson JA, Musk AW, Robinson B. Replication-restricted vaccinia as a cytokine gene therapy vector in cancer: persistent transgene expression despite antibody generation. Cancer Gene Ther 2000; 7:663–670.

120. McCormick F. Interactions between adenovirus proteins and the p53 pathway: the development of ONYX-015. Semin Cancer Biol 2000; 10:453–459.

121. Bischoff JR, Kim DH, Williams A, Heise C, Horn S, Muna M, Ng L, Nye JA, Sampson-Johannes A, Fattaey A, McCormick F. An adenovirus mutant that replicates selectively in p53-deficient human tumor cells. Science 1996; 274:373–376.

122. Heise C, Sampson-Johannes A, Williams A, McCormick F, Von Hoff DD, Kim DH. ONYX-015, an E1B gene-attenuated adenovirus, causes tumor-specific cytolysis and antitumoral efficacy that can be augmented by standard chemotherapeutic agents. Nat Med 1997; 3:639–645.

123. Nemunaitis J, Cunningham C, Buchanan A, Blackburn A, Edelman G, Maples P, Netto G, Tong A, Randlev B, Olson S, Kim D. Intravenous infusion of a replication-selective adenovirus (ONYX-015) in cancer patients: safety, feasibility and biological activity. Gene Ther 2001; 8:746–759.

124. Weinacker AB, Vaszar LT. Acute respiratory distress syndrome: physiology and new management strategies. Annu Rev Med 2001; 52:221–237.

125. Dennehy KC, Bigatello LM. Pathophysiology of the acute respiratory distress syndrome. Int Anesthesiol Clin 1999; 37:1–13.

126. Rogy MA, Auffenberg T, Espat NJ, Philip R, Remick D, Wollenberg GK, Copeland EM 3rd, Moldawer LL. Human tumor necrosis factor receptor (p55) and interleukin 10 gene transfer in the mouse reduces mortality to lethal endotoxemia and also attenuates local inflammatory responses. J Exp Med 1995; 181:2289–2293.

127. Xing Z, Ohkawara Y, Jordana M, Graham FL, Gauldie J. Adenoviral vector-mediated interleukin-10 expression in vivo: intramuscular gene transfer inhibits cytokine responses in endotoxemia. Gene Ther 1997; 4:140–149.

128. Conary JT, Parker RE, Christman BW, Faulks RD, King GA, Meyrick BO, Brigham KL. Protection of rabbit lungs from endotoxin injury by in vivo hyperexpression of the prostaglandin G/H synthase gene. J Clin Invest 1994; 93:1834–1840.

129. von der Leyen HE, Gibbons GH, Morishita R, Lewis NP, Zhang L, Nakajima M, Kaneda Y, Cooke JP, Dzau VJ. Gene therapy inhibiting neointimal vascular lesion:

in vivo transfer of endothelial cell nitric oxide synthase gene. Proc Natl Acad Sci USA 1995; 92:1137–1141.

130. Kumasaka T, Quinlan WM, Doyle NA, Condon TP, Sligh J, Takei F, Beaudet Al, Bennett CF, Doerschuk CM. Role of the intercellular adhesion molecule-1 (ICAM-1) in endotoxin-induced pneumonia evaluated using ICAM-1 antisense oligonucleotides, anti–ICAM-1 monoclonal antibodies, and ICAM-1 mutant mice. J Clin Invest 1996; 97:2362–2369.

131. Lee JC, Laydon JT, McDonnell PC, Gallagher TF, Kumar S, Green D, McNulty D, Blumenthal MJ, Heys JR, Landvatter SW, et al. A protein kinase involved in the regulation of inflammatory cytokine biosynthesis. Nature 1994; 372:739–746.

132. Inoue S, Suzuki M, Nagashima Y, Suzuki S, Hashiba T, Tsuburai T, Ikehara K, Matsuse T, Ishigatsubo Y. Transfer of Heme Oxygenase 1 cDNA by a replication-deficient adenovirus enhances interleukin 10 production from alveolar macrophages that attenuates lipopolysaccharide-induced acute lung injury in mice. Hum Gene Ther 2001; 12:967–979.

133. Factor P. Role and regulation of lung Na,K-ATPase. Cell Mol Biol 2001; 47:347–361.

134. Factor P, Dumasius V, Saldias F, Brown LA, Sznajder JI. Adenovirus-mediated transfer of an Na+/K+-ATPase betal subunit gene improves alveolar fluid clearance and survival in hyperoxic rats. Hum Gene Ther 2000; 11:2231–2242.

135. Stern M, Ulrich K, Robinson C, Copeland J, Griesenbach U, Masse C, Cheng S, Munkonge F, Geddes D, Berthiaume Y, Alton E. Pretreatment with cationic lipid-mediated transfer of the Na+K+-ATPase pump in a mouse model in vivo augments resolution of high permeability pulmonary oedema. Gene Ther 2000; 7:960–966.

136. Suissa S, Ernst P. Inhaled corticosteroids: impact on asthma morbidity and mortality. J Allergy Clin Immunol 2001; 107:937–944.

137. Lee NA, Gelfand EW, Lee JJ. Pulmonary T cells and eosinophils: coconspirators or independent triggers of allergic respiratory pathology? J Allergy Clin Immunol 2001; 107:945–957.

138. Lack G, Renz H, Saloga J, Bradley KL, Loader J, Leung DY, Larsen G, Gelfand EW. Nebulized but not parenteral IFN-gamma decreases IgE production and normalizes airways function in a murine model of allergen sensitization. J Immunol 1994; 152:2546–2554.

139. Schwarze J, Hamelmann E, Cieslewicz G, Tomkinson A, Joetham A, Bradley K, Gelfand EW. Local treatment with IL-12 is an effective inhibitor of airway hyper-responsiveness and lung eosinophilia after airway challenge in sensitized mice. J Allergy Clin Immunol 1998; 102:86–93.

140. Dow SW, Schwarze J, Heath TD, Potter TA, Gelfand EW. Systemic and local interferon gamma gene delivery to the lungs for treatment of allergen-induced airway hyperresponsiveness in mice. Hum Gene Ther 1999; 10:1905–1914.

141. Hogan SP, Foster PS, Tan X, Ramsay AJ. Mucosal IL-12 gene delivery inhibits allergic airways disease and restores local antiviral immunity. Eur J Immunol 1998; 28:413–423.

142. Mathieu M, Gougat C, Jaffuel D, Danielsen M, Godard P, Bousquet J, Demoly P.

The glucocorticoid receptor gene as a candidate for gene therapy in asthma. Gene Ther 1999; 6:245–252.

143. Metzger WJ, Nyce JW. Oligonucleotide therapy of allergic asthma. J Allergy Clin Immunol 1999; 104:260–266.

144. Alton EW, Griesenbach U, Geddes DM. Gene therapy for asthma: inspired research or unnecessary effort? Gene Ther 1999; 6:155–156.

145. Fonseca C, Abraham D, Black CM. Lung fibrosis. Springer Semin Immunopathol 1999; 21:453–474.

146. Sime PJ, O'Reilly KM. Fibrosis of the lung and other tissues: new concepts in pathogenesis and treatment. Clin Immunol 2001; 99:308–319.

147. Nakao A, Fujii M, Matsumura R, Kumano K, Saito Y, Miyazono K, Iwamoto I. Transient gene transfer and expression of Smad7 prevents bleomycin-induced lung fibrosis in mice. J Clin Invest 1999; 104:5–11.

148. Kolb M, Margetts PJ, Galt T, Sime PJ, Xing Z, Schmidt M, Gauldie J. Transient transgene expression of decorin in the lung reduces the fibrotic response to bleomycin. Am J Respir Crit Care Med 2001; 163:770–777.

149. Sisson TH, Hattori N, Xu Y, Simon RH. Treatment of bleomycin-induced pulmonary fibrosis by transfer of urokinase-type plasminogen activator genes. Hum Gene Ther 1999; 10:2315–2323.

150. Epperly M, Bray J, Kraeger S, Zwacka R, Engelhardt J, Travis E, Greenberger J. Prevention of late effects of irradiation lung damage by manganese superoxide dismutase gene therapy. Gene Ther 1998; 5:196–208.

151. Ziesche R, Hofbauer E, Wittmann K, Petkov V, Block LH. A preliminary study of long-term treatment with interferon gamma-lb and low-dose prednisolone in patients with idiopathic pulmonary fibrosis. N Engl J Med 1999; 341:1264–1269.

152. Kuwano K, Hagimoto N, Kawasaki M, Yatomi T, Nakamura N, Nagata S, Suda T, Kunitake R, Maeyama T, Miyazaki H, Hara N. Essential roles of the Fas-Fas ligand pathway in the development of pulmonary fibrosis. J Clin Invest 1999; 104:13–19.

153. Mal H, Dehoux M, Sleiman C, Boczkowski J, Leseche G, Pariente R, Fourier M. Early release of proinflammatory cytokines after lung transplantation. Chest 1998; 113:645–651.

154. Ward S, Muller NL. Pulmonary complications following lung transplantation. Clin Radiol 2000; 55:332–339.

155. Cassivi SD, Cardella JA, Fischer S, Liu M, Slutsky AS, Keshavjee S. Transtracheal gene transfection of donor lungs prior to organ procurement increases transgene levels at reperfusion and following transplantation. J Heart Lung Transplant 1999; 18:1181–1188.

156. Boasquevisque CH, Mora BN, Boglione M, Ritter JK, Scheule RK, Yew N, Debruyne L, Qin L, Bromberg JS, Patterson GA. Liposome-mediated gene transfer in rat lung transplantation: A comparison between the in vivo and ex vivo approaches. J Thorac Cardiovasc Surg 1999; 117:8–14.

157. Itano H, Zhang W, Ritter JH, McCarthy TJ, Mohanakumar T, Patterson GA. Adenovirus-mediated gene transfer of human interleukin 10 ameliorates reperfusion injury of rat lung isografts. J Thorac Cardiovasc Surg 2000; 120:947–956.

158. Schmid RA, Stammberger U, Hillinger S, Gaspert A, Boasquevisque CH, Mali-
      piero U, Fontana A, Weder W. Fas ligand gene transfer combined with low dose
      cyclosporine A reduces acute lung allograft rejection. Transpl Int 2000; 13(Suppl
      1):S324–328.
159. Mora BN, Boasquevisque CH, Boglione M, Ritter JM, Scheule RK, Yew NS, De-
      bruyne L, Qin L, Bromberg JS, Patterson GA. Transforming growth factor-beta1
      gene transfer ameliorates acute lung allograft rejection. J Thorac Cardiovasc Surg
      2000; 119:913–920.
160. Suda T, D'Ovidio F, Daddi N, Ritter JH, Mohanakumar T, Patterson GA. Recipient
      intramuscular gene transfer of active transforming growth factor-beta1 attenuates
      acute lung rejection. Ann Thorac Surg 2001; 71:1651–1656.

15
Artificial Chromosomes

Jonathan Black and Jean-Michel Vos*

University of North Carolina at Chapel Hill, Chapel Hill, North Carolina, U.S.A.

I. INTRODUCTION

Artificial chromosomes are DNA moieties created in vitro that behave like natural chromosomes in vivo; for example, independent replication and segregation to daughter cells. Artificial chromosomal technology provides a powerful biotechnological tool with many broad applications including cloning of large DNA fragments, characterization of genomes by functional and physical mapping, and generation of commercially important transgenic animals and plants. Although the potentials for artificial chromosomes are numerous, it is beyond the scope of this chapter to discuss them all. Therefore, we will limit our discussion to the essential components of the first artificial chromosome, to four types of artificial chromosomes, and to the uses of artificial chromosomes in the areas of physical mapping and animal and plant transgenesis. Furthermore, we will outline the mammalian artificial episomal chromosomal engineering strategy developed by our laboratory.

II. ESSENTIAL COMPONENTS OF ARTIFICIAL CHROMOSOMES

A. Origins

It has been theorized that artificial chromosomes must contain at least three essential elements to function de novo. The first, termed an origin [1], serves as the

---

* Deceased.

location for special replication initiator protein binding and is thus the site for initiation of replication for chromosomal DNA. Origins are usually specifically defined sequences that often contain a region of mostly A-T base pairings presumably for easier unwinding of the DNA helix. The ideal artificial chromosome would possess an origin that acts ubiquitously in nature. Origin discovery and mapping is an active area of research with many dozens mapped from both prokaryotes and eukaryotes to date. For the purpose of brevity, we will list six documented mammalian origins [2]. These origins are called ADA, lamin B^2, beta (B) globin, DHFR, rhodopsin, and RPS14. They span 0.5 Kb–2.0 Kb of DNA and require the modular sequence elements outlined in Table 1 for initiation of replication. Although DNA replication is essential for cellular proliferation, the mechanisms of and regulation by cis-acting elements is poorly understood [2]. However, origin sequence may be less important than chromosomal structure at specific genomic regions [3]. Furthermore, at least one nonorigin has been identified in eukaryotes [4] that may be utilized as origin for artificial chromosomal replication.

B. Telomeres

The second component(s) necessary for artificial chromosome function are a pair of telomeres [5]. Telomeres are stable structural elements located at the ends of chromosomes and are required for the complete replication of the chromosome. Telomeres consist of short repetitive sequences of up to 10,000 bp. The repeat can vary in length from 4 to 8 bp, and in all cases one strand is usually rich in T and G. In mammals, the consensus telomeric sequence is 5′ TTAGGG(n) 3′ in arrays up to 1 Kb in length. It has not been determined whether telomeres are absolutely required for all types of artificial chromosomes.

C. Centromeres

Finally, artificial chromosomes need a centromere. Centromeric regions of eukaryotic chromosomes contain large regions of highly repetitive sequence. The sizes of these repeats range from 15 bp to 350 bp depending on the species. There are usually thousands of copies of these sequences adjacent to a centromere, and an undefined number is required for proper mitotic replication [6]. Although there appears to be consensus in human and primate centromeric regions, other mammalian species show substantial divergence [7]. Interestingly, some specific DNA regions adjacent to but not associated with the centromere may substitute for one in nonpermissive species when no centromere is present [7a]. Consequently, there appears to be much latitude in the centromeric sequences required for artificial chromosomal replication.

Table 1 Six Well-Described Eukaryotic Replication Origins, Their Consensus Sequences, and Identified DNA Unwinding Elements (DUE) Necessary for Replication Initiation

Elements	Consensus sequences	Mapped replication origins
DUE (DNA unwinding elements)	Inherently low helix (>200)	DHFR, C-MYC, rhopsin
SAR (scaffold attachment regions)	AAT(A)₃Y(A)₃WADAWAYAWWTTWTWTTW(T)₃ WWTTDTT(W)₃	DHFR, C-MYC, rhopsin
ARS (autonomous replicating site)	W(T)₃AYR(T)₃W	DHFR, C-MYC, rhopsin
PYR (pyrimidine tract)	(Y)₍₎₁₁	DHFR, C-MYC, rhopsin
CIE (consensus initiation elements)	WAWTTDD(W)₃DHWGWHNAWTT	DHFR, C-MYC, rhopsin
ᵐCpG (methylated CpG clusters)	ᵐCpG clusters (>1 mCpG/40bp)	DHFR, rhopsin, RPS14, ADA

Note: R=A or G; W = A or T; Y = C or T; D = A or G or T; N = A or C or T.
Source: Adapted from Refs. 46 and 47.

III. FOUR TYPES OF ARTIFICIAL CHROMOSOMES

A. Yeast Artificial Chromosomes

Yeast artificial chromosomes (YACs) were first constructed in the budding yeast *Saccharomyces cerevisiae* [8] by successfully combining the previously identified essential minimal eukaryotic elements; for example, an origin, a pair of telomeres, and a centromere as a large circular plasmid. Originally, it was assumed that the core size of a YAC needed to be at least 10 Mb in length or about one-fifth the size of the smallest human chromosome for proper de novo function. Although the original YAC [8] was mitotically stable and could carry mega-base sized inserts, they were very cumbersome. In 1987, large DNA fragments up to 2 Mb were successfully cloned into a newly developed linear YAC of just 50 Kb [9] and was still stably maintained and segregated at mitosis. Therefore, these second-generation YACs provided a novel system for an in-depth study of entire eukaryotic gene loci including all elements required for transcriptional and posttranscriptional regulation in budding yeast. Another plus of this new system was that, like most eukaryotic cells, yeast possess an inherent homologous recombination system which offered a precise and relatively easy avenue for modification of any introduced gene similar to that experienced with prokaryotic cloning systems. Also, genotypic modification was easily and quickly observable by phenotype in each individual recombinant yeast cell. The essential components of a YAC are outlined in Table 2. The origin of replication (ARS1) was isolated from budding yeast, the telomeres (TEL) were derived from *Tetrahymena* sp., and a ubiquitous eukaryotic centromere (CEN3) separates the vector into long and short arms. Although not shown, a yeast selectable marker like LEU2 is always added for selective propagation. The long arm of a YAC can accommodate a large (2 Mb) insert. YACs are usually linearized upon entry into host cells and may or may not integrate into the host genome.

B. Bacterial Artificial Chromosomes

Bacterial artificial chromosomes (BACs) are bacterial plasmid-based vectors first introduced in 1989 [10] as an alternative to YAC use. The logic for their development was driven by a desire to have an artificial chromosome that could easily be propagated in bacteria to obtain large amounts of the replicon. BACs were thus designed to operate in both prokaryotic and eukaryotic cells. The essential components of a BAC are depicted in Table 2 [11]. The quintessential BAC, pBeloBAC11, was designed to operate under an *Escherichia coli* F-factor plasmid replicon. The F-factor plasmid replicon operates together with the partitioning regulatory genes *parA*, *parB*, and *parC* ensuring plasmid copy numbers to as little as 1 in eukaryotic cells and as high as 50 in prokaryotic cells. In BACs, the *oriS* and *repE* genes control unidirectional plasmid replication like a eukaryotic centromere. Selection in prokaryotic cells is guaranteed by the inclu-

Table 2 Four Types of Artificial Chromosomes, Essential Components of Each, and Established Functions of Each Feature

Type	Key features	Key function
YAC	Telomere (TEL)	Complete replication of entire chromosome
	Origin (ARS1)	DNA initiation of replication site
	Centromere (CEN3)	Required for proper mitotic replication
	Insert	Insert size up to 2 Mb
BAC	F-factor replicon	Maintains eukaryotic copy number to
	oriS, repE	Unidirectional plasmid replication
	parA/B/C	Partitioning and regulation
	cosN	Bacteriophage λ terminase restriction site
	loxP	Cre recombinase enzyme restriction site
	lacZ MCS	Blue/white screening; insert up to 350 Kb
PAO	P1 bacteriophage replicon	Maintains copy number to one
	P1 lytic replicon	Permits multicopy propagation of PAC plasmids
	SacBII	Suicide gene
	loxP	Cre recombinase enzyme restriction site
	pUC MCS	Insert up to 350 kb
MAEC	oriP	Latent origin of replication
	EBNA1	Episomal retention, replication, and segregation
	repE	Unidirectional plasmid replication
	parA/B	Partitioning and regulation
	cosN	Bacteriophage λ terminase restriction site
	loxP	Cre recombinase enzyme restriction site
	MCS	Insert up to 350 Kb

sion of a chloramphenicol resistance gene while a hygromycin cassette can be inserted for selection in mammalian cells. Furthermore, three other cloning sites are available for targeted downstream manipulations. The first, cosN, is specific for the restriction enzyme bacteriophage λ terminase, the second is the Cre recombinase recognition site loxP, and the third is a lacZ multicloning site for blue/white screening. Like YACs, BACs can carry large DNA inserts of up to 350 Kb [12–15] and are speculated to be able to carry an insert as large as the entire *E. coli* chromosome [16]. BACs are typically supercoiled plasmids that resist mechanical shearing making purification easier, and they may or may not integrate in the host genome.

C. P1-Derived Artificial Chromosomes

Pl-based artificial chromosomes (PACs) were first designed in 1990 [17]. They were introduced as an alternative cloning vector to BACs and their essential components are illustrated in Table 2. Although they are fundamentally similar to

BACs, PAC replication is driven by both a P1 bacteriophage replicon and P1 lytic replicon. The P1 bacteriophage replicon maintains a plasmid copy number of one in eukaryotic cells. However, P1 lytic replication is driven by an isopropyl β-D-thiogalactoside (IPTG)–inducible lac promoter [17]; for example, under IPTG selection, multicopy plasmids are available for vector purification once a desired clone has been identified. In addition, PACs carry a multilcloning site in the sacBII cassette. The sacBII gene is a conditional suicide gene that is toxic to cells in its unadulterated form. Therefore, any recombination is easily demonstrated by cell survival. Finally, PACs are equipped with a kanamycin resistance gene for rapid screening in prokaryotic cells [18]. Like BACs, PACs carry up to 300 Kb of insert DNA and can be easily transformed into *E. coli*. PACs may or may not integrate into the host genome.

D. Mammalian Artificial Chromosomes

Mammalian artificial chromosomes (MACs) are more difficult to define than other artificial chromosomes, because components of YACs, BACs, or PACs can be used to generate a MAC. Therefore, MAC usage still seeks to guarantee nuclear stability but can be altered to match the research agenda accordingly. By definition, MACs are vectors derived from components of mammalian chromosomes and are designed for ubiquitous use in all mammalian cell types. The impetus to develop MACs was driven by the desire to keep mammalian genomes free of immune-responsive viral and bacterial proteins. Although many types of MACs are theoretically possible, in general, only two types are under development. These two MACs can either possess only eukaryotic replication elements or they can possess both eukaryotic and prokaryotic replication elements. In addition, MACs can also be engineered to be chromosomally integrative or chromosomally nonintegrative depending on the research goal. For over a decade, we have successfully engineered nonintegrative episomal MACs (MAEC) containing both eukaryotic and viral elements that act like artificial chromosomes de novo. Therefore, we will limit our discussion to MAEC development. However, we will describe the strategies for developing both the eukaryotic and prokaryotic MACs later in this chapter.

1. Episomal EBV/BAC–Based MACs

The episomal MAC designed by us and used in our laboratory was generated from the bare essential components of the latently episomal human herpes virus type 4 Epstein–Barr virus (EBV) coupled to a BAC backbone (MAEC). Its more salient features are described in Table 2. This MAEC is equipped with the oriP/ EBNA1 paradigm. In this scenario the latent origin of replication, oriP, is bound by the viral encoded transactivator protein EBNA1 at two specific sites to guarantee unidirectional episomal replication once per cell cycle [19]. Therefore, the copy number is typically one in mammalian cells. Furthermore, EBNA1 is also responsible for nuclear retention and segregation of the latently replicating episome to daughter cells and may have a role in immune escape of MAEC-transfected cells [19]. Multiple copies of this MAEC in bacteria are guaranteed by the inclusion of a chloramphenicol cassette. Selection in mammalian cells is achieved through the hygromycin cassette. In addition, our MAEC possesses BAC elements repE, parA/B, CosN, and loxP, as already described in BAC Section III. B. MAECs do not integrate into the host genome.

IV. APPLICATIONS OF ARTIFICIAL CHROMOSOMES

Artificial chromosomes provide a powerful tool for cloning large DNA fragments. They are engineered to be genetically stable, exhibit autonomous replication, and segregate like natural chromosomes de novo. These vectors have the capacity for in-depth study of functional biology and development, because they can contain an entire eukaryotic gene loci complete with all the cis-regulatory elements necessary for precise spatial orientation and function. Furthermore, the transgene can be inserted and expressed in a wide variety of cultured cells, in transgenic animals, in transplantable cells, and in human somatic tissues. It is envisioned that mammalian artificial chromosomes will revolutionize basic chromosomal research ranging from function and gene expression to transgene production of therapeutics to somatic/germline gene therapy. Rather than discuss them all, we will limit our review to artificial chromosomal, research in physical mapping and transgenesis. In addition, we will outline our "top-down" and "bottom-up" strategies for engineering artificial chromosomes.

A. Physical Mapping

Physical maps can be constructed by a myriad of techniques (as reviewed in Ref. 20). Until the introduction of artificial chromosomes, cosmids were used to construct physical maps. Cosmids are virus-based vectors capable of carrying approximately 25 Kb of insert DNA. The resulting cosmid-based libraries generated a multitude of unrelated contig maps. These maps were informative in that they showed the linear order, distance, and crossing over points of any two mapped genes along the contig. However, the individual contig maps were difficult to align, making whole genomic mapping studies tedious. The advent of large artificial chromosomal cloning systems made whole genomic mapping projects attainable. Artificial chromosomes are ideally suited to map chromosomes, because they can carry the large regions (Mb+) of fragmented chromosomal DNA, permitting easier ordering of the large overlapping regions of fragmented DNA.

In fact, artificial chromosomal cloning systems allowed for the physical linkage of previously unlinked cosmid-based contig maps. Libraries have now been constructed for many species including human, mouse, *Arabidopsis*, tomato, maize, barley, and rice [9,21–28] using YAC, BAC, and PAC technology. Although all forms of artificial chromosomes have been proven to be invaluable in generating the large-insert deep-coverage libraries required for compiling physical maps, most libraries are now constructed using either a BAC or a PAC. BACs are very stable in *E. coli*, making them ideal vectors when constructing physical maps either by restriction enzyme digestion DNA fingerprinting [29] or by sequence-tagged connector series (STS) contig mapping [30], and the two procedures need not be mutually exclusive. DNA fingerprinting can be an effective screening method to identify positive clones, because their restriction patterns will be different and somewhat predictable. The accuracy of physical mapping based on STS technology is dependent on the depth of the clone coverage or deep-coverage large-insert libraries. Generally, there should be enough cloned DNA to cover the genome five times over while keeping the insert size to less than 1 Mb. Primers are designed for the specific STS sites in the clones and polymerase chain reaction (PCR) is used to identify all positive recombinants. Finally, fluorescence in situ hybridization (FISH) of positive recombinant clones on a metaphase spread will show the location of the gene of interest in or around individual chromosomes. This type of approach has been termed reverse genetics, because it discovers and clones a gene before its true function is known. Choosing a vector to generate genomic libraries is not determined by the size of the insert but instead by the integrity of the clone backbone. Although YACs have a cloning capacity of up to 2 Mb, they are not easy to handle, they have a low transformation efficiency, and their "recombination machinery" can result in unwanted chimeric YAC clones. BAC libraries overcome many of the disadvantages of YAC libraries [11]. BACs have been developed that are easily transformed into bacteria, can be purified away from bacteria with little or no DNA shearing, can lower the recombination frequency, and can lower the occurrence of unwanted chimerism.

B. Animal Transgenesis

The ability to create transgenic organisms has developed into one of the most powerful biotechnological tools available today. Transgenic technology has revolutionized the way research is conducted in a variety of scientific disciplines and has linked geneticists, biochemists, and biologists as never before. MAC technology has been used to generate a variety of transgenic animals ranging from commercially important livestock species expressing recombinant proteins in milk [31,32] to mouse disease models mimicking human disorders (33–40). These transgenic organisms have allowed for unparalleled deep investigation of the basic mechanisms of gene regulation, function, and evaluation of therapeutics so crucial for gene therapy. Transgenic organisms are created by the stable introduction of foreign DNA into the host genome. Perhaps a more appropriate definition of transgenicity is retention of transgenic DNA in the host nuclear membrane, because a transgene can be stably replicated either by chromosomal integration or by episomal persistence. Although both approaches will result in transgenic animals and plants there are some inherent disadvantages of the former. Integration of foreign DNA at preselected chromosomal sites is not always feasible and can result in no integration, single integration at an undesirable site, or even multiple integrations. Furthermore, since most virus-based vectors cannot carry even the average-sized gene loci, they must deliver cDNA and thus rely on ectopic expression of the transgene. Finally, chromosomal integration can lead to unwanted recombination events and transgenic sequence instability. Our MAEC cloning system offers an attractive alternative for generating transgenic organisms, because it (1) is able to carry large intact genes, (2) offers ease of interspecies shuttling, (3) replicates once per cell cycle, (4) segregates to daughter cells, (5) maintains long-term episomal persistence, (6) displays sequence stability, (7) eliminates position effects, and (8) reduces the risk of mutational insertions.

V. MAC ENGINEERING TECHNOLOGY

A. MAC Top-Down Assembly

MACS can be engineered following the top-down (tdMAC) strategy as outlined in Figure 1. The tdMAC strategy is accomplished by progressively fragmenting a natural mammalian chromosome by successive targeted deletions; for example, targeted chromosome fragmentation (TCF) [41,42]. In theory, tdMAC technology results in the discovery of minichromosomes containing the minimal essential elements required for stable chromosomal replication and segregation. There is little doubt that this technology offers a unique experimental approach for studying the functional organization of large mammalian chromosomes in an alternative animal model. Indeed, in these two studies [41,42], rodent cells were used to generate a series of human mini-X and mini-Y chromosomes 7–8 Mb and 2.5–4.0 Mb in size, respectively. In addition, the TCF/tdMAC–derived fragmented mini-X and mini-Y chromosomes displayed long-term persistence in the absence of selective pressure, at least in tissue culture cells, indicating the preservation of efficient replication and cis-genomic segregation activities. Although this experimental approach is appealing, there are potential drawbacks. Not all target cells may have the ability to assemble faithfully the MAC components in vivo, natural chromosomes may be resistant to size reduction, there may be unforeseen inherent chromosomal instability, and the TCF/tdMACs are less likely to be available for in vitro manipulation.

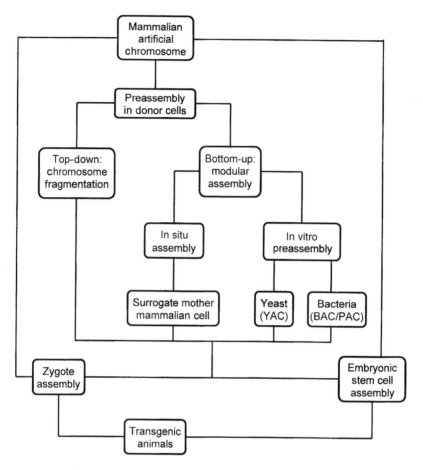

Figure 1   The "top-down" and "bottom-up" strategies used to generate transgenic animals.

B.   MAEC Bottom-Up Assembly

MAECs can also be generated following a bottom-up (buMAEC) strategy outlined in Figure 1. The ideal buMAEC strategy is accomplished by stable preassembly of a functional MAEC in vitro using a surrogate single-cell microorganism followed by transfer into target mammalian cells without any sequence reorganization. Hence, like YACs, MAECs can be assembled using endogenous chromosomal elements derived from mammalian genomes as outlined in Figure 2. In addition, because buMAECs are episomal, they are expected to avoid the negative effects associated with telomeric ends on some linear artificial chromo-

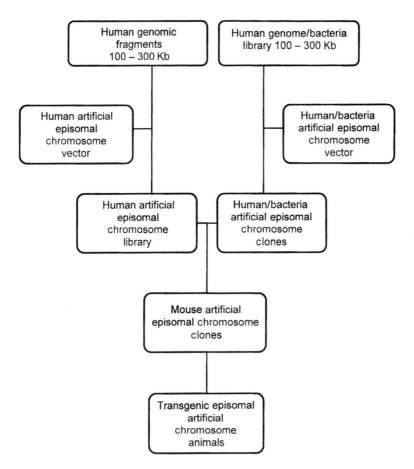

Figure 2 Flowchart of MAEC assemblage using endogenous chromosomal elements derived from mammalian genomes to generate transgenic animals.

somes. This strategy seeks to engineer MAECs with all the essential cis elements necessary for autonomous replication and segregation of episomal DNA independent of host cellular replication processes. In addition, a buMAEC can be generated following faithful in situ assembly of MAEC components by a cellular host much like the tdMAEC strategy outlined previously. Although the in situ method has some of the same drawbacks as described above, it offers researchers more flexibility when designing experiments. Other positives associated with buMAEC engineering are (1) the elements derived from episomal viruses already carry replication and segregation activities functional in mammalian cells; (2) the elements are compactly conjoined, and regulation is accomplished by as little as one single viral transactivator protein; and (3) because the preassembled components are easily manipulated, they offer versatility in developing chimeric MAECs containing smaller and more stable therapeutics.

C. Our EBV/MAC–Based Artificial Chromosome Design

As a proof-of-concept, we designed a chimeric MAEC composed of the essential components of the type 4 latently episomal Epstein–Barr virus (EBV) and a BAC to generate a MAEC for the delivery of large human DNA inserts as outlined in Figure 3. Under this experimental approach, we demonstrated transgene expression of human DNA using this oriP/EBNA1–based MAEC [19]. This large circular MAEC can be manipulated in vitro, is easily microinjected into zygotes by any method of naked DNA delivery, and offers ubiquitous shuttling between mammalian species. Although it appears to be controversial whether an oriP/EBNA1–based plasmid is lost from rodent cells without selective pressure [43], it has been reported that oriP/EBNA1–based plasmids can be retained under certain circumstances [44]. Furthermore, we have successfully demonstrated episomal maintenance and transgene expression of an oriP/EBNA1–based plasmid in mouse fibroblast cell culture without selection for 3 months [44a]. Our laboratory is actively investigating the efficacy of targeting specific disease cell types by therapeutic MAECs. We are engineering MAECs that can be carried in either herpes simplex virus (HSV) "helper-free" capsules or in an EBV "helper-dependent" manner. In theory, the helper-free virions are packaged via viral signals for sequence "a," whereas the helper-dependent MAECs are packaged using viral signals for the terminal repeat (TR) sequence [45] (J. Wang and J.-M. Vos unpublished results; G. Roosen and J.-M. Vos unpublished results). Both systems offer promise in the area of cancer research; that is, targeted cell death, albeit by different mechanisms. We predict that further development of EBV-based MAECs will have many applications once the poorly understood epigenic replication and segregation activities of this type of MAEC are critically analyzed. Once these problems are solved, the commercial applications are numerous, particularly in systems involving large or multiple genes.

VI. SUMMARY

The development of all forms of MACs is expected to be useful in many areas of biotechnological and biomedical endeavors outside the few described in this chapter. The use of MACs as vectors of therapeutic agents offers an exciting alternative treatment paradigm for curing human disease. Rather than continuous administration of chemical agents to alleviate the disease phenotype, MACs can

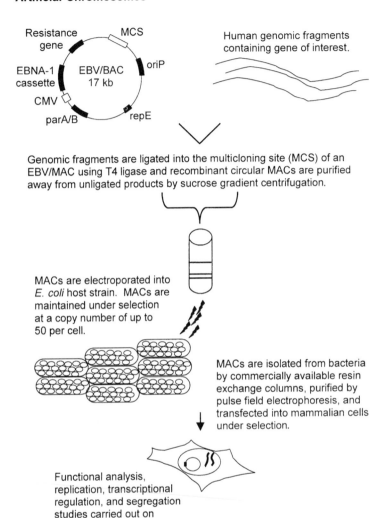

Resistance gene

MCS

EBNA-1 cassette

EBV/BAC 17 kb

oriP

CMV

parA/B

repE

Human genomic fragments containing gene of interest.

Genomic fragments are ligated into the multicloning site (MCS) of an EBV/MAC using T4 ligase and recombinant circular MACs are purified away from unligated products by sucrose gradient centrifugation.

MACs are electroporated into *E. coli* host strain. MACs are maintained under selection at a copy number of up to 50 per cell.

MACs are isolated from bacteria by commercially available resin exchange columns, purified by pulse field electrophoresis, and transfected into mammalian cells under selection.

Functional analysis, replication, transcriptional regulation, and segregation studies carried out on human gene insert.

Figure 3 Protocol outline used by our laboratory for generating a chimeric MAEC.

theoretically deliver and modify specific genes resulting in curing of disease after a single dosage. MACs will be used to study all types of diseases; that is, neurodegenerative, malignant, vascular, immune system, arthritis, and acquired diseases. MAC technology is an active area of research, and some technological hurdles need to be addressed. Specifically, artificial chromosomes need to be further engineered for ubiquitous use and ease of interspecies shuttling, therapeutic MACs need to be developed that can be tested in complementing disease models, the transgenic efficiency of MAECs by zygotic microinjection needs to be evaluated more completely, and the identity of the cis-elements required for MAC replication and segregation needs to be elucidated. Other hurdles not specifically associated with MAC function and development need to be addressed as well. These include gene identification, transfer technology, sufficient understanding of the pathogenesis of the disorder, and thorough knowledge of the disease pathways. Once these technological hurdles are overcome, the biotechnology and biomedical industries will benefit from the genomic libraries, transgenic animals, and therapeutic agents created by MAC technologies.

In Memoriam: Jean-Michel Vos

We are saddened by the death of Jean-Michel Vos on November 31, 2000, who succumbed to cancer at the early age of 44 years. Jean-Michel came to the University of North Carolina in 1987, where he was an active member of the Department of Biochemistry and Biophysics, the Lineberger Comprehensive Cancer Center, the Program in Molecular Biology and Biotechnology, and the Curriculum in Genetics. He was internationally recognized as an expert in the field of artificial chromosome technology and gene therapy in which he was a pioneer. Jean-Michel will not only be missed as a scientist, but also as a fun-loving and good human being. He was committed to his family—his wife Eliane and his two daughters Natasha and Olivia. He left us too soon, and I am honored to have known him. He will be missed by all of us in the Vos Laboratory.

REFERENCES

1. Struhl K, Stinchcomb DT, Scherer S, Davis RW. High-frequency transformation of yeast: autonomous replication of hybrid DNA molecules. Proc Natl Acad Sci USA 1979; 76:1035–1039.
2. DePamphilis ML. Origins of DNA replication. In: DNA Replication in Eukaryotic Cells. Cold Spring Harbor, NY: Cold Spring Harbor Laboratory Press, 1996, pp 45–87.
3. Gilbert DM. Replication origins in yeast versus metazoa: separation of the haves and the haves nots. Curr Opin Genet Dev 1998; 8:194–199.
4. Gilbert DM, Miyazawa H, DePamphilis ML. Site-specific initiation of DNA replication in Xenopus egg extract requires nuclear structure. Mol Cell Biol 1995; 15:2942–2954.
5. Szostak JW, Blackburn EH. Cloning yeast telomeres on linear plasmid vectors. Cell 1982; 29:245–55.

6. Clarke L, Carbon J. Isolation of a yeast centromere and construction of functional small circular chromosomes. Nature 1980; 287:504–509.

7. Grimes B, Cooke H. Engineering mammalian chromosomes. Hum Mol Genet 1998; 10:1635–1640.

7a. Choo KA. Centromere DNA dynamics: latent centromeres and neocentromere formation. Am J Hum Genet 1997; 61:1225–1233.

8. Murray AW, Szostak JW. Construction of artificial chromosomes in yeast. Nature 1983; 305:189–193.

9. Burke DT, Carle GF, Olson MV. Cloning of large segments of exogenous DNA into yeast by means of artificial chromosome. Science, 1987; 236:806–812.

10. O'Connor M, Peifer M, Bender W. Construction of large DNA segments in Escherichia coli. Science 1989; 244:1307–1312.

11. Shizuya H, Birren B, Kim UJ, Mancino V, Slepak T, Tachiiri Y, Simon M. Cloning and stable maintenance of 300-kilobase-pair fragments of human DNA in Escherichia coli using an F-factor-based vector. Proc Natl Acad Sci USA 1992; 89:8794–8797.

12. Sun T-Q, Fenstermacher D, Vos J-M. Human artificial episomal chromosomes for cloning large DNA in human cells. Nat Genet 1994; 8:33–41.

13. Sun T-Q, Livanos E, Vos J-M. Engineering a mini-herpes virus as a general strategy to transduce up to 180 kb of functional self-replicating human mini-chromosomes. Gene Ther 1996; 3:1081–1088.

14. Banerjee S, Livanos E, Vos J-M. Therapeutic gene delivery in human B-lymphoblastoid cells by engineering non-transforming Epstein-Barr virus. Nat Med 1995; 1: 1303–1308.

15. Kelleher ZT, Fu H, Livanos E, Wendelburg B, Gulino S, Vos J-M. Epstein-Barr–based episomal chromosomes shuttle 100 kb of self-replicating circular human DNA in mouse cells. Nat Biotechnol 1998; 16:762–768.

16. Kornberg A, Baker TA. DNA Replication. New York: Freeman, 1991.

17. Sternberg N. Bacteriophage P1 cloning system for the isolation, amplification, and recovery of DNA fragments as large as 100 kilobase pairs. Proc Natl Acad Sci USA 1990; 87:103–107.

18. Ioannou PA, Amemiya CT, Garnes J, Kroisel PM, Shizuya H, Chen C, Batzer MA, de Jong PJ. A new bacteriophage P1-derived vector for the propagation of large human DNA fragments. Nat Genet 1994; 6:84–89.

19. Vos J-MH. Therapeutic mammalian artificial episomal chromosomes. Curr Opin Mol Ther 1999; 2:204–215.

20. Herman GE. Physical mapping of the mouse genome. Methods 1998; 14:135–151.

21. Chartier FL, Keer JT, Sutcliffe MJ, Henriques DA, Mileham P, Brown SD. Construction of a mouse yeast artificial chromosome library in a recombinant-deficient strain of yeast. Nat Genet 1992; 1:132–136.

22. Ward ER, Jen GC. Isolation of single-copy-sequence clones from a yeast artificial chromosome library of randomly-sheared *Arabidopsis thaliana* DNA. Plant Mol Biol 1990; 14:561–568.

23. Grill E, Somerville CR. Construction and characterization of a yeast artificial chromosome library of Arabidopsis which is suitable for chromosome walking. Mol Gen Genet 1991; 226:484–490.

24. Martin G, Ganal M, Tanksley SD. Construction of a yeast artificial chromosome library of tomato and identification of cloned segments linked to two disease resistance loci. Mol Gen Genet 1992; 233:25–32.

25. Budiman MA, Mao L, Wood TC, Wing RA. A deep-coverage tomato BAC library and prospects toward developing of an STC framework for genome sequencing. Genome Res 2000; 10:129–136.

26. Edwards KJ, Thompson H, Edwards D, de Saizieu A, Sparks C, Thompson JA, Greenland AJ, Eyers M, Schuch W. Construction and characterization of a yeast artificial chromosome library containing three haploid maize genome equivalents. Plant Mol Biol 1992; 19:299–308.

27. Kleine M, Michalek W, Graner A, Herrmann RG, Jung C. Construction of a barley (Hordeum vulgare L.) YAC library and isolation of a Hor1-specific clone. Mol Gen Genet 1992; 240:265–272.

28. Umehara Y, Tanoue H, Kurata N, Ashikawa I, Minobe Y, Sasaki T. An ordered yeast artificial chromosome library covering over half of rice chromosome 6. Genome Res 1996; 6:935–942.

29. Marra MA, Kucaba TA, Dietrich NL, Green ED, Brownstein D, Wilson RK, McDonald KM, Hillier LW, McPherson DL, Waterson RH. High throughput fingerprint analysis of large-insert clones. Genome Res 1997; 7:1072–1082.

30. Venter JC, Smith HO, Hood L. A new strategy for genome sequencing. Nature 1996; 381:364–366.

31. Niemann H, Kues WA. Transgenic livestock: premises and promises. Anim Reprod Sci 2000; 60–61:277–293.

32. Zuelke KA. Transgenic modification of cows milk for value-added processing. Reprod Fertil Dev 1998; 10:671–676.

33. Co DO, Borowski AH, Leung JD, van der Kaa J, Hengst S, Platenburg GJ, Pieper FR, Perez CF, Jirik FR, Drayer JI. Generation of transgenic mice and germline transmission of a mammalian artificial chromosome introduced into embryos by pronuclear microinjection. Chrom Res 2000; 8:183–191.

34. Hodgson JG, Agopyan N, Gutekunst CA, Leavitt BR, LePaine F, Singaraja R, Smith DJ, Bissada N, McCutcheon K, Nasir J, Jamot L, Li XJ, Stevens ME, Rosemond E, Roder JC, Phillips AG, Rubin EM, Hersch SM, Hayden MR. A YAC mouse model for Huntington's disease with full-length mutant huntingtin, cytoplasmic toxicity, and selective striatal neurodegeneration. Neuron 1999; 23:181–192.

35. Leavitt BR, Wellington CL, Hayden MR. Recent insights into the molecular pathogenesis of Huntington disease. Semin Neurol 1999; 19:385–395.

36. Duff K. Transgenic models for Alzheimer's disease. Neuropathol Applied Neurobiol 1998; 24:101–103.

37. Probst FJ, Fridell RA, Raphael Y, Saunders TL, Wang A, Liang Y, Morell RJ, Touchman JW, Lyons RH, Noben-Trauth K, Friedman TB, Camper SA. Correction of deafness in shaker-2 mice by unconventional myosin in a BAC transgene. Science 1998; 280:1444–1447.

38. Sauer B. Inducible gene targeting in mice using the Cre/lox system. Methods 1998; 14:381–392.

39. Bruggemann M, Taussig MJ. Production of human antibody repertoires in transgenic mice. Curr Opin Biotechnol 1997; 8:455–458.

40. Peterson KR, Clegg CH, Li Q, Stamatoyannopoulos G. Production of transgenic mice with yeast artificial chromosomes. Trends Genet 1997; 13:61–66.

41. Farr CJ, Bayne RA, Kipling D, Mills W, Critcher R, Cooke HJ. Generation of a human X-derived minichromosome using telomere-associated chromosome fragmentation. EMBO J 1995; 14:5444–5454.

42. Heller R, Brown KE, Burgtorf C, Brown WR. Mini-chromosomes derived from the human Y chromosome by telomere directed chromosome breakage. Proc Natl Acad Sci USA 1996; 93:7125–7130.

43. Yates JL, Warren N, Sugden B. Stable replication of plasmids derived from Epstein-Barr virus in various mammalian cells. Nature 1985; 313:812–815.

44. Tsukamoto H, Wells D, Brown S, Serpente P, Strong P, Drew J, Inui K, Okada S, Dickson G. Enhanced expression of recombinant dystrophin following intramuscular injection of Epstein-Barr virus (EBV)–based mini-chromosome vectors in mdx mice. Gene Ther 1999; 6:1331–1335.

44a. Black J, Vos JM. Establishment of an oriP/EBNA1-based episomal vector transcribing human genomic beta-globin in cultured murine fibroblasts. Gene Therapy 2002, Nov. 9(21):1447–1454.

45. Wang S, Vos J-MH. A hybrid herpesvirus infective vector based on the Epstein-Barr virus and herpes simplex virus type I for gene transfer into human cells in vitro and in vivo. J Virol 1996; 70:8422–8430.

46. Dobbs DL, Shaiu W-L, Benbow RM. Modular sequence elements associated with origin regions in eukaryotic chromosomal DNA. Nucleic Acids Res 1994; 22:2479–2489.

47. Rein T, Zorbas H, DePamphilis M. Active mammalian replication origins are associated with a high-density cluster of ^{m}CPG dinucleotides. Mol Cell Biol 1997; 17:416–426.

Index

415